Engineering Applications of Computational Methods

Volume 6

The book series Engineering Applications of Computational Methods addresses the numerous applications of mathematical theory and latest computational or numerical methods in various fields of engineering. It emphasizes the practical application of these methods, with possible aspects in programming. New and developing computational methods using big data, machine learning and AI are discussed in this book series, and could be applied to engineering fields, such as manufacturing, industrial engineering, control engineering, civil engineering, energy engineering and material engineering.

The book series Engineering Applications of Computational Methods aims to introduce important computational methods adopted in different engineering projects to researchers and engineers. The individual book volumes in the series are thematic. The goal of each volume is to give readers a comprehensive overview of how the computational methods in a certain engineering area can be used. As a collection, the series provides valuable resources to a wide audience in academia, the engineering research community, industry and anyone else who are looking to expand their knowledge of computational methods.

More information about this series at http://www.springer.com/series/16380

You-He Zhou

Wavelet Numerical Method and Its Applications in Nonlinear Problems

Springer

You-He Zhou
Department of Mechanics and Engineering Science
College of Civil Engineering and Mechanics
Lanzhou University
Lanzhou, Gansu, China

ISSN 2662-3366 ISSN 2662-3374 (electronic)
Engineering Applications of Computational Methods
ISBN 978-981-33-6642-8 ISBN 978-981-33-6643-5 (eBook)
https://doi.org/10.1007/978-981-33-6643-5

This Springer imprint is published by the registered company Springer Nature Singapore Pte Ltd.
The registered company address is: 152 Beach Road, #21-01/04 Gateway East, Singapore 189721, Singapore

Foreword

The methods for numerical solutions to mechanics problems, especially strongly nonlinear problems, have been a challenging subject, compared to linear problems to which the solution methods are relatively mature both in formulation and implementations. Many numerical methods have been established for different types of nonlinear problems, but almost of all these methods have some limitations, and there is no single method that can work well for all different kinds of problems. Searching for more effective, unique, and/or alternative methods for different types of problems has motivated many researchers. Wavelet methods are among such methods for nonlinear initial-value and boundary-value problems in engineering and sciences.

Wavelet theory and methods were proposed in the 1980s, this relatively new mathematical tool has been played an important role in signal and picture processing. For initial-value and boundary-value problems, wavelet base functions can be also used and can offer unique features. Following Coifman's wavelets, the author of this text and his research group have made a significant progress for solving nonlinear problems and developed the generalized Coiflets, with which many nonlinear boundary-value problems can be well solved by the wavelet-based methods with high accuracy and efficiency.

This book well summarizes the achievements of the wavelet-based methods for nonlinear problems conducted in Prof. Zhou's research group during the past 20 years. It has a balanced description of mathematic rigor and implementation techniques, and hence is a very good reference for researchers engaged in solving those complex nonlinear problems using wavelet methods. It will also be a valuable reference book for both beginners and engineers.

In the past decades, I have visited China many times and conducted research works together with Chinese scholars in China. From them, I have learned about impressive work by Prof. You-He Zhou and his group. His former Ph.D. student and current colleague, Dr. Xiao-Jing Liu, visited my laboratory at the University of Cincinnati as a postdoctorate researcher and conducted the researches with me during 2017–2019. I have an excellent chance to learn and study the wavelet-based solution methods. From the cooperation, we have successfully extended the wavelet-based solution method to those problems with irregular domain and problems with high gradient fields and hence local refinement is needed. Chapter 11 of the book introduces the main achievements of this extension. This collaboration has demonstrated that a combination of different numerical techniques can be of advantages. I hope that this book can also facilitate the researchers and engineers working on other numerical methods to consider using wavelet methods as a possible alternative or further enrichment.

Cincinnati, OH, USA
November 2020

Gui-Rong Liu

Preface

Nonlinear problems are extensively and widely existed in almost all fields of scientific disciplines and engineering. To find their quantitative solutions except for those simple problems that can be fortunately found by some analytically closed solutions, most of them should be conducted on the basis of a kind of series expansion. As the computer technology and relevant software are developed and supported, to gain quantitative or numerical solutions for a nonlinear problem has become a significant way in the recent 60 years. In theory, such series expansion is inherently infinite but the numerical solutions based on the series expansion should be finitely truncated in the practical calculation, which directly leads to the so-called truncation errors corresponding to the numerical solutions. It is obvious that such truncation errors control the accuracy of the numerical solutions. In order to reduce the effect of truncation errors on the accuracy of numerical solutions, a large amount of effort has been paid by adding several calculation techniques. Due to that, the truncated numerical solutions are dependent on the truncation errors or vise verse, i.e., the numerical solutions are not closed in the conventional methods, such kinds of numerical solution methods are usually suitable to the weak nonlinear problems. When they are applied to solve the strong nonlinear problems if no additional technique is employed, however, the accuracy of the solutions is difficultly ensured as same as that they are used to the weak nonlinear problems even when the same governing equations are employed. When some additional techniques are employed to enhance the solution accuracy, the calculations increase fast with the nonlinearity stronger. Such a situation makes that the numerical solution methods for nonlinear problems are still in immaturity in theory relative to that for the linear problems although they have been used in solving nonlinear problems.

This text is aiming to provide a comprehensive understanding of the efforts what the wavelet-based numerical solution method for nonlinear problems has the closed property proposed by the author and his former Ph.D. students or current colleagues in the past 20 years, containing the main background of numerical methods, mathematical framework of the wavelet method employed, essentials to the numerical solutions, boundary extension techniques and error estimation, wavelet-based Fourier transform and Laplace transform, extended wavelet method for irregular domain and local refinement, and a lot of typical examples of solving nonlinear problems in physics, control and dynamics, solid mechanics, and fluid mechanics, etc. In the final Chap. 12, a brief introduction is given to display the effectiveness of other applications done by those other research groups when the wavelet or relevant method proposed by the author's group was used. Following them, one can gradually enter this research area and can find the significant and powerful ability of the wavelet-based solution method with both high accuracy and low calculation.

Since the 1980s, the author has conducted researches on solving nonlinear problems. For example, early works were concentrated on the von Kármán equations of thin plates with geometrical nonlinearity, then gradually extended to the coupling nonlinear problems of electromagnetic solid mechanics and wind blowing sand movements. At present, areas I am interested in are mainly concentrated on those more complex problems of multi-fold nonlinear mechanics of superconducting materials and relevant magnets, and dust/sand storm movements, etc. In such nonlinear problems, we always get into trouble that the strong nonlinear problems should be numerically solved. Hence, how to get a high accuracy solution with low calculation is so important to a researcher if one hopes to well conduct the theoretical investigations when he or she is in the face of a nonlinear problem, which motivates us to find a suitable method for solving the nonlinear problems. As we have been known, the numerical solutions for nonlinear problems are also strongly dependent on the set of base functions selected in the series expansion, and no set of conventional base functions can be used to support a closed property of their numerical solutions in solving a nonlinear problem. In such situation, we need to find other kinds of base functions if we hope to realize it. This chance was meted to the author when the author visited the University of Tokyo during 1994–1995, from where I accidentally known the newborn wavelet methods and their powerful applications in signal and picture processing, which was originated from the requirements of signal processing application in earthquake and chemistry reaction occurred within a short time. From the wavelets, I have known that many mathematicians such as Daubechies entered this area to establish the framework of wavelets, from where different wavelets can be and have been proposed by following different properties requested. At the same time, many researchers in different areas used the wavelet methods to conduct their investigations, including solving differential equations. The early such works for solving differential equations were mainly concentrated on the second-order ordinary linear differential equations. During the development of recent 20 and more years, the wavelet method has been used to solve more complex differential equations, either locally or globally. After I came back to Lanzhou University in 1995, I went down to do the possible applications of the wavelet methods to solve the bending problems of thin beam and plate structures, where the highest derivative order is 4 such that the wavelets employed should be reformed by increasing the order of derivative function continuous of the base scaling functions. In order to promote the efficiency of calculation in the decomposition coefficients, we also tried to use Gaussian integral such that each decomposition coefficient contains only one sampling point of the decomposed function. Further, when we knew that Daubechies established out the Coiflets which have the property of one sampling point for the signal processing, we reformed and reestablished the Coiflets, which is recently called the generalized Coiflets, such that the wavelet method is suitable to solve the nonlinear problems with high accuracy and low calculation.

During 2000–2004 when I was selected and appointed as a distinguished professor of the Cheung Kong Scholars Programme of the Chinese Ministry of Education, the embryo of such wavelet-based solution method was substantially achieved. Especially in 2003 and later, the author has found that the generalized Coiflet-based solution method has the closed property to any nonlinear problems, which provides us a feasible way to solve a nonlinear equation from weak to strong nonlinearity without additional techniques. After that, we use this wavelet-based method to solve different nonlinear problems for demonstrating its efficiency of accuracy and low calculation, while the mathematic framework of the method is also simultaneously promoted forward. The author deeply appreciates his early Ph.D. student and current colleague, Prof. Jizeng Wang, now a distinguished professor of the Cheung Kong Scholars Programme of the Chinese Ministry of Education, for his hard work in the pioneering foundation of this method together with the author. At the same time, I also thank my former Ph.D. students, Dr. Xiaoming Wang, Dr. Xiaojing Liu, and Dr. Lei Zhang, for their subsequent works in this area, in which Dr. Xiaojing Liu is now an associated professor as my colleague and he paid an effort to extend the wavelet method into an arbitrarily irregular region and the local refinement when he visited the University of Cincinnati to cooperate with Prof. Gui-Rong Liu during 2017–2019. Without their efforts in the researches in this area, it is impossible for me to write this textbook with the present relatively mature version. In addition, I have to give my deep appreciation to my wife, Dr. Prof. Xiaojing Zheng, an academician of the Chinese Academy of Sciences, and my daughter, Dr. Xi Zhou, for their lasting encouragement and accompanying in my career development. Since my father, Xietang Zhou, an ordinary farmer in the Hubei province of China, passed away 25 years ago, his magnanimous attitude of life and positive aggression of unbending life has made an indelible impression in my mind. Although the living condition was not good in my growth environment, he had still encouraged and supported my study in schools until I finished my university education in 1982. I strongly hope that this textbook can provide me to deeply appreciate his indoctrination and support for me and also to deeply memorate 25 years of his leaving from us. Finally, I sincerely appreciate the Springer Nature to give me a chance for publishing this textbook and Dr. Mengchu Huang for selecting the topic of this textbook and his help during its publication program.

Lanzhou, Gansu, China
October 2020

Dr. Prof. You-He Zhou
zhouyh@lzu.edu.cn

Acknowledgments

I am deeply indebted to the lasting partial supports of the Natural Science Foundation of China with either a general grant, or a key grant, or a grant for the Outstanding Young Researchers, or a grant for Innovation Research Team (No. 90405005; 10025208; 1097209554; 11032006; 11121202; 11202087; 113278021; 11421062), the National Key Basic Research and Development Fund of the Ministry of Science and Technology of China (No. G20000048702), and the National Key Project of Magneto-Constrained Fusion Energy Development Program of the Ministry of Science and Technology of China (No. 2013GB110002), as well as the Fund of Ministry of Education of China for the Program of Changjiang Scholars and Innovation Team in Universities (No. IRT0628) to the relevant researches of this textbook during past 20 years.

I am most grateful to my colleagues, Prof. Jizeng Wang and Associate Prof. Xiaojing Liu, for their help in revising and checking many versions of this manuscript. Without their dedication, the content of this book and its production would have taken a great deal longer.

Contents

About the Author

Dr. Prof. You-He Zhou received his bachelor and master degrees at Huazhong University of Science and Technology in 1982 and 1984, respectively, and Ph.D. degree at Lanzhou University in 1989. Since 1996, he has severed as a full professor of mechanics at Lanzhou University. In 1999, he was selected and appointed as a distinguished professor of the Cheung Kong Scholars Program by the Ministry of Education of China. He served as the dean of the College of Civil Engineering and Mechanics at Lanzhou University, a standing member of the Council of the Chinese Associate of Theoretical and Applied Mechanics, an associate editor of *Acta Mechanica Solida Sinica* (Chinese Edition), and the head of the Key Laboratory of Mechanics for Disaster and Environment in Western China (Lanzhou University) of the Ministry of Education of China. At present, he serves as the head of the Superconducting Mechanics Research Institute at Lanzhou University, head of Solid Mechanics Committee of the Chinese Associate of Theoretical and Applied Mechanics, an associate editor of *Chinese Journal of Theoretical and Applied Mechanics*, and an associate editor of*Theoretical and Applied Mechanics Letters*, etc.

His research interests are mainly concentrated in the areas of nonlinear solid mechanics, electromagnetic solid mechanics, mechanics of smart structures and dynamic controls, mechanics of superconductor and superconducting magnets, and aeolian sand/dust environmental mechanics, containing their theoretical modeling, quantitative analysis methods, and experiments. He has published over 350 peer-reviewed

English journal papers, 2 academic books, and 1 textbook. His research contributions were awarded two National Natural Science Prizes (second grade) and one National Scientific and Technology Development Prize (second grade) by the Chinese Central Government during 2007–2018. His two serial academic papers received *the Von Duzer Prize* awarded by the IEEE Superconducting Council in 2008. In 2019, his leading research on mechanics of superconducting materials and magnet structures was awarded the first grade of Technical Invention Prize of the Ministry of Education of China. He holds honors including the Well-Known Teacher of Higher-Education awarded by the Ministry of Education of China in 2008, and the Outstanding Contributor for the Innovation of Western China awarded by the Chinese Association of Science and Technology in 2006, etc. His teaching contributions were awarded one National Teaching Achievement Prize (second grade) by the Ministry of Education of China in 2009, and two Ph.D. dissertations supervised by him won the National Outstanding Ph.D. Dissertations of China awarded by the Ministry of Education of China in 2010 and the Outstanding Ph.D. Dissertation of Mechanics in China awarded by the Chinese Society of Theoretical and Applied Mechanics in 2017, respectively.

Chapter 1
Introduction

1.1 Brief Review of Solution Methods for Linear Systems

Since Isaac Newton published his famous book of the *Mathematical Principles of Natural Philosophy* 300 years ago, the Newton classical mechanics has been recognized as an open of modern science through rigorous logical reasoning, precise mathematical tools, and accurate calculation results [1–3]. After that, the investigation in both science and engineering have been mainly conducted under Newton system of studying, i.e., to develop theoretical models, quantitative solutions, and experiments or their combination. Based on the experiments or observations, the natural laws can be abstractly recognized or formulated in mathematics for a natural or engineering problem, while the algebraic, differential, or/and integral equations are established to a practical problem. By means of those methods of solving the equations, quantitative or predictive results are obtained. Once some predicting characteristics can be demonstrated by further experiments, we can judge that the established theoretical framework of the problems is reasonable and reliable. Such theory provides us a powerful advantage in a large context to predict future developing processes and trace back the developed process in history. Except for the establishment of theoretical equations, or theoretical modeling, and the essential experiments, it is obvious that the theoretical researches are highly dependent on the numerical methods what one uses. As the theories of differential and/or integral equations are developed and progressed, the solution methods are also developed and still developing for solving different kinds of such equations. Due to the different characteristics of such equations existed, in general, the solving approaches are always selected differently. To the differential equations, for example, we know that the earliest one is the linearly ordinary differential equation(s) with constant coefficients, then it is extended to ones with variable coefficients, nonlinear ordinary differential equations, and partial differential equations with either constant coefficients or variable coefficients. In solid mechanics, for example, all the problems of vibrations, bending, and stability of beam, plate, and shell structures can be expressed by the differential equations

Y.-H. Zhou, *Wavelet Numerical Method and Its Applications in Nonlinear Problems*,
Engineering Applications of Computational Methods 6,
https://doi.org/10.1007/978-981-33-6643-5_1

either with initial-value conditions or boundary-value conditions or both. In fluid mechanics, the most typical one is the Navier–Stokes (N–S) equations which can describe behaviors of almost all fluid flows on the earth. Before computer technology was used to take the solving performance, the quantitative results of solving a set of equations were fully conducted by hand and dependent on the analytical derivations, from which fewer equations can be solved quantitatively. After the computers were put into the calculations replacing artificial hands about 70 years ago, the computer technology and the associated software have provided us a powerful tool to solve more and more equations what are almost cannot be conducted by hand. Except for the logical structures of calculations requested in a computer, it has been known that such software for solving a set of equations is strongly dependent upon some solving methods employed, containing the estimations of convergence, accuracy, and computational capacity, etc.

The present numerical methods employed for solving differential equations are originated from the solution method of linearly ordinary differential equations with constant coefficients, which is the simplest case and put a basic solid foundation of train of solving thought. At present, one knows that the solution method for linear differential equations with constant coefficients is fully mature. For such N-order (i.e., the highest derivative of nonzero term is N) ordinary differential equation, i.e.,

$$\sum_{i=0}^{N} a_i \frac{d^{N-i} f(x)}{dx^{N-i}} = g(x) \quad a_0 \neq 0, \tag{1.1}$$

where $g(x)$ is a known function and $f(x)$ is the unknown function, the solution method is performed by the following main steps (see Chap. 3 for details):

(1) *to find the N independent basic solutions* $\{f_i(x); i = 1, 2, \ldots, N\}$

The basic solutions can be gained by an analytic approach and they are independent. Here, the independent basic functions mean that each of them cannot be expressed by a linear composition of others, i.e.,

$$f_i(x) \neq \sum_{j=1}^{i-1} a_j f_j(x) + \sum_{j=i+1}^{N} a_j f_j(x), \tag{1.2}$$

where $a_j (j \neq i)$ are arbitrary constants.

(2) *to construct the general solution of the differential equation*

Based on the basic solutions obtained, the general solution of the original differential equation can be expressed by the linear composition of the basic solutions associated with one specific solution of the original nonhomogeneous differential equation. That is,

$$f(x) = \sum_{i=1}^{N} c_i f_i(x) + f^*(x). \tag{1.3}$$

Here, the coefficients c_i are constants to be determined and $f^*(x)$ is a specific solution of Eq. (1.1).

(3) *to get the solution of Eq. (1.1) with either boundary conditions or initial conditions.*

Substituting the general solution of Eq. (1.3) into those N conditions, one gets a set of N algebraic equations linearly varying with the unknowns $c_i (i = 1, 2, \ldots, N)$. By solving the linear algebraic equation, one obtains the solution of $c_i (i = 1, 2, \ldots, N)$. Thus, the solution of Eq. (1.1) with a set of suitable conditions is gained by substituting the obtained $c_i (i = 1, 2, \ldots, N)$ into Eq. (1.3).

The above solution method obviously has the following characteristics:

Definition 1.1 (*Linear system or operator*) To a system which is expressed by a set of equations on the unknown function(s), if all such terms in the equation(s) linearly vary with the unknown function(s), then the system is called a linear system. In mathematics, we can express it in a linear operator $L[\cdot]$.

Property 1.1 (*Linear combination principle*) To arbitrary different functions $f_1(x)$ and $f_2(x)$, and arbitrary nonzero constants b_1 and b_2, the linear operator $L[\cdot]$ always has the following operation of linear combination principle:

$$L[b_1 f_1(x) + b_2 f_2(x)] = b_1 L[f_1(x)] + b_2 L[f_2(x)]. \tag{1.4}$$

There is no difficulty for one to demonstrate this property when $L[f(x)] = \sum_{i=0}^{N} a_i \frac{d^{N-i} f(x)}{dx^{N-i}}$. In fact, this property is always true for any linear system or problem and vise versa.

Property 1.2 If $f_1(x)$ and $f_2(x)$ are different solutions of the operator equation $L[\cdot] = 0$, then, $b_1 f_1(x) + b_2 f_2(x)$ is also a solution of $L[\cdot] = 0$ for arbitrary nonzero coefficients b_1 and b_2.

Proof According to the conditions, we know $L[f_2(x)] = 0$ and $L[f_2(x)] = 0$. Then following Property 1.1, we have

$$L[b_1 f_1(x) + b_2 f_2(x)] = b_1 L[f_1(x)] + b_2 L[f_2(x)] = 0, \tag{1.5}$$

which shows $b_1 f_1(x) + b_2 f_2(x)$ is a solution of $L[\cdot] = 0$. The proof is finished.

For the nonhomogeneous equation of linear system, we have the following superposition principle for its solutions:

Property 1.3 (*Superposition principle*) To a linear system of nonhomogeneous differential equation(s), e.g., $L[f(x)] = g(x)(\neq 0)$, if $f_1(x)$ and $f_2(x)$ are the solutions of the system corresponding to the non-homogeneous terms $g(x) = g_1(x)$, and $g(x) = g_2(x)$, respectively, then $b_1 f_1(x) + b_2 f_2(x)$ is also a solution of the system when $g(x) = b_1 g_1(x) + b_2 g_2(x)$ for arbitrary nonzero constants b_1 and b_2.

Proof According to the conditions in this property, one can write.

$$L[f_1(x)] = g_1(x) \quad \text{and} \quad L[f_2(x)] = g_2(x). \tag{1.6}$$

Following Property 1.1, we have

$$L[b_1 f_1(x) + b_2 f_2(x)] = b_1 L[f_1(x)] + b_2 L[f_2(x)] = b_1 g_1(x) + b_2 g_2(x). \tag{1.7}$$

which shows that $b_1 f_1(x)+b_2 f_2(x)$ is the solution of equation $L[f(x)] = b_1 g_1(x)+ b_2 g_2(x)$. Thus, the proof of this property is finished.

Property 1.3 is so important to a linear system for finding its solutions in more complex cases by an available way. For example, when a basic solution $h(x, x')$ is gained for the case $L[f(x)] = \delta_{x,x'} = \begin{cases} 1 & x = x' \\ 0 & x \neq x' \end{cases}$, the solution of equation $L[f(x)] = g(x)$ under the same boundary condition (s) can be calculated by the form:

$$f(x) = \int_{\Omega} h(x, x')g(x')dx'. \tag{1.8}$$

Such solving method is called the *influence linear method*. For 2D and 3D linear boundary-value problems, such method constitutes a base of the boundary-element method.

Definition 1.2 (*Independent functions*) If each one of the functions $\{f_i(x) : i = 0, 1, \ldots, N\}$ cannot be expressed by a linear combination of remaining functions in the set, we call the functions are independent of each other.

Definition 1.3 (*Base functions*) If $f_i(x)$ $(i = 0, 1, \ldots, N)$ are independent and when they are selected in solving arithmetic, we call them as a set of base functions. When N is a finite integer, the base functions are referred to a set of finite base functions. Otherwise, when $N \to \infty$, the base functions constitute a set of infinite base functions.

Definition 1.4 (*Inner product or integration*) To arbitrary two different functions $f_1(x)$ and $f_2(x)$, the definite integration of their multiply product is named by the inner integration, and denoted it by $< f_1(x), f_2(x) >$. That is,

$$< f_1(x), f_2(x)>= \int_{-\infty}^{\infty} f_1(x) f_2(x) dx \tag{1.9}$$

Definition 1.5 (*Standard orthogonal base functions*) To the base functions $f_i(x)$, if the following property is satisfied, i.e.,

$$< f_i(x), f_j(x)>=\delta_{ij}, \tag{1.10}$$

we call such base functions as the standard orthogonal base functions. Here $\delta_{ij} = 1$ for $i = j$ and $\delta_{ij} = 0$ for $i \neq j$.

When the differential equation is more difficult than the linear ordinary one with constant coefficients, only a few of them can be solved by the close form of finitely analytical program like above. For example, the linear ordinary differential equation with variable coefficient as

$$\frac{d^2 f(x)}{dx^2} + b(\sin \omega x) f(x) = 0 \quad x > 0, \tag{1.11}$$

one can find that its solution cannot be expressed by a finite set of basic solutions like the ordinary differential equation with unchanged constant(s).

Definition 1.6 (*Expansion of one function in base functions*) For a function $f(x)$ and a set of selected base functions $f_i(x)$ defined in a domain, the following equation

$$f(x) = \sum_{i=0}^{\infty} c_i f_i(x) \tag{1.12}$$

is called the expansion of $f(x)$ under the base functions $f_i(x)$ or abbreviated by function expansion, and c_i are referred to the expansion coefficients.

In the signal processing of a signal $f(x)$, the calculation of c_i is called decomposition of the signal, while the calculation of summation is referred to the reconstruction of the signal. In mathematics, we know that feasibility of Eq. (1.12) is dependent on the characteristic of both the expanded function $f(x)$ and the selected base functions, e.g., continuous and convergence of infinite summation at each point in the domain. To these theoretical problems, here, we do not give their more introduction due to the limit of our concentration in this book. In the following, the function expansion like

Eq. (1.12) implies their availability except for specific indication if it is necessary. To those problems that their closed analytical solution cannot be directly gained, their quantitative solutions are always performed by the function expansion form of Eq. (1.12).

Example 1.1 The simplest two sets of base functions:

(1) power functions $f_i(x) = x^i (i = 0, 1, \ldots, \infty)$ are not orthogonal;
(2) trigonometric functions $f_i(x) = sin\, i\pi x, \cos i\pi x\ (i = 0, 1, \ldots, \infty)$ are orthogonal.

Property 1.4 When the base functions are selected standardly and orthogonally, i.e., Eq. (1.10) is satisfied, then the calculation for expansion coefficients in Eq. (1.12) or the decomposition operation become the form:

$$c_i = < f(x), f_i(x) > . \tag{1.13}$$

To take the inner product operation of Eq. (1.12) using the standard orthogonal base functions, there is no difficulty for one to get Eq. (1.13). It is obvious that the decomposition calculation is the simplest when a set of standard orthogonal base functions are selected.

Definition 1.7 When a (set of) differential equation(s) constitute a determined problem under a set of boundary conditions or initial conditions or some combination of them, then their differential equation(s) associating with their whole conditions constitute a boundary-value or initial-value or initial-boundary-value problem, respectively.

In practical calculations for solving a determinant problem expressed by differential equation(s), once the function expansion of Eq. (1.12) is employed, one can find the summation of infinite terms should be truncated by a set suitable finite terms, denoted by N so as to the approximate solution can be formulated by the form:

$$f(x) \approx \sum_{i=0}^{N} c_i f_i(x). \tag{1.14}$$

To an unknown function $f(x)$ in differential equation(s), here, the base functions $f_i(x)$ are preselected and the unknown coefficients c_i are determined by a suitable or selected numerical method, e.g., finite element method (FEM), finite difference method (FDM), and weighted residual method or Galerkin's Method, etc.. Such solution programs of replacing a continuous problem with infinite freedom with a set of finite freedom are attributed to a discretization of the problem. This discretization method has been extensively used in the numerical methods for solving those determinant problems, especially for boundary-value and initial-boundary-value problems.

1.2 Origination of Nonlinear Science and Some Challenges

In real world, the practical problems all are nonlinear. Even for the earlier era of established Newtonian mechanics, for example, one can find that the nonlinear term is appeared in the differential or governing equation(s) for characterizing the celestial motions. In fluid mechanics, the typical nonlinear governing equations are the Navier–Stokes (N–S) equations, which were fully established in the 1840s. In solid mechanics, the pioneering work of thin plate structures was done in 1910 by von Kármán considering a set of nonlinear geometric relations, to which the governing equations are referred to the von Kármán's equation with the simplest nonlinearity. After that, similar nonlinear governing equations have been found in various areas when the existed nonlinear effects were taken into account in their corresponding linear systems, e.g., nonlinear optics, post-buckling and buffeting of structures, etc. Before the computer added calculations occurred in the 1960s, the solving methods for getting quantitative solution of a nonlinear governing equation had been mainly conducted by hand derivation formulas. With recognizing the importance of nonlinear governing equations both in theory and in practical engineering, the earlier efforts for solving them were mainly attributed to the regular or irregular perturbation methods, iteration methods, and series expansion methods, etc. For example, Ludwig Prandtl proposed an irregular perturbation method in 1904 to deal with the turbulent flow near a solid boundary, from which the named *boundary layer theory* was formed after the complex flows in different engineering applications have successfully been solved by it. After that, the perturbation methods had become a main mathematical tool to find the solution of a nonlinear problem till to 1980s. As the emergence and progress of computing added calculations, numerical methods have been substantially achieved in applications for obtaining those quantitative solutions of both linear and nonlinear problems except for those qualitative theories of solvable, stable, and error analyses etc., of the methods in mathematics. Before the 1960s, people had believed that the governing equation(s) for a determinant dynamic problem like the second Newtonian law, no matter what it is linear or nonlinear, can be used to predict its future and trace back its history [1]. With the accuracy of numerical integrals is promoted on the basis of some refined difference of time deviations, e.g., the Runge–Kutta methods of fourth order, some more complex phenomena, e.g., bifurcation and chaos with strange attractors, were displayed in quantitative results. In 1961, when Lorentz studied atmospheric circulation movement and found that two groups of initial conditions, with a difference of only 1000th, produced completely different evolution processes [4]. From this finding, Lorentz [4, 5] concluded that any physical system that exhibits aperiodic behavior will be unpredictable, completely subverting the conclusion of determinism and marking the birth of a new scientific concept. After that, nonlinear science with chaos theory, fractal geometry, and soliton theory as the main principles was substantially developing and developed. According to the nonlinear theory, it has been known that a determinant dynamic problem with nonlinearity may be not to predict its future and/or to trace back its history under some conditions. Such inherent characteristic implied in a nonlinear dynamic system opens a new era of recognition that a determinant dynamic system without any random effect may exhibit some random features, which constitutes the third greatly scientific revolution in the twentieth century, following the theory of relativity and the quantum mechanics.

At present, the concepts and solution methods in nonlinear science have been extended into almost all scientific fields. A large number of complex nonlinear phenomena have been observed, such as the surging airflow in nature, the undulating lands, the floating smog, the development of embryos in life sciences, the pulsation of the heart, the activities of the nerves, and even the fluctuations in the supply and demand of commodities in economics and society [3–7]. In engineering, nearly all real systems contain various different nonlinear factors [8], such as gaps and dry friction in mechanical systems, large deformations in structural systems, constitutive relationships among nonlinear materials, nonlinear control strategies in control systems, and turbulence in fluid motion [7–10]. In such systems, disregarded nonlinear factors occasionally cause unacceptable errors in analysis and calculation, and may even lead to fundamental cognitive errors [4–10]. For example, in the study of huge magnetostrictive materials, the "flip phenomenon" observed in experiments can only be theoretically predicted when nonlinear constitutive relations are considered [11]. Recent researchers have realized that only nonlinear science can truly interpret the variety and complexity of the natural world; only nonlinear science can guide the continuously developing and increasingly sophisticated practice of engineering in its pursuit of perfection. As stated by scientists, linear science represents the early years of scientific development, while nonlinear science marks the maturity of science and represents the future of scientific development. However, nonlinear problems are currently one of the choke points that restricts the further development of natural science and engineering practice, and they will be among the key scientific issues that humans must face and solve in the twenty-first century [12], where the main challenge is to develop a universal solution method with high accuracy and low calculations.

Definition 1.8 (*Nonlinear problems*) When the essential equation(s) associated with boundary or initial conditions governing a system has at least one nonlinear term in terms of the unknowns, such system is referred to the nonlinear problem.

Example 1.2 N–S equations in fluid mechanics. Without losing generality for our purpose, we restrict our attention to the incompressible viscous flows in 3D space $\mathbf{x} \in \mathbb{R}^3$. Then, we can write the N–S equations as follows:

$$\begin{array}{c} \rho\frac{\partial \mathbf{u}(\mathbf{x},t)}{\partial t} + \rho\mathbf{u}(\mathbf{x},t)\cdot\nabla\mathbf{u}(\mathbf{x},t) = \nu\Delta\mathbf{u}(\mathbf{x},t) - \nabla p(\mathbf{x},t) + \mathbf{g}(\mathbf{x},t) \\ \nabla\cdot\mathbf{u}(\mathbf{x},t) = 0, \quad \mathbf{x}\in\mathbb{R}^3, t > 0 \end{array}. \tag{1.15}$$

Here, the velocity $\mathbf{u} = \mathbf{u}(\mathbf{x}, t)$ and the pressure $p = p(\mathbf{x}, t)$ are the unknowns; ρ is the mass density of the fluid; ν is the viscosity; $\mathbf{g}(\mathbf{x}, t)$ stands for a given, externally applied body force (e.g., gravity); ∇ and $\Delta(= \nabla\cdot\nabla)$ are the vector gradient operator and the Laplacian operator in the space, respectively. The nonlinear part is appeared in the second term on the left hand of the first equation of Eq. (1.15). Such nonlinearity constitutes very high complexity for obtaining its solutions.

It has been found that the N–S equations may be used to model the ocean currents, water flow in a pipe, waves on a lake or ocean, weather, the air's flow around a wing, turbulent air currents following a flight in a modern jet. The N–S equations in their full and simplified forms help with the study of blood flow, the analysis of pollution, the design of power stations, train, and cars, et al. Hence, mathematicians and physicists believe that an explanation for the prediction of both their laminar and turbulence flows can be found through an understanding of solutions to the N–S equations. Due to the complexity induced from the inherent nonlinearity, till today, no perfect solution method has been established by a well-rounded mathematical way, e.g., how to deal with the essential feature of closure necessarily considered in the solution methods is still one of the open keys to find out exactly when the equations work and under what conditions they break down.

The effort for solving the N–S equation on turbulence can be traced back to the Reynolds' pioneer work, where an averaging decomposition method was proposed by the summation of average and turbulent flows. As displayed in Sect. 3.6.4, such decomposition is not closure in mathematics no matter what average form is employed, including large eddy simulation (LES) recently developed. The closure concept can be traceable to a system of differential equations. When the number of the partial differential equations (PDFs) is m, then the closure for such governing equations requests only m unknown functions. Under this closure concept, one can judge that the N–S equations are closure. However, the governing equations for average flows, gained from Reynolds's average decomposition method, cannot be closure in mathematics because the induced governing equations for the average flows are dependent on the nonzero and nonlinear terms of turbulent flow in the N–S equations. In order to close the governing equations for average flows, some experience relationships to formulate the nonzero turbulent part in terms of the average flow unknowns have been proposed. Each one experience relationship is called a turbulent model in literature. Due to such turbulent models all are given by experience, no turbulent model is universally suitable to whole turbulent flows. Meanwhile, the solutions of the N–S equations for a turbulent flow are sensitive to those solution methods what we choose in literature without universal closure in mathematics, which results in the understanding still minimal. Such challenge promotes the substantial progress toward a mathematical theory of solution methods which will unlock the secrets hidden in the N–S equations.

1.3 Main Solution Methods for Nonlinear Problems

Different from the linear problems which have relatively mature solution methods, till today, no universal method has been established for solving different nonlinear problems from weak to strong nonlinearity. Even for a nonlinear problem of mechanics without a closed analytic solution, for example, its quantitative solution should be taken to each magnitude of applied force because the composition principle is false in this case and the solution methods are not closure (see Chap. 3), which directly generates the induced problems relevant to efficient accuracy of solution methods when the nonlinearity becomes strong. As Werner Heisenberg, a Nobel laureate in physics, pointed out "The development of physics is heavily dependent on advances in nonlinear mathematical and nonlinear equation solving methods…, and therefore we can learn by comparing different nonlinear problems." Strictly speaking, most of the recent solution methods for nonlinear problems are suitable to the weakly nonlinear case. While the nonlinear problems become strong as the external actions increase, some special techniques should be added for solving each strongly nonlinear problem one-by-one. In addition, a nonlinear system exhibits complex properties, such as high sensitivity to initial values and having multiple stable states. In the past century, although scientists have exerted considerable effort in studying the two types of solution for nonlinear differential equations, namely, analytical and numerical methods, most existing methods can effectively deal with weak nonlinear problems, but frequently exhibit difficulty when they work for strong nonlinear problems [13, 14]. Here, we give some progress of them.

1.3.1 Analytical Methods

Analytical methods can obtain "exact" or approximate solutions in analytical forms for equations. Such study for the influences of various physical quantities of a system, particularly those for optimizing the parameters of the system, becomes convenient. Various analytical methods have been proposed to study nonlinear problems. They are primarily divided into two categories. The first category comprises methods that use special functions or transformation techniques to obtain explicit solutions for nonlinear equations. Examples of these methods include series expansion, trial and error [15, 16], variable transformation method [17, 18], and variational method [19, 20]. However, the versatility of such methods is very poor and their application scope is extremely narrow. To date, only a few equations have been solved using these methods [21, 22]. The second category includes techniques represented by the perturbation method [23–25] or relevant iteration methods [26].

Consider one of the classical nonlinear problems in mechanics, namely, the large deflection problem of thin circular plates, as an example. von Kármán established a system of nonlinear differential equations that characterize this problem in 1910 [27]. Way [28] first studied the large deflection of thin circular plates under a uniform load by using a power series method. His research was followed by Vincent [29], who used the perturbation method to solve the same problem by regarding load as the small perturbation parameter. However, the two methods have a narrow application scope and only problems with a relatively small load can be solved well. Qian and Ye [30, 31] effectively expanded the applicability of the perturbation method to solve this problem by adopting central deflection as the perturbation parameter. However, when load is increased to a certain extent, this method, which uses the linear bending solution as the initial perturbation solution, give the results being not well when the nonlinear effect increases to a context. To address this issue, Qian [32] used the film solution as the initial perturbation solution and obtained the solution for a large deflection of a circular plate within the vicinity of the film solution. However, an interval still exists outside the applicable range of the two perturbation methods using different initial perturbation solutions. Therefore, the two perturbation techniques fail to work because out-of-plane bending and in-plane stretching play nearly equivalent roles. This phenomenon is also known as the "transition of thin plate to film" [33]. This problem was not solved completely until the 1990s. Zheng and Zhou et al. [26, 34–41] proposed a construction method for the "exact" solution of a large deflection of thin circular plates and the function structures of solutions. They presented the "interpolation iteration method" for dealing with the "transition problem" and completed the proof of convergence. From the research history of the large deflection problem of thin circular plates, we can determine that the applicable range of solutions can vary considerably when the same method is used to study the same problem because of different auxiliary parameters or initial conditions. This deficiency is also common among many analytical methods; that is, these methods are too dependent on the technique used and lack versatility and uniformity. Strong nonlinear problems are frequently not addressed if special techniques are not used [42]. However, finding similar processing techniques is difficult for most strong nonlinear problems encountered in natural science and engineering technology.

Since von Kármán proposed nonlinear differential equations for describing the large deflection of a thin circular plate in the beginning of the twentieth century, nearly a century of hard work had passed to completely solve this problem when Zheng and Zhou et al. finally found an "exact" solution for this equation; nevertheless, this problem had become considerably simpler compared with other complex systems that developed by the end of the twentieth century [26–41]. This situation reflects the difficulty of solving nonlinear differential equations. Moreover, Zheng and Zhou et al. only provided the "exact" solution for the 1D nonlinear problem of thin circular plate bending [26, 34–41]. Meanwhile, the large deflection bending problem of a rectangular plate or a plate with a complicated irregular shape is nearly impossible to solve using various analytical methods. This issue is another limitation of analytical methods, which are typically only applicable to low dimensions and regular regions. For problems involving high dimensions or irregular shapes that are frequently encountered in engineering practice, analytical methods are often ineffective. Only a variety of numerical methods can be used for these problems.

1.3.2 Numerical Methods

Since the 1940s, numerical methods have been applied to solve differential equations. On the one hand, numerical solution techniques are urgently required due to the increasingly complex problems. On the other hand, a firm hardware foundation is provided for numerical calculations with the rapid development of computer technology [43]. Therefore, numerical solution techniques for differential equations have been actively developed for more than half a century. Numerical methods can be classified into major two categories depending on the object of application. The first category comprises time integration methods for initial-value problems; examples include the Runge–Kutta [44], linear multistep [45], Newmark [46], and Wilson–θ [47] methods. These methods are all recursive; that is, the next step can be calculated on the basis of the results of the previous step or a few previous steps. Therefore, the calculation number for each step is nearly the same, and the total amount of calculations required is approximately linear with the size of the research area. Although most of such methods are efficient for nonlinear dynamic problems, however, the inherent accumulate error may lead to inefficient when time persists very long, and the system is sensitive to some factors. Since this book is mainly focused on the numerical research of boundary-value problems, especially for nonlinear problems, here, we do not introduce more detailed progress in this category.

The other category in the numerical researches is for boundary-value relevant problems in natural science and engineering technology. In general, after a differential equation is discretized into a set of algebraic equations, either linearly for linear problems or nonlinearly for nonlinear problems, the approximate solution for a problem is obtained by solving the algebraic equations. In contrast with initial-value problems that require only solutions to meet the initial constraints at one end of the solution region, the solutions for boundary-value problems should satisfy the constraints at all boundaries of the region, considerably increasing the difficulty of solving these problems. At present, the classical common numerical methods for nonlinear boundary-value problems are mainly used by the finite difference method (FDM), weighted residual method, and finite element method (FEM), etc.

(1) *FDM*: The basic idea of FDM is to replace the differentials in the original differential equation with the differences of the unknown function values at discrete grid points. Thus, the differential equation is discretized into a series of algebraic equations. The value of a function at discrete grid points is obtained by solving the algebraic equations. Lastly, the interpolation method is used to obtain the approximate solution in the entire region [48]. Therefore, FDM can be regarded as an interpolation polynomial for approximating the solution of the original differential equation. As one of the classical methods for solving nonlinear differential equations, FDM is widely used in solving linear and nonlinear differential equations. For example, Delfour et al. [49] solved the famous nonlinear Schrödinger equation using FDM. Chen [50] studied the beam fracture problem based on this method. Brian [51] studied the 3D heat conduction problem using FDM. Beam and Warming [52] successfully introduced the finite difference method into the analysis of compressible

fluids. Narasimhan and Witherpoon [53] used the same method to study the seepage problem in porous media. Yuste and Quintana–Murillo [54] recently solved the diffusion equation with fractional derivatives using FDM. From the preceding examples, FDM has been widely and successfully applied to solve linear and nonlinear differential equations [48–54]. However, the approximation of higher order derivatives is extremely cumbersome, and precision is poor due to the inherent defects of numerical differentiation techniques. FDM is typically difficult to apply to solutions for higher order differential equations. For nonlinear problems, the accuracy of a solution will decrease significantly as the nonlinearity of a system increases [48–55]. Thus, such directly dealing with strong nonlinear problems becomes difficult.

(2) *Weighted residual method* (WRM): The basic idea of the weighted residual method is to use a series of base functions with undetermined coefficients, i.e., the trial function, as the approximate solution for a problem. Evidently, the approximate solution cannot fully satisfy the differential equation in the whole region due to the existence of residuals of each base function. Accordingly, a set of algebraic equations about the undetermined coefficients can be obtained by suitably selecting a set of weight functions to make the residual value equal to zero in the meaning of the weighted average. Then the approximate solution for the original problem can be obtained by solving this set of algebraic equations [56]. The weighted residual method can be classified as an internal, boundary, or hybrid method depending on whether the trial function satisfies the differential equation or boundary conditions. In accordance with the choice of weight functions, the weighted residual method can be classified into the least squares, collocation, Galerkin, subdomain, and moment methods [57]. As a common numerical method, the achievements of the weighted residual method in the field of solving differential equations in the past half-century have been considerable. For example, Bramble and Schatz [58] solved an even-order boundary-value problem on the basis of the least square method. Xiu and Hesthaven [59] used the collocation method to study a class of stochastic differential equations. Cockburn and Shu [60] investigated a convective diffusion system using the Galerkin method. Wang et al. [61] solved a shallow water wave equation by applying the subdomain method. Feng and Neilan [62] examined second-order nonlinear differential equations through the moment method. A series of meshless and boundary element methods was developed on the basis of the basic idea of the weighted residual method [63–67]. These traditional processing techniques directly substitute the approximate solution into a nonlinear term [58–68] to obtain the approximate expression. Studies have shown that these methods can effectively solve linear and weak nonlinear problems; however, their accuracy is highly sensitive to nonlinear systems [56–68], and using them to directly solve strong nonlinear problems is frequently difficult.

(3) *FEM*: The FEM is currently the most widely used numerical method in the field of natural science and engineering technology. The basic idea of the FEM is to transform complex problems into a series of simple sub-problems by dividing the object area into small meshes. Then, the problems are solved on the basis of the variation principle or the weighted residual method [69]. FEM has a uniform and fixed solution format, and most engineering problems can be solved by using the same

set of calculation programs, making this method highly versatile [70]. FEM is nearly perfect for solving linear problems. For strong nonlinear problems, however, FEM must be combined with numerical tracking techniques; that is, a nonlinear problem is transformed into a series of linear or weak nonlinear problems [69–71]. Especially, the accumulated error existing in the tracking process from weak to strong nonlinear problem becomes possibly unacceptable except for convergence. In fact, the accuracy of FEM is also highly sensitive to the nonlinearity of a system. Moreover, in solving large deformation problem mechanics, FEM frequently fails due to the excessive distortion of the meshes [72]. In addition, such solution efficiency is low due to the large number of large-scale matrix operations that should be performed in FEM. At the same time, since its calculations take a long time, such method is difficultly applied to real-time control systems [73].

In addition to the three aforementioned classic and widely used numerical methods, other methods for solving nonlinear boundary-value problems are also available, including the shooting [74] and differential quadrature [75] methods, which are built on solving techniques for the initial-value problem. In general, many existing numerical algorithms are effective for some, or even many, weak nonlinear problems. However, their accuracy is frequently significantly reduced down low as the nonlinearity of a system increases, leading to high-precision approximate solutions that are difficult to obtain when studying strong nonlinear problems. Usually, the existing methods cannot provide a unified format of solution method for different problems with from weak to strong nonlinearity. Of course, once some special techniques employed to enhance convergence and accuracy of a numerical method in accordance with a specific nonlinear problem, one can get its numerical solution from weak to strong nonlinearity of the problem, such as an iterative interpolation technique for large deflection of circular plates [26, 34–41].

1.3.3 Examples of Main Program of Solution Methods for Nonlinear Problems

(1) *Analytical methods.*

Example 1.3 To solve axisymmetric *von Kármán's plate equations.* In nonlinear solid mechanics, the von Kármán's plate equations are typical and famous. After the dimensionless quantities $y = r^2/a^2$, $W(y) = \sqrt{3(1-\nu^2)}w(y)/h$, $\varphi(y) = ydW(y)/dy$, $S(y) = 3(1-\nu^2)a^2yN_r/Eh^3$, and $p = \left[3(1-\nu^2)\right]^{3/2}a^4q/Eh^4$ are introduced, the von Kármán's axisymmetric equations for circular plate under uniform pressure can be reduced into the form [41]

$$\begin{aligned} y^2\tfrac{d^2\varphi}{dy^2} &= \varphi(y)S(y) + y^2p \\ y^2\tfrac{d^2S}{dy^2} &= -\tfrac{1}{2}\varphi^2(y) \quad 0 < y < 1 \end{aligned} \tag{1.16}$$

and the boundary conditions

$$
\begin{aligned}
&y = 0 : \varphi(y) = S(y) = 0 \\
&y = 1 : \varphi(y) = \frac{\lambda}{\lambda - 1}\frac{d\varphi}{dy}; \; S(y) = \frac{\mu}{\mu - 1}\frac{dS}{dy}.
\end{aligned} \tag{1.17}
$$

Here, r, w, N_r, q are the axial coordinate, deflection, axial inner force, and magnitude of applied pressure, respectively; and E, ν, a represent Young's modulus, Poisson's ratio, and radius of the circular plate. λ and μ are the coefficients corresponding to the supported form of the outer boundary of the plate.

(1.1) *Series expansion method*

When the plate is subjected to uniform pressure, we can demonstrate that the set of solution functions is of the form $\{y^i; i = 1, 2, \ldots, \infty\}$ such that the solution can be expressed by the form:

$$
\varphi(y) = \sum_{i=1}^{\infty} A_i y^i, \quad S(y) = \sum_{i=1}^{\infty} B_i y^i \tag{1.18}
$$

The "exact" solution is taken by substituting Eq. (1.18) into the governing equations of Eq. (1.16), i.e.,

$$
\begin{aligned}
&\sum_{i=2}^{\infty} i(i-1)A_i y^i = \sum_{i=1}^{\infty}\sum_{j=1}^{\infty} A_i B_j y^{i+j} + y^2 p \\
&\sum_{i=2}^{\infty} i(i-1)B_i y^i = -\tfrac{1}{2}\sum_{i=1}^{\infty}\sum_{j=1}^{\infty} A_i A_j y^{i+j}
\end{aligned}. \tag{1.19}
$$

By comparing the same terms of $y^i (i = 2, 3, \ldots)$, then a set of coefficient recurrence relationships of $A_i, B_i (i = 2, 3, \ldots)$ in terms of A_1, B_1 can be obtained in the form:

$$
\begin{aligned}
&A_2 = \tfrac{1}{2}(A_1 B_1 + p); \; B_2 = \tfrac{1}{2}A_1^2 \quad i \geq 3 \\
&A_i = \tfrac{1}{i(i-1)}\sum_{j=1}^{i-1} A_j B_{i-j}; \; B_i = -\tfrac{1}{2i(i-1)}\sum_{j=1}^{i-1} A_j A_{i-j}
\end{aligned}. \tag{1.20}
$$

It is obvious that the first two boundary conditions of Eq. (1.17) are auto-satisfied by Eq. (1.18). After that, substituting the obtained recurrence relationships of Eq. (1.20) into the boundary conditions of Eq. (1.17), one gets a set of nonlinear algebraic equations on the independent unknown coefficients A_1 and B_1. By solving the algebraic equations, A_1 and B_1 are gained for a pregiven load p, then the solution of the von Kármán's plate equations are gained. It is evident that such solution is strongly dependent on the selection of truncated terms since the algebraic equations on A_1 and B_1 are relevant to infinite coefficients expanded. In the theory of solutions, such solution obtained is referred to strong solution because the differential equation on the region and boundary conditions are strictly satisfied when the expansion is approaching infinite. However, the infinite expansion is impossible in practical calculations. In such case, the expansion of Eq. (1.18) is always truncated by a finite form such as

$$\varphi(y) = \sum_{i=1}^{N} A_i y^i; \quad S(y) = \sum_{i=1}^{N} B_i y^i, \tag{1.21}$$

or the coefficient recurrences in Eq. (1.20) are truncated at $i = N \gg 1$. Except for the truncation errors of such solution existed in practical finite calculations, one can find that the truncated approximate solutions in Eq. (1.21) do not fully or equally appeared in the two sides of the differential equations of Eq. (1.16) since the highest order of power series in the nonlinear terms should be no higher than the highest order of them for linear terms. Thus, there are some errors in such solution in practical calculations, which are generated from both the finite truncation and the inherent unequal status of the approximate solution itself.

(1.2) *Regular perturbation method*

The other analytic method is based on a perturbation with a small perturbation parameter ε preselected. Here, we restrict our attention to the regular perturbation method, where the solution is expressed by the form:

$$\varphi(y) = \sum_{i=1}^{\infty} \varphi_i(y)\varepsilon^i, \quad S(y) = \sum_{i=1}^{\infty} S_i(y)\varepsilon^i, \tag{1.22}$$

in which, $\varphi_i(y)$ and $S_i(y)$ $(i = 0, 1, 2, \ldots, \infty)$ are the unknown functions to be determined by the perturbation program. Substituting Eq. (1.22) into Eqs. (1.16)–(1.17) and comparing the samsse order of ε^i, one gets the following basic equations:

$$\varepsilon^1 : y^2\frac{d^2\varphi_1}{dy^2} = y^2 p, \quad \frac{d^2 S_1}{dy^2} = 0 \quad \text{, i.e., linear for small deflection}$$

$$\varepsilon^2 : y^2\frac{d^2\varphi_2}{dy^2} = \varphi_1(y)S_1(y); \quad y^2\frac{d^2 S_2}{dy^2} = -\frac{1}{2}\varphi_1^2(y)$$

$$\varepsilon^3 : y^2\frac{d^2\varphi_3}{dy^2} = \varphi_1(y)S_2(y) + \varphi_2(y)S_1(y); \quad y^2\frac{d^2 S_3}{dy^2} = -\varphi_1(y)\varphi_2(y)$$

$$\varepsilon^4 : \begin{array}{l} y^2\frac{d^2\varphi_4}{dy^2} = \varphi_1(y)S_3(y) + \varphi_2(y)S_2(y) + \varphi_3(y)S_1(y); \\ y^2\frac{d^2 S_4}{dy^2} = -[\varphi_1(y)\varphi_3(y) + \frac{1}{2}\varphi_2(y)\varphi_2(y)] \end{array}$$

$$\cdots$$

$$\varepsilon^n : y^2\frac{d^2\varphi_n}{dy^2} = \sum\nolimits_{j=1}^{n-1}\varphi_j(y)S_{n-j}(y); \quad y^2\frac{d^2S_n}{dy^2} = -\frac{1}{2}\sum\nolimits_{j=1}^{n-1}\varphi_j(y)\varphi_{n-j}(y)$$

$$\cdots$$

According to the above perturbation solution program, it is evident that the nonlinear von Kármán's equations are reduced into a set of linear governing equations with infinity recurrence associated with their same boundary equations of Eq. (1.17). The first-order solution can be easily gained by the linear case of the plate with small deflection. After that, substituting the gained first-order solution into the two-order equations, there is also no difficulty for one to get its solution $\varphi_2(y)$ and $S_2(y)$. To take this recurrence step by step, one can get arbitrary order of the perturbation solution. The quantitative results comparing with its experiment data reveal that the accuracy of the perturbation solution is dependent on the selection of the perturbation parameter, e.g., the results of Qian's perturbation solution, where the central deflection is selected as the parameter, are much better than those of Vincent's perturbation solution where the pressure is chosen as the perturbation parameter.

The author of this book and his colleagues demonstrated that (1) the unknown functions in the above perturbation program all belong to the power function set $\{y^i; i = 1, 2, \ldots, \infty\}$ such that the perturbation program can be performed by computer recurrence; (2) the solution convergence of both the series expansion and the perturbations above were demonstrated when the applied force is not large enough, and the convergence is uniform in the domain $y \in [0, 1][0, 1]$. When the load becomes larger enough, the magnitudes of nonlinear terms in von Kármán's equations increase as the deflection increases, which the governing equations enter into the case of strong nonlinearity. In such case of strong nonlinearity, the quantitative results of the above analytical solution methods are always inefficient if no special technique is used. By the way, the demonstration of convergence for above analytical methods of the von Kármán's equations with axisymmetric deformation and some quantitative solutions of them were also given by the author and his colleague.

(2) *Numerical Methods*:

No matter what numerical methods, e.g., FED, FEM, and WRM, are mentioned above, their approximate solutions can be uniformly expressed in the form like Eq. (1.14), i.e.,

$$f(x) = \sum_{i=0}^{N} c_i f_i(x) + \eta_N(x) \approx \sum_{i=0}^{N} c_i f_i(x), \tag{1.23}$$

in which $\eta_N(x)$ is an error function when a set of base functions are selected. In the conventional program of numerical methods, the base functions $f_i(x)$ can be generated by a suitable format after the region is meshed by finite grids. Usually, the base functions do not satisfy differential equations of the problem. In the conventional numerical method, the approximate solution like the form of Eq. (1.23) is directly substituted into the equations of the boundary-value problem. In the following, we use the typical nonlinear Batra equation to display how they are working.

Example 1.4 *To solve the Batra equation.* The Batra equation with two-point boundary conditions is written as

$$\frac{d^2u}{dx^2} + \lambda e^u = g(x) \quad x \in (0, 1), \tag{1.24}$$

$$u(0) = u(1) = 0.$$

(2.1) *FDM*

In the FDM, the derivations at the nodes x_j are replaced by node values of unknown functions after the region is meshed by N grids, e.g., for the derivation of second order in Eq. (1.24), the central difference is always employed in the form

$$\left(\frac{d^2u}{dx^2}\right)|_{x_j} = \frac{u_{j+1} - 2u_j + u_{j-1}}{h^2} + \frac{1}{12}h^2\left(\frac{d^4u}{dx^4}\right)|_{x_j} + \cdots = \frac{u_{j+1} - 2u_j + u_{j-1}}{h^2} + O_j(h^2). \tag{1.25}$$

Here, $h = 1/N$. Substituting Eq. (1.25) into Eq. (1.24), one gets

$$\frac{u_{j+1} - 2u_j + u_{j-1}}{h^2} + \lambda e^{u_j} + O_j(h^2) = g_j \quad j = 1, 2, \ldots, N - 1. \tag{1.26}$$

Here, $u_j = u(x_j)$ and $g_j = g(x_j)$, and x_j are the node coordinates. Adding the two boundary conditions, Eq. (1.26) constitute a system of nonlinear algebraic equations on the node values of the unknown function. Due to the residuals $O_j(h^2)$ cannot be known, $O_j(h^2)$= 0 are forced to be set in the practical calculations since they are considered by a set of small quantities, which is the main source of error of such approximation solution. Especially when the magnitude of the applied term increases to a large value, such error may lead the approximate solution being false.

(2.2) *WRM or Galerkin's method*

In the WRM, the approximation of the unknown function is taken under a set of preselected standard orthogonal base functions $u_i(x)$, like Eq. (1.12), i.e.,

$$u(x) = \sum_{i=0}^{N} c_i u_i(x) + \eta_N(x). \tag{1.27}$$

Substituting Eq. (1.27) into Eq. (1.24), one gets

$$\sum_{i=0}^{N} c_i \frac{d^2 u_i}{dx^2} + \eta_N(x) + \lambda \exp\left(\sum_{i=0}^{N} c_i u_i(x) + \eta_N(x)\right) = g(x). \tag{1.28}$$

Since the unknown coefficients c_i are located in the power term shown in Eq. (1.28), the further calculations under such direction substituting become high difficult no matter what calculation arithmetic is employed. Usually, the nonlinear function e^u should be expanded by Tayler's series at the point $u = 0$, i.e.,

$$e^{u(x)} = 1 + u(x) + \frac{1}{2}u^2(x) + \frac{1}{6}u^3(x) + \cdots. \tag{1.29}$$

For the simplicity of description, here, only the first three terms are taken into account in the calculation although such truncation may lead to some errors. Then, Eq. (1.28) becomes the form:

$$\sum_{i=0}^{N} c_i \frac{d^2 u_i}{dx^2} + \lambda \left\{ 1 + \sum_{i=0}^{N} c_i u_i(x) + \frac{1}{2} \sum_{i=0}^{N} \sum_{j=0}^{N} c_i c_j u_i(x) u_j(x) \right\} + \eta(x) = g(x). \tag{1.30}$$

Here, $\eta(x) = \frac{d^2 \eta_N(x)}{dx^2} + \eta_N(x) + 2\lambda \eta_N(x) \sum_{i=0}^{N} c_i u_i(x) + \lambda \eta_N^2(x)$. By using Galerkin's method to Eq. (1.30), that is to take the calculations of $<u_i(x), u_k(x)>$, $<1, u_j(x)>$, $<g(x), u_j(x)>$, $<u_i(x)u_j(x), u_k(x)>$, and $<d^2u_i(x)/dx^2, u_k(x)>$, here, $<u_i(x), u_k(x)> = \int_0^1 u_i(x) u_k(x) dx$, a set of nonlinear algebraic equations on the unknown c_i can be gained after $\eta(x) \equiv 0$ is forcedly selected. By solving the induced nonlinear algebraic equations associated with the boundary conditions, one gets the unknowns c_i and further a solution for the nonlinear boundary-value problem is gained for an applied $g(x)$. With the application changing, the above solution program should be reconducted again. It is obvious that $<\eta(x), u_k(x)> \neq 0$ in the above program. Hence, such forced $<\eta(x), u_k(x)> = 0$ or $\eta(x) \equiv 0$ may generate a large difference between the true solution and the approximation when the magnitude of application increases.

(2.3) *FEM*

In the FEM, the base functions are usually selected by the Lagrange interpolation functions, while the unknown coefficients are the node values of the unknown function. To the governing equation of (1.24), the approximation can be expressed by the form:

$$u(x) = \sum_{i=0}^{N} u_i N_i(x) + R(x) \tag{1.31}$$

in which $N_i(x)$ and $R(x)$ are the formal functions and error function, respectively. For example, they can be formulated by

$$N_i(x) = \prod_{\substack{j=0 \\ j \neq i}}^{N-1} \frac{x - x_j}{x_i - x_j}, \quad R(x) = \frac{1}{(N+1)!} f^{(n+1)}(\xi) \prod_{j=0}^{N} (x - x_j). \tag{1.32}$$

It is evident that $N_i(x_j) = \delta_{ij}(i, j = 0, 1, 2, \ldots, N)$. There is the same problem displayed in Eq. (1.28), i.e., the approximation in Eq. (1.31) cannot be directly substituted into Eq. (1.24); otherwise, the calculations cannot be performed. As done in Eq. (1.30) for Eq. (1.32), one gets

$$\sum_{i=0}^{N} c_i \frac{d^2 N_i}{dx^2} + \lambda \left\{ 1 + \sum_{i=0}^{N} c_i N_i(x) + \frac{1}{2} \sum_{i=0}^{N} \sum_{j=0}^{N} c_i c_j N_i(x) N_j(x) \right\} + \eta(x) = g(x), \tag{1.33}$$

where $\eta(x) = \frac{d^2 R(x)}{dx^2} + R(x) + 2\lambda R(x) \sum_{i=0}^{N} c_i u_i(x) + \lambda R^2(x)$. By the Galerkin FEM, we have

$$\sum_{i=0}^{N} c_i < \frac{d^2 N_i}{dx^2}, N_k(x) > + \lambda \left\{ \int_0^1 N_k(x) dx + \sum_{i=0}^{N} c_i < N_i(x), N_k(x) > \right.$$
$$\left. + \frac{1}{2} \sum_{i=0}^{N} \sum_{j=0}^{N} c_i c_j < N_i(x) N_j(x), N_k(x) > \right\} + < \eta(x), N_k(x) > = < g(x), N_k(x) > \tag{1.34}$$

In the FEM, the error terms $<\eta(x), N_k(x)>$ are not considered in the practical calculations since they can be disappeared in mathematics as $<\eta(x), u_k(x)> \neq 0$in WRM or Galerkin's method.

The above examples display that all about approximate solutions for a nonlinear problem cannot make the errors disappeared in mathematics except for they are forced to be zero. Such performance is always implied in the calculations of approximate solution programs, which may be the main reason why they are suitable to the weak nonlinear problems well rather than to a strong nonlinear problem if no specific technique is employed.

In summary, the introduction of the above methods shows that the accuracy of the approximate methods is highly dependent on the strong nonlinearity of a system [23–25, 27–33, 48–77], except for a few analytical methods that can obtain closed-form solutions for nonlinear problems within an extremely narrow scope [15–19]. For the nonlinear problems when no closed form of analytical solution is found, as similar to the case of linearly ordinary differential equation(s) with constant coefficients, in strictly speaking, all solution methods involved a form of infinite calculation steps in theory. When some finite truncations are forced in such inherent infinite calculations, the accuracy of approximate solution is highly dependent on the convergence, closure, and calculation consumption features of solution arithmetic. Even if the accuracy can be promoted finer by a special technique for one nonlinear problem, the corresponding calculations may increase larger. However, one such special technique for one nonlinear problem is always inefficient for other nonlinear problems. This situation forces us to deal with different problems by means of one-by-one special techniques. In other words, there is no universal format for solving weak and strong nonlinear problems, which is a large challenge that we encounter presently. To conquer this challenge, the author and their colleagues recently proposed a wavelet-based method for solving nonlinear problems, which is the main consideration in this book.

1.4 Brief Review of Wavelet Methods

As showing in the title of this textbook, we will mainly introduce the numerical method on the basis of the wavelet method what we reformed. Since the wavelet theory and methods were proposed, most of its applications have been focused on signal and image processing, to which the wavelet methods have emerged or provided a powerful tool in mathematics.

A "wavelet" is an emerging and rapidly developing field in applied mathematics and engineering. It provides a new technology for time–frequency analysis and a new basis for function characterization [78–80]. In addition, as a set of mathematical theories with rich connotations, a wavelet has high application values and broad application prospects in many disciplines [78–80]. Although the wavelet can be traced back to Haar wavelet in 1910, however, a landmark breakthrough in basic theory is the multi-resolution analysis proposed during the 1980s, and then its applications have made remarkable achievements. This analysis unifies many previous wavelet construction methods and provides a unified mathematical framework for wavelet analysis [78–80]. The Haar, Daubechies (DB), Mexican hat, Morlet, and Meyer wavelets are several classic wavelet functions.

The wavelet theory or method is developing and developed from Fourier transform for signal processing. In such signal decomposition, it has been known that the frequencies obtained are constants without varying with time. However, a real signal progress has always a variation of frequency with time. In order to capture this characteristic and those jumping signals in a short time, like earthquakes and chemical reactions, the earlier way is conducted by adding a moving short-time window in the original signal progress, then using the Fourier transform, such signal processing is named by the short-time Fourier transform. Due to the signal in the window equaling to the original one and zero signal being selected outside the window, i.e., there is signal jumps at the ends of window, it is found that there are strong disturbances or noises in the reconstructed signal progress when the short-time Fourier transform is employed, which directly leads to many filter theories for filtering such induced noises as possible as one can. Another disadvantage of the windowed Fourier analysis is that the frequency and time resolution is a constant due to the fixed window length. This is often not the most desired resolution. For low frequencies, often a good frequency resolution is required over a good time resolution. For high frequencies, the time resolution is more important. In order to eliminate these disadvantages, many filtering methods (mathematically called filters) have been established. Further, the signal processing required in earthquakes and chemical reactions motivated the practical establishment of modern wavelet methods. After that, some mathematicians entered this area for establishing a theoretical framework what is a wavelet, and then many new wavelets have been established out in accordance with trade-off of some prerequired properties, e.g., compacted support, orthogonality, smoothness, interpolation, and symmetry, etc., which are directly relevant to the ability and calculation consumption of a wavelet in processing a signal function.

According to the wavelet theoretical framework (see Chap. 2), it has been known that a wavelet consists of a mother's scaling function $\varphi(x)$ and a mother's wavelet function $\psi(x)$, and they are established by satisfying some previous properties under the wavelet framework. Then the base scaling functions $\varphi_{n,k}(x)$ and base wavelet functions $\psi_{n,k}(x)$ at each one resolution level n (integer) can be gained under the mother's scaling function and wavelet function, here integer k represents a translation. For any square integrable function of the signal, $f \in \mathbf{L}^2(\mathbf{R})$, the signal function can be expanded by one set of base scaling functions at one resolution level added the summation of its expansions on the base wavelet functions at all resolution levels from the base scaling function level to infinite. That is,

$$f(x) = P_n f(x) + \sum_{i=n}^{\infty} Q_i f(x) \tag{1.35}$$

in which

$$P_n f(x) \equiv \sum_{k \in \mathbf{Z}} c_{n,k}\varphi_{n,k}(x),\ Q_i f = \sum_{k} d_{i,k}\psi_{i,k}(x). \tag{1.36}$$

The expansion coefficients $c_{n,k}$ and $d_{i,k}$ are also called the decomposition coefficients, and the calculation progress for $c_{n,k}$ and $d_{i,k}$ are referred to the signal decomposition. Once the decomposition coefficients are gained, the summation calculations of Eq. (1.35) or (1.36) are attributed to the reconstruction of the signal function $f(x)$ under some meanings. The wavelet framework tells us that the established wavelet has the following relationship:

$$P_{n+1} f(x) = P_n f(x) + Q_n f(x) \tag{1.37}$$

which provides a foundation of multi-level analysis to the decomposition coefficients. At present, the wavelet framework shows that the infinite summation of Eq. (1.35)

Table 1.1 Comparison of properties of several popular wavelets

Property	Shannon wavelet	Meyer wavelet	Orthogonal spline wavelet	B-spline wavelet	DB wavelet	Interpolating wavelet	Coiflet wavelet
Orthogonality	Yes	Yes	Yes	No	Yes	No	Yes
Compacted support	No	No	No	Yes	Yes	Yes	Yes
Symmetry	Yes	Yes	Yes	Yes[b]	No	Yes	Yes[c]
Analytical expression[a]	Yes	Yes	Yes	Yes	No	No	No
Interpolation	Yes	No	No	No	No	Yes	Yes[d]

[a]Existence in the time or frequency domain is considered
[b]Symmetry of the even order, anti-symmetry of the odd order
[c]Symmetry of the even order, anti-symmetry of the odd order
[d]Symmetry of the even order, anti-symmetry of the odd order

is always convergent in the square integrable space when the wavelet satisfies the conditions of the framework. Except for such conditions to establish a wavelet, we should also add some other conditions according to other different properties required, thus, different wavelets with different properties can be established in the wavelet framework. Table 1.1 displays a comparison of the main properties of several commonly used wavelets (wavelets with vanishing moments are not listed in the table).

Especially for solving a set of determinant differential equations, the properties of orthogonality, compactness, symmetry, vanishing moment, and interpolation are the most important properties in the numerical method. For example, the orthogonality can provide an improved stability to an algorithm, and the strip matrix can be developed through compactness makes performing large-scale analysis and calculation possibly. The linear phase resulting from symmetry can improve calculation accuracy. An appropriate vanishing moment or differentiability is necessary to solve differential equations. Lastly, interpolation provides convenience when applying essential boundary conditions.

The first equation of Eq. (1.36) has the property of low-frequency pass at the level n, while the second term in the right hand of Eq. (1.35) has the property of high-frequency pass with levels equal to n and higher. These properties reveal that the first equation of Eq. (1.36) can be used as an approximation of the original function $f(x)$ with a relatively smooth part, and the second equation of Eq. (1.36) can capture the signals of $f(x)$ with high-frequency part at each resolution level. If Eq. (1.35) is employed in the numerical solution of a boundary-value problem, it is obvious that the part of high-frequency pass becomes an error function in the truncation. In such a situation, we rewrite Eq. (1.35) in the form:

$$f(x) = \sum_{k\in\mathbf{Z}} c_{n,k}\varphi_{n,k}(x) + \sum_{i=n}^{\infty} Q_i f(x) \approx \sum_{k\in\mathbf{Z}} c_{n,k}\varphi_{n,k}(x). \tag{1.38}$$

Once a standard ordinary wavelet is selected, i.e.,

$$<\varphi_{n,k}(x), \varphi_{n,l}(x)>=\delta_{kl}, \quad <\psi_{n,k}(x), \psi_{m,l}(x)>=\delta_{nm}\delta_{kl}, \tag{1.39a,b}$$

$$<\varphi_{n,k}(x), \psi_{m,l}(x)>=0 \quad m \geq n, \tag{1.39c}$$

the coefficients $c_{n,k}$ can be directly calculated by

$$c_{n,k} =< f(x), \varphi_{n,k}(x) >= \int_R f(x)\varphi_{n,k}(x)dx. \tag{1.40}$$

With the extensive developments of wavelet methods applying in the signal or image processing, some efforts have been made for applying the wavelet method in solving those boundary-value problems containing in dynamic problems, fluid mechanics. and solid mechanics, etc. However, the researches did not show a powerful ability more than those conventional numerical methods mentioned above when the wavelet method is employed independently. Here, we use a simple example to show why it is.

For simplicity, we still use the example of the nonlinear problem of Eq. (1.24) solved by the conventional wavelet methods. Here, the standard orthogonal wavelet is selected. In such case, the approximate solution with wavelet base functions are re-written in the form:

$$u(x) = \sum_{k\in\mathbf{Z}} c_{n,k}\varphi_{n,k}(x) + \sum_{i=n}^{\infty} Q_i u(x) \approx \sum_{k\in\mathbf{Z}} c_{n,k}\varphi_{n,k}(x), \tag{1.41}$$

in which, $c_{n,k}$ are the unknowns to be determined by a program of calculations. As the same as done in the conventional numerical methods shown above, the conventional wavelet solution methods were performed by directly substituting Eq. (1.41) into the expanded equation of (1.24) rather than itself with the same reason shown above. Such substitution leads to

$$\sum_k \tilde{c}_{n,k}\varphi_{n,k}(x) + \lambda\left\{1 + \sum_k c_{n,k}\varphi_{n,k}(x) + \frac{1}{2}\sum_k\sum_l c_{n,k}c_{n,l}\varphi_{n,k}(x)\varphi_{n,l}(x)\right\}. \\ + \eta(x) = g(x) \tag{1.42}$$

Here,

$$\eta(x) = \sum_{i=n}^{\infty} Q_i u(x) + \sum_{i=n}^{\infty} Q_i \left\{ \frac{d^2}{dx^2} u(x) \right\} + 2\lambda \left[\sum_k c_{n,k} \varphi_{n,k}(x) \right] \left[\sum_{i=n}^{\infty} Q_i u(x) \right] + \lambda \left[\sum_{i=n}^{\infty} Q_i u(x) \right]^2 \tag{1.43}$$

and the second derivation function is expanded by the form:

$$\begin{aligned} \frac{d^2 u(x)}{dx^2} &= v(x) = \sum_{k \in \mathbf{Z}} \tilde{c}_{n,k} \varphi_{n,k}(x) + \sum_{i=n}^{\infty} Q_i v(x) \\ &= \sum_{k \in \mathbf{Z}} \tilde{c}_{n,k} \varphi_{n,k}(x) + \sum_{i=n}^{\infty} Q_i \left\{ \frac{d^2}{dx^2} u(x) \right\}, \end{aligned} \tag{1 44}$$

In which the expansion coefficients $\tilde{c}_{n,k}$ can be calculated by the coefficients $c_{n,k}$ in the connection coefficient method (see detail in Chap. 2). By means of Galerkin's method to the induced Eq. (1.41), one gets

$$\begin{aligned} &\tilde{c}_{n,j} + \lambda \Big\{ 2^{-n/2} + c_{n,j} + \tfrac{1}{2} \sum_k \sum_l c_{n,k} c_{n,l} \textstyle\int_R \varphi_{n,k}(x) \varphi_{n,l}(x) \varphi_{n,j}(x) dx \\ &+ \textstyle\int_R \eta(x) \varphi_{n,j}(x) dx = \textstyle\int_R g(x) \varphi_{n,j}(x) dx \quad k, j, l \in Z \end{aligned} \tag{1.45}$$

Considering the orthogonal property of Eq. (1.39c), we can get

$$< \varphi_{n,j}(x), \sum_{i=n}^{\infty} Q_i v(x) > = \sum_{i=n}^{\infty} \int_R \varphi_{n,j}(x) Q_i u(x) dx = 0 \tag{1.46}$$

for any function $u(x) \in L^2(R)$. Further, the integration of $\int_R \eta(x)\varphi_{n,j}(x)dx$ in Eq. (1.44) becomes

$$\begin{aligned} \int_R \eta(x) \varphi_{n,j}(x) dx = &\int_R \varphi_{n,k}(x) \\ &\left\{ 2\lambda \left[\sum_k c_{n,k} \varphi_{n,k}(x) \right] \left[\sum_{i=n}^{\infty} Q_i u(x) \right] + \lambda \left[\sum_{i=n}^{\infty} Q_i u(x) \right]^2 \right\} dx \neq 0. \end{aligned} \tag{1.47}$$

In the conventional wavelet numerical calculations for solving a nonlinear boundary-value problem, either the truncated error $\eta(x) \approx 0$ or its averages $< \eta(x), \varphi_{n,j}(x) > \approx 0$ are forcedly selected as same as done in the conventional numerical methods shown above. Such situation of the approximate solution coupled with its truncated errors displayed in all above solution methods is the possible main origination that the methods cannot be directly used in the strong nonlinear problems. In this textbook, this challenge will be conquered by reforming the generation of the generalized Coiflets.

References

1. Newton I (1802) Mathematical principles of natural philosophy. A. Strahan
2. Hartog JP (1948) Mechanics. General Publishing Company, Ontario
3. Wei N (2004) Basis and application of nonlinear science. Science Press, Beijing
4. Nicolis C (1995) Introduction to nonlinear science. Cambridge University Press, Cambridge
5. Lui L (1997) Introduction to nonlinear physics. Springer-Verlag, New York
6. Wilson JR (1981) Nonlinear system theory. The Johns Hopkins University Press, Maryland
7. Scott AC (2007) The Nonlinear universe: Chaos, emergence, life. Springer, Berlin Heidelberg, New York
8. Hu HY (2006) Applied nonlinear dynamics. Aviation Industry Press, Beijing
9. Thompson JMT, Stewart HB (2002) Nonlinear dynamics and chaos. Wiley, Chichester
10. Lin JZ, Ruan XD, Chen BG et al (2005) Fluid mechanics. Tsinghua University Press, Beijing
11. Zheng XJ, Liu XE (2005) A nonlinear constitutive model for Terfenol-D rods. J Appl Phys 97:053901
12. Zhu ZX, Huang Y (1998) Science in the 21st century-nonlinear science. Chin Sci Bull 7:1–2
13. Tadmor E (2012) A review of numerical methods for nonlinear partial differential equations. Bull Am Math Soc 49:507–554
14. Neilan M, Glowinski R, Feng X (2013) Recent developments in numerical methods for fully nonlinear second order partial differential equations. Soc Indus Appl Math Rev 55:205–267
15. Acton JR, Squire PT (1985) Solving equations with physical understanding. Adam Hilger Ltd., Boston
16. Elwakil SA, El-labany SK, Zahran MA (2002) Modified extended tanh-function method for solving nonlinear partial differential equations. Phys Lett A 299:179–188
17. Sedov LI (1982) Similarity methods and dimensional analysis in mechanics. Science Press, Beijing
18. Yan XL (2004) The exact solution of a nonlinear equation. Economic Science Press, Beijing
19. Lu WR (2003) Variational methods in differential equations. Science Press, Beijing
20. Struwe M (2008) Variational methods, applications to nonlinear partial differential equations and Hamiltonian systems. Springer-Verlag, Berlin
21. Li JC, Zhou XC (1998) Asymptotic methods in mathematical physics. Science Press, Beijing
22. Yang BJ, Zhao YF (2003) Methods of advanced mathematical physics. Beijing University of Posts and Telecommunications Press, Beijing
23. Cole JD (1968) Perturbation method in applied mathematics. Blaisdell Publishing Company, Waltham
24. Nayfeh AH (2000) Perturbation method. Wiley, New York
25. Von Dyke M (1975) Perturbation methods in fluid mechanics. The Parabolic Press, Stanford
26. Zheng XJ, Zhou YH (1989) On convergence of interpolated iterative method of geometrically nonlinear equations of circular plates. Sci China A 32:316–327
27. Von Kármán T (1910) Festigkeits problem in maschinenbau. Encyklopädie Der Mathematischen Wissenschaften 4:311–385
28. Way S (1934) Bending of circular plate with large deflection. J Appl Mech 56:627–636
29. Vincent JJ (1931) The bending of a thin circular plate. Phil Mag 12:185–196
30. Chien W (1947) Large deflection of a circular clamped plate under uniform pressure. Chin J Phys 7:102–113
31. Chien W, Yeh K (1954) On the large deflection of circular plates. China Sci 3:405–436

32. Qian WC (1989) A selection of scientific papers by Qian Weichang. Fujian Education Press, Fuzhou
33. Qian WC, Lin HS, Hu HC, Yeh KY (1954) Large deflection of elastic circular thin plate. Science Press, Beijing
34. Zheng XJ (1990) Theory and application of large deflection of circular thin plate. Jilin Science & Technology Publishing House, Changchun
35. Yeh KY, Zheng XJ, Zhou YH (1989) An analytical formula of the exact solution to von Kármán's equations of a circular plate under a concentrated load. Int J Non-Linear Mech 24:551–560
36. Zheng XJ, Lee J (1995) On the convergence of the Chien's perturbation method for von Kármán plate equations. Int J Eng Sci 33:1085–1094
37. Zheng XJ, Zhou YH (1990) Analytical formulas of solutions of geometrically nonlinear equations of axisymmetric plates and shallow shells. Acta Mech Sin 6:69–80
38. Zhou YH, Zheng XJ (1989) On the range of applicability of von Karman plate equations. J Appl Mech 56:724–726
39. Yeh KY, Zheng XJ, Wang XZ (1990) On some properties and calculation of the exact solution to von Kármán's equations of circular plates under a concentrated load. Int J Non-Linear Mech 25:17–26
40. Zheng XJ, Zhou YH (1988) The exact solution of large deflection problem of circular thin plate of elastic foundation under concentrated load. Chin J Theor Appl Mech 20:161–171
41. Zheng XJ, Zhou YH (1987) On exact solution of Karman's equations of rigid clamped circular plate and shallow spherical shell under a concentrated load. Appl Math Mech 8:1057–1068
42. He JH (2006) Some asymptotic methods for strongly nonlinear equations. Int J Mod Phys B 20:1141–1199
43. Chapra SC, Canale RP (2010) Numerical methods for engineers. McGraw-Hill, New York
44. Atkinson KA (1989) An Introduction to numerical analysis. Wiley, New York
45. Lambert JD (1991) Numerical methods for ordinary differential systems: the Initial Value Problem. Wiley, New York
46. Khiyal MSH (2005) Implementation of Newmark's method for second order initial value problems. J Appl Sci 5:402–410
47. Iserles A (1996) A first course in the numerical analysis of differential equations. Cambridge University Press, Cambridge
48. Causon DM, Mingham CG (2010) Introductory finite difference method for PDEs. Ventus Publishing, London
49. Delfour M, Fortin M, Payr G (1981) Finite-difference solutions of a non-linear Schrödinger equation. J Comput Phys 44:277–288
50. Chen YM (1975) Numerical computation of dynamic stress intensity factors by a Lagrangian finite-difference method. Eng Fract Mech 7:653–660
51. Brian PLT (1961) A finite-difference method of high-order accuracy for the solution of three-dimensional transient heat conduction problems. AIChE J 7:367–370
52. Beam RM, Warming RF (1978) An implicit factored scheme for the compressible Navier-Sstokes equations. Am Inst Aeronaut Astronaut J 16:393–402
53. Narasimhan TN, Witherspoon PA (1976) An integrated finite difference method for analyzing fluid flow in porous media. Water Resour Res 12:57–64
54. Yuste SB, Quintana-Murillo J (2012) A finite difference method with non-uniform timesteps for fractional diffusion equations. Comput Phys Commun 183:2594–2600
55. Buckmire R (2004) Application of a Mickens finite-difference scheme to the cylindrical Bratu-Gelfand problem. Numer Methods Partial Differen Equations Int J 20:327–337
56. Qiu JB (1991) Theory and application of weighted residual value method. Aerospace Press, Beijing
57. Wang XZ (2006) Computational mechanics. Lanzhou University Press, Lanzhou
58. Bramble JH, Schatz AH (1971) Least squares methods for 2mth order elliptic boundary-value problems. Math Comput 25:1–32
59. Xiu D, Hesthaven JS (2006) High-order collocation methods for differential equations with random inputs. Soc Indus Appl Math J Sci Comput 27:1118–1139

60. Cockburn B, Shu CW (2006) The local discontinuous Galerkin method for time-dependent convection-diffusion systems. Soc Indus Appl Math J Numer Anal 35:2440–2463
61. Wang P, Yao Y, Tulin MP (1995) An efficient numerical tank for non-linear water waves, based on the multi-subdomain approach with BEM. Int J Numer Meth Fluids 20:1315–1336
62. Feng X, Neilan M (2009) Vanishing moment method and moment solutions for fully nonlinear second order partial differential equations. J Sci Comput 38:74–98
63. Atluri SN, Zhu T (1998) A new Meshless Local Petrov-Galerkin approach in computational mechanics. Comput Mech 22:117–127
64. Zhu T, Zhang J, Atluri SN (1998) A meshless local boundary integral equation method for solving nonlinear problems. Comput Mech 22:174–186
65. Zhang X, Liu Y (2004) Meshless method. Tsinghua University Press, Beijing
66. Sauter SA, Schwab C (2011) Boundary element methods. Springer-Verlag, Berlin
67. Sutradhar A, Paulino GH, Gray LJ (2008) Symmetric Galerkin boundary element method. Springer-Verlag, Berlin
68. Caglar H, Caglar N, Özer M (2008) Fifth-degree B-spline solution for nonlinear fourth-order problems with separated boundary conditions. J Phys Conf Ser 96:012031
69. Wang XC (2004) Finite element method. Tsinghua University Press, Beijing
70. Desai YM, Eldho TI, Shah AH (2011) Finite element method with applications in engineering. Dorling Kindersley Pvt. Ltd, Noida
71. Wang XC, Shao M (1997) Basic principles and numerical methods of finite element method. Tsinghua University Press, Beijing
72. Chen Y, Lee J, Eskandarian A (2006) Meshless methods in solid mechanics. Springer Science & Business Media, Inc., New York
73. Gu ZQ, Kou GM (1997) Active vibration control. National Defense Industry Press, Beijing
74. Ja YH (1983) Using the shooting method to solve boundary-value problems involving nonlinear coupled-wave equations. Opt Quant Electron 15:529–538
75. Bert CW, Malik M (1996) Differential quadrature method in computational mechanics: a review. Appl Mech Rev 49:1–28
76. Ali AH, Al-Saif ASJ (2008) Adomian decomposition method for solving some models of nonlinear partial differential equations. Basrah J Sci A 26:1–11
77. Liao S (2012) Homotopy analysis method in nonlinear differential equations. Springer-Verlag, Berlin
78. Meyer Y (1992) Wavelets and operators. Cambridge University Press, Cambridge
79. Cui JT (1995) Introduction to wavelet analysis. Xi 'an: Xi 'an Jiaotong University Press
80. Daubechies I (1988) Orthonormal bases of compactly supported wavelets. Commun Pure Appl Math 41:909–996

Chapter 2
Mathematical Framework of Compactly Supported Orthogonal Wavelets

After the wavelet method for processing the jumping signal was proposed in the 1970s, its powerful advantages attract the attention of mathematicians, and the mathematical framework of a wavelet has been gradually established and various wavelets with different anticipatable properties are constructed since 1980s. For simplicity, here, we briefly introduce the compactly supported orthogonal wavelets what we will employ, containing their essential concepts, construction, and main properties, etc.

2.1 Essentials of Compactly Supported Orthogonal Wavelets

Compactly supported orthogonal wavelets are most widely used in practice. After the wavelet method was emerged and then promoted by mathematicians such as Daubechies, the wavelet theory has been raised to the level of mathematical frame, including those general priority conditions that should be met for a wavelet, and what additional conditions can be added for an expected wavelet with some special properties, e.g., compacted support, orthogonal, vanishing moments, and continuous, etc. At present, a wavelet is generated by a trade-off of those special properties under the constraints of the framework. Considering the goal of this book focused on to solve differential equations relevant problems, especially nonlinear equations, and ensuring that the wavelet method what we use has the obvious advantages of high accuracy, low computational complexity, and universal applicability for linear, weak nonlinear, and strong nonlinear problems, in this chapter, we pay our attention to the detail introduction for the compactly supported orthogonal wavelet after some appropriate improvements are made by the author and his colleagues, which is the basis of this book and its solution method has not achieved by other wavelets.

From the wavelet framework, we know that any wavelet system is composed of scaling function, wavelet function, and their base functions. Then such wavelet can

Y.-H. Zhou, *Wavelet Numerical Method and Its Applications in Nonlinear Problems*, Engineering Applications of Computational Methods 6,
https://doi.org/10.1007/978-981-33-6643-5_2

be used to decompose or reconstruct any square integrable function $f \in L^2(R)$ (that is, expansion).

Definition 2.1 A function $f(x)$ defined in the real number set $R = (-\infty, +\infty)$ is called a square integrable function, i.e., $f(x) \in L^2(R)$, if there is the relation

$$\int_{-\infty}^{\infty} |f(x)|^2 dx < \infty. \tag{2.1}$$

Definition 2.2 A function $\varphi(x) \in L^2(R)$ is referred to orthogonal scaling or dilation function if it has the following properties or characteristics:

(1) There is a set of real sequence $\{p_k,\ k \in Z\}$ satisfying

$$\varphi(x) = \sum_{k \in Z} p_k \varphi(2x - k), \tag{2.2}$$

(2) A set of base scaling functions, $\left\{\varphi_{n,k}(x) = 2^{n/2}\varphi(2^n x - k),\ n,\ k \in Z\right\}$, satisfy the orthogonal condition

$$\int_R \varphi_{n,k}(x)\varphi_{n,l}(x)dx = \delta_{k,l} = \begin{cases} 0, & k \neq l \\ 1, & k = l \end{cases}, \tag{2.3}$$

(3) For $\forall f \in L^2(R)$, a transformation is defined as

$$P_n f(x) \equiv \sum_{k \in \mathbf{Z}} c_{n,k}\varphi_{n,k}(x), \quad c_{n,k} \equiv \int_R f(x)\varphi_{n,k}(x)dx, \tag{2.4}$$

and the set of function space

$$V_n = \left\{ \sum_{k \in Z} c_{n,k}\varphi_{n,k}(x), n, k \in Z;\ c_{n,k} \in R \right\} \tag{2.5}$$

has the following relationships [1]:

i. $\{0\} \cdots \subset V_{-2} \subset V_{-1} \subset V_0 \subset V_1 \subset \cdots \subset L^2(R)$
ii. $f(x) \in V_n \leftrightarrow f(2x) \in V_{n+1}$
iii. $f(x) \in V_n \leftrightarrow f(x + k2^{-n}) \in V_n$
iv. $\lim_{n\to\infty} V_n = \bigcup_n V_n$, is dense in $L^2(R)$

v. $\lim_{n\to\infty} \cap V_n = \{0\}$

vi. The set $\{\varphi(x-k)\}$ form unconditional bases for V_0, that is, there exist constants A and B with $0 < A \le B < \infty$ such that

$$A\sum_{k\in Z} |c_k|^2 \le \left\| \sum_{k\in Z} c_k\varphi(x-k) \right\|_2^2 \le B\sum_{k\in Z} |c_k|^2$$

for any sequence $\{c_k\} \in l_2$, where l_2 is the space of all square summable sequences, and $Z(=\{k; -\infty < k < \infty\})$ is the set of integers [2].

Definition 2.3 If there is a real sequence $\{q_k,\ k \in Z\}$ such that the function cxprcssion

$$\psi(x) = \sum_{k\in Z} q_k\varphi(2x-k) \tag{2.6}$$

satisfies the following conditions:

(1) The function sequences

$$\psi_{n,k}(x) = 2^{n/2}\psi(2^n x - k), \tag{2.7}$$

are orthogonal, i.e.,

$$\int_R \psi_{n,k}(x)\psi_{m,l}(x)dx = \delta_{m,n}\delta_{k,l}. \tag{2.8}$$

For arbitrary coefficients $d_{n,k} \in l^2$, the function space spanned by the base functions $\psi_{n,k}(x)$

$$W_n = \left\{ \sum_{k\in Z} d_{n,k}\psi_{n,k}(x),\ n,\ k \in \mathrm{Z} \right\} \tag{2.9}$$

has the following properties:

i. $V_{n+1} = V_n \oplus W_n$

ii. For different integers $m,\ n,\ W_n$ is orthogonal to W_m

iii. $\bigoplus_{j\in Z} W_j = L^2(R)$.

Then the function $\psi(x)$ is called a mother wavelet function or a wavelet function, and $\psi_{n,k}(x)$ is called the base wavelet functions.

A function that satisfies all the conditions in Definitions 2.2 and 2.3 is, respectively, called a scaling function or a wavelet function of the wavelet. On this basis, the corresponding base functions with a resolution level n can be used to represent a function $f(x) \in L^2(R)$ by its expansions on either base scaling functions or base wavelet functions or both. The expansion in terms of base scaling functions has been expressed in Eq. (2.4). According to Eq. (2.9), the expansion by using the base wavelet functions can be written as

$$Q_n f(x) = \sum_{k\in Z} d_{n,k}\psi_{n,k}(x), \quad d_{n,k} \equiv \int_R f(x)\psi_{n,k}(x)dx \quad n,k \in Z \tag{2.10}$$

which has the following characteristics:

$$P_{n+1} f(x) = P_n f(x) + Q_n f(x), \tag{2.11}$$

$$f(x) = \lim_{n\to\infty} P_n f(x) = P_n f(x) + \sum_{i=n}^{\infty} Q_i f(x) = P_n f(x) + \tilde{Q}_n f(x) \tag{2.12}$$

Here, the notation

$$\tilde{Q}_n f(x) = \sum_{i=n}^{\infty} Q_i f(x). \tag{2.13}$$

is employed.

As the wavelet method was originally and successfully used in the field of signal processing, the relevant coefficients p_k and q_k had been named as low-pass filter coefficients and high-frequency filter coefficients, respectively. Usually, the scaling function and wavelet function have a compact support, i.e., they are nonzero only in a finite region and are zero outside this region. Thus, there are only a set of finite nonzero coefficients to be determined, which are related to the size of the compactly supported region of the wavelet. At the same time, it can be seen that the scaling function expansion $P_n f(x)$ is the low-frequency band-pass (i.e., smooth and continuous) part of a signal, while $\tilde{Q}_n f(x)$ is the high-frequency band-pass (i.e., jitter) part of the signal function $f(x)$. The calculation of the coefficients in the second equations of Eqs. (2.4) and (2.10) is called the signal decomposition, while the calculation of the first equations in Eqs. (2.4) and (2.10) is called signal reconstruction. Of course, the former reconstructs the smooth part of the signal, while the latter reconstructs the high-frequency jitter part of the signal. It should be noted that the properties given in Definitions 2.2 and 2.3 will be realized by the subsequent construction procedures of wavelets.

2.2 Conditions for Constructing an Orthogonal Wavelet

From the previous introduction of the general framework, we know that once the filter coefficients p_k and q_k in Eqs. (2.2) and (2.6) are determined by adding some conditions and properties what they should satisfy, then we can generate a wavelet. In the following, we give a construction way for general orthogonal wavelets that satisfies those conditions with necessary properties what we expect.

2.2.1 *General Conditions on Filter Coefficients from Orthogonality*

Assume that the compact support region of the scaling function is $\text{Supp}[\varphi(x)] = [0, \tilde{N}]$, where $\tilde{N}$ is a positive odd integer. This means that $\varphi(x) \neq 0$ only in $[0, \tilde{N}]$, and $\varphi(x) \equiv 0$ outside $[0, \tilde{N}]$. Therefore, there are $2(\tilde{N}+1)$filter coefficients,p_k and q_k ($k = 0, 1, 2, \ldots, \tilde{N}$), in Eqs. (2.2) and (2.6) to be determined. When $n = 0$, the orthogonal relation in Eq. (2.3) should be satisfied, that is

$$\int_R \varphi(x-k)\varphi(x-l)dx = \delta_{k,l} \quad \text{or} \quad \int_R \varphi(x)\varphi(x-k)dx = \delta_{0,k} \tag{2.14}$$

which is one of the basic conditions in constructing the scaling function. When q_k are selected by the form [3]

$$q_k = (-1)^k p_{\tilde{N}-k}, \quad k = 0, 1, \ldots, \tilde{N}, \tag{2.15a}$$

Equation (2.6) can be rewritten as

$$\psi(x) = \sum_{k=0}^{\tilde{N}} (-1)^k p_{\tilde{N}-k}\varphi(2x-k). \tag{2.15b}$$

After that, we can verify that the compactly support region of the wavelet function is $\text{Supp}[\psi(x)] = [1 - N_*, N_*]$ with $N_* = (\tilde{N}+1)/2$, and the orthogonal relations are held as follows:

$$\int_R \varphi(x)\psi(x-k)dx = 0, \tag{2.16a}$$

$$\int_R \psi(x)\psi(x-k)dx = \delta_{0,k} \tag{2.16b}$$

$$\int_R \varphi_{n,j}(x)\varphi_{n,k}(x)dx = \delta_{j,k}, \tag{2.16c}$$

$$\int_R \varphi_{n,j}(x)\psi_{m,k}(x)dx = 0 \quad m \geq n, \tag{2.16d}$$

$$\int_R \psi_{n,j}(x)\psi_{m,k}(x)dx = \delta_{n,m}\delta_{j,k} \quad m \geq n. \tag{2.16e}$$

Proof Before we give the proof of Eq. (2.16), let us see the explicit expression of Eq. (2.14). Substituting Eq. (2.2) into Eq. (2.14), one gets

$$\begin{aligned}\delta_{0,k} &= \int_{-\infty}^{\infty} \varphi(x)\varphi(x-k)dx = \sum_{i=0}^{\tilde{N}}\sum_{j=0}^{\tilde{N}} p_i p_j \int_R \varphi(2x-i)\varphi(2x-2k-j)dx \\ &= \frac{1}{2}\sum_{i=0}^{\tilde{N}}\sum_{j=0}^{\tilde{N}} p_i p_j \delta_{i,j+2k} = \frac{1}{2}\sum_{i=0}^{\tilde{N}} p_i p_{i-2k} \quad k \in Z\end{aligned} \tag{2.17}$$

which is one set of the conditions for constructing the wavelets, see Eq. (2.34) in the next subsection. Under the conditions of Eq. (2.17), the proofs are given as follows:

(I) to prove Eq. (2.16a). Substitution of Eqs. (2.2) and (2.15b) into the left of Eq. (2.16a) leads to

$$\begin{aligned}\int_{-\infty}^{\infty} \varphi(x)\psi(x-k)dx &= \sum_{i=0}^{\tilde{N}}\sum_{j=0}^{\tilde{N}} p_i(-1)^j p_{\tilde{N}-j} \int_R \varphi(2x-i)\varphi(2x-2k-j)dx \\ &= \frac{1}{2}\sum_{i=0}^{\tilde{N}}\sum_{j=0}^{\tilde{N}} (-1)^j p_i p_{\tilde{N}-j}\delta_{i,j+2k} \\ &= \frac{1}{2}\sum_{i=0}^{\tilde{N}} (-1)^{i-2k} p_i p_{\tilde{N}-i-2k} = 0 \quad k \in Z\end{aligned} \tag{2.18}$$

which is automatically satisfied for arbitrary k. Here, $p_{\tilde{N}-i-2k} = 0$ is used when $\tilde{N} - i - 2k \notin [0, \tilde{N}]$ and the anti-symmetry is considered in the last summation. For example, when $k = 0$, we always have a pair of $(-1)^i p_i p_{\tilde{N}-i}$ and $(-1)^{\tilde{N}-i} p_{\tilde{N}-i} p_i$. When $\tilde{N}$ is an odd integer, we have $(-1)^i = -(-1)^{\tilde{N}-i}$, further, we have $(-1)^i p_i p_{\tilde{N}-i} + (-1)^{\tilde{N}-i} p_{\tilde{N}-i} p_i = 0$. Thus,

$$\sum_{i=0}^{\tilde{N}} (-1)^{i-2k} p_i p_{\tilde{N}-i} = 0.$$

For the same reason, Eq. (2.18) is true for the case of $k \neq 0$. Hence, Eq. (2.16a) is proved.

(II) to prove Eq. (2.16b). Substituting Eq. (2.15b) into the left of Eq. (2.16b), one gets

$$\begin{aligned}\int_{-\infty}^{\infty} \psi(x)\psi(x-k)dx &= \sum_{i=0}^{\tilde{N}}\sum_{j=0}^{\tilde{N}} (-1)^{i+j} p_{\tilde{N}-i} p_{\tilde{N}-j} \int_R \varphi(2x-i)\varphi(2x-2k-j)dx \\ &= \frac{1}{2}\sum_{i=0}^{\tilde{N}}\sum_{j=0}^{\tilde{N}} (-1)^{i+j} p_{\tilde{N}-i} p_{\tilde{N}-j} \delta_{i,j+2k} \\ &= \frac{1}{2}\sum_{i=0}^{\tilde{N}} (-1)^{2i-2k} p_{\tilde{N}-i} p_{\tilde{N}-i-2k} = \frac{1}{2}\sum_{i=0}^{\tilde{N}} p_i p_{i-2k}.\end{aligned}$$

By using Eq. (2.17), further, one gets that Eq. (2.16b) is held.

(III) to prove Eq. (2.16c). According to Definition 2.2, the base scaling functions are expressed by $\varphi_{n,k}(x) = 2^{n/2}\varphi(2^n x - k),\ n, k \in Z$. Substituting the expression of base scaling functions into the left of Eq. (2.16c), we have

$$\begin{aligned}\int_{-\infty}^{\infty} \varphi_{n,j}(x)\varphi_{n,k}dx &= 2^n \int_R \varphi(2^n x - j)\varphi(2^n x - k)dx \\ &= \int_R \varphi(x-j)\varphi(x-k)dx = \delta_{j,k} \quad j, k \in Z.\end{aligned}$$

That is, Eq. (2.16c) is held.

(IV) to prove Eq. (2.16d). Substituting definitions of the base wavelet functions $\psi_{n,k}(x)$ and the base scaling functions $\varphi_{n,k}(x)$ into the left of Eq. (2.16d), we get

$$A_{j,k}^{n,m} = \int_{-\infty}^{\infty} \varphi_{n,j}(x)\psi_{m,k}dx = 2^{(n+m)/2} \int_R \varphi(2^n x - j)\psi(2^m x - k)dx. \tag{2.19}$$

Denote $r = m - n(\geq 0)$. Then, $A_{j,k}^{n,m} = A_{j,k}^{n,n+r}$. Now, let us use mathematical induction to give the proof for the integers $r \geq 0$. For the case $r = 0$, Eq. (2.19) becomes

$$A_{j,k}^{n,n} = 2^n \int_R \varphi(2^n x - j)\psi(2^n x - k)dx = \int_R \varphi(x - j)\psi x - k)dx$$
$$= \int_R \varphi(x)\psi x - k + j)dx = 0 \quad n, j, k \in Z.$$

That is, $A_{j,k}^{n,m} = A_{j,k}^{n,n+r} = 0$ is true for $r = 0$. Assuming $A_{j,k}^{n,m} = A_{j,k}^{n,n+r} = 0$ for $r = s$ with arbitrary integer $s > 0$. Then for the case of $r = s + 1$, we have

$$A_{j,k}^{n,n+s+1} = 2^{(2n+s+1)/2} \int_R \varphi(2^n x - j)\psi(2^{n+s+1} x - k)dx$$
$$= 2^{(2n+s+1)/2} \int_R \varphi(2^n x - j)\psi(2^{n+s+1} x - k)dx.$$

Substituting Eq. (2.2) with $x = 2^n x' - j$ into the above equation, we obtain

$$A_{j,k}^{n,n+s+1} = 2^{(2n+s+1)/2} \sum_{l \in Z} p_l \int_R \varphi(2^{n+1} x - 2j - l)\psi(2^{n+s+1} x - k)dx.$$

Using the transform of integral variable of $2x$ to x, the above equation is reduced into the form

$$A_{j,k}^{n,n+s+1} = 2^{(2n+s-1)/2} \sum_{l \in Z} p_l \int_R \varphi(2^n x - 2j - l)\psi(2^{n+s} x - k)dx,$$

or

$$A_{j,k}^{n,n+s+1} = 2^{-1/2} \sum_{l \in Z} p_l \int_R \varphi_{n,2j+l}(x)\psi_{n+s,k}(x)dx = 2^{-1/2} \sum_{l \in Z} p_l A_{2j+l,k}^{n,n+s}$$

Due to $A_{2j+l,k}^{n,n+s} = 0$ in the assumption for the case of arbitrary integers $n, s, j, k \in Z$, hence, $A_{j,k}^{n,n+s+1} = 0$. That is $A_{j,k}^{n,n+r} = 0$ for $r = s + 1$. From the mathematical induction, we know that $A_{j,k}^{n,n+r} = 0$ is true for arbitrary integer $r = m - n \geq 0$, or Eq. (2.16d) is held.

(V) to prove Eq. (2.16e). Denote $r = m - n \geq 0$ and

$$B_{j,k}^{n,m} = B_{j,k}^{n,n+r} = \int_{-\infty}^{\infty} \psi_{n,j}(x)\psi_{m,k}dx = 2^{(n+m)/2}\int_{R} \psi(2^n x - j)\psi(2^m x - k)dx \tag{2.20}$$

When $r = 0$, we have

$$B_{j,k}^{n,n} = 2^n \int_{R} \psi(2^n x - j)\psi(2^m x - k)dx = \int_{R} \psi(x - j)\psi(x - k)dx.$$

By means of Eq. (2.16b), one gets $B_{j,k}^{n,n} = \delta_{j,k}$. For $r = 1$, we have

$$B_{j,k}^{n,n+1} = 2^{n+1/2}\int_{R} \psi(2^n x - j)\psi(2^{n+1} x - k)dx \tag{2.21}$$

Applying Eq. (2.2) to $\psi(2^n x - j)$, i.e.,

$$\psi(2^n x - j) = \sum_{l \in Z} q_l \varphi(2^{n+1} x - 2j - l), \tag{2.22}$$

and substituting Eq. (2.22) into Eq. (2.21), one gains

$$\begin{aligned} B_{j,k}^{n,n+1} &= 2^{n+1/2}\sum_{l \in Z} q_l \int_{R} \varphi(2^{n+1}x - 2j - l)\psi(2^{n+1}x - k)dx \\ &= 2^{-1/2}\sum_{l \in Z} q_l \int_{R} \varphi(x - 2j - l)\psi(x - k)dx. \end{aligned}$$

According to Eq. (2.16a), we have

$$\int_{R} \varphi(x - 2j - l)\psi(x - k)dx = \int_{R} \varphi(x)\psi(x - k + 2j - l)dx = 0.$$

That is, $B_{j,k}^{n,n+1} = 0$. Assume $B_{j,k}^{n,n+r} = 0$ for arbitrary integer $r = s(\geq 1)$, i.e., $B_{j,k}^{n,n+s} = 0$. For the case of $r = s + 1$, Eq. (2.20) becomes

$$B_{j,k}^{n,n+r} = 2^{(2n+s+1)/2}\int_{R} \psi(2^n x - j)\psi(2^{n+s+1} x - k)dx.$$

Substitution Eq. (2.22) into the above equation yields

$$B_{j,k}^{n,n+r} = 2^{(2n+s+1)/2} \sum_{l \in Z} q_l \int_R \varphi(2^{n+1}x - 2j - l)\psi(2^{n+s+1}x - k)dx$$
$$= 2^{(s+1)/2} \sum_{l \in Z} q_l \int_R \varphi(x - 2j - l)\psi(x - k)dx = 0.$$

According to the mathematical induction, we get that $B_{j,k}^{n,n+r} = 0$ is held for arbitrary integer $r \geq 1$. Combined this conclusion with $B_{j,k}^{n,n} = \delta_{j,k}$, we gain $B_{j,k}^{n,m} = \delta_{n,m}\delta_{j,k}$, i.e., Eq. (2.16e) is held.

Under the condition of Eq. (2.16a) and the selection of Eq. (2.15a), all properties displayed in Definitions 2.2 and 2.3 are satisfied as proved above. In addition, it should be noted that the property of Eq. (2.16d) tells us that the following equation:

$$< \varphi_{n,j}(x), \tilde{Q}_n f(x) >= \int_R \varphi_{n,j}(x)\tilde{Q}_n f(x)dx = 0 \tag{2.23}$$

is always true for arbitrary $f(x) \in L^2(R)$.

Therefore, to construct a compactly supported orthogonal wavelet, only the filter coefficients in Eq. (2.2) are needed to be determined. To make the constructed wavelet have other additionally required properties, sometimes other conditions can be considerably added. In the following, we introduce the system of algebraic equations to determine the filter coefficients, $p_k(k = 0, 1, \ldots, \tilde{N})$, under the general situation, which constitutes a definite solution problem by adjusting the relevant performance indexes.

2.2.2 Properties on Moments of Scaling and Wavelet Functions

For an orthogonal scaling function with the compact support, $\text{Supp}[\varphi(x)] = [0, \tilde{N}]$, we have

Definition 2.4 The moments M_i of the scaling function $\varphi(x)$ are defined as [4]

$$M_i = \int_{-\infty}^{\infty} x^i \varphi(x)dx = \int_0^{\tilde{N}} x^i \varphi(x)dx, \quad i = 0, 1, \ldots, s \tag{2.24}$$

where the zeroth moment M_0 is directly determined by the normalization condition in the construction of the scaling function, namely, $M_0 = 1$.

Definition 2.5 A wavelet has $r + 1(< [(\tilde{N} + 1)/2 + 1])$ vanishing moments means that its wavelet function satisfies the following conditions:

$$\int_{-\infty}^{\infty} x^i \psi(x)dx = \int_{0}^{\tilde{N}} x^i \psi(x)dx = 0, \quad i = 0, 1, \ldots, r \tag{2.25}$$

where [.] represents the rounding operation.

Property 2.1 When a wavelet function $\psi(x)$ has $r + 1$ vanishing moments of up to r order shown in Eq. (2.25), then the unconditional bases $\{\varphi(x - k)\}$ generated by the scaling function can accurately represent all polynomial functions up to order r. That is, we have the following formula:

$$x^i \equiv \sum_k c(i, k)\varphi(x - k), \quad c(i, k) = \int_{-\infty}^{\infty} x^i \varphi(x - k)dx, \quad k \in Z \tag{2.26}$$

for $i = 0, 1, \ldots, r$.

Proof From Eq. (2.12), we have

$$\begin{aligned} x^i &= P_0 f(x) + \tilde{Q}_0 f(x) = \sum_{k \in Z} c_{0,k}\varphi_{0,k}(x) + \sum_{n=0}^{\infty}\sum_{k \in Z} d_{n,k}\psi_{n,k}(x) \\ &= \sum_{k \in Z} c_{0,k}\varphi(x - k) + \sum_{n=0}^{\infty}\sum_{k \in Z} d_{n,k}\psi_{n,k}(x) \quad i = 0, 1, \ldots, r \end{aligned} \tag{2.27a}$$

where $c_{0,k} = \int_{-\infty}^{\infty} x^i \varphi(x - k)dx = c(i, k)$ and

$$\begin{aligned} d_{n,k} &= \int_{-\infty}^{\infty} x^i \psi_{n,k}(x)dx = 2^{-n/2} \int_{-\infty}^{\infty} x^i \psi(2^n x - k)dx \\ &= 2^{-n(i+3/2)} \int_{-\infty}^{\infty} (x + k)^i \psi(x)dx = 2^{-n(i+3/2)} \sum_{j=0}^{i} C_i^j k^{i-j} \int_{-\infty}^{\infty} x^j \psi(x)dx \end{aligned} \tag{2.27b}$$

Because $j \leq i$ and $\int_{-\infty}^{\infty} x^i \psi(x)dx = 0$ for $0 \leq i \leq r$, one gets $d_{n,k} \equiv 0$ for all $n \geq 0$ and k. Then Eq. (2.26) is exactly held, that is, the conclusion of Property 2.1 is proved.

Property 2.2 If a scaling function $\varphi(x)$ can accurately express the polynomial functions $f(x) = x^j$ $(j = 0, 1, 2, \ldots, r)$, and when

$$M_{2j-1} = \frac{1}{2}\sum_{k=0}^{\tilde{N}} p_k k^{2j-1} = M_1^{2j-1} \quad (j = 1, 2, \ldots, l(< [(r-1)/2])) \tag{2.28}$$

holds, the following relations are also held:

$$M_{2j} = \frac{1}{2}\sum_{k=0}^{\tilde{N}} p_k k^{2j} = M_1^{2j}, \qquad j = 1, 2, \ldots, l(2l \le r). \tag{2.29}$$

We will see later that relations in Eq. (2.28) are part of basic equations in generating orthogonal wavelets. This property tells us that when Eq. (2.28) for the odd number case is selected to be satisfied, we can get their similar relations for the even number case. The detailed proof for this property can be seen in the Appendix of this chapter.

2.2.3 *Generalized Gaussian Integral for Calculating Expansion Coefficients*

Property 2.3 When the moments of a scaling function $\varphi(x)$ satisfy relations of $M_j = M_1^j (j = 1, 2, \ldots, r(= 2l-1))$, the generalized Gaussian integral to $c_{0,0} \equiv \int_0^N f(x)\varphi(x)dx$ has algebraic accuracy of order r. Among the weight coefficients corresponding to all Gaussian points, there is a set of real solutions in which only one weight coefficient is equal to 1 and all other ones are equal to 0. In this set of solutions, the Gaussian point corresponding to the nonzero weight coefficient is $x^* = M_1$, and others can be selected arbitrarily.

Proof To construct a generalized Gaussian integral format with algebraic accuracy of order r, one needs to find a set of weight coefficients A_k and their corresponding Gaussian points $\tilde{x}_k$ $(k = 1, 2, \ldots, l = (r+1)/2)$ to satisfy

$$\int_0^{\tilde{N}} \varphi(x)x^i dx = \sum_{k=1}^{l} A_k \tilde{x}_k^i \quad \text{for} \quad i = 1, 2, \ldots, r(= 2l-1).$$

Using the condition of $M_j = M_1^j$ $(j = 1, 2, \ldots, r)$, the above Gaussian integral equations become

$$
\begin{aligned}
A_1 + A_2 + \cdots + A_l &= M_0, \\
A_1\tilde{x}_1 + A_2\tilde{x}_2 + \cdots + A_l\tilde{x}_l &= M_1, \\
A_1\tilde{x}_1^2 + A_2\tilde{x}_2^2 + \cdots + A_l\tilde{x}_l^2 &= M_1^2, \\
&\cdots \\
A_1\tilde{x}_1^{2l-1} + A_2\tilde{x}_2^{2l-1} + \cdots + A_l\tilde{x}_l^{2l-1} &= M_1^{2l-1}.
\end{aligned}
$$

Obviously, the above $2l$ nonlinear algebraic equations can give at least one set of determinant solutions of the $2l$ unknowns for the weight coefficients and the Gaussian points if their real solutions exist or are found. According to the theory of nonlinear algebraic equations, this system of nonlinear algebraic equations has $2l(2l-1)$ sets of such solutions. Of course, one set of such real solutions is enough for our calculation of the coefficient $c_{0,0}$. From the normalization conditions in Eq. (2.24), we have $M_0 = 1$. Then one set of such solutions can be found by $A_1 = 1$, $A_i = 0$ $(i = 2, 3, \ldots, l)$, $\tilde{x}_1 = M_1 = x^*$, and arbitrarily different $\tilde{x}_i (\neq x^*, i = 2, 3, \ldots, l)$. It is evident that such real solution for the generalized Gaussian integral has algebraic accuracy of order $r(= 2l - 1)$.

Thus, the conclusion of this property is proved.

Deduction 2.1 When the moments of a scaling function $\varphi(x)$ satisfies $M_j = M_1^j (j = 1, 2, \ldots, r(= 2l - 1))$, its generalized Gaussian integral

$$\int_0^{\tilde{N}} f(x)\varphi(x)dx = \sum_{k=1}^{l} A_k f(\tilde{x}_k^i)$$

can be represented by only one Gaussian point (referred to as a single point), namely $x^* = M_1$. Then we have

$$\int_0^{\tilde{N}} f(x)\varphi(x)dx = f(M_1) \tag{2.30}$$

which has the algebraic precision of order r.

Deduction 2.2 When the moments of a scaling function $\varphi(x)$ satisfies $M_j = M_1^j$ $(j = 1, 2, \ldots, r(= 2l-1))$, the decomposition coefficients of $\forall f \in L^2(R)$ under the base scaling functions can be given by a single-point generalized Gaussian integral with rth-order algebraic accuracy, that is,

$$c_{n,k} \equiv \int_R f(x)\varphi_{n,k}(x)dx \approx 2^{-n/2} f\left(\frac{M_1 + k}{2^n}\right) = 2^{-n/2} f_{n,k+M_1}. \tag{2.31}$$

Here, $f_{n,k+M_1} = f(\frac{M_1+k}{2^n})$. Then the reconstruction equation or the expansion in terms of base scaling functions, Eq. (2.4), can be explicitly expressed as

$$P_n f(x) = 2^{-n/2} \sum_{k \in Z} f_{n,k+M_1} \varphi_{n,k}(x) = 2^{-n/2} \sum_{k \in Z} f_{n,k} \varphi_{n,k-M_1}(x)$$
$$= \sum_{k \in Z} f_{n,k} \varphi(2^n x - k + M_1) \tag{2.32}$$

For a finite interval $\Omega = [0, 1] \subset R$ on $f(x)$, denote the corresponding finite summation index set by Z_Ω. Due to that the summation number in Eq. (2.32) at each point x is only dependent on integers in the support region $[0, \tilde{N}]$ of $\varphi(x)$, considering $\varphi(0) = \varphi(\tilde{N}) = 0$ (see in the next section) and the function values of $P_n f(x)$ at the dyadic points $j2^{-n}(j = 0, 1, 2, \ldots, 2^n)$, i.e., $P_n f(j2^{-n}) = \sum_{k \in Z} f_{n,k} \varphi(j - k + M_1)$, one can get $Z_\Omega = \{k; -\tilde{N} + 1 + M_1 \leq k \leq 2^n + \tilde{N} - 1 + M_1\}$. In such situation, Eq. (2.32) can be rewritten by

$$P_n f(x) = 2^{-n/2} \sum_{k \in Z_\Omega} f_{n,k} \varphi_{n,k-M_1}(x) \quad 0 \leq x \leq 1 \tag{2.32a}$$

Since the summation of Eq. (2.32a) at each dyadic point is equal to $2\tilde{N} - 1$ times, hence, the total summation number in Eq. (2.32) at all dyadic points in [0, 1] is equal to $(2\tilde{N} - 1)2^n$ times.

It should be noted that the feasibility of above calculations in Eq. (2.32) or (2.32a) is dependent on whether or not M_1 is an integer. Following the establishment of the compactly supported orthogonal wavelets in the next section, one can see that M_1 can be taken an integer only in the generalized Coiflets.

2.3 Numerical Generation of Orthogonal Wavelets

2.3.1 Determination of Filter Coefficients

In order to achieve the performance of the scaling function and wavelet function, the filter coefficients have firstly been determined by the necessary and desirable conditions. When $\tilde{N}$ in the region of $\mathrm{Supp}[\varphi(x)] = [0, \tilde{N}]$ is an odd integer great than 3, the above conditions for an orthogonal wavelet are as follows:

(1) The normalized equation of the scaling function. That is, $M_0 = \int_{-\infty}^{\infty} \varphi(x) dx = 1$. By substituting Eq. (2.2) into it and considering the support domain, we have

$$\sum_{i=0}^{\tilde{N}} p_i = p_0 + p_1 + \cdots + p_N = 2 \tag{2.33}$$

(2) Orthogonality condition of the scaling function. This requires the orthogonality of the scaling function and its translation function, i.e., Eq. (2.14) or Eq. (2.17) should be satisfied by the equations:

$$\sum_{i=0}^{\tilde{N}} p_i p_{i-2k} = 2\delta_{0,k}, \qquad k = 0, 1, 2, \ldots, [(\tilde{N} - 1)/2] \tag{2.34}$$

(3) Selected vanishing moment equations of wavelet function. Substituting Eq. (2.15b) into the Eq. (2.25), one gets (see in Appendix 2.3 for details)

$$\sum_{j=0}^{\tilde{N}} (-1)^j p_j j^i = 0, \qquad i = 0, 1, 2, \ldots, r - 1 \tag{2.35}$$

Additional conditions of moment relationships. That is, when $M_j = M_1^j$ are held for $j = 1, 2, \ldots 2l(\leq r)$, the necessary conditions are of the form:

$$\sum_{k=0}^{\tilde{N}} p_k k^{2i-1} = 2M_1^{2i-1} \qquad i = 1, 2, \ldots, l. \tag{2.36}$$

In the compact supported orthogonal wavelets, when the order of vanishing moment is taken as r, the index of its supported domain needs to satisfy $\tilde{N} \geq 2r - 1$, and $\tilde{N}$ should be odd integer. In the algebraic equations of Eqs. (2.33)–(2.36), there are $\tilde{N}+1$ unknown coefficients p_k ($k = 0, 1, 2, \ldots, \tilde{N}$), and the number of above equations is $1 + [(\tilde{N} - 1)/2] + 1 + r + l$. To Eqs. (2.33)–(2.35), one can demonstrate that there is one equation is dependent on others [5], that is, $[(\tilde{N} - 1)/2] + 1 + r + l$ equations are independent. In order to ensure them being determinant equations, the number of such independent equations should be equal to the number of the unknowns, i.e., $\tilde{N} + 1 = [(\tilde{N} - 1)/2] + 1 + r + l$ or

$$\tilde{N} = 2(r + l) - 1. \tag{2.37}$$

Such a condition constricts the selection of the total number of Eqs. (2.35) and (2.36). For example, $l = 0$ and $\tilde{N} = 2r - 1$ are the selection in the Daubechies wavelets, while $2l = r$ is the selection in the generalized Coifman wavelets. In general case, such Coiflets requires r to be an even number, further $\tilde{N} = 6l - 1$ ($l \geq 1$). For other cases, we can construct other compact supported orthogonal wavelets different from these two types of wavelets by selecting different r and l under the condition of Eq. (2.37). It can be seen from Property 2.2 that after the order of vanishing moment r is selected, the relationship of the moment equation in Property 2.2 holds only under the conditions $2l \leq r$ such that the conclusion of Property 2.3 holds. Restricted by this condition in Eq. (2.37), Table 2.1 lists the relevant feasible parameters of various compactly supported orthogonal wavelets that can be generated within the region of $5 \leq \tilde{N} \leq 23$.

Once one set of parameters of the compact supported orthogonal for a wavelet are selected in Table 2.1, then a set of filter coefficient solutions p_k $(k = 0, 1, 2, \ldots, \tilde{N})$ can be obtained by solving their corresponding independent nonlinear algebraic equations displayed in Eqs. (2.33)–(2.36). Further, the filter coefficients $q_k(k = 0, 1, 2, \ldots, \tilde{N})$ in Eq. (2.15a) can be gained. In theory, the foundation of such an expected wavelet has been laid after this sept is finished.

2.3.2 *Generation of Scaling and Wavelet Functions*

The numerical generation of the compact supported orthogonal wavelet is mainly carried out by using Eq. (2.2) satisfied by the scaling function after the filter coefficients p_k are gained.

Step 1: to calculate the values of $\varphi(x)$ at each integer point in the support domain $[0, \tilde{N}]$ $(\tilde{N} = 2(r + l) - 1)$. In order to be consistent with the mark used in the previous resolution level n, denote $\varphi_{0k} = \varphi(k)(k = 0, 1, \ldots, \tilde{N})$. Then the following equations can be obtained by Eq. (2.2), i.e.,

$$\varphi_{0,0} = p_0\varphi_{0,0}, \quad \varphi_{0,1} = p_0\varphi_{0,2} + p_1\varphi_{0,1} + p_2\varphi_{0,0},$$
$$\varphi_{0,2} = p_0\varphi_{0,4} + p_1\varphi_{0,3} + p_2\varphi_{0,2} + p_3\varphi_{0,1} + p_4\varphi_{0,0}$$
$$\cdots\cdots\cdots$$
$$\varphi_{0,(\tilde{N}-1)/2} = p_0\varphi_{0,\tilde{N}-1} + p_1\varphi_{0,\tilde{N}-2} + \cdots + p_{\tilde{N}-2}\varphi_{0,1} + p_{N-1}\varphi_{0,0},$$
$$\cdots\cdots\cdots$$
$$\varphi_{0,\tilde{N}-1} = p_{\tilde{N}-2}\varphi_{0,\tilde{N}} + p_{\tilde{N}-1}\varphi_{0,\tilde{N}-1} + p_{\widetilde{N}}\varphi_{0,\tilde{N}-2}, \varphi_{0,\tilde{N}} = p_{\tilde{N}}\varphi_{0,\tilde{N}}$$

From the above first and last equations, it is obvious that $\varphi(0) = 0$ and $\varphi(\tilde{N}) = 0$ since $p_0 \neq 0$ and $p_{\tilde{N}} \neq 0$. After these two equations are omitted, and $\varphi(0) = 0$ and $\varphi(\tilde{N}) = 0$ are eliminated in the remainder equations, the above system of linear algebraic equations with unknowns $\varphi_{0,i}$ $(i = 1, 2, \ldots, \tilde{N} - 1)$ has $\tilde{N} - 1$ equations. It is obvious that such remainder equations are homogeneous. Denote unknowns by column $\mathbf{\Phi}_{\tilde{N}-1} = [\varphi_{0,1}, \varphi_{0,2}, \ldots, \varphi_{0,\tilde{N}-1}]^T$, and their coefficients by matrix $\mathbf{M}_{(\tilde{N}-1)(\tilde{N}-1)}$. Then the above algebraic equations can be abbreviated as in the matrix form

$$(\mathbf{M}_{(\tilde{N}-1)(\tilde{N}-1)} - \mathbf{I}_{(\tilde{N}-1)(\tilde{N}-1)})\mathbf{\Phi}_{\tilde{N}-1} = \mathbf{0}_{\tilde{N}-1} \tag{2.38}$$

where $\mathbf{I}_{(\tilde{N}-1)(\tilde{N}-1)}$ is the identity matrix of order $\tilde{N} - 1$, and

Table 2.1 Main parameters and attributes of compact supported orthogonal wavelets

Support region $\tilde{N}$(Odd)	Number of filter coefficients $(\tilde{N}+1)$	Number of independent orthogonal equations $(\tilde{N}+1)/2$	Number of vanishing moment equations $r \le (\tilde{N}+1)/2$	Number of moment equations l	Maximum moment order $l \le r/2$	Order of algebraic precision in Gaussian integral $2l \le r$
3	4	2	2	0		
5	6	3	3	0		
			2	1	2	2
7	8	4	4	0		
			3	1	2	2
9	10	5	5	0		
			4	1	2	2
11	12	6	6	0		
			5	1	2	2
			4	2	4	4
13	14	7	7	0		
			6	1	2	2
			5	2	4	4
15	16	8	8	0		
			7	1	2	2
			6	2	4	4
17	18	9	9	0		
			8	1	2	2
			7	2	4	4
			6	3	6	6
19	20	10	10	0		
			9	1	2	2
			8	2	4	4
			7	3	6	6
21	22	11	11	0		
			10	1	2	2
			9	2	4	4
			8	3	6	6
23	24	12	12	0		
			11	1	2	4
			10	2	4	4
			9	3	6	6
			8	4	8	8
....						

Note The wavelets in the table corresponding to $l = 0$ is the Daubechies wavelets, and that corresponding to $2l = r$ is for the generalized Coifman wavelets

$$\mathbf{M}_{(\tilde{N}-1)(\tilde{N}-1)} = \begin{bmatrix} p_1 & p_0\ 0\ 0\ 0 \cdots 0 & 0 & 0 & 0 & 0 \\ p_3 & p_2\ p_1\ p_0\ 0 \cdots 0 & 0 & 0 & 0 & 0 \\ \cdots\cdots & \cdots & & & & \\ 0 & 0\ 0\ 0\ 0 \cdots 0 & p_{\tilde{N}} & p_{\tilde{N}-1} & p_{\tilde{N}-2} & p_{\tilde{N}-3} \\ 0 & 0\ 0\ 0\ 0 \cdots\cdots 0 & 0 & 0 & p_{\tilde{N}} & P_{\tilde{N}-1} \end{bmatrix}. \quad (2.39)$$

As pointed out previously, we have Rank $[\mathbf{M}_{(\tilde{N}-1)(\tilde{N}-1)}] = \tilde{N}-2$. In such case, we should add one nonhomogeneous equation without adding new unknown to Eq. (2.38) such that the nonzero solution of $\mathbf{\Phi}_{\tilde{N}-1} \neq \mathbf{0}$ can be gained. From the solution theory, one knows that Eq. (2.38) has nonzero solution $\mathbf{\Phi}_{\tilde{N}-1} \neq \mathbf{0}$ in one-dimensional solution space. In other words, such nonzero solutions varying with one unknown when it is nonzero. In order to obtain a determinant solution, we add the following standardized condition to the nonzero solution with one freedom, i.e.,

$$\varphi_{0,1} + \varphi_{0,2} + \cdots + \varphi_{0,\tilde{N}-1} = 1 \quad (2.40)$$

By solving the linear algebraic equations of Eqs. (2.38) and (2.40), the values of $\varphi(x)$ at each integer point in the support domain, $\varphi_{0,i} (i = 1, 2, \ldots, \tilde{N} - 1)$, can be numerically obtained.

Step 2: to calculate the values of function $\varphi(x)$ at all dyadic points $j2^{-n}$ in support region $[0, \tilde{N}]$ $(j = 0, 1, 2, \ldots, \tilde{N}2^n)$. Since the values of function $\varphi(x)$ at the integer points in the supported region have been obtained in step 1 previously, which are in the dyadic points, rest of the function values at dyadic points can be obtained by the iteration through Eq. (2.2). After a resolution level $n(\geq 1)$ is specified, firstly, let $x = (2j-1)2^{-1}$ $(j = 1, 2, \ldots, \tilde{N} - 1)$ and substitute it into Eq. (2.2), then we have

$$\varphi(j2^{-1}) = \sum_{k=0}^{j} p_k \varphi(j-k). \quad (2.41)$$

Here, $\varphi(x) = 0$ outside the support domain has been considered. Since the right side of Eq. (2.41) is the obtained values of $\varphi(x)$ at the integer points in the support region, hence, the values of $\varphi(x)$ at the dyadic point $j2^{-1}$ $(j = 1, 2, \ldots, \tilde{N}2)$ all can be directly obtained by Eq. (2.41). Following the same way, let $x = (2^2 j - 3)2^{-2}$, $x = (2^3 j - 2^2 - 1)2^{-3}$, $\cdots$, $x = (2^n j - 2^{n-1} - 1)2^{-n}$, and respectively, substitute each one set of dyadic point coordinates into Eq. (2.2) in orderly. Then the values of $\varphi(x)$ at the dyadic points $x = (2^{i+1} j - 2^{i+1} - 1)2^{-(i+1)}$ can be gained from the values of $\varphi(x)$ at the dyadic points $x = (2^i j - 2^i - 1)2^{-i}$. Finally, the values of $\varphi(x)$ at the dyadic points $j2^{-n} (j = 0, 1, 2, \ldots, \tilde{N}2^n)$ in the support region are obtained at the resolution level n. That is, $\varphi(j/2^n) = \sum_{k=0}^{N-1} p_k \varphi(j2^{-(n-1)} - k)$ $(j = 0, 1, 2, \ldots, \tilde{N}2^n)$.

Further, the values of the base scaling functions $\varphi_{n,k}(x)$ at the dyadic points $x_j = j2^{-n} (j \in Z)$ in R at the resolution level n can be gained by the definition of the base scaling function and the above value of $\varphi(x)$ at the same dyadic points. That is,

$$\varphi_{n,k}(x) = 2^{n/2}\varphi(2^n x - k) \quad \text{or} \quad \varphi_{n,k}(j2^n) = 2^{n/2}\varphi(j - k) \quad j, k \in Z$$

Finally, the base scaling functions $\varphi_{n,k}(x)$ are numerically generated at the dyadic points with resolution level n.

Step 3: to numerical generation of base wavelet functions at the dyadic point with resolution level n if they are needed.

The generation of the base wavelet functions $\psi_{n,k}(x)$ are the same as one of $\varphi_{n,k}(x)$. After $p_k(k = 0, 1, \ldots, \tilde{N})$ are gained, the coefficients $q_k(k = 0, 1, \ldots, \tilde{N})$ can be given by Eq. (2.15a). Then, the values of the wavelet function $\psi(x)$ at the integers in its support region can be obtained by the two-scale Eq. (2.15b) based on the scaling function $\varphi(x)$. After that, similar to the previous steps shown in the scaling function by successively taking the higher resolution level, the values of the wavelet function $\psi(x)$ at the dyadic points $j2^{-n}$ under the resolution index n can be obtained. Finally, the base wavelet functions $\psi_{n,k}(x)$ $(k \in Z)$ can be numerically obtained at the dyadic points $j2^{-n}(j \in Z)$.

2.3.3 Examples of Compactly Supported Orthogonal Wavelets

Here, we show the construction of the Daubechies wavelet and the Coifman wavelet as examples, which are used widely in the application of signal processing.

Example 2.1 *Daubechies wavelets*. This kind of wavelets was established by the famous mathematician Daubechies in the early 1980s, and it has the characteristic of the largest vanishing moment for a given compact support. In this kind of wavelet, $l = 0$, and $\tilde{N} = 2r - 1$. Here, $l = 0$ means that the moment relationships of Eq. (2.36) are not considered. In the simplest case $r = 2$, the four independent equations in (2.33)–(2.35) become explicitly

$$p_0 + p_1 + p_2 + p_3 = 2$$

$$p_0^2 + p_1^2 + p_2^2 + p_3^2 = 2$$

$$p_0 - p_1 + p_2 - p_3 = 0$$

$$-p_1 + 2p_2 - 3p_3 = 0$$

It is lucky that two sets of analytical solutions for the filter coefficients p_i $(i = 0, 1, 2, 3)$ are getting by the form:

$$p_0 = (1 \pm \sqrt{3})/4, p_1 = (3 \pm \sqrt{3})/4, p_2 = (3 \mp \sqrt{3})/4, p_3 = (1 \mp \sqrt{3})/4$$

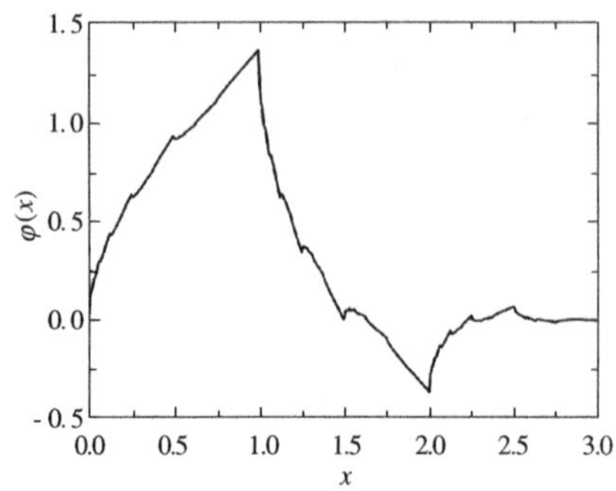

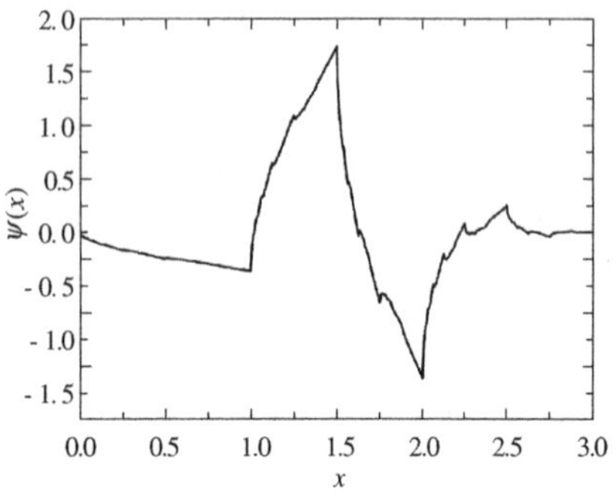

Fig. 2.1 Daubechies scaling function $\varphi(x)$ and wavelet functions $\psi(x)$ (support [0, 3]), $n = 7$. The $\psi(x)$ image here is obtained by translating the previously introduced $\psi(x)$ to the right by $N_* = (\tilde{N} - 1)/2$ on its support domain. This translation does not change all the characteristics of the wavelet function

In the case of taking the former set of the solution, applying steps 1–3, the scaling function $\varphi(x)$ and wavelet function $\psi(x)$ can be gained numerically. Figure 2.1 displays such generated $\varphi(x)$ and $\psi(x)$.

In addition, Table 2.2 displays the filter coefficients p_k of Daubechies wavelet at $\tilde{N} = 3$, 5, and 7, respectively. And Fig. 2.2 exhibits the scaling function $\varphi(x)$ and wavelet function $\psi(x)$ of Daubechies wavelet when $\tilde{N} = 7$.

It should be noted that since the moment relationship of Eq. (2.36) is not used in the construction of Daubechies wavelet, thus, the Daubechies orthogonal wavelets does not have the relevant conclusions displayed in Property 2.2 to Deduction 2.2 introduced above. In this way, the Gaussian integral calculation of the decomposition coefficients $c_{n,k}$ under the base scaling functions has no algebraic accuracy of order r.

Example 2.2 Coifman wavelet and its Generalization. Coifman suggested new class of wavelets with compact support based on the Daubechies's orthogonal wavelet [3], by adding the moment equations of the scaling function, e.g., Eq. (2.36). In the

Table 2.2 Daubechies wavelet filter coefficients p_k

k	$\tilde{N} = 3$	$\tilde{N} = 5$	$\tilde{N} = 7$
0	0.68301270189222	0.47046720778405	0.32580342805130
1	1.18301270189222	1.14111691583131	1.01094571509183
2	0.31698729810778	0.65036500052554	0.89220013824676
3	−0.18301270189222	−0.19093441556797	−0.03957502623564
4		−0.12083220831036	−0.26450716736904
5		0.04981749973164	0.04361630047418
6			0.04650360107098
7			−0.01498698933036

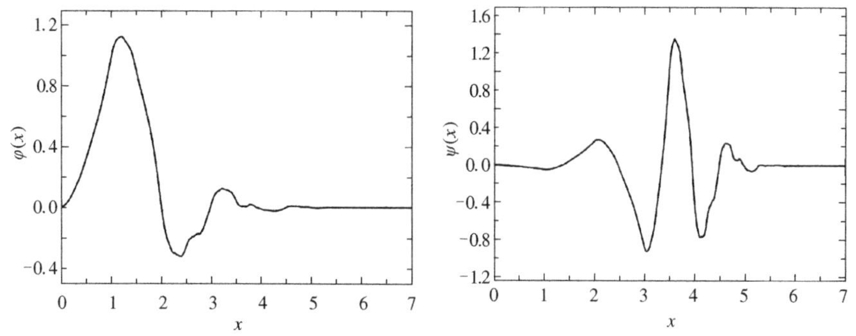

Fig. 2.2 Daubechies scaling function $\varphi(x)$ and wavelet functions $\psi(x)$ (support [0, 7]), $n = 7$. The $\psi(x)$ image here has been shifted to the right by $N_* = (\tilde{N} - 1)/2$

simplest case $\tilde{N} = 5$, we have $r = 2$ and $l = 1$. From Eq. (2.36), one finds that M_1 can be freely selected. Hence, the wavelet is dependent on a selected M_1, which can improve some expected property of the wavelet applying in different situations. Here, we introduce the original Coiflets. When $M_1 = 4$, the independent equations of Eqs. (2.33)–(2.36) are of the form

$$p_0 + p_1 + p_2 + p_3 + p_4 + p_5 = 2$$

$$p_0^2 + p_1^2 + p_2^2 + p_3^2 + p_4^2 + p_5^2 = 2$$

$$p_0 p_2 + p_1 p_3 + p_2 p_4 + p_3 p_5 = 0$$

$$p_0 p_4 + p_1 p_5 = 0$$

$$-p_1 + 2p_2 - 3p_3 + 4p_4 - 5p_5 = 0$$

$$p_1 + 2p_2 + 3p_3 + 4p_4 + 5p_5 = 8$$

Two sets of analytical solutions of the above equations can be obtained by the form:

$$p_0 = (-3 \mp \sqrt{15})/16, \quad p_1 = (1 \pm \sqrt{15})/16, \quad p_2 = (3 \pm \sqrt{15})/8,$$

$$p_3 = (3 \mp \sqrt{15})/8, \quad p_4 = (13 \mp \sqrt{15})/16, \quad p_5 = (9 \pm \sqrt{15})/16$$

Such the simplest wavelet generated by the above solution is called Coiflets, i.e., Coifman wavelet. It was suggested by Coifman, and Daubechies realized its construction. It is obvious that such Coiflets have all the characteristics mentioned in the previous sections. Although the Coiflets are successfully applied in the signal processing, however, it is found that their relevant algebraic accuracy is not high enough to solve a boundary-value problem due to the calculations of the Gaussian integral for the decomposition coefficients with a relative low order of algebraic precision. In this case, we have to look for other Coiflets with larger support region. According to Eqs. (2.33)–(2.36), one finds that the order of nonlinear algebraic equations increases as the support region increases, which leads to many difficulties in solving the equations. In addition, even when the support region is unchanged, the generalized Coiflets are different each other when M_1 is taken by different integers. Hence, different Coiflets can be gained by selecting different support regions and M_1 although their mathematical framework is full same. Of course, such different generalized Coiflets are strongly dependent on that at least one set of real solutions for the filter coefficients can be gained from the nonlinear algebraic equations expressed by Eqs. (2.33)–(2.36) when M_1 and $\tilde{N}$ are chosen. If one expects to get a relevant good Coiflet in a support region, usually, it is better that the integer M_1 is selected in the support region of $\varphi(x)$, i.e., $0 < M_1 < \tilde{N}$. In this situation, there are still many different generalized Coiflets in a fixed support region $[0, \tilde{N}]$ when different M_1 is chosen.

According to the knowledge of solution of boundary-value problems either linear or nonlinear, we know that when a set of orthogonal base scaling functions are used to expand an unknown function, the contradiction between the accuracy and the calculation efficiency will be encountered. That is, when the accuracy is improved, the efficiency will be decreased. In order to solve this contradiction, after the high-frequency filtering characteristics are taken into account in the generalized Coiflets, i.e., the higher algebraic accuracy of the generalized Gaussian integral is achieved by a suitably increasing of the support domain, this contradiction can be solved well. For such purpose of the support region, by means of the numerical trade-off to the accuracy and the calculations, we find that the generalized Coiflets is good for most cases when $\tilde{N} = 17$, $r = 6$, and $l = 3$ as well as $M_1 = 7$. As the support domain increases, it is necessary to properly select an integer value of M_1 to obtain the solution of the nonlinear equations of the filter coefficients. Here, we select $M_1 = 7$. Table 2.3 lists the values of the filter coefficients p_k for the case of $\tilde{N} = 5$ with $M_1 = 4$, and other two cases of $\tilde{N} = 11, 17$ with $M_1 = 7$. Figures 2.3 and 2.4 display these two scaling functions and wavelet functions of the original Coiflets with $\tilde{N} = 5$ and the generalized Coiflets with $\tilde{N} = 17$, respectively.

2.3.4 Analysis for Decomposition and Reconstruction Calculations

So far, the scaling function $\varphi(x)$ and wavelet function $\psi(x)$ are obtained under a resolution level n, then the base scaling functions $\{\varphi_{n,k}(x) = 2^{n/2}\varphi(2^n x - k),\ k \in Z\}$

Table 2.3 Coiflets wavelet filter coefficients p_k

k	$\tilde{N}=5$	$\tilde{N}=11$	$\tilde{N}=17$
0	0.054561459137964	0.001689380907696	−0.002392638657280
1	−0.179561459137964	−0.018166392820735	−0.004932601854180
2	−0.109122918275927	0.035078620626054	0.027140399711399
3	0.859122918275927	0.070743940368093	0.030647555946200
4	1.054561459137960	−0.219708291581175	−0.139310237070997
5	0.320438540862036	−0.101311830407117	−0.080606530717800
6		0.806759341910244	0.645994543293994
7		1.061135780078050	1.116266213257990
8		0.396844803880349	0.538189055707998
9		−0.010479864874492	−0.099615433862400
10		−0.020663855743163	−0.079923139434800
11		−0.001921632058008	0.051491462932400
12			0.012388695657060
13			−0.015831780392559
14			−0.002717178600540
15			0.002886948664020
16			0.000630499394708
17			−0.000305833973596

Note $M_1 = 4$ corresponds to $\tilde{N} = 5$, and $M_1 = 7$ corresponds to $\tilde{N} = 11$ and $\tilde{N} = 17$

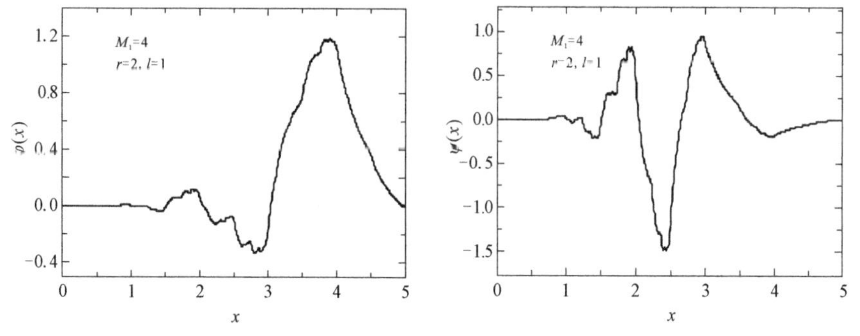

Fig. 2.3 Scaling function $\varphi(x)$ and wavelet functions $\psi(x)$ of the original Coiflets (support [0, 5]), $n = 7$

and the base wavelet functions $\{\psi_{n,k}(x) = 2^{n/2}\psi(2^n x - k),\ k \in Z\}$ can be obtained accordingly. The variable x here takes values at dyadic points associated with the resolution index n, that is $x = j2^{-n}$. The integer j should make x to satisfy $x(= j2^{-n}) \in \Omega$. Obviously, when Ω is a finite domain, then j must be a finite integer. The integer k should make $2^n x = j - k$ be located in the support domain of $\varphi(x)$ and

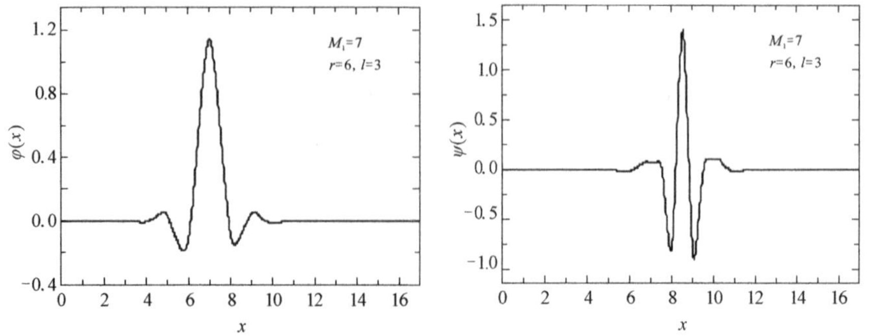

Fig. 2.4 Scaling function $\varphi(x)$ and wavelet functions $\psi(x)$ of the Generalized Coiflets (support $[0, 17]$), $n = 7$

$\psi(x)$, respectively. And the range of variation, or the number of selectable k, is the same as that of the available integer points on the support domain, which changes with j. Therefore, the account of calculations is finite for the decomposition and reconstruction of a function $f(x)$ defined on a finite domain. For the reconstruction based on the base scaling functions, since the generalized Gaussian integral method, Eqs. (2.30) or (2.32), is applied, then the relevant decomposition coefficients can be directly characterized by the function values of the decomposed function at the dyadic points, so that the integral operation of the decomposition coefficients can be disappeared. Following this way, for each specified dyadic point, $x = j2^{-n}$, under the reconstruction of Eq. (2.32) based on the generalized Gaussian integral, the number of calculations of the convolution summation is at most $\tilde{N} - 1$. Consider a function $f(x)$ defined in a finite domain $\Omega = [a, b]$ $(b > a)$, we can transform the variable x into $\tilde{x} = (x - a)/(b - a)$, then $x \in \Omega$ can be transformed into $\tilde{x} \in [0, 1]$. After that, we still use the notations of $f(x)$ and $x \in \Omega = [0, 1]$ without losing generality. When $\Omega = [a, b]$ rather than $\Omega = [0, 1]$ for a function $f(x)$, the variable transform is implicitly used. In such case, once the solution of $f(\tilde{x})$ is obtained in $\Omega = [0, 1]$, the solution of $f(x)$ in $\Omega = [a, b]$ can be easily gained by using its inverse transform, i.e., $\tilde{x} = x(b - a) + a$, to the solution $f(\tilde{x})$. In the following introduction related to problems of finite domain, they all are dealt with by $[0, 1]$. As pointed out previously, the total summations in the reconstruction are $(\tilde{N} - 1)2^n$ for a one-dimensional problem.

2.4 Spectrum Characteristics of the Orthogonal Wavelets

In this section, we use the Fourier analysis method to investigate the spectrum characteristics of the scaling functions and its reconstruction of Eq. (2.32). In fact, some establishing process of wavelet theory has been developed based on the Fourier analysis. From the Fourier analysis method introduced, later on, we can know that, for any square integrable function, $f(x) \in L^2(R)$, in the time domain of $(-\infty, \infty)$, the frequency obtained by using Fourier analysis is invariable with time, which is inconsistent with many practical situations. For example, when a singer sings a song, not only the loudness of the sound can be changed but also its frequency can always be varied as needed (i.e., the frequency varies with time). This feature of signals cannot be captured by the classical Fourier analysis. In the real world, there are many similar signals that need to be processed to obtain their frequency spectrum that vary with time. To solve this issue, in the early stage of signal processing, the so-called windowed Fourier analysis is used. That is, the original signal is intercepted at a certain time range and then processed by Fourier analysis. This is mathematically equivalent to multiplying the original signal function with a window function of fixed height 1 and width Δx, and the window is moved with time. When the width of the window function is reduced to a certain extent, the frequency obtained by using Fourier analysis can behave for the transient frequency. That is, when the window function is moved with time, the frequency varying with time for the whole signal can be obtained. Hence, the signal analysis on time-varying frequency can be realized by such windowed Fourier analysis. This windowed signal processing is called the short-time Fourier analysis. However, the problem that appeared in the short-time Fourier analysis is that there are some noises near the ends of the window, due to that the signal function is zero-processed outside the window and the original signal is selected in the window in this method. In theory, such a signal process may lead to the first type of discontinuity at the ends of the window. According to the principle of Fourier analysis, we know that such first type of discontinuity or non-smooth points may generate the Gibbs phenomenon of oscillation in the reconstructed signal in a small region containing these discontinuity points. Hence, the original signal cannot be accurately reconstructed near these points by the short-time Fourier analysis if no additional technique is employed. In order to eliminate this type of noise possibly, many filtering methods (mathematically called filters) have been established. With the requirements to process some short-time high-frequency signals in earthquakes and chemical engineering, the wavelet method was gradually developed and established to overcome the disadvantages of the short-time Fourier analysis methods, including the fixed window length and noise. From the point of view of function continuity, we know that the low-frequency part of a function is corresponding to the information of those relevant smooth of the function, while its high-frequency part is connected to that of the high-order continuity or jittering of the function. Under this understanding, the spectrum analysis can provide us another way to highlight the characteristics in the wavelets.

2.4.1 Essentials of Spectrum Analysis

According to Fourier analysis, the Fourier transform, and its inverse transform are, respectively, defined by

$$\hat{f}(\omega) = \int_{-\infty}^{+\infty} f(x)e^{-i\omega x}dx, \quad f(x) = \frac{1}{2\pi}\int_{-\infty}^{+\infty} \hat{f}(\omega)e^{i\omega x}d\omega \tag{2.42}$$

which are the information of a signal that appeared in frequency and time domain, respectively. Using this pair of relationships, there is no difficulty for one to prove

$$f(ax) \leftrightarrow a^{-1}\hat{f}(\omega/a) \tag{2.43}$$

At the following, one will find that the spectrum analysis is dependent on M_1 in the generalized Coiflets. Hence, we omit it in the corresponding formulae for simplicity. In fact, such omission implies that the spectrum analysis is conducted for any compactly supported orthogonal wavelets containing the generalized Coiflets. Once M_1 is needed, we will return to their formulae.

Definition 2.6 $\Delta_{\hat{f}}$ is called the Fourier spectrum domain of a function $f(x)$, which mathematically means

$$\hat{f}(\omega) = \begin{cases} \neq 0 & \omega \in \Delta_{\hat{f}} \\ \equiv 0 & \omega \notin \Delta_{\hat{f}} \end{cases}. \tag{2.44}$$

For the expansion of $f(x)$ under the base scaling functions, i.e., Eq. (2.4), denote $f_*(x) = P_n f(x) \equiv \sum_{k\in Z} c_{n,k}\varphi_{n,k}(x)$, then there are $f_* \leftrightarrow \hat{f}_*$ and $\varphi_{n,k}(x) \leftrightarrow \hat{\varphi}_{n,k}(\omega)$. The question now is whether or not $\Delta_{\hat{f}_*} \subset \Delta_{\hat{\varphi}_{n,k}}$ must be true? We will answer this question below.

Definition 2.7 The Fourier energy spectrum for a function $f(x)$ is defined by the product of the Fourier spectrum function of $f(x)$ and its conjugate function, i.e.,

$$\left\|\hat{f}(\omega)\right\| = \hat{f}(\omega)\overline{\hat{f}(\omega)} \tag{2.45}$$

which is abbreviated by $\left\|\hat{f}\right\|$. Correspondingly, its energy spectrum domain is denoted by $\Delta_{\|\hat{f}\|}$.

Definition 2.8 For all spectral functions $\hat{f}(\omega)$ in the spectral domain $\Delta_{\hat{f}}$, the set of spectral functions $\hat{V}_{\hat{f}}$ is defined by

$$\hat{V}_{\hat{f}} = \{\hat{f}(\omega);\, \omega \in \Delta_{\hat{f}}\}. \tag{2.46}$$

Lemma 2.1 The spectrum domain of the product of two different Fourier spectral functions is the intersection of the spectrum domains of the two spectral functions, and the spectrum domain of the product of functions is also a subset of each spectrum domain of the two spectral functions.

Proof Let $\hat{g}(\omega) = \hat{f}_1(\omega)\hat{f}_2(\omega)$ with $\hat{f}_1(\omega) \neq \hat{f}_2(\omega)$. When $\omega \in \Delta_{\hat{f}_1}$ and $\omega \in \Delta_{\hat{f}_2}$ simultaneously, we have $\hat{g}(\omega) = \hat{f}_1(\omega)\hat{f}_2(\omega) \neq 0$, i.e., $\omega \in \Delta_{\hat{g}}$. Otherwise, if $\omega \in \Delta_{\hat{g}}$, we know that $\hat{g}(\omega) \neq 0$. Further, we should have both $\hat{f}_1(\omega) \neq 0$ and $\hat{f}_2(\omega) \neq 0$, then $\omega \in \Delta_{\hat{f}_1}$ and $\omega \in \Delta_{\hat{f}_2}$ simultaneously. Combining these two cases, we get $\Delta_{\hat{g}}=\Delta_{\hat{f}_1} \cap \Delta_{\hat{f}_2}$. Accordingly, we have

$$\Delta_{\hat{g}} \subseteq \Delta_{\hat{f}_1} \quad \text{and} \quad \Delta_{\hat{g}} \subseteq \Delta_{\hat{f}_2}. \tag{2.47}$$

Hence, the proof of Lemma 2.1 is finished.

Theorem 2.1 The Fourier spectrum domain is the full same as its energy spectrum domain, that is, $\Delta_{\|\hat{f}\|} = \Delta_{\hat{f}}$.

Proof Since the conjugate function of a Fourier spectral function does not change its spectrum domain, i.e., $\Delta_{\bar{\hat{f}}} = \Delta_{\hat{f}}$, then, from the definition $\left\|\hat{f}(\omega)\right\| = \hat{f}(\omega)\overline{\hat{f}(\omega)}$ and Lemma 2.1, we know $\Delta_{\|\hat{f}\|} \subseteq \Delta_{\hat{f}}$. On the other hand, for $\forall \omega \in \Delta_{\hat{f}}$, we have $\hat{f}(\omega) \neq 0$, further $\left\|\hat{f}(\omega)\right\| \neq 0$, i.e., $\omega \in \Delta_{\|\hat{f}\|}$. Hence, $\Delta_{\hat{f}} \subseteq \Delta_{\|\hat{f}\|}$. Combining these two cases, one gets $\Delta_{\hat{f}} = \Delta_{\|\hat{f}\|}$. Thus, the proof is finished.

Consider the window function $h(x) = \begin{cases} 1 & x \in [x_1, x_1 + \Delta x] \\ 0 & x \notin [x_1, x_1 + \Delta x] \end{cases}$, where Δx is the window width. For $\forall f(x) \in L^2(R)$, the convolution integral with the window function is defined as

$$f*(x) = \int_{-\infty}^{\infty} h(y) f(x-y) dy, \tag{2.48}$$

further, we get

$$\hat{f}*(\omega) = \hat{h}(\omega)\hat{f}(\omega) \quad \text{and} \quad \left\|\hat{f}*(\omega)\right\| = \left\|\hat{h}(\omega)\right\| \cdot \left\|\hat{f}(\omega)\right\|. \tag{2.49}$$

In the usual case, $\Delta_{\hat{f}} \in (-\infty, \infty)$, that is, $f(x)$ can contain all frequencies of signals. After the signal is processed by the window function, $\Delta_{\hat{f}*} \subset \Delta_{\hat{h}}$ can be obtained by Lemma 2.1. That is, the signal frequency of $f_*(x)$ obtained by the processing contains those in the frequency domain of the window function, which is the main principle of a filter designed in signal analysis.

Since such a simple window function in the time domain will produce noise, in actual filter design, the ideal case is to choose $h(x)$ such that its spectrum has the form:

$$\hat{h}(\omega) = \begin{cases} 1 & \omega \in [\omega_1, \omega_2] \\ 0 & \omega \notin [\omega_1, \omega_2] \end{cases} \quad (\omega_2 > \omega_1). \tag{2.50}$$

Then $\hat{f}*(\omega) = \begin{cases} \hat{f}(\omega) & \omega \in [\omega_1, \omega_2] \\ 0 & \omega \notin [\omega_1, \omega_2] \end{cases}$. This is an ideal filter in theory, which has not been achieved in practice. Such an ideal filter provides us a representative goal for designing the practical filters to approach the ideal case possibly.

From Theorem 2.1 and the definition of Eq. (2.46) for the set of spectrum domain, there is no difficulty for one to give the following conclusions.

Deduction 2.3 For all energy spectral functions $\left\|\hat{f}(\omega)\right\|$ in the spectrum domain $\Delta_{\left\|\hat{f}\right\|}$, $\hat{V}_{\left\|\hat{f}\right\|} = \hat{V}_{\hat{f}}$ holds, where $\hat{V}_{\left\|\hat{f}\right\|} = \{\left\|\hat{f}(\omega)\right\|; \omega \in \Delta_{\left\|\hat{f}\right\|}\}$.

Deduction 2.4 For two spectral function sets $\hat{V}_{\left\|\hat{f}\right\|}$ and $\hat{V}_{\|\hat{g}\|}$, if their two spectrum domains are the same, that is, $\Delta_{\left\|\hat{f}\right\|} = \Delta_{\|\hat{g}\|}$, then they have the same spectral function set, that is, $\hat{V}_{\left\|\hat{f}\right\|} = \hat{V}_{\|\hat{g}\|}$ holds, and vice versa.

It can be seen that the bandwidth of the Fourier spectrum and its spectral function set are the same as the bandwidth of the energy spectrum and its spectral function set, respectively. Therefore, we will use this result to directly give their relevant statements without strictly distinguishing them in the following discussions.

According to Eq. (2.42), one can get the following deductions:

Deduction 2.5 For $\forall f(x) \in L^2(R)$ and any real constant $a(\neq 0)$, when the spectrum function of $f(x)$ belongs to $\hat{V}_{\left\|\hat{f}\right\|}$, then the spectrum domain of $af(x)$ is equal to $\Delta_{\left\|\hat{f}\right\|}$. In other words, a nonzero multiplier to a function $f(x)$ does not affect the set and the domain of the original spectrum function $\hat{f}(\omega)$.

Deduction 2.6 For any two different functions of $\forall f_1(x) \neq f_2(x) \in L^2(R)$, the spectrum of their summation is equal to the summation of their spectrums. That is when $g(x) = f_1(x) + f_2(x)$, we have $\hat{g}(\omega) = \hat{f}_1(\omega) + \hat{f}_2(\omega)$ and $\Delta_{\hat{g}} = \Delta_{\hat{f}_1} \cup \Delta_{\hat{f}_2}$.

Deduction 2.7 For $\forall f(x) \in L^2(R)$ and any nonzero real constant b, the energy spectrum function of $f(x \pm b)$ is equal to the energy spectrum function of $f(x)$, i.e., $\|\hat{g}(\omega)\| = \left\|\hat{f}(\omega)\right\|$, here $g(x) = f(x \pm b)$. Further, $\Delta_{\hat{g}} = \Delta_{\hat{f}}$.

2.4.2 *Spectrum Characteristics of Compactly Supported Orthogonal Wavelets*

Based on the Fourier analysis introduced above, here, we discuss the spectrum characteristics of the compact supported orthogonal scaling function and wavelet function constructed in the previous section such that we can obtain the main characteristics of the wavelet approximation in the spectrum space. For such situation, we define

$$\varphi_{n,k}^*(x) = 2^{n/2}\varphi_{n,k}(x) = 2^n\varphi(2^n x - k), \quad k \in Z \tag{2.51a}$$

$$\psi_{n,k}^*(x) = 2^{n/2}\psi_{n,k}(x) = 2^n\psi(2^n x - k), \quad k \in Z \tag{2.51b}$$

It is evident that such $\{\varphi_{n,k}^*(x)\}$ and $\{\psi_{n,k}^*(x)\}$ have the same orthogonality as the standard orthogonal base functions $\{\varphi_{n,k}(x)\}$ and $\{\psi_{n,k}(x)\}$, respectively, except for their normalization condition. According to Deductions 2.6 and 2.7, moreover, we know that their corresponding frequency domain and spectral function set are also the same. In this case, we can use the characteristics of $\{\varphi_{n,k}^*(x)\}$ and $\{\psi_{n,k}^*(x)\}$ to conduct the analysis of the expansion $P_n f(x)$ and $Q_n f(x)$. Due to the reason that the filters of $\{\varphi_{n,k}^*(x)\}$ and $\{\psi_{n,k}^*(x)\}$ approach to a low-frequency band-pass and high-frequency band-passes, respectively, here, their main parts of the band-passes are discussed. That is, there is obvious no zero spectrum in the main part of the spectrum domain and almost zero outside the main pain part.

Theorem 2.2 The spectrum domains of base scaling functions $\{\varphi_{n,k}(x)\}$ and base wavelet functions $\{\psi_{n,k}(x)\}$ are both independent of the translation index, k, and increase with increasing the resolution level n. That is, they satisfy the following relationships:

$$\Delta_{\hat{\varphi}_{n,k}} = \Delta_{\hat{\varphi}_{n,k}^*} = 2^n\Delta_{\hat{\varphi}_{\tilde{N}}}, \quad \Delta_{\hat{\psi}_{n,k}} = \Delta_{\hat{\psi}_{n,k}^*} = 2^n\Delta_{\hat{\psi}_{\tilde{N}}} \tag{2.52}$$

and the following energy spectrum relationship:

$$\left\|\hat{\varphi}_{n,k}(\omega)\right\| = 2^{-n}\left\|\hat{\varphi}_{\tilde{N}}(\omega/2^n)\right\|, \left\|\hat{\psi}_{n,k}(\omega)\right\| = 2^{-n}\left\|\hat{\psi}_{\tilde{N}}(\omega/2^n)\right\| \tag{2.53}$$

$$\left\|\hat{\varphi}_{n,k}^*(\omega)\right\| = \left\|\hat{\varphi}_{N}(\omega/2^n)\right\|, \quad \left\|\hat{\psi}_{n,k}^*(\omega)\right\| = \left\|\hat{\psi}_{\tilde{N}}(\omega/2^n)\right\| \tag{2.54}$$

where the subscript $\tilde{N}$ in $\varphi_{\tilde{N}}(x) - \varphi(x)$ and $\psi_{\tilde{N}}(x) = \psi(x)$ is used only to identify the support domain.

Proof Firstly, we give the conclusion of this theorem on the base scaling functions. For the scaling function $\varphi(x)$ and its basis functions $\varphi_{n,k}(x)$, we have

$$\varphi_{\tilde{N}}(x) \leftrightarrow \hat{\varphi}_{\tilde{N}}(\omega),$$

$$\varphi_{n,k}(x) = 2^{n/2}\varphi_{\tilde{N}}(2^n x - k) \leftrightarrow \hat{\varphi}_{n,k}(\omega) = 2^{-n/2}\hat{\varphi}_{\tilde{N}}(\omega/2^n)e^{i\omega k/2^n}.$$

Then their energy spectrum relationships can be written in the form:

$$\left\|\hat{\varphi}_{n,k}(\omega)\right\| = 2^{-n}\left\|\hat{\varphi}_{\tilde{N}}(\omega/2^n)\right\|, \quad \left\|\hat{\varphi}_{n,k}^*(\omega)\right\| = \left\|\hat{\varphi}_{\tilde{N}}(\omega/2^n)\right\|$$

Thus, the former equations in Eqs. (2.53) and (2.54) are achieved. It is obvious that the energy spectrum of the base scaling function $\varphi_{n,k}(x)$ is independent of k.

From the relationships between the obtained energy spectrums of the base scaling functions $\{\varphi_{n,k}(x)\}$ and $\{\varphi^*_{n,k}(x)\}$, and that of the scaling function $\varphi_{\tilde{N}}(x)$, it is obvious that the first expression of the spectrum domain relationship of Eq. (2.52) holds.

To the relations of the spectral domains and sets in Eqs. (2.52)–(2.54) for the wavelet function and its base functions, the conclusion can be proved by using the same method as the scaling function, which is omitted here for saving space. Hence, the conclusion of this theorem is proved.

Since the spectrum domains of the base scaling and wavelet functions are only related to the resolution level n, but not to the translation index k, then following Deductions 2.6 and 2.7, we can obtain the following theorem:

Theorem 2.3 For $\forall f(x) \in L^2(R)$, denote $f_n^*(x) = P_n f(x) \equiv \sum_{k\in Z} c_{n,k}\varphi^*_{n,k}(x)$ and $g_n^*(x) = Q_n f(x) \equiv \sum_{k\in Z} c_{n,k}\psi^*_{n,k}(x)$, where $c_{n,k}, d_{n,k} \in R$. Then the following relationships:

$$\Delta_{\hat{f}_n^*} = \Delta_{\hat{\varphi}_{n.k}} = \Delta_{\hat{\varphi}^*_{n,k}}, \quad \Delta_{\hat{g}_n^*} = \Delta_{\hat{\psi}_{n.k}} = \Delta_{\hat{\psi}^*_{n,k}} \tag{2.55}$$

$$\hat{f}_n^*(\omega) \in \hat{V}_{\hat{f}_n^*} = \hat{V}_{\hat{\varphi}_{n,k}} = \hat{V}_{\hat{\varphi}^*_{n,k}}, \quad \hat{g}_n^*(\omega) \in \hat{V}_{\hat{g}_n^*} = \hat{V}_{\hat{\psi}_{n,k}} = \hat{V}_{\hat{\psi}^*_{n,k}} \tag{2.56}$$

are held.

From Theorem 2.3 and the wavelet decomposition relation of Eq. (2.11), i.e.,$P_{n+1}f(x) = P_n f(x) + Q_n f(x)$ $(n \geq 0)$, there is no difficulty for one to obtain the following theorem.

Theorem 2.4 For any resolution level $n(\geq 0)$, we have

$$\Delta_{\widehat{\varphi}_{n+1,k}} = \Delta_{\widehat{\varphi}_{n,k}} \cup \Delta_{\hat{\psi}_{n,k}}, \tag{2.57}$$

where $\Delta_{\widehat{\varphi}_{0,k}} = \Delta_{\widehat{\varphi}_{\tilde{N}}}$.

From the conclusion of Theorem 2.4, we have the following deduction.

Deduction 2.8 For compact supported orthogonal wavelets, the spectrum bandwidth of the base scaling functions $\varphi_{n,k}(x)$ satisfies the following relationships:

$$\Delta_{\hat{\varphi}_{n+1,k}} = \Delta_{\hat{\varphi}_{\tilde{N}}} \cup \Delta_{\hat{\psi}_{1,k}} \cup \Delta_{\hat{\psi}_{2,k}} \cup \cdots \cup \Delta_{\hat{\psi}_{n,k}}, \quad n \geq 0, \tag{2.58}$$

$$\Delta_{\hat{\varphi}_{\tilde{N}}} \subset \Delta_{\hat{\varphi}_{1,k}} \subset \Delta_{\hat{\varphi}_{2,k}} \subset \cdots \subset \Delta_{\hat{\varphi}_{n,k}} \subset \Delta_{\hat{\varphi}_{n+1,k}} \cdots \tag{2.59}$$

The result displayed in Deduction 2.8 tells us that with the increase of resolution index n, the spectral domain of $P_n f(x) \equiv \sum_{k\in Z} c_{n,k}\varphi_{n,k}(x)$ is increased by combining each spectral bandwidth of the scaling functions $\varphi_{r,k}(x)$ with resolution level r from 0 to n.

Theorem 2.5 When the resolution level n is finite, the spectrum bandwidth of any compactly supported orthogonal scaling function $\varphi_{\hat{N}}(x)$ must be finite too, and $\omega = 0$ should be a point in its frequency domain.

Proof Without loss of generality and considering the symmetry of frequency domain about $\omega = 0$, hereafter, we only consider the case of $\omega \geq 0$. Denote $\Delta_{\hat{\varphi}_{\tilde{N}}} = [\omega_1, \omega_2]$ $(\omega_1 < \omega_2)$. Let us firstly prove $\omega_1 = 0 \in \Delta_{\hat{\varphi}_{\tilde{N}}}$. If it not true, i.e., $\omega_1 > 0$, according to the first formula of (2.52), we have $\Delta_{\hat{\varphi}_{n,k}} = [2^n\omega_1, 2^n\omega_2]$. Denote $\Delta_{\min} = \min\{\omega_1, \Delta_{\hat{\varphi}_{\tilde{N}}}\}$. It is obvious $\omega_1 + \Delta_{\min} \leq 2\omega_1$, thus $[\omega_1, \omega_1 + \Delta_{\min}) \subset \Delta_{\hat{\varphi}_{\tilde{N}}}$. Then for the case of $\forall \omega \in [\omega_1, \omega_1 + \Delta_{\min}) \subset \Delta_{\hat{\varphi}_{\tilde{N}}}$, we have $\omega_1 + \Delta_{\min} < 2^n\omega_1 (n \geq 1)$. Hence, $\omega \notin [2^n\omega_1, 2^n\omega_2]$ $(n \geq 1)$. That is $[\omega_1, \omega_1 + \Delta_{\min}) \not\subset \Delta_{\hat{\varphi}_{n,k}} (n \geq 1)$, which is contradictory to the inclusion relations of Eqs. (2.58) and (2.59). Therefore, $\omega_1 = 0 \in \Delta_{\hat{\varphi}_{\tilde{N}}}$ should be held.

Now, let us prove that ω_2 is finite. If it is not true, we have $\omega_2 \to \infty$, i.e., $\Delta_{\hat{\varphi}_{\tilde{N}}} = [0, \infty)$. In this case, the first expression of Eq. (2.52) tells us that $\Delta_{\hat{\varphi}_{n,k}} = [0, \infty)$ is held for any $n \geq 1$. This implies that

$$\Delta_{\hat{\varphi}_{\tilde{N}}} = \Delta_{\hat{\varphi}_{1,k}} = \Delta_{\hat{\varphi}_{2,k}} = \cdots = \Delta_{\hat{\varphi}_{n,k}} = \Delta_{\widehat{\phi}_{n+1,k}} \cdots$$

holds. It is obvious that this is contradictory to the inclusion relationship of (2.59). Therefore, ω_2 should be finite. Thus, the proof for this theorem is finished.

Denote $\Delta_{\hat{\varphi}_{\tilde{N}}} = [0, \omega_*]$ (ω_* is a finite real number), then the scaling function $\varphi_{\hat{N}}(x)$ constitutes a low-frequency band-pass filter with the energy spectrum $\left\|\hat{h}_{\varphi_{\tilde{N}}}(\omega)\right\|$ in the frequency spectrum domain $[0, \omega_*]$, and its energy spectrum is

$$\left\|\hat{h}_{\varphi_{\tilde{N}}}(\omega)\right\| = \left\|\hat{\varphi}_{\tilde{N}}(\omega)\right\|, \quad \omega \in \Delta_{\hat{\varphi}_{\tilde{N}}} = [0, \omega_*] \tag{2.60}$$

where we have used subscript in $\left\|\hat{h}(\omega)\right\|$ to identify its corresponding which functions energy spectrum is employed. Further, from the former formula of (2.52) in Theorem 2.2, we have the following deductions.

Deduction 2.9 If the spectrum domain of a scaling function $\varphi_{\tilde{N}}(x)$ is $\Delta_{\hat{\varphi}_{\tilde{N}}} = [0, \omega_*]$, then for any resolution level $n(\geq 1)$, its base scaling functions $\{\varphi^*_{n,k}(x)\}$ constitutes a low-pass filter with the energy spectrum $\left\|\hat{h}(\omega)\right\|$ in the spectral domain $\Delta_{\hat{\varphi}^*_{n,k}} = [0, 2^n\omega_*]$, and its energy spectrum is expressed by

$$\left\|\hat{h}_{\varphi^*_{n,k}}(\omega)\right\| = \left\|\hat{\varphi}^*_{n,k}(\omega)\right\| = \left\|\hat{\varphi}_{\tilde{N}}(\omega/2^n)\right\|, \quad \omega \in \Delta_{\varphi^*_{n,k}} = [0, 2^n\omega_*]. \tag{2.61}$$

Deduction 2.10 For $\forall f(x) \in L^2(R)$, its expansion or projection on the base scaling functions $\{\varphi_{n,k}(x)\}$ or $\{\varphi^*_{n,k}(x)\}$ under any resolution level $n(\geq 1)$

$$f^*_n(x) = P_n f(x) \equiv \sum_{k \in \mathbf{Z}} c_{n,k}\varphi_{n,k}(x)$$

$$= 2^{-n/2}\sum_{k\in Z} c_{n,k}\varphi^*_{n,k}(x) = 2^{n/2}\sum_{k\in Z} c_{n,k}\varphi(2^2 x - k) \tag{2.62}$$

is also low-frequency band-pass, i.e., $\Delta_{\hat{f}^*_n} = [0, 2^n\omega_*]$. In this frequency domain, the energy spectrum of its low-frequency band-pass can be characterized by Eq. (2.61).

Such results display that the expansion $P_n f(x)$ with any resolution level $n(\geq 1)$ is obtained by filtering $f(x)$ through the filter $\left\|\hat{\varphi}^*_{n,k}(\omega)\right\| = \left\|\hat{\varphi}_{\tilde{N}}(\omega/2^n)\right\|$ ($\omega \in [0, 2^n\omega_*]$) characterized by Eq. (2.61). Since this filter has a low band-pass characteristic, such filtering or expansion of a function can capture the smooth part of $f(x)$. As the resolution level $n(\geq 1)$ increases by each 1, the bandwidth will double the former domain, such that the relevant higher frequency parts can be contained in expansion with doubled low-frequency band-pass. In fact, according to Theorem 2.2, the energy spectrum curve of the low-pass filter $\{\varphi^*_{n,k}(x)\}$ characterized by Eq. (2.61) under a specified resolution level $n(\geq 1)$ can be determined by the corresponding energy spectrum curve of the low-pass filter of the scaling function $\varphi_{\tilde{N}}(x)$ (expressed by Eq. (2.52)) to be extended 2^n times along ω coordinate direction. Therefore, as long as we obtain the energy spectrum curve of the $\varphi_{\tilde{N}}(x)$, we can easily obtain the energy spectrum curve of the $\{\varphi^*_{n,k}(x)\}$ with low-pass filter.

Theorem 2.6 The energy spectrum of the wavelet function $\psi_{\tilde{N}}(x)$ must be a finite spectrum domain without $\omega = 0$, that is, $\Delta_{\hat{\psi}_{\tilde{N}}} = [\omega_0, \omega_1]$, $\omega_1 > \omega_0 > 0$.

Proof We firstly prove that ω_1 is a nonzero finite value. From Theorems 2.3 and 2.4, we have $f_1^*(x) = f_0^*(x) + g_0^*(x)$ with $\hat{f}_0^* \in \hat{V}_{\hat{f}_0^*} = \hat{V}_{\hat{\varphi}_{\tilde{N}}}$, $\hat{g}_0^* \in \hat{V}_{\hat{g}_0^*} = \hat{V}_{\hat{\psi}_{\tilde{N}}}$, and $\hat{f}_1^* \in \hat{V}_{\hat{f}_1^*} = \hat{V}_{\hat{\varphi}^*_{1,k}}$. According to Eq. (2.11), further, we have $\hat{V}_{\hat{\varphi}^*_{1,k}} = \hat{V}_{\hat{\varphi}_{\tilde{N}}} \cup \hat{V}_{\hat{\psi}_{\tilde{N}}}$, and $\Delta_{\hat{\varphi}_{1,k}} = \Delta_{\hat{\varphi}_{\tilde{N}}} \cup \Delta_{\hat{\psi}_{\tilde{N}}}$. Then following Deduction 2.9 and Theorem 2.5, both $\Delta_{\hat{\varphi}_{1,k}}$ and $\Delta_{\hat{\varphi}_{\tilde{N}}}$ are bounded. Therefore, $\Delta_{\hat{\psi}_{\tilde{N}}}$ must be bounded too. That is, ω_1 must be a nonzero finite value.

Next, we use the proof of contradiction for $\omega_0 \neq 0$. Otherwise, if $\omega_0 = 0$, we know that for any function $f(x)$ with $\omega_0 = 0$, its expansion on the base wavelet functions $\{\psi_{\tilde{N}}(x-k)\}$ has to contain the nonzero part corresponding to $\omega_0 = 0$. Without losing generality, we take the function $f(x) \equiv 1$ in a finite domain $[x_1, x_2]$. Obviously, its frequency is always $\omega_0 = 0$. From the fact that wavelet function $\psi_{\tilde{N}}(x)$ has up to jth-order vanishing moment, the following Property 2.1, $f(x) \equiv 1$ can be accurately expressed by the corresponding base scaling functions $\{\varphi_{\tilde{N}}(x-k)\}$. Due to the orthogonality of the base scaling functions $\{\varphi_{\tilde{N}}(x-k)\}$ and the base wavelet functions $\{\psi_{\tilde{N}}(x-k)\}$, one can obtain that $f(x) \equiv 1$ cannot have any nonzero component in the space spread by $\{\psi_{\tilde{N}}(x-k)\}$, which implies that no information of a function $f(x)$ with $\omega_0 = 0$ is contained in $\Delta_{\hat{\psi}_{\tilde{N}}} = [0, \omega_1]$. This contradictory conclusion indicates that $\omega_0 \neq 0$ should be held to the base wavelet functions $\{\psi_{\tilde{N}}(x-k)\}$. Thus, this theorem holds.

The result of Theorem 2.6 tells us that the wavelet function $\psi_{\tilde{N}}(x)$ constitutes the energy spectrum $\left\|\hat{h}(\omega)\right\|$ of a band-pass filter in the spectral domain $\Delta_{\hat{\psi}_{\tilde{N}}} = [\omega_0, \omega_1]$ ($\omega_1 > \omega_0 > 0$), and its energy spectrum is

$$\left\|\hat{h}_{\psi_{\tilde{N}}}(\omega)\right\| = \left\|\hat{\psi}_{\tilde{N}}(\omega)\right\|, \quad \Delta_{\hat{\psi}_{\tilde{N}}} = [\omega_0, \omega_1]. \tag{2.63}$$

Furthermore, according to Theorems 2.2 and 2.6, we can obtain the spectrum of the band-pass filter corresponding to the base wavelet functions $\{\psi^*_{n,k}(x)\}$ as follows.

Deduction 2.11 For any resolution level $n(\geq 1)$, the base wavelet functions $\{\psi^*_{n,k}(x)\}$ is a band-pass filter with the energy spectrum $\left\|\hat{h}(\omega)\right\|$ in the frequency domain $\Delta_{\hat{\psi}^*_{n,k}} = [2^n\omega_0, 2^n\omega_1]$, and its energy spectrum can be obtained from the energy spectrum of the wavelet function $\psi_{\tilde{N}}(x)$ filter by extending 2^n times along ω coordinate direction and then translating $(2^n - 1)\omega_0$ along this direction. That is,

$$\left\|\hat{h}_{\psi^*_n}(\omega)\right\| = \left\|\hat{\psi}^*_{n,k}(\omega)\right\| = \left\|\hat{\psi}_{\tilde{N}}(\omega/2^n)\right\|, \quad \omega \in \Delta_{\widehat{\psi}^*_{n,k}} = (2^n\omega_0, 2^n\omega_1] \tag{2.64}$$

It is obvious that with the increase of resolution level $n(\geq 1)$, the frequency band of the filter $\left\|\hat{h}_{\psi^*_n}(\omega)\right\|$ is gradually moved and expanded to the high-frequency region.

Deduction 2.12 For $\forall f(x) \in L^2(R)$, its expansion or projection on the base wavelet functions $\{\psi^*_{n,k}(x)\}$ under any resolution level $n(\geq 1)$, or

$$\begin{aligned} g^*_n(x) = Q_n f(x) &\equiv \sum_{k\in\mathbf{Z}} d_{n,k}\psi_{n,k}(x) \\ &= 2^{-n/2}\sum_{k\in Z} d_{n,k}\psi^*_{n,k}(x) = 2^{n/2}\sum_{k\in Z} d_{n,k}\psi(2^2x - k) \end{aligned} \tag{2.65}$$

is high-frequency band-pass, i.e., $\Delta_{\hat{g}^*_n} = (2^n\omega_0, 2^n\omega_1]$. In this frequency band, the energy spectrum of its band-pass filter is characterized by Eq. (2.64).

According to Deduction 2.10, with the increase of resolution level $n(\geq 1)$, the expansion under the scaling functions of Eq. (2.62) can capture the information of the relevant high frequency of function $f(x)$ into the low-pass band step by step. Then according to Eq. (2.13), we have the following deduction.

Deduction 2.13 For $\forall f(x) \in L^2(R)$ and any specified resolution level $n(\geq 1)$, the projection function $\tilde{g}^*_n(x) = \tilde{Q}_n f(x) = \sum_{i=n}^{\infty} Q_i f(x)$ captures the part of high-frequency band-pass of the function $f(x)$ in frequency domain $(2^n\omega_0, \infty)$.

Theorem 2.7 The energy spectrum of the orthogonal scaling function $\varphi_{\tilde{N}}(x)$ and wavelet function $\psi_{\tilde{N}}(x)$ makes the following formulae

$$\int_0^{\omega_*} \left\|\hat{\varphi}_{\tilde{N}}(\omega)\right\| d\omega = \pi \quad \text{and} \quad \int_{\omega_0}^{\omega_1} \left\|\hat{\psi}_{\tilde{N}}(\omega)\right\| d\omega = \pi \tag{2.66}$$

hold.

Proof From the Parseval identity and the orthogonal normalization conditions of $\varphi_{\tilde{N}}(x)$ and $\psi_{\tilde{N}}(x)$, we have

$$1 = \int_{-\infty}^{\infty} \varphi_{\tilde{N}}^2(x)dx = \frac{1}{2\pi}\int_{-\infty}^{\infty} |\hat{\varphi}_{\tilde{N}}(\omega)|^2 d\omega,$$

$$1 = \int_{-\infty}^{\infty} \psi_{\tilde{N}}^2(x)dx = \frac{1}{2\pi}\int_{-\infty}^{\infty} |\hat{\psi}_{\tilde{N}}(\omega)|^2 d\omega.$$

According to Eqs. (2.60) and (2.61) and considering their symmetries with respect to $\omega = 0$, we have

$$\int_{-\infty}^{\infty} \left|\hat{\varphi}_{\tilde{N}}(\omega)\right|^2 d\omega = \int_{-\omega_*}^{\omega_*} \left\|\hat{\varphi}_{\tilde{N}}(\omega)\right\| d\omega = 2\int_{0}^{\omega_*} \left\|\hat{\varphi}_{\tilde{N}}(\omega)\right\| d\omega = 2\pi$$

$$\int_{-\infty}^{\infty} \left|\hat{\psi}_{\tilde{N}}(\omega)\right|^2 d\omega = \int_{-\omega_1}^{\omega_1} \left\|\hat{\psi}_{\tilde{N}}(\omega)\right\| d\omega = 2\int_{\omega_0}^{\omega_1} \left\|\hat{\psi}_{\tilde{N}}(\omega)\right\| d\omega = 2\pi$$

That is, Eq. (2.66) in Theorem 2.7 holds.

From Theorems 2.2 and 2.7, there is no difficulty for one to get the following deduction.

Deduction 2.14 For any resolution level $n(\geq 1)$, the energy spectrum of the standard orthogonal base scaling functions $\{\varphi_{n,k}(x)\}$ and the standard base wavelet functions $\{\psi_{n,k}(x)\}$ satisfy the following relationship:

$$\int_{0}^{2^n\omega_*} \left\|\hat{\varphi}_{n,k}(\omega)\right\| d\omega = \int_{0}^{\omega_*} \left\|\hat{\varphi}_{\tilde{N}}(\omega)\right\| d\omega. = \pi \tag{2.67}$$

$$\int_{2^n\omega_0}^{2^n\omega_1} \left\|\hat{\psi}_{n,k}(\omega)\right\| d\omega = \int_{\omega_0}^{\omega_1} \left\|\hat{\psi}_{\tilde{N}}(\omega)\right\| d\omega. = \pi \tag{2.68}$$

The results of Deduction 2.14 tell us that the energy spectrum of the base scaling functions $\{\varphi_{n,k}(x)\}$ and the base wavelet functions $\{\psi_{n,k}(x)\}$ contain the same area in their respective spectrum domains, and they are independent of the resolution level $n(\geq 1)$. Therefore, no matter how n increases, the spectral domain of the base scaling functions and the base wavelet functions increases with 2^n, while the energy spectrum intensity decreases with 2^n, such that the total energy spectrum areas remain unchangeable.

Theorem 2.8 For any resolution level $n(\geq 1)$, the energy spectrum $\left\|\hat{\varphi}^*_{n,k}(\omega)\right\|$ of the base scaling functions $\{\varphi^*_{n,k}(x)\}$ is equal to the summation of the energy spectrum $\left\|\hat{\varphi}^*_{n-1,k}(\omega)\right\|$ of the base scaling functions $\{\varphi^*_{n-1,k}(x)\}$ and the energy spectrum $\left\|\hat{\psi}^*_{n-1,k}(\omega)\right\|$ of its corresponding base wavelet functions $\{\psi^*_{n-1,k}(x)\}$, namely,

$$\left\|\hat{\varphi}^*_{n,k}(\omega)\right\| = \left\|\hat{\varphi}^*_{n-1,k}(\omega)\right\| + \left\|\hat{\psi}^*_{n-1,k}(\omega)\right\|, \quad \omega \in \Delta_{\hat{\varphi}^*_{n,k}} = [0, 2^n\omega_*]. \qquad (2.69)$$

Further, when $\Delta_{\hat{\varphi}^*_{n,k}} = [0, 2^n\omega_*]$ and $\Delta_{\hat{\psi}^*_{n,k}} = (2^n\omega_0, 2^n\omega_1]$, we have $\omega_0 \leq \omega_*$ and $\omega_1 = 2\omega_*$.

Proof According to the wavelet decomposition of Eq. (2.11) for $\forall f \in L^2(R)$ and any specified resolution level $n(\geq 1)$, we always have

$$P_n f(x) = P_{n-1} f(x) + Q_{n-1} f(x)$$

to be true. Then, following Eq. (2.60) of the Deduction 2.7 and Eq. (2.65) of the Deduction 2.12, we know that $P_n f(x)$ is the decomposition on the low-pass filter $\{\varphi^*_{n,k}(x)\}$, and the energy spectrum of the low-pass filter is $\left\|\hat{\varphi}^*_{n,k}(\omega)\right\|$ $(\omega \in \Delta_{\hat{\varphi}^*_{n,k}} = [0, 2^n\omega_*])$, while $P_{n-1} f(x)$ and $Q_{n-1} f(x)$ are respectively the decomposition on the filters $\{\varphi^*_{n-1,k}(x)\}$ and $\{\psi^*_{n-1,k}(x)\}$, and their corresponding energy spectrums are $\left\|\hat{\varphi}^*_{n-1,k}(\omega)\right\|$ $(\omega \in \Delta_{\hat{\varphi}^*_{n-1,k}} = [0, 2^{n-1}\omega_*])$ and $\left\|\hat{\psi}^*_{n-1,k}(\omega)\right\|$ $(\omega \in \Delta_{\widehat{\psi}^*_{n-1,k}} = (2^{n-1}\omega_0, 2^{n-1}\omega_1])$, respectively. Furthermore, according to Eq. (2.57) of Theorem 2.4, we get

$$\Delta_{\hat{\varphi}^*_{n,k}} = [0, 2^n\omega_*] = [0, 2^{n-1}\omega_*] \cup (2^{n-1}\omega_0, 2^{n-1}\omega_1]. \qquad (2.70)$$

Firstly, we prove $\omega_0 \leq \omega_*$. Otherwise, if $\omega_0 > \omega_*$, one gets

$$[0, 2^{n-1}\omega_*] \cap (2^{n-1}\omega_0, 2^{n-1}\omega_*] = \emptyset.$$

In such case, to $\hat{f}(\omega) \in \Delta_{\hat{\varphi}^*_{n,k}}$ in $(2^{n-1}\omega_*, 2^{n-1}\omega_0)$, its corresponding information of the function $f(x)$ cannot be captured by both the base scaling functions $\{\varphi^*_{n-1,k}(x)\}$ and the base wavelet functions $\{\psi^*_{n-1,k}(x)\}$, which is obviously contradicted to Eq. (2.11) for $\forall f \in L^2(R)$. Therefore, $\omega_0 \leq \omega_*$ must be held. Then, we have

$$[0, 2^{n-1}\omega_*] \cup (2^{n-1}\omega_0, 2^{n-1}\omega_1] = [0, 2^{n-1}\omega_*]$$

$$[0, 2^{n-1}\omega_*] \cap (2^{n-1}\omega_0, 2^{n-1}\omega_1] = (2^{n-1}\omega_0, 2^{n-1}\omega_*]$$

being the sum of spectrum domains of $\{\varphi^*_{n-1,k}(x)\}$ and $\{\psi^*_{n-1,k}(x)\}$. From Eq. (2.70) and above former equation, one gets $\omega_1 = 2\omega_*$.

Now we prove that Eq. (2.69) holds for any resolution level n. In such case, it is OK when Eq. (2.67) is held by using Eq. (2.69). According to Eqs. (2.67) and (2.68) and the definitions of the base functions $\{\varphi^*_{n-1,k}(x)\}$ and $\{\psi^*_{n-1,k}(x)\}$, we have

$$\int_0^{2^{(n-1)}\omega_*} \left\|\hat{\varphi}^*_{n-1,k}(\omega)\right\| d\omega = 2^{(n-1)}\pi, \quad \int_{2^{(n-1)}\omega_0}^{2^{(n-1)}\omega_*} \left\|\hat{\psi}^*_{n-1,k}(\omega)\right\| d\omega = 2^{(n-1)}\pi \tag{2.71}$$

Since $\left\|\hat{\varphi}_{n-1,k}(\omega)\right\| = 2^{-(n-1)}\left\|\hat{\varphi}^*_{n-1,k}(\omega)\right\|$ and $\left\|\hat{\psi}_{n-1,k}(\omega)\right\| = 2^{-(n-1)}\left\|\hat{\psi}^*_{n-1,k}(\omega)\right\|$, then, substituting them into Eq. (2.69), one gets

$$\begin{aligned} \int_0^{2^n\omega_*} \left\|\hat{\varphi}^*_{n,k}(\omega)\right\| d\omega &= \int_0^{2^n\omega_*} \left(\left\|\hat{\varphi}^*_{n-1,k}(\omega)\right\| + \left\|\hat{\psi}^*_{n-1,k}(\omega)\right\|\right) d\omega \\ &= \int_0^{2^{(n-1)}\omega_*} \left\|\hat{\varphi}^*_{n-1,k}(\omega)\right\| d\omega + \int_{2^{(n-1)}\omega_)}^{2^{(n-1)}\omega_*} \left\|\hat{\psi}^*_{n-1,k}(\omega)\right\| d\omega = 2^n\pi \end{aligned} \tag{2.72}$$

Substituting $\left\|\hat{\varphi}_{n,k}(\omega)\right\| = 2^{-n}\left\|\hat{\varphi}^*_{n,k}(\omega)\right\|$ into Eq. (2.72), we obtain

$$\int_0^{2^n\omega_*} \left\|\hat{\varphi}_{n,k}(\omega)\right\| d\omega = 2^{-n}\int_0^{2^n\omega_*} \left\|\hat{\varphi}^*_{n,k}(\omega)\right\| d\omega = \pi$$

That is, Eq. (2.67) is true under the summation operation of Eq. (2.69). Thus, Eq. (2.69) is true. The proof is finished.

To clearly understand Theorem 2.8, we can rewrite Eq. (2.69) into a piecewise form

$$\left\|\hat{\varphi}^*_{n,k}(\omega)\right\| = \begin{cases} \left\|\hat{\varphi}^*_{n-1,k}(\omega)\right\| & \omega \in [0, 2^{n-1}\omega_0] \\ \left\|\hat{\varphi}^*_{n-1,k}(\omega)\right\| + \left\|\hat{\psi}^*_{n-1,k}(\omega)\right\| & \omega \in (2^{n-1}\omega_0, 2^{n-1}\omega_*] \\ \left\|\hat{\psi}^*_{n-1,k}(\omega)\right\| & \omega \in (2^{n-1}\omega_*, 2^n\omega_*] \end{cases} \tag{2.73}$$

Considering the filter spectrum characteristics of the above base scaling functions and base wavelet functions, we know that once the scaling function $\varphi_{\tilde{N}}(x)$ is obtained in a support domain, then the curve of its energy spectrum expressed in Eq. (2.60) can be gained. Further, using the above-mentioned relevant conclusions, the energy spectrum curves corresponding to all its base scaling and wavelet functions can be obtained from the energy spectrum curve of (2.73). The main steps are as follows:

Step 1: After $\varphi_{\tilde{N}}(x)$ is generated according to the scaling function generation program in the previous section, the energy spectrum $\|\hat{\varphi}_{\tilde{N}}(\omega)\|$ and the spectrum domain $\omega \in [0, \omega_*]$ of Eq. (2.60) can be obtained by using the Fourier analysis method;

Step 2: Extending the energy spectrum $\|\hat{\varphi}_{\tilde{N}}(\omega)\|$ of the scaling function along ω coordinate direction of the spectrum domain to twice the original spectrum domain, we obtain the energy spectrum $\|\hat{\varphi}^*_{1,k}(\omega)\| = \|\hat{\varphi}_{\tilde{N}}(\omega/2)\|$ with $\omega \in [0, 2\omega_*]$;

Step 3: Substituting the results of the above two steps into Eq. (2.73), the energy spectrum of wavelet function $\psi_{\tilde{N}}(x)$ is obtained by the form

$$\begin{aligned}\left\|\hat{\psi}_{\tilde{N}}(\omega)\right\| &= \left\|\hat{\psi}^*_{0,k}(\omega)\right\| = \left\|\hat{\varphi}^*_{1,k}(\omega)\right\| - \left\|\hat{\varphi}^*_{0,k}(\omega)\right\| \\ &= \left\|\hat{\varphi}^*_{1,k}(\omega)\right\| - \left\|\hat{\varphi}_{\tilde{N}}(\omega)\right\|, \quad \omega \in (\omega_0, 2\omega_*]\end{aligned}$$

Here, ω_0 is the largest root of the following equation:

$$\left\|\hat{\varphi}^*_{1,k}(\omega_0)\right\| - \left\|\hat{\varphi}_{\tilde{N}}(\omega_0)\right\| = 0$$

Step 4: After $\|\hat{\varphi}_{\tilde{N}}(\omega)\|$ and $\left\|\hat{\psi}_{\tilde{N}}(\omega)\right\|$ are stretched to $2^n (n \geq 2)$ times of the original spectral domain along the ω coordinate direction of the spectral domain, and the energy spectrum of such extending $\left\|\hat{\psi}_{\tilde{N}}(\omega)\right\|$ is translated to the right along the ω coordinate direction by $(2^n - 1)\omega_0$, then the energy spectrums of $\|\hat{\varphi}_{n,k}(\omega)\|$ and $\left\|\hat{\psi}_{n,k}(\omega)\right\|$ under the resolution level n are, respectively, obtained by the form:

$$\left\|\hat{\varphi}^*_{n,k}(\omega)\right\| = \left\|\hat{\varphi}_{\tilde{N}}(\omega/2^n)\right\|, \quad \omega \in [0, 2^n\omega_*]$$

$$\left\|\hat{\psi}^*_{n,k}(\omega)\right\| = \left\|\hat{\psi}_{\tilde{N}}(\omega/2^n)\right\|, \quad \omega \in [2^n\omega_0, 2^{n+1}\omega_*]$$

On the one hand, the generation of each energy spectrum provides a test criterion for checking the generation accuracy of each set of base scaling and wavelet functions of a wavelet. On the other hand, in the wavelet-based signal analysis, as long as the low-pass spectrum of the scaling function filter is obtained, the low-pass and band-pass filter spectra under each resolution level n (> 1) can be obtained according to the above steps. Then, we can directly use them to evaluate the samples of a signal or function characteristic under the base scaling functions and base wavelet functions, respectively.

2.4.3 Spectrum Characteristics of Ideal Wavelets

Although an ideal wavelet approaching to an ideal filter cannot be gained in a finite support domain or are very few in practice, but through this discussion, we can understand the spectrum characteristics of the ideal wavelet when it is realized, which provides a way to evaluate the effectiveness of actual wavelet characteristics on the spectrum space.

Definition 2.9 If the scaling function $\varphi_{\tilde{N}}(x)$ of a wavelet has $\left\|\hat{\varphi}_{\tilde{N}}(\omega)\right\|\,|| = 1$ in its frequency domain, we call this wavelet as an ideal wavelet.

Using the discussion introduced in this section, it is not difficult to find that the following theorem holds.

Theorem 2.9 If a wavelet is an ideal wavelet, then its frequency domain should be $[0, \pi]$, and $\left\|\hat{\varphi}_{\tilde{N}}(\omega)\right\| = 1$ for $\omega \in [0, \pi]$ and $\left\|\hat{\psi}_{\tilde{N}}(\omega)\right\| = 1$ for $\omega \in (\pi, 2\pi]$ are the low- and high-frequency band-passes of ideal filters $\varphi_{\tilde{N}}(x)$ and $\psi_{\tilde{N}}(x)$, respectively.

Theorem 2.10 If a wavelet is an ideal wavelet, then the spectrum of the base scaling functions $\{\varphi^*_{n,k}(x)\}$ and the base wavelet functions $\{\psi^*_{n,k}(x)\}$ at any resolution level $n(\geq 1)$, respectively, constitute an ideal low-pass filter, i.e.,

$$\left\|\hat{\varphi}^*_{n,k}(\omega)\right\| = 1 \quad \forall\omega \in [0, 2^n\pi]$$

and an ideal band-pass filter, i.e.,

$$\left\|\hat{\psi}^*_{n,k}(\omega)\right\| = 1 \quad \forall\omega \in (2^n\pi, 2^{n+1}\pi]$$

2.4.4 Spectrum Characteristic of the Generalized Coiflets

Using the previous discussion on wavelet spectrum characteristics, we can evaluate the characteristics of any compactly supported orthogonal wavelets. Considering the reason that the generalized Coiflets is employed in our solution methods for either linear or nonlinear boundary-value problems in this book, here, we only introduce its relevant spectrum characteristic quantitatively.

Figure 2.5 exhibits the distribution curve of the energy spectrum $\left\|\hat{\varphi}_{\tilde{N}}(\omega)\right\|$ of the scaling function $\varphi_{\tilde{N}}(x)$ of the generalized Coiflets in the finite domains $\tilde{N} = 5, 11, 17$ shown in Table 2.3. From this figure, one can see that with the increase of $\tilde{N}$, $\left\|\hat{\varphi}_{\tilde{N}}(\omega)\right\|$ approaches to the energy spectrum of the ideal wavelet in $[0, \pi)$. In addition, there is a region outer $[0, \pi]$, i.e., $(\pi, \omega_*]$ in its energy spectrum domain. For the case $\tilde{N}= 5$, for example, $\omega_* \approx 5(> \pi)$. Demote $[0, \omega*]$ is the domain of the ideal spectrum part

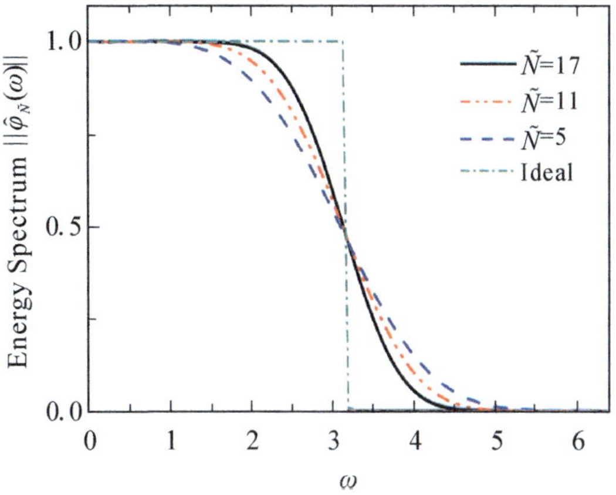

Fig. 2.5 The energy spectrum curve $\left\|\hat{\varphi}_{\tilde{N}}(\omega)\right\|$ of the generalized Coifman scaling function $\varphi_{\tilde{N}}(x)$ under different support domain $[0, \tilde{N}]$

with $\left\|\hat{\varphi}_{\tilde{N}}(\omega)\right\| = 1$. Obviously, $\omega* < \pi$. As $\tilde{N}$ increases, this ideal spectral domain or $\omega*$ increases, while ω_* decreases. When the region of the support domain approaches to infinity, i.e., $\omega*$, it is found that the domain of $\left\|\hat{\varphi}_{\tilde{N}}(\omega)\right\| = 1$ approaches to $[0, \pi]$ or both $\omega*$ and $\omega_* \to \pi$. In other words, when $\tilde{N} \to \infty$,$\varphi_{\tilde{N}}(x)$ approaches to an ideal wavelet. In addition, to each curve of $\left\|\hat{\varphi}_{\tilde{N}}(\omega)\right\|$ in $[0, \omega_*]$ for a specified finite support domain, we have $\pi - \omega* \approx \omega_* - \pi$, or $\omega* \approx 2\pi - \omega_*$.

According to the generation septs, one can gets $\left\|\hat{\varphi}^*_{n-1,k}(\omega)\right\|$ and $\left\|\hat{\psi}^*_{n-1,k}(\omega)\right\|$ which satisfies $\left\|\hat{\varphi}^*_{n-1,k}(\omega)\right\| + \left\|\hat{\psi}^*_{n-1,k}(\omega)\right\| \le 1$. To the $\left\|\hat{\varphi}_{\tilde{N}}(\omega)\right\|$ shown in Fig. 2.5 for the generalized Coiflets, we have

$$\left\|\hat{\varphi}^*_{n,k}(\omega)\right\| = \begin{cases} 1 & \omega \in [0, 2^n\omega^*] \\ < 1 & \omega \in (2^n\omega^*, 2^n\omega_*] \end{cases}. \tag{2.74}$$

When $\tilde{N} = 17$, Fig. 2.6 displays the energy spectrum curve $\left\|\hat{\varphi}^*_{n,k}(\omega)\right\|$ of the generalized Coiflets at the resolution level n from 0 to 7. Here, the ω coordinate is drawn by the compression with $\ln\omega$, while the coordinate annotation still adopts

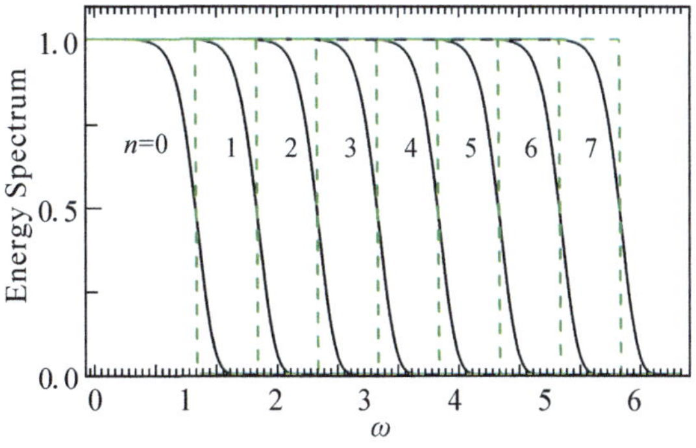

Fig. 2.6 The Energy spectrum curve $\left\|\hat{\varphi}^*_{n,k}(\omega)\right\|$, $\tilde{N} = 17$, of the generalized Coifman scaling function bases $\varphi^*_{n,k}(x)$ under different resolution index n

the value of ω. When $n = 0$, $\left\|\hat{\varphi}^*_{0,k}(\omega)\right\| = \left\|\hat{\varphi}_{\tilde{N}}(\omega)\right\|$ shown in Fig. 2.5 where $\omega_* < 4.5(> \pi)$ and $\omega* = 2\pi - \omega_* \approx 1.78(> \pi/2)$. Following this result, when the resolution level $n \geq 1$, the region of the ideal spectrum (i.e., $\left\|\hat{\varphi}^*_{n,k}(\omega)\right\| = 1$) on the low-pass spectrum is $[0, 2^n\omega*]$ and $2^n\omega* > 2^{n-1}\pi$. Thence, the results in Fig. 2.6 confirm that when the resolution index n is equal to 4, 5, 6, and 7, respectively, where the mesh number of x in [0, 1] is equal to 16, 32, 64, and 128, respectively, their bandwidths $2^n\omega*$ of the ideal part $\left\|\hat{\varphi}^*_{n,k}(\omega)\right\| = 1$ in $[0, 2^n\omega*]$ are, respectively, greater than 28, 56, 112, and 224.

In addition, there is some request of the resolution level n in the reconstruction of an approximation under the base scaling functions when the approximation is used to a boundary-value problem in finite domain. As pointed out previously, such finite domain can be always reduced into [0, 1]. Since all calculations are taken at the dyadic points $j2^{-n}$ $(j = 0, 1, 2, \ldots, 2^n)$ in [0, 1] except for those extended ones, and at each dyadic point, the summation number in the reconstruction is equal to $\tilde{N} - 1$ when the support region of a compactly supported orthogonal wavelet is $[0, \tilde{N}]$. This means that at least $\tilde{N} - 1$ dyadic points should be located in [0, 1] such that the efficiency of such calculations is ensured better. Thus, the necessary condition for the resolution level n is $2^n \geq \tilde{N} - 1$, or

$$n \geq [\ln(\tilde{N} - 1)]. \tag{2.75}$$

Denote $n_{\min} = [\ln(\tilde{N} - 1)]$. When $\tilde{N} = 17$ for the generalized Coiflets, for example, we have $n_{\min} = 4$ and $n \geq 4$.

2.5 Calculations for Derivatives, Integrations, and Connection Coefficients of the Orthogonal Base Scaling Functions

As displayed above, both the base scaling functions and the base wavelet functions are numerically generated when a compactly supported orthogonal wavelet is employed. In practical cases, it is unavoidable that there are some calculations of derivatives and integrations on the base scaling functions and their multiply production, especially in solving differential equations. Although there are many mutual fitting or interpolation methods for such calculations after the original unknown function is solved out, however, there are unavoidable errors in such fitting calculations. With the development of wavelets in application, it is fortunate for us that some algebraic approaches, similar to the generation of scaling function, have been proposed in literature for their calculations such that the calculations have high accuracy almost without loss of accuracy appeared in the calculations. In this subsection, we introduce them as follows for a preparation when they are needed.

2.5.1 Calculation of Derivatives of Scaling Function

Denote the mth-order derivative of $\varphi(x)(=\varphi_{\tilde{N}}(x))$ by $\varphi^{(m)}(x) = d^m\varphi(x)/dx^m$ ($m \geq 1$), here the subscript $\tilde{N}$ is eliminated for simplicity. Taking the mth-order derivative to Eq. (2.2), we have [6]

$$\varphi^{(m)}(x) = 2^m \sum_{k=0}^{\tilde{N}} p_k \varphi^{(m)}(2x-k), \quad m \geq 1 \tag{2.76}$$

Step 1: to calculate the derivatives $\varphi^{(m)}(i)$ $(i = 0, 1, 2, \ldots, \tilde{N})$ at each integer point in the support domain $[0, \tilde{N}]$. Substituting $x = i(= 0, 1, 2, \ldots, \tilde{N})$ into Eq. (2.76), one gets a system of linear algebraic equations with $\tilde{N}+1$ unknowns $\varphi^{(m)}(i)$ in the form

$$\varphi^{(m)}(0) = 2^m p_0\varphi^{(m)}(0), \varphi^{(m)}(1) = 2^m[p_0\varphi^{(m)}(2) + p_1\varphi^{(m)}(1) + p_2\varphi^{(m)}(0)],$$
$$\varphi^{(m)}(2) = 2^m[p_0\varphi^{(m)}(4) + p_1\varphi^{(m)}(3) + p_2\varphi^{(m)}(2) + p_3\varphi^{(m)}(1) + p_4\varphi^{(m)}(0)]$$
$$\cdots\cdots$$
$$\varphi^{(m)}((\tilde{N}-1)/2) = 2^m[p_0\varphi^{(m)}(\tilde{N}-1) + p_1\varphi^{(m)}(\tilde{N}-2) + \cdots + p_{\tilde{N}-2}\varphi^{(m)}(1) + p_{N-1}\varphi^{(m)}(0)],$$
$$\cdots\cdots$$
$$\varphi^{(m)}(\tilde{N}) = 2^m[p_{\tilde{N}-2}\varphi^{(m)}(\tilde{N}) + p_{\tilde{N}-1}\varphi^{(m)}(\tilde{N}-1) + p_{\tilde{N}}\varphi^{(m)}(\tilde{N}-2)],$$
$$\varphi^{(m)}(\tilde{N}) - 2^m p_{\tilde{N}}\varphi^{(m)}(\tilde{N}).$$

Due to $p_0 \neq 0$ and $p_{\tilde{N}} \neq 0$, one can get $\varphi^{(m)}(0) = 0$ and $\varphi^{(m)}(\tilde{N}) = 0$. In such case, the above algebraic equations can be compressed in the matrix form, i.e.,

$$(\mathbf{M}_{(\tilde{N}-1)(\tilde{N}-1)} - 2^{-m}\mathbf{I}_{(\tilde{N}-1)(\tilde{N}-1)})\mathbf{\Phi}^{(m)}_{\tilde{N}-1} = \mathbf{0}_{N-1} \tag{2.77}$$

where the column $\mathbf{\Phi}^{(m)}_{\tilde{N}-1} = [\varphi^{(m)}(1), \varphi^{(m)}(2), \cdots, \varphi^{(m)}(\tilde{N}-1)]^T$ and the matrix $\mathbf{M}_{(\tilde{N}-1)(\tilde{N}-1)}$ is the same as Eq. (3.39).

It is obvious that Eq. (2.77) cannot give a set of unique solutions. In order to get a set of unique solutions, an algebraic equation should be added. To this end, we consider r vanishing moments in a compactly supported orthogonal wavelet. It has been known that the base scaling functions $\{\varphi(x-k)\}$ can accurately represent x^m ($1 \leq m \leq r$). When $\tilde{N} = 17$ and $r = 6$ of the generalized Coiflets, for example, we have $1 \leq m \leq 6$, which implies that the m-order derivatives of x^m ($1 \leq m \leq 6$) can be accurately formulated by the base scaling functions $\{\varphi(x-k)\}$. It is evident that the highest order 6 of derivative is sufficient for solving differential equations of most boundary-value problems.

For a given mth-order derivative ($1 \leq m \leq r$), we select $(x - M_1)^m$ which can also be accurately expanded by $\{\varphi(x-k)\}$. According to Eqs. (2.124) and (2.118), we have

$$(x - M_1)^m \equiv \sum_{k \in Z} k^m \varphi(x - k), \quad x \in R \tag{2.78}$$

Further, one gains

$$\sum_{k \in Z} k^m \varphi^{(m)}(x - k) = m!, \quad x \in R \tag{2.79}$$

Without loss of generality, we take $x = 0$ to Eq. (2.79), then one obtains a supplementary algebraic equation of the form

$$\sum_{k=1}^{\tilde{N}-1} (-k)^m \varphi^{(m)}(k) = m! \tag{2.80}$$

Thus, a set of unique solutions for $\mathbf{\Phi}^{(m)}_{\tilde{N}+1} = [\varphi^{(m)}(1), \varphi^{(m)}(2), \ldots, \varphi^{(m)}(\tilde{N} - 1)]^T$ can be obtained by solving the simultaneous equations consisting of Eqs. (2.77) and (2.80) since $\text{Rank}(\mathbf{M}_{(\tilde{N}-1)(\tilde{N}-1)} - 2^{-m}\mathbf{I}_{(\tilde{N}-1)(\tilde{N}-1)}) = \tilde{N} - 2$ as the same reason as pointed out to Eq. (2.32).

Step 2: to calculate the derivatives $\varphi^{(m)}_{n,k}(j2^{-n})(j = 0, 1, 2, \ldots, 2^n)$. After the values of $\mathbf{\Phi}^{(m)}_{\tilde{N}-1} = [\varphi^{(m)}(1), \varphi^{(m)}(2), \ldots, \varphi^{(m)}(\tilde{N} - 1)]^T$ are obtained, according to the definition of base scaling functions $\varphi_{n,k}(x) = 2^{n/2}\varphi(2^n x - k)$, the mth-order derivative $\varphi^{(m)}_{n,k}(x)$ at the dyadic points $x = j2^{-n}$, $j \in Z$, i.e., $\varphi^{(m)}_{n,k}(j2^n) = 2^{n/2}\varphi^{(m)}(j - k)$, can be gained by $\varphi^{(m)}(j - k)$, where $j \in Z$, $j < k < j + \tilde{N} - 1$.

2.5.2 *Calculation of Integral of Scaling Function*

Denote $\theta_m(x)$ as the m multiple integral of the scaling function $\varphi(x)$ in the form

$$\theta_m(x) = \int_0^x \left(\int_0^{y_m} \cdots \int_0^{y_2} \varphi(y_1)\, dy_1 dy_2 \cdots dy_{m-1} \right) dy_m. \tag{2.81}$$

Then we have $\theta_m(x) = \int_0^x \theta_{m-1}(y)dy$. When $x \le 0$, it is evident that $\theta_m(x) \equiv 0$ since $\varphi(x) \equiv 0$ in $x \le 0$. Taking the above multiple integration expressed in Eq. (2.81) to Eq. (2.2), we obtain

$$\theta_m(x) = 2^{-m} \sum_{k=0}^{\tilde{N}} p_k \theta_m(2x - k) \tag{2.82}$$

Step 1: to calculate the values of $\theta_m(x)$ at the integer points outside the support region $[0, \tilde{N}]$. Since $\theta_m(x) \neq 0$ when $x \geq \tilde{N}$, the calculation of $\theta_m(x)$ at the dyadic points $x = j2^{-n}(j = 0, 1, 2, \ldots, 2^n\tilde{N}, \ldots)$ will be different from the calculation of $\varphi(x)$ at the dyadic points in its support region. Due to $\int_0^{\tilde{N}} \varphi(x)dx = 1$ for $x \geq \tilde{N}$, and considering the support region of $\varphi(x)$, we have

$$\theta_1(x) = \int\limits_0^{\tilde{N}} \varphi(x)dx + \int\limits_{\tilde{N}}^{x} \varphi(x)dx = 1$$

$$\theta_2(x) = \int\limits_0^{\tilde{N}} \theta_1(x)dx + \int\limits_{\tilde{N}}^{x} \theta_1(x)dx = \theta_2(\tilde{N}) + x - \tilde{N}$$

$$\theta_3(x) = \int\limits_0^{\tilde{N}} \theta_2(x)dx + \int\limits_{\tilde{N}}^{x} \theta_2(x)dx = \theta_3(\tilde{N}) + \theta_2(\tilde{N})(x - \tilde{N}) + \frac{1}{2!}\theta_1(\tilde{N})(x - \tilde{N})^2.$$

By mathematical induction, we can get

$$\theta_m(x) = \sum_{j=0}^{m-1} \frac{(x - \tilde{N})^j}{j!}\theta_{m-j}(\tilde{N}) \quad , \quad m \geq 1 \quad , \quad x \geq \tilde{N} \tag{2.83}$$

where $\theta_1(\tilde{N}) = 1$, and $\theta_i(\tilde{N})$ $(i = 2, 3, \ldots, m)$ are still unknown. In order to determine them, let us return to the two-scale Eq. (2.82) and take $x = \tilde{N}$, namely,

$$\theta_m(\tilde{N}) = 2^{-m} \sum_{k=0}^{\tilde{N}} p_k \theta_m(2\tilde{N} - k). \tag{2.84}$$

Since $2\tilde{N} - k \geq \tilde{N}$ in the summation on the right side of the above formula, we can substitute Eq. (2.83) into Eq. (2.84), then the right term $\theta_m(2\tilde{N} - k)$ is calculated by the form

$$\theta_m(2\tilde{N} - k) = \sum_{j=0}^{m-1} \frac{(\tilde{N} - k)^j}{j!}\theta_{m-j}(\tilde{N}). \tag{2.85}$$

Substituting Eq. (2.85) into Eq. (2.84), we obtain

$$\theta_m(\tilde{N}) = 2^{-m} \sum_{k=0}^{\tilde{N}} p_k \sum_{j=0}^{m-1} \frac{(\tilde{N} - k)^j}{j!}\theta_{m-j}(\tilde{N})$$

$$= (2^{-m}\sum_{j=0}^{\tilde{N}} p_k)\theta_m(\tilde{N}) + 2^{-m}\sum_{k=0}^{\tilde{N}} p_k \sum_{j=1}^{m-1} \frac{(\tilde{N}-k)^j}{j!}\theta_{m-j}(\tilde{N}).$$

From Eq. (2.33), we know $\sum_{j=0}^{\tilde{N}} p_k = 2$. Thus, above equation becomes

$$\theta_m(\tilde{N}) = \frac{1}{2^m - 2}\sum_{k=0}^{\tilde{N}} p_k \sum_{j=1}^{m-1} \frac{(\tilde{N}-k)^j}{j!}\theta_{m-j}(\tilde{N}). \tag{2.86}$$

According to Eq. (2.86), we can get the values of $\theta_m(\tilde{N})$ $(m \geq 2)$ successively from $\theta_1(\tilde{N}) = 1$. Substituting the induced results into Eq. (2.83), the values of $\theta_m(x)$ in the region $x \geq \tilde{N}$ are obtained. Of course, such obtained values of $\theta_m(x)$ contain those values when x is located at the integer points in this region.

Step 2: to calculate the values of $\theta_m(x)$ at the integer points in the support region. After the value of $\theta_m(x)$ at the integer points in the region $x \geq \tilde{N}$ is gained, let us first calculate $\theta_m(x)$ at the integer points in $\text{Supp}\varphi(x) = [0, \tilde{N}]$, that is, $\theta_m(i)$ $(i = 1, 2, \ldots, \tilde{N} - 1)$. For such case, at both ends of the support region, $\theta_m(0) = 0$ and $\theta_m(\tilde{N})$ have been gained by Eq. (2.86). Denote

$$\Theta^m_{\tilde{N}-1} = [\theta_m(1), \theta_m(2), \ldots, \theta_m(\tilde{N}-1)]^T, \tag{2.87}$$

$$\mathbf{C}^m_{\tilde{N}-1} = [c^m_1, c^m_2, \ldots, c^m_{\tilde{N}-1}]^T \tag{2.88}$$

where $c^m_i = \sum_{\substack{0<2i-k<\tilde{N}=1 \\ k=0,1\ldots\tilde{N}}} 2^{-m} p_k \theta_m(2i-k)(1 \leq i \leq \tilde{N}-1)$. Then, the calculations of two-scale equation of Eq. (2.82) at $\theta_m(i)$ $(i = 1, 2, \ldots, \tilde{N} - 1)$ can be abbreviated by the following matrix equation:

$$(\mathbf{I}_{(\tilde{N}-1)\times(\tilde{N}-1)} - 2^{-m}\mathbf{M}_{(\tilde{N}-1)\times(\tilde{N}-1)})\Theta^m_{\tilde{N}-1} = \mathbf{C}^m_{\tilde{N}-1}. \tag{2.89}$$

Solving this nonhomogeneous system of linear algebra equations, we can obtain $\theta_m(i)$ $(i = 1, 2, \ldots, \tilde{N} - 1)$.

Step 3: to calculate the values of $\theta_m(x)$ at the dyadic points $x_j = j2^{-n}$ $(j = 0, 1, 2, \ldots)$. In this case, the two-scale Eq. (2.82) of $\theta_m(x)$ can be used successively, and then the values of $\theta_m(x)$ at the dyadic points $x = j2^{-n}(j = 0, 1, 2, \ldots, 2^n\tilde{N}, \ldots)$ under the resolution index n is obtained. Finally, the values of integral function corresponding to the base scaling functions $\varphi_{n,k}(x)$ at the dyadic points are obtained by $\theta_m(j2^{-n})$. For example, when $m = 1$, we get

$$\Phi_1(x) = \int_0^x \varphi_{n,k}(y_1)dy_1 = 2^{n/2}\int_0^x \varphi(2^n y_1 - k)dy_1$$

$$= 2^{-n/2}\int_{-k}^{x2^n-k} \varphi(y_1)dy_1 = 2^{-n/2}\int_0^{x2^n-k} \varphi(y_1)dy_1 = 2^{-n/2}\theta_1(2^n x - k)$$

further, $\Phi_1(j2^{-n}) = 2^{n/2}\theta_1(j-k),\ j = 0, 1, 2, \ldots, 2^n\tilde{N}, \ldots$.

2.5.3 Calculation of Connection Coefficients

When the wavelet Galerkin method is employed to solve the boundary-value problems, it is often necessary to calculate the connection coefficients of the product of the base scale functions and their derivatives, which is introduced at follows [7]. Denote

$$\Gamma_{k,j}^{n,m} = \int_0^1 \varphi_{n,k}^{(m)}(x)\varphi_{n,j}(x)dx. \tag{2.90}$$

Here, we introduce a system of algebraic equations to get the values of $\Gamma_{k,j}^{n,m}$ in Eq. (2.90) such that high precision can be ensured without using a numerical integral.

Step 1: To Calculate $\int_0^{\tilde{N}} \varphi^{(m)}(x-k)\varphi(x)dx$

For the scaling function $\varphi(x)$ defined on the support domain $[0, \tilde{N}]$, denote

$$\Gamma_k^m = \int_{-\infty}^{\infty} \varphi^{(m)}(x-k)\varphi(x)dx = \int_0^{\tilde{N}} \varphi^{(m)}(x-k)\varphi(x)dx. \tag{2.91}$$

It is obvious that when $k \geq \tilde{N}$ and $k \leq -\tilde{N}$, we have $\Gamma_k^m = 0$. Furthermore, we only need to calculate Γ_k^m in the region $-\tilde{N} < k < \tilde{N}$. Substituting Eqs. (2.2) and (2.76) into the right side of Eq. (2.91), one gets

$$\Gamma_k^m = 2^m \sum_{j=0}^{\tilde{N}}\sum_{i=0}^{\tilde{N}} p_i p_j \int_0^{\tilde{N}} \varphi^{(m)}(2x - 2k - j)\varphi(2x - i)dx$$

$$= 2^{m-1} \sum_{j=0}^{\tilde{N}}\sum_{i=0}^{\tilde{N}} p_i p_j \int_0^{\tilde{N}} \varphi^{(m)}(y + i - 2k - j)\varphi(y)dy$$

$$= 2^{m-1} \sum_{j=0}^{\tilde{N}} \sum_{i=0}^{\tilde{N}} p_i p_j \Gamma^m_{2k+j-i}. \tag{2.92}$$

By transforming the index of the connection coefficient in the summation on the right side of Eq. (2.92), we can rewrite Eq. (2.92) in the form:

$$\Gamma^m_k = 2^{m-1} \sum_{l=1-\tilde{N}}^{\tilde{N}-1} \sum_{j=0}^{\tilde{N}} p_j p_{2k+j-l} \Gamma^m_l \tag{2.93}$$

Abbreviated Eq. (2.93) into the matrix form, we have

$$(\mathbf{H} - \mathbf{I})\mathbf{\Gamma}^{(m)} = \mathbf{0} \tag{2.94}$$

where $\mathbf{H}$ is a square matrix of order $2(\tilde{N} - 1) \times 2(\tilde{N} - 1)$, and its elements are

$$H_{ij} = 2^{m-1} \sum_{l=0}^{\tilde{N}} p_l p_{2i+l-j}, \quad i, j = 1 - \tilde{N},\ 2 - \tilde{N}, \ldots, \tilde{N} - 1 \tag{2.95}$$

and the column $\mathbf{\Gamma}^{(m)} = [\Gamma^m_{1-\tilde{N}}, \Gamma^m_{2-\tilde{N}}, \ldots, \Gamma^m_{\tilde{N}-1}]^T$. It can demonstrate that $\text{Rank}[\mathbf{H}] = 2(\tilde{N} - 1) - 1$. Hence, we need to add an equation to find a set of unique solutions for unknowns in $\mathbf{\Gamma}^{(m)}$. For this purpose, we take the operations that the Eq. (2.79) is firstly multiplied by $\varphi(x)$, then the induced equation is taken by the integral on $[0, \tilde{N}]$. Finally, we get

$$\sum_{k=-\infty}^{\infty} k^m \int_0^{\tilde{N}} \varphi^{(m)}(x - k)\varphi(x)dx = m! \int_0^{\tilde{N}} \varphi(x)dx = m!. \tag{2.96}$$

That is,

$$\sum_{k=1-\tilde{N}}^{\tilde{N}-1} k^m \Gamma^m_k = m! \tag{2.97}$$

By solving the $2(\tilde{N} - 1)$-order-independent linear algebraic equations of (2.94) and (2.97) simultaneously, we can obtain the solution of the connection coefficient $\Gamma^m_k(k)(k = 1 - \tilde{N}, 2 - \tilde{N}, \ldots, \tilde{N} - 2, \tilde{N} - 1)$.

Step 2: to calculate $\int_0^x \varphi^{(m)}(y - k)\varphi(y)dy$. Define

$$\Gamma_k^m(x) = \int_0^x \varphi^{(m)}(y-k)\varphi(y)dy. \tag{2.98}$$

Then from the support domain $[0, \tilde{N}]$ of the wavelet function (i.e., $\varphi(x) = 0$ outside the domain), we have

$$\Gamma_k^m(x) = \Gamma_k^m(\tilde{N}) = \Gamma_k^m, \quad x \geq \tilde{N} \tag{2.99a}$$

$$\Gamma_k^m(x) = 0, \quad x \leq 0 \,\text{or}\, x \leq k \tag{2.99b}$$

$$\Gamma_{-k}^m(\tilde{N}) = (-1)^m \Gamma_k^m(\tilde{N}) = (-1)^m \Gamma_k^m \tag{2.99c}$$

$$\Gamma_{-k}^m(x) = \Gamma_{-k}^m(\tilde{N}) = (-1)^m \Gamma_k^m(\tilde{N}) = (-1)^m \Gamma_k^m, \quad x + k \geq \tilde{N} \tag{2.99d}$$

$$\Gamma_{-k}^m(k) = 0 \quad k = 0, 1, 2, \ldots, \tilde{N}. \tag{2.99e}$$

Substituting the two-scale equations of Eqs. (2.2) and (2.74) into (2.98), we have

$$\begin{aligned}\int_0^x \varphi^{(m)}(y-k)\varphi(y)dy &= 2^m \sum_{j=0}^{\tilde{N}} \sum_{i=0}^{\tilde{N}} p_j p_i \int_0^x \varphi^{(m)}(2y-2k-j)\varphi(2y-i)dx \\ &= 2^{m-1} \sum_{j-0}^{\tilde{N}} \sum_{i-0}^{\tilde{N}} p_j p_i \int_0^{2x-i} \varphi^{(m)}(y+i-2k-j)\varphi(y)dy\end{aligned} \tag{2.100}$$

or

$$\Gamma_k^m(x) = 2^{m-1} \sum_{j=0}^{\tilde{N}} \sum_{i=0}^{\tilde{N}} p_j p_i \Gamma_{2k+j-i}^m(2x-i). \tag{2.101}$$

Next, we take the integer points of x in the support domain, namely $x = l$ $(1 \leq l \leq \tilde{N} - 1)$, one by one to Eq. (2.101) with $-(\tilde{N} - 1) + l < k < l$. This result displays that for any specified $x = l$, the number of k is only $\tilde{N} - 1$ desirable integers. Therefore, Eq. (2.101) at the integer points may become

$$\Gamma_k^m(l) = 2^{m-1} \sum_{j=0}^{\tilde{N}} \sum_{i=0}^{\tilde{N}} p_j p_i \Gamma_{2k+j-i}^m(2l-j). \tag{2.102}$$

According to the nonzero condition of (2.99a) for $\Gamma_k^m(x)$ and $\Gamma_{2k+j-i}^m(2l-j)$, we have, $j < 2l$ and $i > 2(l+k-j)$.Thus, the summation range in Eq. (2.102) can be reduced into the form

$$\Gamma_k^m(l) = 2^{m-1}\sum_{j=0}^{2l-1}\sum_{i=2(k+j-l)}^{\tilde{N}} p_j p_i \Gamma_{2k+j-i}^m(2l-j). \tag{2.103}$$

When $2l-1 \geq \tilde{N}$, we have $2l-j \geq \tilde{N}$ items. The conditions in this situation are $l \geq (\tilde{N}+1)/2$ and $j \leq 2l-\tilde{N}$. Under these conditions, the right side of the above formula becomes $\Gamma_{2k+j-i}^m(2l-j) = \Gamma_{2k+j-i}^m$. And when $2k+j-i < 0$, that is $i > 2k+j$, there is $\Gamma_{2k+j-i}^m(2l-j) = (-1)^m\Gamma_{i-(2k+j)}^m$. Therefore, Eq. (2.103) can be further rewritten as

$$\begin{aligned}\Gamma_k^m(l) &= 2^{m-1}\sum_{j=0}^{2l-1}\sum_{i=2(k+j-l)}^{2k+j} p_j p_i \Gamma_{2k+j-i}^m(2l-j)\\ &+ 2^{m-1}\sum_{j=0}^{2l-1}\sum_{i=2k+j}^{\tilde{N}} p_j p_i (-1)^m \Gamma_{2k+j-i}^m.\end{aligned} \tag{2.104}$$

Corresponding to it, the calculation range for the summation index on the right side of Eq. (2.101) can be given too.

Step 3: to calculate $\Gamma_{k,j}^{n,m}$

Now let us give the calculations of the connection coefficients $\Gamma_{k,j}^{n,m}$. By the definition of Eq. (2.90), here we rewrite it as

$$\Gamma_{k,j}^{n,m} = \int_0^1 \varphi_{n,k}^{(m)}(x)\varphi_{n,j}(x)dx = 2^{(m+1)n}\int_0^1 \varphi^{(m)}(2^n x-k)\varphi(2^n x-j)dx.$$

By means of variable transform, the above equation can be reduced by the form

$$\Gamma_{k,j}^{n,m} = 2^{mn}\int_{-j}^{2^n-j}\varphi^{(m)}(x+j-k)\varphi(x)dx. \tag{2.105}$$

Thus, we get

$$\Gamma_{k,j}^{n,m} = \begin{cases} 2^{mn}\Gamma_{k-j}^m(2^n-j), & j \geq 0 \\ 2^{mn}\Gamma_{k-j}^m(2^n-j) - \Gamma_{k-j}^n(-j), & j < 0 \end{cases}. \tag{2.106}$$

From the previous step, the values on the right side of Eq. (2.106) have been obtained. Hence, the values of the connection coefficients $\Gamma_{k,j}^{n,m}$ can be obtained too by Eq. (2.106). According to the definition formula of $\Gamma_{k,j}^{n,m}$ in Eq. (2.105), we can get $\Gamma_{k,j}^{n,m} = \Gamma_{j,k}^{n,m}$, i.e.,

$$\begin{aligned}\Gamma_{k,j}^{n,m} &= 2^{(m+1)n}\int_0^1 \varphi^{(m)}(2^n x - k)\varphi(2^n x - j)dx\\ &= 2^{(m+1)n}\int_0^1 \varphi^{(m)}(2^n x - j)\varphi(2^n x - k)dx = \Gamma_{j,k}^{n,m}\end{aligned}$$

where the variable transform in integration is employed.

Appendix 2.1 Moment Relationships of Orthogonal Scaling Functions

From the description of Property 2.2, we know that the expression

$$M_i = \frac{1}{2}\sum_{k=0}^{\tilde{N}} p_k k^i \ (i = 0, 1, 2, \ldots, r)$$

and Property 2.2 should be proved, which will give in Appendix 2.2.

According to Definition 2.4 of the moment of the scaling function, we have

$$M_i = \int_{-\infty}^{\infty} x^i\varphi(x)dx = \int_0^{\tilde{N}} x^i\varphi(x)dx \qquad 0 \le i \le s. \tag{2.107}$$

where $M_0 = 1$ is prechosen in the establishing wavelet. Substituting Eq. (2.2) into Eq. (2.107), one gets

$$\begin{aligned}M_i &= \sum_{k=0}^{\tilde{N}} p_k \int_{-\infty}^{\infty} x^i\varphi(2x - k)dx = \frac{1}{2}\sum_{k=0}^{\tilde{N}} p_k \int_{-\infty}^{\infty}\left(\frac{x+k}{2}\right)^i \varphi(x)dx\\ &= \frac{1}{2^{i+1}}\sum_{k=0}^{\tilde{N}} p_k \sum_{j=0}^{i} C_i^j k^j \int_{-\infty}^{\infty} x^{i-j}\varphi(x)dx = \frac{1}{2^{i+1}}\sum_{k=0}^{\tilde{N}} p_k \sum_{j=0}^{i} C_i^j k^j M_{i-j}\end{aligned} \tag{2.108}$$

in which $C_i^j = \frac{j!(i-j)!}{i!}$ is the expansion coefficients of $(x + k)^i$. Denote

$$m_j = \frac{1}{2}\sum_{k=0}^{\tilde{N}} p_k k^j. \tag{2.109}$$

and consider $C_i^0 = 1 = k^0$, then, Eq. (2.108) can be rewritten by the form

$$M_i = \frac{1}{2^i - 1}\sum_{j=1}^{i} C_i^j m_j M_{i-j}. \tag{2.110}$$

Appendix 2.2 Moment Relationships of Scaling Functions of Coiflets

Conclusion 2.1. To a compactly supported orthogonal wavelet with up to s-order vanishing moment, when the relationships

$$M_j = M_1^j, \quad j = 1, 2, \ldots, s. \tag{2.111}$$

are established, then we have

$$(x - M_1)^j = \sum_{k \in Z} k^j \varphi(x - k), \quad j = 1, 2, \ldots, s, \tag{2.112}$$

$$M_i = m_i, \quad i = 1, 2, \ldots, s. \tag{2.113}$$

Proof

(I) To prove Eq. (2.112). Since the wavelet has up to s-order vanishing moment, following Property 2.1, we have

$$(x - M_1)^j \equiv \sum_{k \in Z} d_k^j \varphi(x - k) \quad , \quad x \in R, \quad 0 \le j \le s, \tag{2.114}$$

$$d_k^j = \int_{-\infty}^{\infty} (x - M_1)^j \varphi(x - k) dx. \tag{2.115}$$

Taking the transform of the integral variable to Eq. (2.115), one gets

$$d_k^j = \int_{-\infty}^{\infty} (x + k - M_1)^j \varphi(x) dx$$

$$= \sum_{i=0}^{j} C_j^i k^i \int_{-\infty}^{\infty} (x - M_1)^{j-i} \varphi(x) dx. \tag{2.116}$$

Since $C_j^j = 1$ and $\int_{-\infty}^{\infty} \varphi(x)dx = 1$, Eq. (2.116) can be reduced into the form:

$$\begin{aligned} d_k^j &= k^j + \sum_{i=0}^{j-1} C_j^i k^i \int_{-\infty}^{\infty} (x - M_1)^{j-i} \varphi(x) dx \\ &= k^j + \sum_{i=0}^{j-1} C_j^i k^i \sum_{l=0}^{j-i} C_{j-i}^l (-1)^l M_1^l \int_{-\infty}^{\infty} x^{j-i-l} \varphi(x) dx. \\ &= k^j + \sum_{i=0}^{j-1} C_j^i k^i \sum_{l=0}^{j\ i} C_{j-i}^l (-1)^l M_1^l M_{j-i-l} \end{aligned} \tag{2.117}$$

Substitution Eq. (2.111) into Eq. (2.117) leads to

$$\begin{aligned} d_k &= k^j + \sum_{i=0}^{j-1} C_j^i k^i \sum_{l=0}^{j-i} C_{j-i}^l (-1)^l M_1^{j-i} \\ &= k^j + \sum_{i=0}^{j-1} C_j^i k^i M_1^{j-i} \sum_{l=0}^{j-l} C_{j-i}^l (-1)^l. \\ &= k^j + \sum_{i=0}^{j-1} C_j^i k^i M_1^{j-i} (1 - 1)^{j-i} = k^j \end{aligned} \tag{2.118}$$

Then, substituting the above result of Eq. (2.118) into Eq. (2.114), we get Eq. (2.112)

(II) To prove Eq. (2.113). Here, we use mathematical induction for this proof. When $i = 1$, according to Eq. (2.110), it is obvious that $M_1 = \frac{1}{2^1-1} C_1^1 m_1 M_0 = m_1$ is held since $C_1^1 = 1 = M_0$. Hence, Eq. (2.113) holds for $i = 1$. Assume Eq. (2.113) holds for $i = r (1 \le r < s)$, that is, the equations

$$M_i = m_i \quad , \quad i = 1, 2, \ldots, r \tag{2.119}$$

are held. For the case of $i = r + 1$, Eq. (2.110) can be expressed by

$$M_{r+1} = \frac{1}{2^{r+1} - 1} \sum_{j=1}^{r+1} C_{r+1}^j m_j M_{r+1-j} = \frac{1}{2^{r+1} - 1} \left\{ m_r + \sum_{j=1}^{r} C_{r+1}^j m_j M_{r+1-j} \right\}. \tag{2.120}$$

Substitution of Eq. (2.119) into Eq. (2.120) leads to

$$M_{r+1} = \frac{1}{2^{r+1} - 1}\left\{m_{r+1} + \sum_{j=1}^{r} C_{r+1}^{j} M_j M_{r+1-j}\right\}. \tag{2.121}$$

According to the conditions of Eqs. (2.111) and (2.121) is reduced in the form

$$M_{r+1} = \frac{1}{2^{r+1} - 1}\left\{m_{r+1} + \sum_{j=1}^{r} C_{r+1}^{j} M_1^{r+1}\right\} = \frac{1}{2^{r+1} - 1}\left\{m_{r+1} + M_{r+1}\left(\sum_{j=0}^{r+1} C_{r+1}^{j} - 2\right)\right\}$$

Since $\sum_{j=0}^{r} C_{r+1}^{j} = (1+1)^{r+1} = 2^{r+1}$, the above equation becomes into

$$M_{r+1} = \frac{1}{2^{r+1} - 1}\{m_{r+1} + M_{r+1}(2^{r+1} - 2)\}$$

Further, one gets $M_{r+1} = m_{r+1}$, i.e., Eq. (2.113) holds for $i = r + 1$. From the mathematical method of induction, we know that Eq. (2.113) holds for all integers of $1 \le i \le s$. Hence, the proof of conclusion (2.107) is finished.

Combining Eqs. (2.109) and (2.113), we have the expression of moments in terms of the filter coefficients p_k of the form

$$m_j = \frac{1}{2}\sum_{k \in Z} p_k k^j = \frac{1}{2}\sum_{k=0}^{\tilde{N}} p_k k^j \quad , \qquad j = 1, 2, \ldots, s \tag{2.122}$$

which is the first part of Property 2.2.

(III) Proof of the second part in Property 2.2:

That is, when

$$M_{2j-1} = \frac{1}{2}\sum_{k=0}^{\tilde{N}} p_k k^{2j-1} = M_1^{2j-1}, \quad j = 1, 2, \ldots, l(< [(r-1)/2]) \tag{2.123}$$

are prechosen in the establishment of a wavelet, then, we should prove the following equations:

$$M_{2j} = \frac{1}{2}\sum_{k=0}^{\tilde{N}} p_k k^{2j} = M_1^{2j} \quad , \quad j = 1, 2, \ldots, l(2l \le r) \tag{2.124}$$

holding. Here, the mathematical method of induction is employed in this proof.

When $j = 1$, Eq. (2.124) becomes $M_2 = \frac{1}{2}\sum_{k=0}^{\tilde{N}} p_k k^2$. According to Eqs. (2.114) and (2.115) and considering $c_k = \int_{-\infty}^{\infty} \varphi(x-k)dx = \int_{-\infty}^{\infty} \varphi(x)dx = 1$, we have

$$\sum_{k\in Z} \varphi(x-k) = 1 \tag{2.125}$$

$$\sum_{k\in Z} k\varphi(x-k) = x - M_1. \tag{2.126}$$

Denote

$$\kappa_k = \int_{-\infty}^{\infty} x\varphi(x)\varphi(x-k)dx. \tag{2.127}$$

Then

$$\begin{aligned}\kappa_{-k} &= \int_{-\infty}^{\infty} x\varphi(x)\varphi(x+k)dx = \int_{-\infty}^{\infty} (x-k)\varphi(x-k)\varphi(x)dx \\ &= \int_{-\infty}^{\infty} x\varphi(x-k)\varphi(x)dx - k\int_{-\infty}^{\infty} \varphi(x-k)\varphi(x)dx. \\ &= \int_{-\infty}^{\infty} x\varphi(x-k)\varphi(x)dx - k\delta_{0,k} = \kappa_k \end{aligned} \tag{2.128}$$

Thus, we have $\sum_{k\in Z} k\kappa_k = \sum_{k=1}^{\infty} k\kappa_k + \sum_{k=-1}^{-\infty} k\kappa_k = 0$. Multiplying $x\varphi(x)$ to Eq. (2.126), and integrating the induced equation respective to x in $(-\infty, \infty)$, one obtains

$$\begin{aligned}0 &= \sum_{k\in Z} k\kappa_k = \sum_{k\in Z} k\int_{-\infty}^{\infty} x\varphi(x)\varphi(x-k)dx \\ &= \int_{-\infty}^{\infty} x^2\varphi(x)dx - M_1\int_{-\infty}^{\infty} x\varphi(x)dx = M_2 - M_1^2.\end{aligned}$$

Thus, Eq. (2.124) holds for $j = 1$.

When $j = p\ (< l)$, we assume that the following equations:

$$M_{2j} = M_1^{2j}, \qquad j = 2, 3, \ldots, p \tag{2.129}$$

are held. Then for $j = p + 1$, the equation $M_{2(p+1)} = M_1^{2(p+1)}$ should be proved. According to Eq. (2.112), we have

$$(x - M_1)^{2p+1} = \sum_{k \in Z} k^{2p+1} \varphi(x - k). \tag{2.130}$$

Taking the calculations of Eq. (2.130) as same as ones of Eq. (2.126), we get

$$\int_{-\infty}^{\infty} (x - M_1)^{2p+1} x\varphi(x)dx = \sum_{k \in Z} k^{2p+1} \int_{-\infty}^{\infty} x\varphi(x)\varphi(x - k)dx = \sum_{k \in Z} k^{2p+1} \kappa_k = 0. \tag{2.131}$$

To the left side of Eq. (2.131), we have

$$\begin{aligned}
\int_{-\infty}^{\infty} (x - M_1)^{2p+1} x\varphi(x)dx &= \sum_{j=0}^{2p+1} C_{2p+1}^j (-1)^j M_1^j \int_{-\infty}^{\infty} x^{2p-j+2} \varphi(x)dx \\
&= \sum_{j=0}^{2p+1} C_{2p+1}^j (-1)^j M_1^j M_{2p-j+2} \\
&= M_{2p+2} + \sum_{j=1}^{2p+1} C_{2p+1}^j (-1)^j M_1^j M_{2p-j+2}.
\end{aligned} \tag{2.132}$$

Since $2p + 2 - j \leq 2p + 1$ $(1 < j \leq 2p + 1)$ in the summation of Eq. (2.132), from Eqs. (2.123) and (2.129), one gets

$$M_{2p-j+1} = M_1^{2p-j+1}, \qquad j = 1, 2, \ldots, 2p + 1. \tag{2.133}$$

Substitution of Eq. (2.133) into (2.132) yields

$$\begin{aligned}
\int_{-\infty}^{\infty} (x - M_1)^{2p+1} x\varphi(x)dx &= M_{2p+2} + M_1^{2p+2} \sum_{j=1}^{2p+1} C_{2p+1}^j (-1)^j \\
&= M_{2p+2} - M_1^{2p+2} + M_1^{2p+2} \sum_{j=0}^{2p+1} C_{2p+1}^j (-1)^j \\
&= M_{2p+2} - M_1^{2p+2}.
\end{aligned} \tag{2.134}$$

Substituting Eq. (2.146) into Eq. (2.131), one gets

$$M_{2p+2} = M_1^{2p+2}.$$

According to the mathematical method of induction, we know that Eq. (2.124) holds for all integers of $j = 1, 2, \ldots, l(2l \leq r)$. Thus, the proof of Property 2.2 is finished.

Appendix 2.3 Condition on Filer Coefficients from Vanishing Moments

In Definition 2.5, we give the general expression of vanishing moments of a compact supported orthogonal wavelet, see Eq. (2.25). Here, we give their explicit expressions in terms of the filter coefficients $p_k(k = 0, 1, 2, \ldots, \tilde{N})$. Substituting Eq. (2.15b) into Eq. (2.25), one gets

$$\begin{aligned}
0 &= \sum_{k=0}^{\tilde{N}} (-1)^k p_k \int_R x^i \varphi(2x-k)dx = \frac{1}{2^{i+1}} \sum_{k=0}^{\tilde{N}} (-1)^k p_k \int_R (x+k)^i \varphi(x)dx \\
&= \frac{1}{2^{i+1}} \sum_{k=0}^{\tilde{N}} (-1)^k p_k \sum_{l=0}^{i} C_i^l k^l \int_R x^{i-l} \varphi(x)dx = \frac{1}{2^{i+1}} \sum_{k=0}^{\tilde{N}} (-1)^k p_k \sum_{l=0}^{i} C_i^l k^l M^{i-l} \\
&= \frac{1}{2^{i+1}} \sum_{k=0}^{\tilde{N}} (-1)^k p_k \sum_{l=0}^{i} C_i^l k^l M_1^{i-l}
\end{aligned}$$

i.e.,

$$\sum_{k=0}^{\tilde{N}} (-1)^k p_k \sum_{l=0}^{i} C_i^l k^i M_1^{i-l} = 0 \quad , \quad i = 0, 1, \ldots, r-1. \tag{2.135}$$

When $i = 0$, we have

$$\sum_{k=0}^{\tilde{N}} (-1)^k p_k = 0. \tag{2.136}$$

Further, when $i = 1$ and Eq. (2.136) is considered, Eq. (2.135) becomes

$$\begin{aligned}
&\sum_{k=0}^{\tilde{N}} (-1)^k p_k (C_1^0 k^0 M_1^1 + C_1^1 k^1 M_1^0) \\
&= \sum_{k=0}^{\tilde{N}} (-1)^k p_k + \sum_{k=0}^{\tilde{N}} (-1)^k p_k k = \sum_{k=0}^{\tilde{N}} (-1)^k p_k k = 0
\end{aligned}$$

i.e., $\sum_{k=0}^{\tilde{N}}(-1)^k p_k k = 0$. Assume $\sum_{k=0}^{\tilde{N}}(-1)^k p_k k^i = 0$ held for $i = 2, \ldots, s(< r-1)$. Then, for $i = s+1(\leq r)$, Eq. (2.135) can be rewritten in the form:

$$0 = \sum_{k=0}^{\tilde{N}}(-1)^k p_k \sum_{l=0}^{s+1} C_{s+1}^l k^l M_1^{s+1-l}$$

$$= \sum_{k=0}^{\tilde{N}}(-1)^k p_k (M_1^{s+1} + C_{s+1}^1 k M_1^s + C_{s+1}^2 k^2 M_1^{s-1} + \cdots + C_{s+1}^s k^s M_1^1 + k^{s+1} M_1^0)$$

$$= M_1^{s+1}\sum_{k=0}^{\tilde{N}}(-1)^k p_k + C_{s+1}^1 M_1^s \sum_{k=0}^{\tilde{N}}(-1)^k p_k k + C_{s+1}^2 M_1^{s-1}\sum_{k=0}^{\tilde{N}}(-1)^k p_k k^2 + \cdots$$

$$+ C_{s+1}^s M_1 \sum_{k=0}^{\tilde{N}}(-1)^k p_k k^s + \sum_{k=0}^{\tilde{N}}(-1)^k p_k k^{s+1}. \tag{2.137}$$

According to the previous assumption and results, we know that

$$\sum_{k=0}^{\tilde{N}}(-1)^k p_k k^i = 0 \quad , \quad i = 0, 1, \ldots, s \tag{2.138}$$

are held. Substitution such results of Eq. (2.138) into Eq. (2.137) yields

$$\sum_{k=0}^{\tilde{N}}(-1)^k p_k k^{s+1} = 0$$

being held too. By means of the mathematical induction, we get that Eq. (2.138) is true for arbitrary integer $i = 0, 1, \ldots, r-1$.

References

1. Mallat S (1989) Multiresolution approximations and wavelet orthonormal bases of $L_2(R)$. Trans Am Math Soc 315:69–87
2. Burt PJ, Adelson EH (1983) The Laplacian pyramid as a compact image code. IEEE. Trans Commun 31:532–540
3. Mallat S (1998) A wavelet tour of signal processing. Academic Press, New York
4. Sweldens W, Piessens R (1994) Quadrature formulae and asymptotic error expansions for wavelet approximations of smooth functions. SIAM J Numer Anal 31:1240–1264
5. William J, Amaratunga K (1994) Introduction to wavelet in engineering. Int J Numer Meth Eng 37:2365–2388
6. Chen MQ, Wang CH, Shih YP (1996) The computation of wavelet-Galerkin approximation on a bounded interval. Int J Numer Methods Eng 39:2921–2944
7. Zhang T, Tian YC, Tade MO, Utomo J (2007) Comments on The computation of Wavelet-Galerkin approximation on a bounded interval. Int J Numer Meth Eng 72:244–251

Chapter 3
Essentials to Solving Nonlinear Boundary-Value Problems

In this chapter, we first briefly introduce the general mathematical knowledge on solving boundary-value problems with differential equations, and the solution strategy based on the generalized Coiflets as proposed in the previous chapter such that the readers can have an overall understanding on how the book will solve various differential equations. For the initial-value problems, we will give a specific introduction somewhere later once it is encountered.

It is well-known that only a few kinds of boundary-value problems can be solved by exact solutions to be expressed as summations of finite terms of basic functions [1]. As displayed in Chap. 1, in most situations, such solution methods for a boundary-value problem are developed by an infinite series of base functions in theory. In the practice of numerical computations, it is obvious that such infinite series should be truncated by its finite form. After that, the truncated finite form of the infinite series is referred to an approximation solution or numerical solution to the exact solution of the problem, while the part thrown away in the infinite series is called the residual or error. In a solution method for a nonlinear problem, in general, as shown in Chap. 1, we have known that almost no one solution method can make the residual being zero at every point. A common and effective way is to make the residual be zero in weight average by using a set of appropriate weight functions. In theory, such weighted error zero is forced rather than exactly for the nonlinear terms no matter what solution method is employed. When the weight functions are selected by the base functions (also called trial functions) used in the solution method, the corresponding approach is called as the Galerkin method. The Galerkin method is the most popular one for solving boundary-value problems due to the optimal balance between accuracy, efficiency, and stability [2, 3]. Therefore, the Galerkin method will be mainly adopted in this book.

In the following, we first review the two-point boundary-value problems in one-dimensional cases [4]. The so-called one-dimensional problem means that there is only one independent variable, and the governing equations are also called as ordinary differential equations. Then, the Galerkin method based on the generalized

Y.-H. Zhou, *Wavelet Numerical Method and Its Applications in Nonlinear Problems*, Engineering Applications of Computational Methods 6,
https://doi.org/10.1007/978-981-33-6643-5_3

Coiflets established in Chap. 2 is proposed for the solution of the one-dimensional problems, especially for nonlinear problems. From the introduction, one will find that the prominent advantage in the present wavelet-based method is that all weighted errors are naturally zero. Finally, the proposed wavelet-based method is successfully extended to the cases of the two-dimensional and three-dimensional boundary-value problems with the same features as a one-dimensional situation.

3.1 Governing Equations of 1D Nonlinear Boundary-Value Problems

Without loss of generality, we consider the 1D boundary-value problem in the region [0, 1] of the form

$$L[f(x)] + N[f(x)] = q(x), \quad 0 < x < 1 \tag{3.1}$$

$$x = 0: \; L_{B0}[f(x)] + N_{B0}[f(x)] = q_{B0} \tag{3.2}$$

$$x = 1: \; L_{B1}[f(x)] + N_{B1}[f(x)] = q_{B1}, \tag{3.3}$$

where $L[\cdot]$ is the linear differential operator of the form.

$$L[\cdot] = \sum_{i=0}^{m} a_i(x) \frac{d^{m-i}}{dx^{m-i}}. \tag{3.4}$$

Here, $a_0(x) \neq 0$, $a_m(x) \neq 0$, and $N[\cdot]$ is a nonlinear operator in terms of the unknown function $f(x)$. In Eqs. (3.2) and (3.3), the subscript "B" is used to represent the operator for the boundary conditions. If the derivative is included in the boundary conditions, the order of the derivative must be, at least, one order lower than the order of Eq. (3.4). The subscripts "0" and "1" are used to indicate that the boundary conditions at $x = 0$ and $x = 1$, respectively. For practical problems, these given boundary conditions can be obtained by measurement. For many mechanical, physical, and engineering problems, the order of Eq. (3.4) usually does not exceed 4, such as the Bernoulli equation for beam bending.

According to the theory of ordinary differential equations [5], we know that a necessary and sufficient condition for a boundary-value problem of Eqs. (3.1)–(3.4) existed and a definite solution is that the number of boundary conditions is equal to the highest order of the derivatives in the differential equation. When the boundary conditions are represented by operators, we have assumed that such a requirement has been satisfied automatically.

Obviously, all linear differential operators should satisfy the property of the superposition principle. That is, for any two different functions, $f(x)$ and $g(x)$, and arbitrary two different real constants, a and b, the following operation is held, i.e.,

$$L[af(x) + bg(x)] = aL[f(x)] + bL[g(x)]. \tag{3.5}$$

For nonlinear operators, this property no longer exists. It is this main difference to make that finding a solution of nonlinear problems is much more difficult than that of linear problems. Generally speaking, the theory and method of solving linear problems have been basically mature, while for nonlinear problems, most of them rely on empirical methods and most of them are only effective in solving weak nonlinear problems. For strong nonlinear problems, although there are some methods used in practical solution, the accuracy of the solution is often difficultly not guaranteed, because the contradiction between the calculation accuracy and the amount of calculations is particularly prominent in the empirical methods.

3.2 Solution Methods from 1D Linear Ordinary Differential Equations

In this section, we give a general description of the original solution methods of the linear differential equation problems, which will involve some basic approaches and concepts about the solution approach such that one can find how the modern numerical method is developed, and when the relevant methods are extended to solve the nonlinear problems, as well as what limitations and difference exist.

Definition 3.1 To a system expressed by an initial- or boundary-value problem represented by differential equations, when all nonlinear terms in terms of the unknown function $f(x)$ are zero, that is, $N[\cdot] \equiv 0$ and $N_{B0}[\cdot] = 0$ and $N_{B1}[\cdot] = 0$ in Eqs. (3.1)–(3.3), then this system is called a linear system.

For linear boundary-value problems, if the boundary conditions are nonhomogeneous, that is, if $q_{B0} \neq 0$ or $q_{B1} \neq 0$, we can always find a function transformation to make them homogeneous, and the nonhomogeneous part on the boundary conditions are converted to the nonhomogeneous part $q(x)$ in the differential equation. Therefore, the equations for the linear boundary-value problem can be written as in the general form

$$L[f(x)] = q(x), \quad 0 < x < 1 \tag{3.6}$$

$$x = 0: \ L_{B0}[f(x)] = 0 \tag{3.7}$$

$$x = 1: \ L_{B1}[f(x)] = 0. \tag{3.8}$$

It should be pointed out that, if it is an initial-value problem, the boundary condition of Eq. (3.8) disappeared since the independent variable region is $0 < x < \infty$, and the corresponding Eq. (3.7) is called the initial-value condition.

Definition 3.2 For the ordinary differential equation (3.6), if all parameters in its differential operator, a_i $(i = 1, 2, \ldots, m)$, are real constants independent of the variable x, then we call it the linear ordinary differential equation with constant coefficients. Otherwise, if these are at least one parameter dependent on x in them, we refer it to a linear ordinary differential equation with variable coefficients.

In practice, the boundary-value problem of ordinary differential equations with variable coefficients is corresponding to the so-called "non-uniform" problems, and the corresponding initial-value problem is the so-called "parametric excitation" dynamic problem. At present, there is no perfectly complete mathematical method for solving the problems of the variable coefficient differential equation, but there are some approximate methods.

3.2.1 General Solutions of the Ordinary Differential Equations with Constant Coefficients

For any $m(\geq 2)$ order of ordinary differential equations with constant coefficients in Eq. (3.6), its differential operator is

$$L[\cdot] = a_0\frac{d^m}{dx^m} + a_1\frac{d^{m-1}}{dx^{m-1}} + \cdots + a_{m-1}\frac{d}{dx} + a_m. \tag{3.9}$$

And we call the differential equation in the domain

$$L[f(x)] = a_0\frac{d^m f}{dx^m} + a_1\frac{d^{m-1} f}{dx^{m-1}} + \cdots + a_{m-1}\frac{df}{dx} + a_m f(x) = 0 \tag{3.10}$$

as the homogeneous differential equation being associated with the nonhomogeneous differential equation (3.6) when $q(x) \neq 0$.

Step 1: To find the general solution $f^*(x)$ of the homogeneous differential equation. Let $f^*(x) = ce^{rx}$, where r is called the eigenvalue, and c is an arbitrary constant. Substituting $f^*(x) = ce^{rx}$ into the homogeneous differential equation (3.10), we obtain the algebraic equation for eigenvalue r as

$$a_0r^m + a_1r^{m-1} + \cdots + a_{m-1}r + a_m = 0. \tag{3.11}$$

This is an mth order of nonlinear algebraic equation on the eigenvalue r. According to the fundamental theorem of algebra, there must be m roots in the complex domain. And those complex roots are conjugated for each one pair. Of course, there could be multiple roots. We denote the roots as r_i $(i = 1, 2, 3 \ldots, m)$. After the roots are obtained by solving Eq. (3.11), for each r_i, the basic solution of the homogeneous differential equation can be given by

$$f_i(x) = e^{r_i x}, \quad i = 1, 2, 3 \ldots, m. \tag{3.12}$$

For the case with multiple roots, the form of the basic solution needs to be improved on the above basis, otherwise it is impossible to obtain the independent basic solutions of its geometric space in this algebraic space. At the same time, when the complex roots appear in pairs, the corresponding complex basic solutions (3.12) can also be converted into relevant basic solutions in real space. Here, we omit these contents and do not introduce them in detail. For details, one can see the textbooks on solving ordinary differential equations. Without losing generality, here, we assume that the m roots are real and different from each other, so their fundamental solutions can be given by Eq. (3.12). Then according to the superposition principle of the linear differential operator, that is Eq. (3.5), the general solution of the homogeneous differential equation (3.10) can be formulated by a linear combination of its m basic solutions, that is

$$f^*(x) = c_1 e^{r_1 x} + c_2 e^{r_2 x} + \cdots + c_m e^{r_m x} = \sum_{i=1}^{m} c_i e^{r_i x}, \tag{3.13}$$

where $c_i (i = 1, 2, \ldots, m)$ are real and arbitrary constants, which can be determined by either the boundary conditions for boundary-value problems or initial conditions of initial-value problems.

Step 2: To find a particular solution to the inhomogeneous differential equation $L[f(x)] = q(x)$. Denote the particular solution to the inhomogeneous differential equation $L[f(x)] = q(x)$ by $\hat{f}(x)$ such that the differential equation $L[\hat{f}(x)] = q(x)$ is satisfied. Usually, to find such particular solutions is not difficult, either by direct observation or by constant variation. In principle, as long as one of them is found, which depends on the specific problem to implement, the general solution for the inhomogeneous differential equation can be easily expressed. Of course, sometimes it is not easy to find a particular solution for a complex problem, but it is always feasible in theory.

Step 3: The general solution of the original inhomogeneous ordinary differential equation and the definite solution of the boundary-value problem. By adding a particular solution to the general solution of the homogeneous differential equation, that is $f(x) = f^*(x) + \hat{f}(x)$, one can find that the obtained solution satisfies the original inhomogeneous differential equation, that is, $L[f^*(x) + \hat{f}(x)] = q(x)$. Hence, the general solution of the original differential equation can be formulated by the form

$$f(x) = f^*(x) + \hat{f}(x) = \sum_{i=1}^{m} c_i e^{r_i x} + \hat{f}(x), \tag{3.14}$$

where $c_i (i = 1, 2, \ldots, m)$ are arbitrary constants. For definite solution problems, on the basis of differential equations in the domain, either initial-value conditions or boundary conditions are needed to determine these constants c_i. Substituting the general solution of Eq. (3.14) into the m boundary-value conditions, we get a set of linear algebraic equations about the undetermined coefficients as follows

$$\begin{bmatrix} A_{11} & \cdots & A_{1m} \\ \cdots & \cdots & \cdots \\ A_{m1} & \cdots & A_{mm} \end{bmatrix} \begin{bmatrix} c_1 \\ \vdots \\ c_m \end{bmatrix} = \begin{bmatrix} q_1 \\ \vdots \\ q_m \end{bmatrix}, \tag{3.15}$$

where the coefficient matrix $\mathbf{A} = [A_{ij}]_{m \times m}$, and the column $\mathbf{q} = [q_1, q_2, \ldots, q_m]^T$ is associated with the boundary conditions, nonhomogeneous terms, and the basic solutions. $\mathbf{c} = [c_1, c_2, \ldots, c_m]^T$ is the column of unknown constants. By solving the linear algebra equations of Eq. (3.15), one can get the solution of $\mathbf{c} \neq \mathbf{0}$. As long as the determinant of the coefficient matrix is nonzero, that is $\det \mathbf{A} \neq 0$ or Rank$\mathbf{A} = m$, the solution $\mathbf{c} \neq \mathbf{0}$ is unique. Substituting the obtained $\mathbf{c} = [c_1, c_2, \ldots, c_m]^T$ into Eq. (3.14), one gains the definite solution of Eqs. (3.6)–(3.8), then the solution approach is finished.

The solution of either the initial- or the boundary-value problems of linear ordinary differential equation with constant coefficients introduced above has been very classical, and the obtained solution is also called an exact solution. From the above introduction, one can see that the core problem in such a solution approach is to find the roots of the m-order nonlinear algebraic equation as shown in Eq. (3.11). Once these roots are obtained in a practical problem, the exact solution is immediately gained for it.

3.2.2 *Solution Method of Homogeneous Linear Ordinary Differential Equations with Constant Coefficients*

In the initial- and boundary-value problems of such kind of differential equations, there is $q(x) \equiv 0$ in the domain. The dynamic problem represented by these initial-value equations are corresponding to the free vibrations, while the boundary-value problems are corresponding to the buckling instability in solid mechanics. They all constitute the eigenvalue problems mathematically.

In this kind of problem, the general solution of the differential equation is expressed by Eq. (3.13). For the initial-value problems, all eigenvalues appear by a pair of conjugate complex roots, in which its imaginary part of the roots is related to the period of vibrations, and the real part corresponds to the increase or decay of the amplitude. When the real part is greater than zero, the amplitude increases with time, which corresponds to the so-called "flutter phenomenon." Conversely, when the real part is less than zero, the amplitude decays with time. When the real part is equal to zero, that is, the eigenvalue is a pure imaginary, it is a simple harmonic oscillation of free vibration, that is, the sine or cosine vibration of trigonometric functions. Here, the initial conditions are used to determine the initial state of the free vibration. For undamped free vibration, the initial-value condition can determine the amplitude and initial phase. For the mechanics problem of buckling instability, in the case where both the differential equation and the boundary conditions are homogeneous, it is clear that $f(x) \equiv 0$ is a solution to its boundary-value problem. Thus, the study of the so-called buckling instability problem is to find the conditions under which nonzero solutions $f(x) \neq 0$ exists. In this way, substitution of the general solution of Eq. (3.13) into the boundary conditions, one gets the m homogeneous linear algebraic equations with respect to the undetermined coefficients $c_i (i = 1, 2, \ldots, m)$ of the form

$$\begin{bmatrix} A_{11} & \cdots & A_{1m} \\ \cdots & \cdots & \cdots \\ A_{m1} & \cdots & A_{mm} \end{bmatrix} \begin{bmatrix} c_1 \\ \vdots \\ c_m \end{bmatrix} = \begin{bmatrix} 0 \\ \vdots \\ 0 \end{bmatrix}. \tag{3.16}$$

In Eq. (3.16), the elements A_{ij} in the coefficient matrix changes with the system parameter a_i, that is.

$$A_{ij} = A_{ij}(a_1, a_2, \ldots, a_m) \quad (i, j = 1, 2, \ldots, m). \tag{3.17}$$

Obviously, the nonzero solution of $f(x) \neq 0$ corresponds to $\mathbf{c} \neq \mathbf{0}$ solution of the homogeneous linear algebraic equations in Eq. (3.16). From the knowledge of the solution, we know that the necessary and sufficient condition of such nonzero solution is $\det \mathbf{A} = 0$ or Rank$\mathbf{A} < m$, which leads to

$$\begin{vmatrix} A_{11} & \cdots & A_{1m} \\ \cdots & \cdots & \cdots \\ A_{m1} & \cdots & A_{mm} \end{vmatrix} = 0. \tag{3.18}$$

Equation (3.18) constitutes a system of nonlinear algebraic equations for the parameters a_i, especially containing some loads or other factors contained in the system. From Eq. (3.18), one can get the solutions of changeable parameters a_i. In theory, there are m sets of such solutions in Eq. (3.18). Each one of them gives a buckling or instability condition. With the process of system parameters changing, e.g., load parameter increasing, once a set of the changeable parameter makes one of the above condition be satisfied first, and such result for the changeable parameter is called the critical condition of buckling instability. In the researches of buckling or instability, the main work is to search the solutions of parameters under Eq. (3.18).

After the parameter solutions are gained, then each set of them is substituted into Eq. (3.16) to determine its corresponding nonzero $\mathbf{c}$. Due to Rank$\mathbf{A} < m$, therefore, corresponding to each set of such parameter solution, the solution space of $\mathbf{c} = [c_1, c_2, \ldots, c_m]^T$ is at least one dimension, that is, at least one undetermined constant can be arbitrarily valued. However, once these parameters that can be arbitrarily valued are preselected, the remaining coefficients can be obtained by solving the algebraic equations in Eq. (3.16). Further, $\mathbf{c} = [c_1, c_2, \ldots, c_m]^T \neq \mathbf{0}$ is obtained for the set of substituted parameters. Following this way, the m sets of nonzero solutions are finally gained. Although the absolute value of each $\mathbf{c}$ cannot be uniquely determined, their relative values are unique. Usually, a certain normalization method is used to determine all relative values in each $\mathbf{c}$. In the buckling instability theory, we call such obtained $\mathbf{c}$ a buckling mode with corresponding to the parameter value obtained previously. In many cases, this type of change of system parameters is the single-parameter change. In this case, the parameter values obtained from Eq. (3.18) are also called the eigenvalues of buckling instability. And each buckling mode is corresponding to each of these eigenvalues. Obviously, since there are m eigenvalues in Eq. (3.18), the buckling problem has m bucking modes corresponding to m eigenvalues, respectively. In this way, there is a problem which buckling mode will occur, which is related to the path of parameter changes. The parameter corresponding to the minimum energy that causes buckling is called the critical parameter, and the corresponding buckling mode is the critical buckling mode.

3.3 Essentials to Approximate Solutions in Mathematics

The above two sections briefly introduce the solutions to the problems of linear ordinary differential equations with constant coefficients, and they also constitute the mathematical basis for finding solutions to those complex systems. For example, for the initial- and boundary-value problems of linear ordinary differential equations with variable coefficients, except for a few special classes (such as Euler-type ordinary differential equations) that can be transformed into constant coefficient linear ordinary differential equations by appropriate transformation, and then the analytical solutions expanded by a set of finite basic solutions can be given. For a more complex case, to get such exactly finite and analytic solutions are almost impossible. Especially for a nonlinear system with ordinary differential equation(s), the proposition principle held in linear systems is not held, which yields much more difficulty in the solution arithmetic. Even in the linear situation, when the coefficients are relatively complex where the previous solving process is not easy to execute, it is also necessary to establish approximate methods, such as finite element method, finite difference method, and variational methods. As displayed in Chap. 1 and the following introduction, such conventional methods have no property of closure in the approximation methods, especially for nonlinear problems. Under such a situation, it is necessary to establish an effective approximate solution to effectively realize its quantitative analysis in a uniform way, no matter of linear or weak and strong nonlinear problems, which is our aim.

In this kind of approximate calculation, generally speaking, although its calculation ideas are similar to those introduced before, but it is quite different. The similarities are achieved by expanding the unknown functions on a set of base functions, while the difference is that the selected base functions introduced in Chap. 2 make the approximation method closed. In general, the selected base functions cannot be obtained by the basic solutions of the problem similar to the linear ordinary differential equation, then the selected base functions are consisted by a set of finite terms. Obviously, there are more than one type of base functions to be preselected, which leads to the problem of how to select a set of base functions so as to make the approximate solution approach to "exact" one quickly. Of course, this must be done on the premise that the approximation converges.

Definition 3.3 For a n-dimensional function space, if its n unrelated functions $\{\varphi_i(x)\}$ can represent any function $f(x)$ in the space, then this set of functions $\varphi_i(x)(i = 1, 2, \ldots, n)$ are called a set of base functions (or function basis) of this n-dimensional function space. When $n \to \infty$, if there is still the linearly independent property in this set of functions, then this set of functions are called a set of function bases in the infinite-dimensional function space.

The so-called linear independence means that any function in $\{\varphi_i(x)\}$ cannot be characterized by other $n - 1$ functions using linear combination. For example, for the power functions, x^i $(i = 0, 1, 2, \ldots, n)$, and the trigonometric functions, $e^{i\omega x} = \cos\omega x + j\sin\omega x$ $(i = 0, 1, 2, \ldots, n)$, where $j = \sqrt{-1}$, each of them are a set of function bases. Obviously, when $n \to \infty$, these two sets of base functions are still linearly independent, so each of them can be selected as a set of base functions in the infinite-dimensional function space. Therefore, a function can be expanded by such function bases as

$$f(x) = \sum_{i=1}^{\infty} c_i\varphi_i(x). \tag{3.19}$$

Since the calculations are with infinite terms in practical calculations, we need to truncate the above expansion by a finite form. In general, such truncation can be taken by the first N items, i.e.,

$$f_N(x) = \sum_{i=1}^{N} c_i\varphi_i(x). \tag{3.20}$$

It is called the approximate characterization of $f(x)$, that is, $f_N(x) \approx f(x)$. The convergence of Eq. (3.19) on a set of base functions is conditional, that is, there are requirements on the characteristics of both the function $f(x)$ and the base functions. In theory, the function characteristics of $f_N(x)$ represented by Eq. (3.20) are consistent with the characteristics of its base functions. That is to say, the approximation converges in the space of the base functions (for details, the readers can refer to those textbooks of real variable function and functional analysis, etc.), we do not introduce this too much here, since we mainly focus on the relevant content of truncation error.

Definition 3.4 When a function $f(x)$ can be represented by the infinite expansion of Eq. (3.19) in a function space (which means that its convergence and convergence into its function space have been guaranteed), and $f(x)$ is approximated by the finite term of Eq. (3.20), the removed part is called the truncation error $\eta_f(x)$ of function $f(x)$ in the approximation by $f_N(x)$, that is.

$$\eta_f(x) = f(x) - f_N(x) = \sum_{i=N+1}^{\infty} c_i\varphi_i(x) \tag{3.21}$$

where the subscript "f" represents the truncation error $\eta(x)$ of the function $f(x)$ in cases of Eqs. (3.19) and (3.20).

Therefore, for an unknown function expanded on any infinite base functions, Eq. (3.19) can be rewritten as

$$f(x) = f_N(x) + \eta_f(x). \tag{3.22}$$

It is evident that the representation in Eq. (3.22) is always feasible no matter which approximate solution is adopted. In fact, for the differential equation, in addition to its undetermined function $f(x)$ needs to be represented by the approximate solution and truncation error according to Eq. (3.22) to judge its progress, all the equation terms also need to be judged according to this. When an approximate solution is effective for the definite solution of a differential equation, it means that the truncation errors generated in each term of the definite solution problem of the differential equation expressed by Eq. (3.22) are very small relative to its approximate solution. Only at this time, we think that the approximate solution has enough accuracy.

Definition 3.5 We call the solution of the boundary-value problem of differential equations being closed, which means the truncated approximate solution being independent of its truncation error, and vice versa.

According to the definition of closedness of the approximate solution, we can judge that the solutions to the initial- and boundary-value problems of ordinary differential equations with constant coefficients are closed. In mathematics, when its solution only has finite terms exactly, it is said to be an exact solution.

Theorem 3.1 For any linear differential operator $L[\cdot]$ acting on a function $f(x)$, the number of terms after it acts on the finite approximation $f_N(x)$ truncated from $f(x)$ is as the same as that of $f_N(x)$, and the truncation error of $L[f(x)]$ is $\eta_L(x) = L[\eta_f(x)]$.

Proof Substituting Eqs. (3.19) into (3.10), and applying Eq. (3.5), we obtain.

$$L[f(x)] = L[\sum_{i=0}^{N} c_i\varphi_i(x) + \sum_{i=N}^{\infty} c_i\varphi_i(x)] = \sum_{i=0}^{N} c_i L[\varphi_i(x)] + \sum_{i=0}^{N} c_i L[\varphi_i(x)].$$

Let $g(x) = L[f(x)]$, then

$$g_N(x) = \sum_{i=0}^{N} c_i L[\varphi_i(x] = L[f_N(x)]$$

was established. Obviously, the number of items in $g_N(x)$ is the same as the number of items in $f_N(x)$. Secondly, according to the definition of truncation error, for a linear operator acting on $f(x)$, the truncation error of $L[\cdot]$ is

$$\begin{aligned} \eta_L(x) = \eta_g(x) &= L[f(x)] - L[f_N(x)] \\ &= L[f(x) - f_N(x)] = L[\eta_f(x)]. \end{aligned} \tag{3.23}$$

Thus, the theorem is completed.

It is worth noting that, in the above proof, the interchangeability of infinite summation operation and derivative operation is used, which is mathematically conditional, that is, the function $f(x)$ is required to be uniformly continuous in the definite region and uniformly converge on the infinite series.

With the conclusion of Theorem 3.1, after the approximate solution of Eq. (3.20) is substituted into the linear differential equation, no matter what algorithm we use to obtain the first N coefficients $c_i (i = 0, 1, 2, \ldots, N)$ of the approximate solution $f_N(x)$ from the differential equation, we will find that the determination of these undetermined coefficients has nothing to do with the truncation error. Once the truncation errors are equal to zero under an average approach, such approximate solution will be closed for the linear systems.

It is worth pointing out that the solution closedness and solution convergence are two different concepts. For linear systems, even if the solution is closed, the solution may be not necessarily convergent. For example, the finite element method does not converge under all meshes and arbitrarily selected interpolation functions. The same situation holds for the finite difference method, and its convergence depends on the choice of difference in grid or difference in format for different problems. There are similar problems in other cases. However, once the approximate solution converges, it means that the truncation error of the solution tends to zero.

3.4 Closedness Concepts of Approximate Solutions for Nonlinear Boundary-Value Problems

Since the nonlinear initial-value problem can be solved by a numerical integration to obtain a high-precision numerical solution, here, we omit its introduction for solving the nonlinear initial-value problem by using a wavelet-based method. When the wavelet method was first used to solve linear initial-value problems, one had not found a strong advantage in the wavelet method over the conventional ones. In such a situation, we will focus our main attention on the wavelet-based closed solution method of solving nonlinear boundary-value problems, from which one can find the high efficiency and high accuracy of the wavelet-based closed method in solving such nonlinear problems, especially for strong nonlinear problems.

For nonlinear boundary-value problems, there is $N[\cdot] \neq 0$ in the domain of an unknown function $f(x)$. In this case, except for very few simple problems that have found exact solutions with finite representations, the remaining all others should be given by an approximation method of infinite summation expression [6]. Such as the above-mentioned series expansion method, regular perturbation method, singular perturbation method, and various forms of iterative methods. When modern numerical methods such as finite element method (FEM), finite difference method (FDM), and finite volume method (FVM) are developed to solve different kinds of complex boundary-value problems, it is found that their solutions are not closed in solving nonlinear problems. The reader will valuate that some of such some approximate methods are always not closed even when they are employed to a linear boundary-value problem after the closed concepts are introduced. In order to make them applicable to the solution of nonlinear problems, they are often combined with a tracking technique (such as piecewise linearization or homotopy tracking method), but the problem in such tracking arithmetic has the accumulated error after multistep tracking. Due to the nonlinear situation, the superposition principle in a linear system is no longer valid. In this way, for each system factor change (such as mechanical loading), the solution must be recalculated. In a linear system, as long as the solution of a certain type of load, i.e., unit magnitude is obtained at a certain size, it is not necessary to calculate the solution again according to the superposition principle when the load is changed. Therefore, for nonlinear systems, the difficulty of calculations during the solution process is much greater than that of linear systems. In theory, there is no general method which has been found to solve all nonlinear differential equation systems in literature. Especially in the strong nonlinear problem, there is no perfectly effective method in theory. Because of the lack of closed solution method, these conventional solution methods are effective only in solving weak nonlinear problems.

Before we give their relevant concepts, let us look at a few simple nonlinear examples.

3.4.1 *Examples of Non-closed Solutions to Nonlinear Problems in Approximation*

In order to show that the approximate solution of the nonlinear problem $N[f(x)] \neq 0$ is not closed, let us use the simplest nonlinear term $N[f(x)] = f^2(x)$ as an example.

Case 1: The power function $x^i (i = 0, 1, 2, \ldots)$ is selected to the base functions. In such case, the approximations for a function $f(x)$ can be expressed by

$$f(x) = f_N(x) = \sum_{i=0}^{N} c_i x^i + \eta_f(x), \quad \eta_f(x) = \sum_{i=N+1}^{\infty} c_i x^i. \tag{3.24}$$

At the same time, we can also expand $f^2(x)$ on the same base functions, that is,

$$f^2(x) = f_N^2(x) = \sum_{i=0}^{N} d_i x^i + \eta_{f^2}(x), \quad \eta_{f^2}(x) = \sum_{i=N+1}^{\infty} d_i x^i. \tag{3.25}$$

In this approximation, the calculation formulae of its coefficients are often obtained by comparing the same order of power x^i in the equation. Hence, after Eq. (3.24) is substituted into $N[f(x)] = f^2(x)$, and let the induced results be equal to Eq. (3.25), then, the formulae for calculating the expansion coefficient d_i in terms of the expansion coefficient c_i can be gained as follows:

$$d_0 = c_o^2,$$
$$d_1 = 2c_0c_1, d_2 = 2c_0c_2 + c_1^2, \ldots\ldots$$

$$d_N = \sum_{j=0}^{N} c_j c_{N-j},$$
$$d_{N+1} = c_0c_{N+1} + c_1c_N + c_2c_{N-1} + \cdots + c_Nc_1 + c_{N+1}c_0, \cdots\cdots$$

which obviously display that $\eta_{f^2}(x)$ varies with the approximate solution $f_N(x)$.

Case 2: Take the trigonometric sine and cosine functions as base functions, that is 1, $\sin i\omega x, \cos i\omega x$ $(i = 1, 2, \ldots)$. Let

$$f(x) = f_N(x) + \eta_f(x) = b_0 + \sum_{i=1}^{N} (a_i \sin i\omega x + b_i \cos i\omega x) + \eta_f(x),$$

$$\eta_f(x) = \sum_{i=N+1}^{\infty} (a_i \sin i\omega x + b_i \cos i\omega x)$$

$$f^2(x) = f_N^2(x) + \eta_{f^2}(x) = d_0 + \sum_{i=1}^{N} (c_i \sin i\omega x + d_i \cos i\omega x) + \eta_{f^2}(x),$$

$$\eta_{f^2}(x) = \sum_{i=N+1}^{\infty} (c_i \sin i\omega x + d_i \cos i\omega x).$$

Then substituting the expansion of $f(x)$ into $f^2(x)(= [f(x)]^2)$, since the square of the trigonometric function should be converted to the base functions in order to compare the coefficients on the same base functions, e.g., $\cos^2 i\omega x =$

$(1 + \cos 2i\omega x)/2$ and $\sin^2 i\omega x = (1 - \cos 2i\omega x)/2$, we have

$$d_0 = \frac{1}{2}[b_0^2 + \sum_{i=1}^{N}(a_i^2 + b_i^2) + \sum_{i=N+1}^{\infty}(a_i^2 + b_i^2)],$$

$$d_{N+1} = b_0 b_{N+1} + \frac{1}{2}\sum_{j=1}^{N} b_j b_{N+1-j} + \cdots$$

It is evident that approximation $f_N^2(x)$ contains d_0 that is dependent on the truncation error $\eta_f(x)$, and the truncation error $\eta_{f^2}(x)$ contains d_{N+1} that is dependent on the coefficients $a_1, \ldots, a_N$ and $b_0, b_1, \ldots, b_N$ appeared in the approximation $f_N(x)$. If more expansion coefficients d_i are explicitly given in terms of a_j, b_j, then, in fact, one can find more terms of the similar coupling dependence and that the approximation $f_N^2(x)$ is dependent on the truncation error $\eta_f(x)$ and the truncation error $\eta_{f^2}(x)$ is varied with $f_N(x)$.

Case 3: Regular perturbation method. In this method, the solution is represented as a power series of any small parameter ε preselected. For the nonlinear problem with square term, we have

$$f(x) = f_N(x) + \eta_f(x) = \sum_{i=1}^{N} u_i(x)\varepsilon^i + O(\varepsilon^{N+1}).$$

Substituting it into $f^2(x)$, one gets

$$f^2(x) = f_N^2(x) + O(\varepsilon^{2N}) = \sum_{i=1}^{N}\sum_{j=1}^{i} u_j(x)u_{i-j}(x)\varepsilon^i + O(\varepsilon^{N+1}).$$

As shown in Chap. 1 as an example, one finds that for any order perturbation solution $u_i(x)$, the function index in $\sum_{j=1}^{i} u_j(x)u_{i-j}(x)$ corresponding to the same ε^i power in its nonlinear term is always lower than the index of $u_i(x)$. This shows that the perturbed solution is obtained from the solution of $u_1(x)$ from the linear boundary-value problems corresponding to the original problem by ε^1. Then, $u_1(x)$ is substituted into the induced linear problem corresponding to ε^2 for obtaining $u_2(x)$. Step-by-step liking this iteration up to ε^N, $u_N(x)$ can be gained by its induced linear governing equations by substituting those gained $u_i(x)$ $(i = 1, 2, \ldots, N-1)$. After that, the approximate solution $f_N(x)$ is obtained in the perturbation program. In such program, there is always relations of the form

$$u_i(x) = g_i(u_1(x), u_2(x), \ldots, u_{i-1}(x)), i > 1.$$

That is, the solution $u_i(x)$ of the i-th order perturbation solution is always associated with the previous perturbation solutions $u_j(x)$ $(j = 1, 2, \ldots, i-1)$. At the same time, it is obvious that the truncation error $\eta_f(x)$ also contains $u_j(x)(j = 1, 2, \ldots, N)$, i.e.,.

$$\eta_f(x) = \sum_{i=N+1}^{\infty} u_i(x)\varepsilon^i = \tilde{g}(u_1(x), u_2(x), \ldots, u_{i-1}(x))$$

is strongly dependent on the approximate solution truncated.

In practice, almost all the current approximation methods cannot make an independence between an approximate solution and its truncation errors in the solution process in solving nonlinear problems, containing tracking methods. This fact may reveal that most of the numerical solution methods often lose their efficiency in solving strong nonlinear problems.

3.4.2 Nonlinear Problems with Non-integer Power Nonlinearity

Another difficulty what the conventional solution methods met is that they cannot be directly applied to solve the nonlinear problems with non-integer power nonlinearity of the unknown function. In fact, there are many such nonlinear terms in practice like logarithmic nonlinear term, exponential function nonlinear term, trigonometric function nonlinear term, transcendental function, etc. In such a situation, when one directly substitutes the approximate expansion of the unknown function into the nonlinear terms, one will find that the calculations cannot be directly conducted to the approximation of the nonlinear terms. To solve this difficulty, the practical process for such direct substitution is that before such substitution is conducted, Taylor's expansion is applied to the non-integer nonlinear term $N[f(x)]$on unknown function $f(x)$ at $N[f(x)]_{f(x)\equiv 0}$ such that $N[f(x)]$ is reduced into a power series of unknown function, i.e., the form of integer power nonlinearity. It is obvious that such expansion adds some new calculation errors. Usually, the lowest two or three nonlinear terms are truncated in Taylor's expansion. In theory, there is some restriction in the truncated Taylor's expansion, e.g., the magnitude order $||f(x)||$ of unknown function is always restricted in small region relative to $O(1)$, which is attributed to the category of weak nonlinear problems. Therefore, such treatment for a non-integer power nonlinearity will be inefficient to solve a strong nonlinearity with large $||f(x)||$ when the external application in a nonlinear problem increases to a large value.

Example 1 Fractional nonlinear term. For example, $\sqrt{f(x)}$, when the expansion approximate solution of Eq. (3.19) is directly substituted into it, one gains.

$$\sqrt{f(x)} = \sqrt{\sum_{i=1}^{N} c_i \varphi_i(x) + \eta_f(x)}.$$

Then, the calculations for the unknown coefficients c_i cannot be directly calculated by the present numerical techniques.

Example 2 Logarithmic nonlinear term, such as $\ln f(x)$. If the expansion approximate solution of Eq. (3.19) is adopted and directly substituted into it, one gets.

$$\ln f(x) = \ln[\sum_{i=1}^{N} c_i \varphi_i(x) + \eta_f(x)].$$

Similar to the above in $\sqrt{f(x)}$, such formula is still difficult to transform the above equation into a system of explicit algebraic equation about the unknown constants c_i. Thus, the calculations for the unknown coefficients c_i cannot be again performed by the present techniques.

Example 3 The nonlinear term of sinusoidal function, such as $\sin f(x)$. Similarly, we have.

$$\sin f(x) = \sin[\sum_{i=1}^{N} c_i \varphi_i(x) + \eta_f(x)]$$

to which the problems that appeared in the above two examples still existed.

There are many examples of these kinds of nonlinear terms that we omit here to save space. How to deal with this kind of nonlinearity terms such that the accuracy can be ensured by the same expansion in an approximation method? In order to make it possible, this kind of nonlinear function is transformed into a power series of unknown function (i.e., the nonlinearity of integer power) by Taylor expansion. Since this expansion is an infinite series, its convergence is related to a radius defined in function norm, which is also called the radius of convergence. Within this radius of convergence, the first several terms are intercepted as the approximation of the original nonlinear term. From the mathematical analysis and computational analysis, we know that even for exponential nonlinear functions, although the Taylor expansion has an infinite radius of convergence, when some finite terms is intercepted, the approximation will be seriously deviated when the magnitude of unknown function is larger than a criterion. In this way, due to the limit of the convergence radius of Taylor expansion, there may still be a large deviation when the finite terms are intercepted. That is, the validity scope of application in this expansion method to deal with the non-integer nonlinear terms is greatly limited, because the approximate solution will be no longer valid when the magnitude of unknown function in the solution is over the convergence radius, and the truncation error may become large for the strong nonlinear problems.

3.4.3 Concepts of Strong Nonlinearity and Weak Nonlinearity

In order to illustrate the effect of the concept of closedness of solution to numerical methods, we firstly give the concepts of weak nonlinearity and strong nonlinearity in a nonlinear problem. From the previous preliminary discussion, we can find that when the solution of a nonlinear problem increases with external factors, the norm $||f(x)||$ of its approximate solution will also increase. For the linear part, this increasing is a linear relationship of the external factors. But for the nonlinear part, this growth relationship will be nonlinear. Therefore, in this growth process, when $||f(x)||$ is small, the influence of its nonlinear part is also relatively small compared to the linear part. But as $||f(x)||$ reaches a certain level, the magnitude $||N[f(x)]||$ of the nonlinear term will approach to or even over the magnitude $||L[f(x)]||$ of the linear term, and then grow faster than the linear part. Therefore, we can divide a nonlinear system into a weak nonlinear problem and a strong nonlinear problem.

Definition 3.6 For a nonlinear system, if its solution makes the magnitude of the nonlinear term in the governing equation be far smaller than the magnitude of the linear term (generally it can be slightly close to the same magnitude), $||N[f_N(x)]|| \ll ||L[f_N(x)]||$, then the nonlinear system is called a weak nonlinear problem. In contrary, if the solution makes the magnitude of the nonlinearity in the equation reach to or over the magnitude of the linear term, that is, $||N[f_N(x)]|| \geq ||L[f_N(x)]||$, then the nonlinear system is called a strong nonlinear problem.

From Definition 3.6, one can judge that when the Taylor expansion is used to the non-integer power nonlinearity of a problem, the numerical solution methods are suitable only to the weak nonlinear problem of the system.

From the above example, we see that the solution of nonlinear problems is strongly dependent on its truncation errors, or vice versa. Here, we give a general operation of the nonlinear term in the form

$$\sum_{i=0}^{N} d_i\varphi_i(x) + \eta_{N[f]}(x) = g(x) = N[f(x)] = N[\sum_{i=0}^{N} c_i\varphi_i(x) + \eta_f(x)]. \quad (3.26)$$

Since the nonlinear operator in the above equation does not have the interchangeability of the summation operation and the $N[f(x)]$ operator, there is the coupling either between the approximation part $\sum_{i=0}^{N} d_i\varphi_i(x)$and the truncation error $\eta_f(x)$ or between the approximation $\sum_{i=0}^{N} c_i\varphi_i(x)$and the truncation error $\eta_{N[f]}(x)$ or both in Eq. (3.26), when the conventional numerical methods are employed. Such coupling may lead to that there is the invalid possibility in the solution methods for nonlinear problems when the magnitude order of the solution $||f(x)||$ increases over a criterion, that is, the system enters into a strong nonlinear problem because the accompany truncation error $||\eta_{N[f]}(x)||$ may increase faster than $||\eta_f(x)||$ such that the approximations in the governing equation is either non-convergent or infeasible to ignore the truncation errors or both.

When we use this intuitive valuation to a linear problem, one will find that the truncation errors of all terms in the system are in the same order of small quantity such that the solution is always valid except for the convergence. Although the convergence for an approximate method is necessary in theory, in practice, it is always impossibly realized. When the proof of convergence cannot be obtained, the judge is often replaced by a weak condition with the concept of asymptotic behavior to the approximation. In fact, the asymptotic condition is always used even when the convergence of the approximate method is demonstrated for a problem.

Definition 3.7 We call an approximate solution to be effective when the magnitude order of all truncation errors in an approximate method are within a high-order small quantity of magnitude of the approximate solution.

Definition 3.8 We call an approximate solution asymptotic when the approximate solution to the same system is quantitatively solved by taking two differently adjacent truncations in the approximation, and the error between the two truncated approximate solutions is within a pre-specified accuracy range with a small quantity. Otherwise, the approximate solution is not asymptotic and the approximate method is invalid.

From Definition 3.8, we know that the asymptotic judgment of efficiency for applying an approximate method in a problem is weak and not strict. Such evaluation based on the asymptotic behavior is strongly dependent on a specified precision condition preselected. To a nonlinear system, when the experience of numerical tests has been displayed by that the approximation is valid for a selected precision condition, there is a high possibility to invalid in the same arithmetic when the magnitude of the solution increases up to a criterion such that the asymptotic precision condition cannot be satisfied. In such a situation, some techniques for the arithmetic have to be added if one still expects to use the same approximate method. To the static deflection of circular plates of nonlinear von Kármán's equations when the applied load is up to a value, for example, the approximate solution exhibits period-two solution phenomena which are not true in practice. Once we down the increase step of the applied load, which is the simplest added technique, the efficiency of the arithmetic of the original approximate solution method can be still ensured in some regions of the applied region. With increasing of the applied load over the former region, however, such adjustment of the increased step becomes invalid. For such a case of a strong nonlinearity high enough, we need to look for other techniques of calculations, e.g., interpolation iteration method associated with the adjustment of load increasing step. Finally, the solution of the von Kármán's plate equations is realized by the approximate method from the small deflection (i.e., solution for linear case) to enough large deflection approximating to deflection of the corresponding membrane form [7]. Such approximate method for solving strong nonlinear problems is highly dependent on both the nonlinear problem and the skill knowledge that a researcher has. When the kind of nonlinear problems is changed by another one, usually, the mature approximate method for the former kind of nonlinear problems established by a researcher is often invalid to the latter kind of nonlinear problems, containing the change of the defined region form of the same kind of governing equations. That is to say that approximate solution methods should be taken into account for different kinds of nonlinear problems one-by-one, which leads to the lack of a unified method for different kinds of nonlinear problems. During the past 20 years, the author and his students or colleagues pay their attention on this area to establish a unified method for solving different kinds of nonlinear problems with both high accuracy and low computations by means of the generalized Coiflets displayed in Chap. 2.

In addition, the efficiency of an approximate method is also dependent on the number of the finite truncated terms in approximation. To it, there is no difficulty for one to obtain the following theorem.

Theorem 3.2 If an approximate solution is valid, it means that there must be a truncated integer number N^* in the approximate solution such that when $N \geq N^*$, the magnitudes of truncation errors in the expansion process all are far smaller than the magnitude of approximate solution $f_N(x)$, that is,

$$||\eta_f(x)||, ||\eta_{L[f]}(x)||, ||\eta_{N[f]}(x)|| \ll ||f(x)||.$$

To an effective truncated approximate solution, as pointed out above, its effectiveness is related to a preselected precision condition with enough accuracy. Hence, the truncated number N^* is dependent on the preselected precision condition. In general, as the selection precision increases, the truncated number N^* increases too if the approximate method is efficient.

As introduced previously, when an efficient approximate method for the linear system is extended to its corresponding nonlinear ones, usually, they are effective in the case of their weak nonlinear problems, which is judged by the asymptotic behavior of the extended approximate solution rather than by a relatively strict theory. In this case, there is not a relative "standard solution" and "standard quantitative solution" for the solution of strong nonlinear problems in the strict meaning except for a few simple problems that have closed analytical solutions. From the closed solutions, one can find that such solutions are all either exact with no error or in approximation being independent on its truncation errors. Due to the reason that such closed approximation is independent on its truncation errors, hence, the efficiency of the method is dependent only on the truncated number N^*. In other words, when N^* is larger enough, such closed approximate method is still efficient no matter what $||f(x)||$ is small or larger.

Therefore, how to establish an effective solution method with closed property for nonlinear problems becomes a key challenge problem in the scientific committee for solving strong nonlinear problems. To approach this goal, we find that the generalized Coiflets has the closed property in solution method for nonlinear problems, which will be introduced in the next section and thereafter.

3.5 Closed Wavelet-Based Solution for Solving 1D Nonlinear Boundary-Value Problems

In this section, we introduce the closed solution of the boundary-value problem with nonlinear ordinary differential equations using the generalized Coifman wavelet method established in the previous chapter, and then we can give a unified solution method for differently weak and strong nonlinear problems. On the basis of it, the solution method is successfully extended to solving the 2D and 3D nonlinear boundary-value problems in the next section. In the literature, there are also other existing wavelet methods to solve nonlinear problems, such as solving the one-dimensional Burger's equation of fluid mechanics [8, 9], which is mainly realized by the Galerkin method after the wavelet approximation is used for the spatial part of the initial-boundary-value problem, while the time part is still carried out by the traditional numerical integral method with high accuracy enough. From it, however, the readers can see that the wavelet-based approximate solution without closed property is still directly substituted into the nonlinear term as done in the traditional way for the processing of nonlinear term. In such a case, there are no advantages over those conventional methods to be found in such wavelet-based method since the solution method is still not closed.

Now, let us introduce the Generalized Coifman's wavelet-based solution with close property to the nonlinear boundary-value problems of Eqs. (3.1)–(3.3). Here, we mainly use Eqs. (2.12), (2.24), (2.31) and (2.32), and rewrite them as follows:

$$f(x) = \lim_{n\to\infty} P_n f(x) = P_n f(x) + \sum_{i=n}^{\infty} Q_i f(x) = P_n f(x) + \tilde{Q}_n f(x) \tag{3.27}$$

$$P_n f(x) = 2^{-n/2} \sum_{k\in Z_\Omega} f_{n,k}\varphi_{n,k-M_1}(x) = \sum_{k\in Z_\Omega} f_{n,k}\varphi(2^n x - k + M_1) \tag{3.28}$$

in which the support domain of scaling function $\varphi(x)(=\varphi_{\tilde{N}}(x))$ is $[0, \tilde{N}]$, $f_{n,k} = f(k2^{-n})$, and the summation set is $Z_\Omega = \{k; -\tilde{N}+1+M_1 \le k \le 2^n + \tilde{N} - 1 + M_1\}$. Since the summation in Eq. (3.28) is selected from $Z = \{k; -\infty < k < \infty\}$ and the values of $P_n f(x)$ in [0, 1] are fully the same as that summation in Eq. (3.28) is conducted in Z. For this reason, at the following introduction, we do not differ them strictly and the all properties are ensured. In the next chapter, the error estimation is introduced in detail to the approximation under Eq. (3.28).

3.5.1 *Expansion of Nonlinear Operator Terms*

According to the previous introductions, one has known that the key problem of an approximate method for a nonlinear problem is how to expand the nonlinear term in the basic equations of the problem. For our wavelet-based method, first, let us introduce the expansion method of nonlinear operator terms based on Eqs. (3.27) and (3.28). To any nonlinear term $N[f(x)]$ in the basic equation of the problems, we use a function transform of $g(x) = N[f(x)]$. Usually, $g(x) \in L^2(R)$when the unknown function $f(x) \in L^2(R)$. Hence, the expansion of Eq. (3.28) can be used to $g(x)$, i.e.,

$$\begin{aligned} P_n N[F(x)] &= P_n g(x) = 2^{-n/2} \sum_{k\in Z_\Omega} g_{n,k}\varphi_{n,k-M_1}(x) \\ &= 2^{-n/2} \sum_{k\in Z_\Omega} N[f_{n,k}]\varphi_{n,k-M_1}(x). \end{aligned} \tag{3.29}$$

From Eq. (3.29), one finds that the expression of the nonlinear operator $N[f(x)]$ has the same form as that of the unknown function $f(x)$ with a truncation under the resolution level n, and its expansion coefficients $g_{n,k}$ in Eq. (3.29) is easily formulated by

$$g_{n,k} = g(k2^{-n}) = N[f_{n,k}], \tag{3.30}$$

which is a full form of the original nonlinear function. Equation (3.29) reveals that its operations of summation and nonlinear operator can be exchanged as done in the expansion of a linear operator. Applying the same notations in Eq. (3.27) to the nonlinear operator, then, one gets

$$N[f(x)] = P_n N[f(x)] + \tilde{Q}_n N[f(x)] = P_n N[f(x)] + \eta_{N[f]}(x), \tag{3.31}$$

where $\eta_{N[f]}(x) = \tilde{Q}_n N[f(x)]$ is the truncation error of the approximation $P_n N[f(x)]$. Thereafter, the truncation error in Eq. (3.27) is rewritten by $\eta_f(x) = \tilde{Q}_n f(x)$.

To the examples of nonlinear terms shown in Sect. 3.3 using Eq. (3.29) or (3.31), one can immediately get their expansion as follows:

$$f^2(x) = 2^{-n/2} \sum_{k\in Z_\Omega} f^2\left(\frac{M_1+k}{2^n}\right)\varphi_{n,k}(x) + \tilde{Q}_n f^2(x) = P_n f^2(x) + \eta_{f^2}(x)$$

$$\sqrt{f(x)} = 2^{-n/2} \sum_{k\in Z_\Omega} \sqrt{f\left(\frac{M_1+k}{2^n}\right)}\varphi_{n,k}(x) + \tilde{Q}_n \sqrt{f(x)} = P_n \sqrt{f(x)} + \eta_{\sqrt{f}}(x)$$

$$\ln f(x) = 2^{-n/2} \sum_{k\in Z_\Omega} \ln\{f\left(\frac{M_1+k}{2^n}\right)\}\varphi_{n,k}(x) + \tilde{Q}_n \ln f(x) = P_n \ln f(x) + \eta_{\ln f}(x)$$

$$\sin f(x) = 2^{-n/2} \sum_{k\in Z_\Omega} \sin\{f\left(\frac{M_1+k}{2^n}\right)\}\varphi_{n,k}(x) + \tilde{Q}_n \sin\{f(x)\} = P_n \sin f(x) + \eta_{\sin f}(x).$$

3.5.2 Wavelet-Based Solution Program of 1D Nonlinear Boundary-Value Problems

Before we conduct it, we introduce the notation of the form to a given function $q(x)$ in Eq. (3.1), i.e.,

$$P_n q(x) =< q(x), \varphi_{n,k}(x) >= 2^{-n/2} \sum_{k \in Z_\Omega} q_{n.k} \varphi_{n,k-M_1}(x), \tag{3.32}$$

where $q_{n.k} = q(k2^{-n})$ is known. To the linear operator $L[f(x)]$, its expansion is conducted as done as $N[f(x)]$, which is written in Eq. (3.34) later.

With these preparations, we mathematically show that this approximate solution method is closed when using the expansion of scaling functions in Eq. (3.27) to solve the boundary-value problems (3.1)–(3.3).

For simplicity, here, we rewrite the expansions of all terms in the nonlinear differential equation of Eq. (3.1) with uniform notations as follows:

$$f(x) = P_n f(x) + \tilde{Q}_n f(x) = P_n f(x) + \eta_f(x) \tag{3.33}$$

$$L[f(x)] = P_n L[f(x)] + \tilde{Q}_n L[f(x)] = P_n L[f(x)] + \eta_{L[f]}(x) \tag{3.34}$$

$$N[f(x)] = P_n N[f(x)] + \tilde{Q}_n N[f(x)] = P_n N[f(x)] + \eta_{N[f]}(x) \tag{3.35}$$

$$q(x) = P_n q(x) + \tilde{Q}_n q(x) = P_n q(x) + \eta_q(x), \tag{3.36}$$

where $\eta_f(x) = \tilde{Q}_n f(x), \eta_{L[f]}(x) = \tilde{Q}_n L[f(x)], \eta_{N[f]}(x) = \tilde{Q}_n N[f(x)]$, and $\eta_q(x) = \tilde{Q}_n q(x)$ are the truncation errors of the respective functions on the basis of the base scaling functions of the generalized Coiflets under the resolution level n.

According to Eq. (2.32), we have known that the orthogonal property

$$< \varphi_{n,j}(x), \tilde{Q}_n f(x) >= \int_R \varphi_{n,j}(x) \tilde{Q}_n f(x) dx = 0 \tag{3.37}$$

is always held for any $f(x) \in L^2(R)$, which will be used in the deduction of the solution method we used.

Substituting the expansions of Eqs. (3.33)–(3.37) into the nonlinear differential equation in Eq. (3.1), one gets

$$P_n L[f(x)] + \tilde{Q}_n L[f(x)] + P_n N[f(x)] + \tilde{Q}_n N[f(x)] = P_n q(x) + \tilde{Q}_n q(x). \tag{3.38}$$

By using the Galerkin method to Eq. (3.38), that is, taking each of the base scaling functions $\varphi_{n,k}(x)$ to multiply Eq. (3.38) then conducting the integration of the induced equation, one gets

$$\int_R \{P_n L[f(x)] + P_n N[f(x)] - P_n q(x)\}\varphi_{n,i}(x)dx = 0 \quad i \in Z, \tag{3.39}$$

where orthogonality of Eq. (3.37) is used such that all truncation errors are equal to zero under the average meaning of $L^2(R)$ space with $< \cdot, \varphi_{n.i}(x) >$, i.e., $< \tilde{Q}_n L[f(x)], \varphi_{n,i}(x) >= 0$, $< \tilde{Q}_n N[f(x)], \varphi_{n,i}(x) >= 0$ and $< \tilde{Q}_n g(x), \varphi_{n,i}(x) >= 0$. By using the explicit expressions of the approximations on each term in Eq. (3.39) and the orthogonality $< \varphi_{n,k}(x), \varphi_{n,i}(x) >= \delta_{k,i}$, further, Eq. (3.39) is reduced in the form

$$L[f_{n,i+M_1})] + N[f_{n,i+M_1})] = q_{n,i+M_1}, \quad i = -M_1, -M_1+1, \ldots, 2^n - M_1 \tag{3.40}$$

in which the notation of base scaling functions $\{\varphi_{n,k-M_1}(x)\}$ has been taken into account. Taking a transform of the index $i = k - M_1$, then Eq. (3.40) can be reduced by the form

$$L[f_{n,k})] + N[f_{n,k})] = q_{n,k}, \quad k = 0, 1, \ldots, 2^n, \tag{3.41}$$

which reveals that the induced algebraic equations are independent of the index M_1. Substitution of the explication of the linear operator of Eq. (3.5) and its expansion into Eq. (3.41) leads to

$$\sum_{i=0}^{m} a_{i,n,k} f_{n,k}^{(m-i)} + N[f_{n,k})] = q_{n,k}, k \in \mathrm{Z} \tag{3.42}$$

in which

$$f_{n,k}^{(m-i)} = \frac{d^{m-i} f(x)}{dx^{m-i}}|_{x=k2^{-n}}, \; f_{n,k}^{(0)} = f_{n,k}, \quad \text{and} \quad a_{i,nk} = a_i(x)|_{x=k2^{-n}}. \tag{3.43}$$

Since

$$P_n\{\frac{d^i f(x)}{dx^i}\} = 2^{-n/2} \sum_{j\in Z} f_{n,j}^{(i)} \varphi_{n,j-M_1}(x), \quad 1 \le i \le m \tag{3.44}$$

$$\frac{d^i}{dx^i}\{P_n f(x)\} = 2^{-n/2} \sum_{j\in Z} f_{n,k} \varphi_{n,j-M_1}^{(i)}(x), \quad 1 \le i \le m, \tag{3.45}$$

considering the summations in Eqs. (3.44) and (3.45) at each point x being of finite and the exchange of operator P_n and the derivatives, we have

$$\sum_{j\in Z_\Omega} f_{n,j}^{(i)}\varphi_{n,j-M_1}(x)=\sum_{j\in Z_\Omega} f_{n,j}\varphi_{n,j-M_1}^{(i)}(x),\ 1\le i\le m. \tag{3.46}$$

In Sect. 2.5.3, the values of connection coefficients

$$\Gamma_{k,j}^{n,i}=\int_0^1 \varphi_{n,k}^{(i)}(x)\varphi_{n,j}(x)dx,\quad 1\le i\le m \tag{3.47}$$

has been obtained. Multiplying the scaling functions $\varphi_{n,k-M_1}(x)$ to Eq. (3.46) and taking the integral on x over [0, 1] to the induced equation, we get

$$f_{n,k}^{(i)}=\sum_{j\in Z_k} f_{n,j}\int_0^1 \varphi_{n,j-M_1}^{(i)}(x)\varphi_{n,k-M_1}(x)dx=\sum_{j\in Z_k} f_{n,j}\Gamma_{j-M_1,k-M_1}^{n,i},\quad k\in Z. \tag{3.48}$$

Since $f_{n,k}^{(i)}\approx P_n\{\frac{d^i f(x)}{dx^i}\}_{x=j2^n}=\sum\limits_{j\in Z_k} f_{n,j}^{(i)}\varphi_{\tilde{N}}(k-j+M_1)$ and $\mathrm{Supp}[\varphi_{\tilde{N}}(x)]=[0,\tilde{N}]$, we get $Z_k(=\{i;i=k-\tilde{N}+1+M_1,k-\tilde{N}+2+M_1,\ldots,k-1+M_1\})$ which is a finite subset of integer set Z.

Substitution Eqs. (3.48) into (3.42) induces

$$\sum_{i=1}^{m} a_{i,nk}\sum_{j\in Z_k}\Gamma_{j-M_1,k-M_1}^{n,i} f_{n,j}+a_{m,nk}f_{n,k}+N[f_{n,k})]=q_{n,k},k\in \mathrm{Z}. \tag{3.49}$$

Now let us select a set of finite algebraic equation on unknowns $f_{n,k}$ from Eq. (3.49) with infinite equations. Due to that $f_{n,k}=f(k2^{-n})$ $(k=0,1,2,\ldots,2^n)$ are the values of $f(x)$ at the dyadic points in [0, 1], hence such selected system of algebraic equations should contain these unknowns. Considering the region of each base scaling function transitioned in [0, 1] at the resolution level n as shown in Fig. 3.1, it is better for us to select the algebraic equations with k index in the region of $k=0,1,2,\ldots,2^n$. To such selection, we rewrite Eq. (3.49) in the form

$$\sum_{i=1}^{m} a_{i,nk}\sum_{j\in Z_k}\Gamma_{j-M_1,k-M_1}^{n,i} f_{n,j}+a_{m,nk}f_{n,k}+N[f_{n,k})]=q_{n,k},\ k\in \mathrm{Z}_{[0,1]}, \tag{3.50}$$

where $\mathrm{Z}_{[0,1]}=\{k;k=0,1,2,\ldots,2^n\}$.

To the summation term in Eq. (3.50), we introduce the notations of the form $A_{kj}=\sum\limits_{i=1}^{m} a_{i,nk}\Gamma_{j-M_1,k-M_1}^{n,i}+a_{m,nk}\delta_{j,k}$. Then, Eq. (3.50) can be explicitly rewritten as follows:

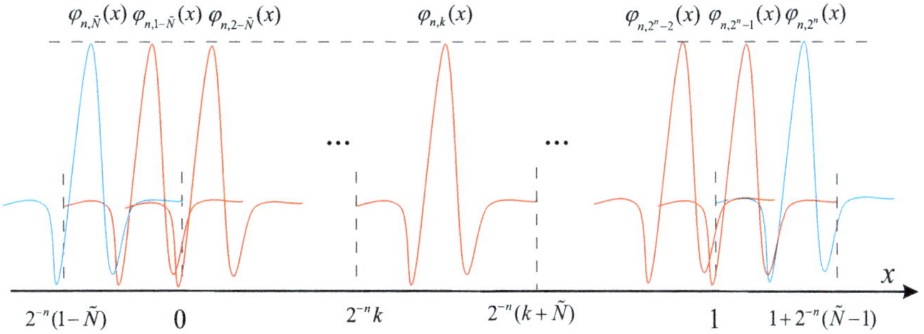

Fig. 3.1 Diagram of base scaling function $\varphi_{n.k}(x)$ translating with index k for the generalized Coiflets when $\tilde{N} = 17$ and $x \in [0, 1]$. To this case, $0 \leq k \leq 2^n$

$$\begin{aligned} &A_{k-\tilde{N}+1+M_1,k} f_{n,k-\tilde{N}+1+M_1,} + A_{k-\tilde{N}+2+M_1,,k} f_{n,k-\tilde{N}+2+M_1,} + \cdots A_{k,k} f_{n,k} \\ &\quad + \cdots + A_{k-1+M_1,k} f_{n,k-1+M_1} + N[f_{n,k}) = q_{n,k}, \end{aligned} \tag{3.51}$$

where $k = 0, 1, 2, \ldots, 2^n$. Denote

$$[\mathbf{f}_L]_0 = [f_{n,-\tilde{N}+1+M_1}, f_{n,-\tilde{N}+2+M_1}, \cdots f_{n,-1}]^T, \tag{3.52}$$

$$[\mathbf{f}_L]_1 = [f_{n,2^n+1}, f_{n,2^n+2}, \ldots, f_{n,2^n+M_1-1}]^T \tag{3.53}$$

$$[\mathbf{f}_L]_{(2^n+1)\times 1} = [f_{n,0}, f_{n,1}, \ldots, f_{n,2^n}]^T, \tag{3.54}$$

$$[\mathbf{q}]_{(2^n+1)\times 1} = \left[q_{n,0}, q_{n,1}, \ldots, q_{n,2^n}\right]^T \tag{3.55}$$

$$[\mathbf{N}(\mathbf{f}_L)]_{(2^n+1)\times 1} = \left[N(f_{n,0}), N(f_{n,1}), \ldots, N(f_{n,2^n})\right]^T \tag{3.56}$$

$$[\mathbf{A}]_{00} = \begin{bmatrix} A_{-\tilde{N}+1+M_1,0} & A_{-\tilde{N}+2+M_1,0} & A_{-\tilde{N}+3M_1,0} & \cdots & A_{-3,0} & A_{-2,0} & A_{-1,0} \\ 0 & A_{-\tilde{N}+2M_1,1} & A_{-\tilde{N}+3M_1,1} & \cdots & A_{-3,1} & A_{-2,\,1} & A_{-1,1} \\ 0 & 0 & A_{-\tilde{N}+3M_1,2} & \cdots & A_{-3,2} & A_{-2,2} & A_{-1,2} \\ \vdots & \vdots & \vdots & \cdots & \vdots & \vdots & \vdots \\ 0 & 0 & 0 & \cdots & A_{-3,\,\tilde{N}-4-M_1} & A_{-2,\,\tilde{N}-4-M_1} & A_{-1,\,\tilde{N}-3-M_1} \\ 0 & 0 & 0 & \cdots & 0 & A_{-2,\,\tilde{N}-3-M_1} & A_{-1,\,\tilde{N}-3-M_1} \\ 0 & 0 & 0 & \cdots & 0 & 0 & A_{-1,\,\tilde{N}-2-M_1} \end{bmatrix}_{(\tilde{N}-1-M_1)\times(\tilde{N}-1-M_1)} \tag{3.57}$$

$$[\mathbf{A}]_{11} = \begin{bmatrix} A_{2^n+1,2^n-M_1+1} & 0 & \cdots & 0 & 0 & 0 \\ A_{2^n+1,2^n+2-M_1} & A_{2^n+2,2^n+2-M_1} & \cdots & 0 & 0 & 0 \\ A_{2^n+1,2^n+3-M_1} & A_{2^n+2,2^n+3-M_1} & \cdots & 0 & 0 & 0 \\ \vdots & \vdots & \vdots & \vdots & \vdots & \\ A_{2^n+1,2^n-2} & A_{2^n+2,2^n-2} & & A_{2^n-2+M_1,2^n-2} & 0 & 0 \\ A_{2^n+1,2^n-1} & A_{2^n+2,2^n-1} & \cdots & A_{2^n-2+M_1,2^n-1} & A_{2^n-1+M_1,2^n-1} & 0 \\ A_{2^n+1,2^n} & A_{2^n+2,2^n} & \cdots & A_{2^n-2+M_1,2^n} & A_{2^n-1+M_1,2^n-1} & A_{2^n+M_1,2^n} \end{bmatrix}_{(M_1-1)\times(M_1-1)} \tag{3.58}$$

$$[\mathbf{A}]=\begin{bmatrix}
A_{0,0} & \cdots & A_{\tilde{N}-1-M_1,0} & 0 & 0 & \cdots & 0 & \cdots & 0 & 0 & 0 & 0 & 0 \\
\vdots & \cdots & \vdots & \times & 0 & \cdots & \vdots & \cdots & \vdots & \vdots & \vdots & \vdots & \vdots \\
A_{0,\tilde{N}-1-M_1} & \cdots & A_{\tilde{N}-1-M_1,\tilde{N}-1-M_1} & \cdots & A_{2\tilde{N}-2-M_1,\tilde{N}-1-M_1} & 0 & 0 & \cdots & \cdots & \cdots & 0 & \cdots & 0 \\
0 & \times & \vdots & \vdots & \vdots & \times & 0 & \cdots & \vdots & \vdots & \vdots & \cdots & \vdots \\
0 & 0 & A_{\tilde{N}-1-M_1,2\tilde{N}-2-M_1} & \cdots & A_{2\tilde{N}-2-M_1,2\tilde{N}-2-M_1} & \cdots & A_{3\tilde{N}-3-M_1,2\tilde{N}-2-M_1} & 0 & \cdots & \cdots & 0 & & 0 \\
\vdots & \vdots & 0 & \times & \vdots & \cdots & \vdots & \times & 0 & \cdots & \vdots & \cdots & \vdots \\
0 & 0 & 0 & 0 & A_{2\tilde{N}-2-M_1,3\tilde{N}-3-M_1} & \cdots & A_{3\tilde{N}-3-M_1,3\tilde{N}-3-M_1} & \cdots & \times & 0 & 0 & \cdots & 0 \\
\vdots & \vdots & \vdots & \cdots & 0 & \times & \vdots & \cdots & \vdots & \times & 0 & & \vdots \\
0 & 0 & 0 & \cdots & 0 & 0 & \times & \cdots & \cdots & \cdots & A_{2^n-M_1,2^n-\tilde{N}-M_1} & 0 & 0 \\
\vdots & \vdots & \vdots & \cdots & \vdots & \cdots & 0 & \times & \cdots & \vdots & \vdots & \times & 0 \\
0 & 0 & 0 & \cdots & 0 & \cdots & 0 & 0 & A_{2^n-M_1,2^n-\tilde{N}-M_1} & \cdots & A_{2^n-M_1,2^n-M_1} & \cdots & A_{2^n,2^n-M_1} \\
\vdots & \vdots & \vdots & \cdots & \vdots & \cdots & \vdots & \cdots & 0 & \cdots & \vdots & \cdots & \vdots \\
0 & 0 & 0 & \cdots & 0 & \cdots & 0 & \cdots & \cdots & 0 & A_{2^n-M_1,2^n} & & A_{2^n,2^n}
\end{bmatrix}. \tag{3.59}$$

It is obvious that the coefficient matrix $[\mathbf{A}]_{(2^n+1)\times(2^n+1)}$ in Eq. (3.58) is a bound-type matrix and those nonzero elements are symmetrically distributed about its diagonal line. Further, denote

$$\left\{[\tilde{\mathbf{A}}]_{00}\right\}_{(2^n+1)\times(\tilde{N}-1-M_1)} = \begin{bmatrix} \{[\mathbf{A}]_{00}\}_{(\tilde{N}-1-M_1)\times(\tilde{N}-1-M_1)} \\ \mathbf{0}_{(2^n+2-\tilde{N}+M_1)\times(\tilde{N}-1-M_1)} \end{bmatrix} \tag{3.60}$$

$$[\tilde{\mathbf{A}}]_{11}=\begin{bmatrix} \mathbf{0}_{(2^n-M_1+2)\times(M_1-1)} \\ \{[\mathbf{A}]_{11}\}_{(M_1-1)\times(M_1-1)} \end{bmatrix}. \tag{3.61}$$

Then, Eq. (3.53) can be rewritten by the matrix form

$$\begin{aligned} &\left\{[\tilde{\mathbf{A}}]_{00}\right\}_{(2^n+1)\times((\tilde{N}-1-M_1)}\{[\mathbf{f}_L]_0\}_{(\tilde{N}-1-M_1)\times 1} + [\mathbf{A}]_{(2^n+1)\times(2^n+1)}[\mathbf{f}_L]_{(2^n+1)\times 1} \\ &\quad +\{[\tilde{\mathbf{A}}]_{11}\}_{(2^n+1)\times(M_1-1)}\{[\mathbf{f}_L]_1\}_{(1-1)\times 1}+[\mathbf{E}]_{(2^n+1)\times(2\tilde{N}+1)}[\mathbf{N}(\mathbf{f}_L)]_{(2^n+1)\times 1} = [\mathbf{q}]_{(2^n+1)\times 1}. \end{aligned} \tag{3.62}$$

Here, $[\mathbf{E}]_{(2^n+1)\times(2\tilde{N}+1)}$ represents an identity matrix.

According to the above deduction process, one finds that the system of nonlinear algebraic equations obtained in Eq. (3.62) for the unknowns $f_{n,k}$ in $[\mathbf{f}_L]_{(2^n+1)\times 1}$ are held for any resolution level n and all truncation errors are automatically disappeared in the process. Thus, the solution of Eq. (3.62) is independent on the truncation errors. According to Definition 3.5, we know that such approximate solution based on (3.62) is closed.

3.5.3 Some Discussions for the Wavelet-Based Approximate Solution Program

(**I**) *Relevant parts that have to be dealt with by a boundary extension technique*

It should be noted that there are some unknowns $\{[\mathbf{f}_L]_0\}_{(\tilde{N}-1-M_1)\times 1}$ and $\{[\mathbf{f}_L]_1\}_{(M_1-1)\times 1}$ of function $f(x)$ outside [0, 1] which are contained in Eq. (3.62). If one expects to get a better solution with satisfied accuracy, such outside values should be suitably selected. For example, the simplest selection is $\{[\mathbf{f}_L]_0\}_{(\tilde{N}-1-M_1)\times 1} = \mathbf{0}$ and $\{[\mathbf{f}_L]_1\}_{(M_1-1)\times 1} = \mathbf{0}$, which may induce some oscillation $P_n f(x)$ near the ends of [0, 1] in theory, liking the Gibbs phenomenon in the Fourier transform. Hence, the better solution is to select $\{[\mathbf{f}_L]_0\}_{(\tilde{N}-1-M_1)\times 1} \neq \mathbf{0}$ and $\{[\mathbf{f}_L]_1\}_{(M_1-1)\times 1} \neq \mathbf{0}$ with a continuous extension technology as done in other numerical methods. In such case, the better selection for M_1 is to make the element numbers of $\{[\mathbf{f}_L]_0\}_{(\tilde{N}-1-M_1)\times 1}$ and $\{[\mathbf{f}_L]_1\}_{(M_1-1)\times 1}$ be nearly equal to each other. As shown in Table 2.3, it is luck for us to find that real solutions of the filter coefficient $p_k(k = 0, 1, 2, \ldots, \tilde{N})$ when $\tilde{N} = 5, 11, 17$ for either $M_1 = 4$ or $M_1 = 7$, respectively.

Here, we give a general description for the boundary extension with a transformed form to formulate $\{[\mathbf{f}_L]_0\}_{(\tilde{N}-1-M_1)\times 1}$ and $\{[\mathbf{f}_L]_1\}_{(M_1-1)\times 1}$ in terms of the unknowns $[\mathbf{f}_L]_{(2^n+1)\times 1}$ in [0, 1]. Denote the matrices of such continuous extension by

$$\begin{gathered}\{[\mathbf{f}_L]_0\}_{(\tilde{N}-1-M_1)\times 1} = [\mathrm{T}_0]_{(\tilde{N}-1-M_1)\times(\tilde{N}-1-M_1)}[\tilde{\mathbf{f}}_L]_0\}_{(\tilde{N}-1-M_1)\times 1},\\ [\tilde{\mathbf{f}}_L]_0\}_{(\tilde{N}-1-M_1)\times 1} \in [\mathbf{f}_L]_{(2^n+1)\times 1}\end{gathered} \tag{3.63}$$

$$\{[\mathbf{f}_L]_1\}_{(M_1-1)\times 1} = [\mathrm{T}_1]_{(M_1-1)\times(M_1-1)}[\tilde{\mathbf{f}}_L]_1\}_{(M_1-1)\times 1},\ [\tilde{\mathbf{f}}_L]_1\}_{(M_1-1)\times 1} \in [\mathbf{f}_L]_{(2^n+1)\times 1}. \tag{3.64}$$

Here, $[\mathrm{T}_0]_{(\tilde{N}-1-M_1)\times(\tilde{N}-1-M_1)}$ and $[\mathrm{T}_1]_{(M_1-1)\times(M_1-1)}$ are the transform matrices relevant to the continuous extension technique selected, while $[\tilde{\mathbf{f}}_L]_0\}_{(\tilde{N}-1-M_1\times 1}$ and $\{[\mathbf{f}_L]_1\}_{(M_1-1)\times 1}$ are the selected unknowns in $[\mathbf{f}_L]_{(2^n+1)\times 1}$ at the extrapolation points. Once an extension technique is selected, the transform matrices can be gained soon afterward. Substituting Eqs. (3.63) and (3.64) into Eq. (3.62), one gets

$$\begin{aligned}&\left\{[\tilde{\mathbf{A}}]_{00}\right\}_{(2^n+1)\times(\tilde{N}-1-M_1)}[\mathrm{T}_0]_{(\tilde{N}-1-M_1)\times(\tilde{N}-1-M_1)}[\tilde{\mathbf{f}}_L]_0\}_{(\tilde{N}-1-M_1)\times 1}\\ &+[\mathbf{A}]_{(2^n+1)\times(2^n+1)}[\mathbf{f}_L]_{(2^n+1)\times 1} + \{[\tilde{\mathbf{A}}]_{11}\}_{(2^n+1)\times(\tilde{N}-1)}[\mathrm{T}_1]_{(M_1-1)\times(M_1-1)}[\tilde{\mathbf{f}}_L]_1\}_{(M_1-1)\times 1}\\ &\quad+[\mathbf{E}]_{(2^n+1)\times(2\tilde{N}+1)}[\mathbf{N}(\mathbf{f}_L)]_{(2^n+1)\times 1} = [\mathbf{q}]_{(2^n+1)\times 1}\end{aligned} \tag{3.65}$$

which evidently displays that only those coefficients in $[\mathbf{A}]_{(2^n+1)\times(2^n+1)}$ corresponding to the extrapolation points employed in the continuous extension technique are reformed, and all unknowns are of $[\mathbf{f}_L]_{(2^n+1)\times 1}$. After that, Eq. (3.65) associated with the boundary conditions in a specified problem can be solved to get the unknowns in $[\mathbf{f}_L]_{(2^n+1)\times 1}$, and finally the approximate solution $f_{n,k} = f(x)|_{x=k2^{-n}}$ $(k = 0, 1, \ldots, 2^n)$ with higher precision can be gained to $f(x)$ in [0, 1].

Obviously, such obtained approximate solution to a boundary-value problem is dependent on the accuracy of the continuous extension method selected. In the next chapter, we will describe its selection method, containing its error estimation combined with the error analysis of the generalized Coiflets. From it, one will find an optimal extension method such that its precision order is equal to that of the generalized Coiflets used in the solving process.

(II) *Treatment of derivatives of unknown function contained in the nonlinear operator*

If the derivative with order lower than m is contained in the nonlinear operator$N[(f(x)]$, denoted by $N[(f(x), f^{(l)}(x)]$ $(1 \le l < m)$, then when the expansion of $N[(f(x), f^{(l)}(x)]$ is applied following the previous program, the above treatment to the $f^{(l)}(x)$ is fully same as Eq. (3.47). After that, the above induced Eq. (3.52) can be replaced by the form

$$\begin{aligned}&\sum_{i=1}^{m} a_{i,nk} \sum_{j\in Z_k} \Gamma^{n,i}_{j-M_1,k-M_1} f_{n,j} + a_{m,nk} f_{n,k} + N[f_{n,k}, \sum_{j\in Z_k} \Gamma^{n,l}_{j-M_1,k-M_1} f_{n,j})] \\ &\quad = q_{n,k},\ k \in \mathrm{Z}_{[0,1]}.\end{aligned} \tag{3.66}$$

In such situation, the coefficient matrix for $[\mathbf{N}(\mathbf{f}_L)]_{(2^n+1)\times 1}$ is a matrix similar to that obtained for $[\mathbf{f}_L]_{(2^n+1)\times 1}$ rather than the identity matrix. For example, when $N[f(x), f^{(l)}(x)] = f(x)f^{(1)}(x)$, one can immediately write

$$N[f_{n,k}, \sum_{j\in Z_k} \Gamma^{n,1}_{j-M_1,k-M_1} f_{n,j})] = f_{n,k} \sum_{j\in Z_k} \Gamma^{n,1}_{j-M_1,k-M_1} f_{n,j}, \quad k = 0, 1, 2, \ldots, 2^n.$$

Denote $B_{kj} = \sum_{j\in Z_k} \Gamma^{n,1}_{j-M_1,k-M_1} f_{n,j}$. One can see that the form of B_{kj} are all fully similar to A_{kj} except for different values between their elements. Finally, the matrix of $[\mathbf{B}] = [B_{kj}]$ is fully similar to $[\mathbf{A}] = [A_{kj}]$ which has been explicitly gained. It should be pointed out that $N[f(x), f^{(l)}(x)] = f(x)f^{(1)}(x)$ should be extended to the outside [0, 1] as done as the unknown function $f(x)$. Here, we omit them for save space.

(III) *Properties of closure and convergence for the wavelet-based solution method*

Definition 3.9 If an approximate solution method makes the system of induced N algebraic equations only on N unknowns appeared in the approximate solution and all the truncation errors in the solution method disappeared automatically, we call such approximate solution method as strongly closed. Otherwise, if only all truncation errors have automatically disappeared by the solution method but there are more than

N unknowns in the method that are contained in the induced N algebraic equations, such approximate solution method is referred to be weakly closed.

Following Definition 3.9, we know that the approximate solution in our method based on the generalized Coiflets is weakly closed in general situation. When the outside values of $\{[\mathbf{f}_L]_0\}_{(\tilde{N}-1-M_1)\times 1}$ and $\{[\mathbf{f}_L]_1\}_{(M_1-1)\times 1}$ are all taken by zero, the number of unknowns in Eq. (3.65) is equal to the number of equations, hence, the solution method is strongly closed. As displayed previously, such strongly closed solution is not good in accuracy. Hence, we usually take $\{[\mathbf{f}_L]_0\}_{(\tilde{N}-1-M_1)\times 1} \neq \mathbf{0}$ and $\{[\mathbf{f}_L]_1\}_{(M_1-1)\times 1} \neq \mathbf{0}$ by means of a continuous extension mentioned above.

Even when such approximate solution method is weakly closed, however, such weakly closed property has not been found in the conventional numerical method for solving a nonlinear problem as shown in Chap. 1, such as FEM and FDM even when they are applied in solving a linear boundary-value problem since the truncation errors in the solution methods are not automatically disappeared. Hence, the conventional numerical methods are not closed such that their approximate solutions are obviously dependent on their truncation errors, which may yield some implicit errors to their approximate solutions. It is this great possibility that such implicit errors in the conventional methods make their efficiency lose when they are directly applied to solve a strong nonlinear problem without adding some specific techniques.

By the way, we should point out that the convergence of the approximate solution method in Eq. (3.65) for solving a boundary-value problem, either linear to nonlinear, is implicitly contained in the establishment of the generalized Coiflets, which has been demonstrated by the relevant mathematician such as Daubechies for the framework of compactly supported orthogonal wavelets [10]. From Definition 2.2, such established wavelets have the property of the form

$$\{0\}\cdots\subset V_{-2}\subset V_{-1}\subset V_0\subset V_1\subset\cdots\subset L^2(R),\quad V_n=\{\sum_k c_{n,k}\varphi_{n,k}(x),\, n,k\in Z\}, \tag{3.67}$$

which means that the approximate solution $f(x)\approx\sum_k c_{n,k}\varphi_{n,k}(x)$ is convergent to $L^2(R)$.

3.6 Wavelet Closed Solution Method for 2D and 3D Nonlinear Boundary-Value Problems

After the wavelet-based closed solution method for solving nonlinear boundary-value problems with ordinary differential equations or 1D boundary-value problems are established, there is no difficulty for us to generalize this closed numerical method into those 2D and 3D boundary-value problems. In such generalization, the key problem is the generalization of base functions in the wavelet on the basis of 1D wavelet established. In this section, we give this generalization of what is done similar to those in literature except for those closed properties.

3.6.1 Two-Dimensional Generalized Coiflets

Without loss of generality, the bounded region for a 2D boundary-value problem is denoted by $(x, y) \in [0, 1] \times [0, 1] \subset R^2$ similar to the one-dimensional situation conducted previously.

Definition 3.10 For $\forall(x, y) \in R^2 = (-\infty, \infty) \times (-\infty, \infty)$, the two-dimensional base scaling functions $\Phi_{n,jk}(x, y)$and the relevant base wavelet functions $\Psi_{n,jk}(x, y)$ are defined as follows:

$$\Phi_{n,jk}(x, y) = \varphi_{n,j}(x)\varphi_{n,k}(y),\ \Psi_{0,n,jk}(x, y) = \psi_{n,j}(x)\psi_{n,k}(y),\ j, k \in Z \tag{3.68}$$

$$\Psi_{1,n,jk}(x, y) = \varphi_{n,j}(x)\psi_{n,k}(y), \quad \Psi_{2,n,jk}(x, y) = \psi_{n,k}(x)\varphi_{n,j}(y), \tag{3.69}$$

where $\varphi_{n,j}(x)$,$\varphi_{n,j}(y)$ and $\psi_{n,j}(x)$, and $\psi_{n,j}(y)$ are the base scaling functions and the base wavelet functions obtained in the one-dimensional case, respectively.

Following the orthogonality of the base functions of the 1D orthogonal wavelets, there is no difficulty for one to give the demonstration of orthogonal properties for the 2D base functions defined in Definition 3.10 as follows:

$$\int_{R^2} \Phi_{n,jk}(x, y)\Phi_{n,rs}(x, y)dxdy = \delta_{rs}^{jk} = \delta_{j,r}\delta_{k,s},\ j, k, r, s \in Z \tag{3.70}$$

$$\int_{R^2} \Psi_{0,n,jk}(x, y)\Psi_{0,m,rs}(x, y)dxdy = \delta_{mrs}^{njk} = \delta_{m,n}\delta_{j,r}\delta_{k,s},\ j, k, r, s \in Z, m \geq n \tag{3.71}$$

$$\int_{R^2} \Phi_{n,jk}(x, y)\Psi_{i,m,rs}(x, y)dxdy = 0,\ j, k, r, s \in Z, m \geq n; i = 0, 1, 2. \tag{3.72}$$

Definition 3.11 For $\forall f(x, y) \in L^2(R^2)$, i.e.,$\int_{-\infty}^{\infty}\int_{-\infty}^{\infty} |f(x, y)|^2 dxdy < \infty$, for integer $n \geq 0$, we define the following projection:

$$P_n f(x, y) = \sum_{j \in Z}\sum_{k \in Z} c_{n,jk}\Phi_{n,jk}(x, y) \tag{3.73}$$

$$Q_{n,i} f(x, y) = \sum_{j \in Z}\sum_{k \in Z} d_{i,n,jk}\Psi_{i,n,jk}(x, y) \quad ,i = 0, 1, 2 \tag{3.74}$$

in which the decomposition coefficients under the base functions defined in Definition 3.10 have the form

$$c_{n,jk} = \int_{R^2} f(x, y)\Phi_{n,jk}(x, y)dxdy, \quad j, k \in Z \tag{3.75}$$

$$d_{i,n,jk} = \int_{R^2} f(x, y)\Psi_{i,n,jk}(x, y)dxdy, \ j, \ k \in Z; i = 0, 1, 2. \tag{3.76}$$

Then the function $f(x, y)$ can be formulated by the form

$$f(x, y) = P_n f(x, y) + \sum_{j=n}^{\infty}\sum_{i=0}^{2} Q_{j,i} f(x, y) = P_n f(x, y) + \tilde{Q}_n f(x, y). \tag{3.77}$$

Here.

$$\tilde{Q}_n f(x, y) = \sum_{j=n}^{\infty}\sum_{i=0}^{2} Q_{j,i} f(x, y). \tag{3.78}$$

After $Q_n f(x, y) = \sum_{i=0}^{2} Q_{n,i} f(x, y)$ is set at the resolution level n, similar to 1D projection relationship, we can get the same formula of 1D case in R^2, i.e.,

$$P_{n+1} f(x, y) = P_n f(x, y) + Q_n f(x, y) \ x, y \in R^2. \tag{3.79}$$

Definition 3.12 The function spaces spread by the base scaling functions $\{\Phi_{n,jk}(x, y)\}$and the base wavelet functions $\{\Psi_{n,jk}(x, y)\}$ are, respectively, denoted by the form.

$$V_n = \{\sum_{j,k\in Z} c_{jk}\Phi_{n,jk}(x, y); \ j, k \in Z; c_{jk} \in R\} \tag{3.80}$$

$$W_n = \{\sum_{j,k\in Z} d_{i,jk}\Psi_{i,n,jk}(x, y); \ j, k \in Z; d_{i,jk} \in R; i = 0, 1, 2\}. \tag{3.81}$$

Then, all properties of projection in the one-dimensional case are also true in 2D generation, that is,

$$V_{n+1} = V_n \oplus W_n, \ P_{n+1} f(x, y) = P_n f(x, y) + Q_n f(x, y) \tag{3.82}$$

and $\{0\} \cdots \subset V_{-2} \subset V_{-1} \subset V_0 \subset V_1 \subset \cdots \subset L^2(R)$, or,

$$f(x, y) = \lim_{n\to\infty} P_n f(x, y) = P_n f(x, y) + \tilde{Q}_n f(x, y), \tag{3.83}$$

which means the approximation $P_n f(x, y)$ is convergent to $L^2(R^2)$ when the resolution level n approaches to infinite.

When the one-dimensional generalized Coiflets with compact support $[0, \tilde{N}]$ is selected to the 2D case, then there is no difficulty for one to get the decomposition coefficients like Eq. (2.31) by the form

$$c_{n,jk} = \int_{R^2} f(x, y)\Phi_{n,jk}(x, y)dxdy = \int_{R^2} f(x, y)\varphi_{n,j}(x)\varphi_{n,k}(y)dxdy = 2^{-n} f(\tfrac{M_1+j}{2^n}, \tfrac{M_1+k}{2^n}) = 2^{-n} f_{n,(j+M_1)(k+M_1)}. \tag{3.84}$$

Here, $f_{n,jk} = f(j2^{-n}, k2^{-n})$. Substituting Eqs. (3.84) into (3.73), one further gets

$$P_n f(x, y) = 2^{-n} \sum_{j\in Z} \sum_{k\in Z} f_{n,jk} \Phi_{n,(j-M_1)(k-M_1)}(x, y). \tag{3.85}$$

When $f(x, y)$ is defined on the finite domain $[0, 1] \times [0, 1]$, as done in an one-dimensional case, the summation in Eq. (3.85) is finite at each one point in the domain. In such situation, once the summation indexes are selected by a suitable region such that the domain of expression of Eq. (3.85) contains $[0, 1] \times [0, 1]$, the approximation of Eq. (3.85) defined in R^2 is almost unchanged for the finite domain except for the summation indexes. That is,

$$\begin{aligned} P_n f(x, y) &= 2^{-n} \sum_{j\in Z_\Omega} \sum_{k\in Z_\Omega} f_{n,jk} \Phi_{n,(j-M_1)(k-M_1)}(x, y) \\ &= 2^{-n} \sum_{j\in Z_\Omega} \sum_{k\in Z_\Omega} f_{n,jk} \varphi_{\tilde{N}}(2^n x - j + M_1)\varphi_{\tilde{N}}(2^n y - k + M_1). \end{aligned} \tag{3.86}$$

For example, to the dyadic points $(x, y) = (r2^{-n}, s2^{-n})$ $(r, s \in [0, 2^n])$, the summation calculation of Eq. (3.86) can be explicitly rewritten as

$$P_n f(r2^{-n}, s2^{-n}) = 2^{-n} \sum_{j\in Z_\Omega} \sum_{k\in Z_\Omega} \varphi_{\tilde{N}}(r - j + M_1)\varphi_{\tilde{N}}(s - k + M_1). \tag{3.87}$$

It can be seen from Eq. (3.87) that to each specified point $(r2^{-n}, s2^{-n})$, the number of summation calculations is $(\tilde{N} - 1)^2$, further the total summation number of the reconstruction of Eq. (3.87) in $[0, 1] \times [0, 1]$ is $(\tilde{N} - 1)^2 2^{2n}$.

For a nonlinear operator of function $f(x, y)$, denot it by $N[f(x, y)]$. By using the formula of Eq. (3.86), similar to one-dimensional expansion, one can immediately gain the approximation and full expansion of $N[f(x, y)]$ as follows:

$$P_n N[f(x,y)]{=}2^{-n}\sum_{j\in Z_\Omega}\sum_{k\in Z_\Omega} N[f_{n,jk}]\Phi_{n,(j-M_1)(k-M_1)}(x,y) \tag{3.88}$$

$$N[f(x,y)] = P_n N[f(x,y)] + \tilde{Q}_n N[f(x,y)] = P_n N[f(x,y)] + \eta_{N[f]}(x,y). \tag{3.89}$$

3.6.2 Three-Dimensional Generalized Coiflets

Following the generalization in the above subsection for 2D base scaling functions and base wavelet functions under the generalized Coiflets, there is no difficulty for one to do a similar program once the 3D base scaling functions and base wavelet functions are defined in 3D generalization, while all properties what the 1D case has are ensured by the generalization. As follows, we directly write the relevant formulae in the conclusions rather than giving their demonstration for saving space.

Here, we extend the base scaling and wavelet functions in a situation similar to that done in the two dimensions. In such generalization, the 3D base scaling functions and base wavelet functions are defined as follows:

$$\Phi_{n,jkl}(x,y,z) = \varphi_{n,j}(x)\varphi_{n,k}(y)\varphi_{n,l}(z),\quad \forall(x,y,z)\in R^3 = (-\infty,\infty)^3 \tag{3.90}$$

$$\Psi_{0,n,jkl}(x,y,z) = \psi_{n,j}(x)\psi_{n,k}(y)\psi_{n,k}(z),\quad \forall(x,y,z)\in R^3 = (-\infty,\infty)^3 \tag{3.91}$$

$$\Psi_{1,n,jkl}(x,y,z) = \varphi_{n,j}(x)\psi_{n,k}(y)\psi_{n,k}(z),\quad \forall(x,y,z)\in R^3 = (-\infty,\infty)^3 \tag{3.92}$$

$$\Psi_{2,n,jkl}(x,y,z) = \psi_{n,j}(x)\varphi_{n,k}(y)\psi_{n,k}(z),\quad \forall(x,y,z)\in R^3 = (-\infty,\infty)^3 \tag{3.93}$$

$$\Psi_{3,n,jkl}(x,y,z) = \psi_{n,j}(x)\psi_{n,k}(y)\varphi_{n,k}(z),\quad \forall(x,y,z)\in R^3 = (-\infty,\infty)^3 \tag{3.94}$$

$$\Psi_{4,n,jkl}(x,y,z) = \varphi_{n,j}(x)\varphi_{n,k}(y)\psi_{n,k}(z),\quad \forall(x,y,z)\in R^3 = (-\infty,\infty)^3 \tag{3.95}$$

$$\Psi_{5,n,jkl}(x,y,z) = \varphi_{n,j}(x)\psi_{n,k}(y)\varphi_{n,k}(z),\quad \forall(x,y,z)\in R^3 = (-\infty,\infty)^3 \tag{3.96}$$

$$\Psi_{6,n,jkl}(x,y,z) = \psi_{n,j}(x)\varphi_{n,k}(y)\varphi_{n,k}(z),\quad \forall(x,y,z)\in R^3 = (-\infty,\infty)^3. \tag{3.97}$$

They have the following orthogonal properties:

$$\int_{R^3}\Phi_{n,jkl}(x,y,z)\Phi_{n,rst}(x,y,z)dxdydz = \delta_{rst}^{jkl} = \delta_{j,r}\delta_{k,s}\delta_{l,t} \tag{3.98}$$

$$\int_{R^3} \Psi_{0,n,jkl}(x,y,z)\Psi_{0,m,rst}(x,y,z)dxdydx = \delta_{mrst}^{njkl} = \delta_{m,n}\delta_{j,r}\delta_{k,s}\delta_{l,t},\ m \geq n \tag{3.99}$$

$$\int_{R^3} \Phi_{n,jkl}(x,y,z)\Psi_{i,m,rst}(x,y,z)dxdydz = 0,\ m \geq n;\ i = 0, 1, \ldots, 6. \tag{3.100}$$

For $\forall f(x, y, z) \in L^2(R^3)$, its expansion can be formulated in the form

$$\begin{aligned} f(x,y) &= P_n f(x,y,z) + \tilde{Q}_n f(x,y,z) \\ &= P_n f(x,y,z) + \eta_f(x,y,z) \end{aligned} \tag{3.101}$$

$$\begin{aligned} P_n f(x,y,z) &= 2^{-3n/2} \sum_{j,k,l \in Z} f_{n,(j+M_1)(k+M_1)(l+M_1)} \Phi_{n,jkl}(x,y,z) \\ &= 2^{-3n/2} \sum_{j,k,l \in Z} f_{n,jkl} \Phi_{n,(j-M_1)(k-M_1)(l-M_1)}(x,y,z) \end{aligned} \tag{3.102}$$

$$\begin{aligned} f_{n,jkl} &= 2^{3n/2} \int_{R^3} f(x,y,z)\Phi_{n,(j-M_1)(k-M_1)(l-M_1)}(x,y,z)dxdydz \\ &= f(j2^n, k2^n, l2^n) \end{aligned} \tag{3.103}$$

$$d_{i,n,jkl} = \int_{R^3} f(x,y,z)\Psi_{i,n,jkl}(x,y,z)dxdydz, \quad i = 0, 1, \ldots, 6 \tag{3.104}$$

$$Q_{n,i} f(x,y,z) = \sum_{j,k,l \in Z} d_{i,n,jk} \Psi_{i,n,jk}(x,y,z), \quad ,i = 0, 1, \ldots, 6 \tag{3.105}$$

$$Q_n f(x,y,z) = \sum_{i=0}^{6} Q_{n,i} f(x,y,z) \tag{3.106}$$

$$\tilde{Q}_n f(x,y,z) = \sum_{j=n}^{\infty} Q_j f(x,y,z) \tag{3.107}$$

$$V_n = \{ \sum_{j,k,l \in Z} c_{jkl} \Phi_{n,jk}(x,y);\ j,k,l \in Z;\ c_{jkl} \in R \} \tag{3.108}$$

$$W_n = \{ \sum_{j,k.l \in Z} d_{i,jkl} \Psi_{i,n,jkl}(x,y);\ j,k,l \in Z;\ d_{i,jkl} \in R;\ i = 0, 1, \ldots, 6 \} \tag{3.109}$$

$$V_{n+1} = V_n \oplus W_n, \quad P_{n+1} f(x,y) = P_n f(x,y) + Q_n f(x,y) \tag{3.110}$$

$$\{0\} \cdots \subset V_{-2} \subset V_{-1} \subset V_0 \subset V_1 \subset \cdots \subset L^2(R) \tag{3.111}$$

$$f(x, y) = \lim_{n \to \infty} P_n f(x, y) = P_n f(x, y) + \tilde{Q}_n f(x, y). \tag{3.112}$$

When the last equality is used in Eqs. (3.102) and (3.103), it implies that the generalized Coiflets have been employed in the above 3D compactly supported orthogonal wavelets.

According to Eq. (3.101) and (3.102), further, the expansion for any nonlinear operator $N[f(x, y, z)]$ can be immediately written by the form

$$\begin{aligned} N[f(x, y, z)] &= P_n N[f(x, y, z)] + \tilde{Q}_n N[f(x, y, z)] \\ &= P_n N[f(x, y, z)] + \eta_{N[f]}(x, y, z) \end{aligned} \tag{3.113}$$

$$P_n N[f(x, y, z)] = 2^{-3n/2} \sum_{j,k,l \in Z} N[f_{n,jkl}] \Phi_{n,(j-M_1)(k-M_1)(l\ M_1)}(x, y, z). \tag{3.114}$$

Similarly, for a function $f(x, y, z) \in L^2(\mathrm{R}^3)$ in $[0, 1]^3$ at each dyadic point$(r2^{-n}, s2^{-n}, t2^{-n})$ when the indexes r, s, t are specified in the region, the number of summation calculations is $(\tilde{N} - 1)^3$. Hence, the total summations are $(\tilde{N} - 1)^3 2^{3n}$ times.

3.6.3 *Closed Spatial Discretization for Initial-Boundary-Value Problems in 3D Space*

Here, we give a closed decomposition approach for the nonlinear initial-boundary-value problems on the basis of the 3D generalized Coiflets, which has all the calculation properties displayed in 1D case, and the corresponding results can be easily degraded to 1D and 2D nonlinear initial-boundary-value problems. When the time variable is considered in the calculations, one can find that all the decomposition coefficients contain the time variable, that is, all decomposition coefficients are varied with time. In such a case, the decomposition for the space variables is closed.

Without loss of generality, we consider a dynamics problem in a 3D finite space. Let the time variable be $t \in [0, \infty)$ and the three-dimensional space coordinates be $(x, y, z) \in [0.1]^3 \subset R^3$. Then, for any nonlinear initial-boundary-value problem in this domain, we can write its differential equations, boundary conditions, and initial-value conditions of the operator forms

$$\frac{\partial^2 f}{\partial t^2} + L[f(x, y, z, t)] + N[f(x, y, z, t)] = q(x, y, z, t), \ 0 < x, y, z < 1, \ t > 0 \tag{3.115}$$

$$x = 0: \ L_{B10}[f(x, y, z, t)] + N_{B10}[f(x, y, z, t)] = q_{B10}(y, z, t) \tag{3.116}$$

$$x = 1: \ L_{B11}[f(x, y, z, t)] + N_{B11}[f(x, y, z, t)] = q_{B11}(y, z, t) \tag{3.117}$$

$$y = 0: \ L_{B20}[f(x, y, z, t)] + N_{B20}[f(x, y, z, t)] = q_{B20}(x, z, t) \tag{3.118}$$

$$y = 1: \ L_{B21}[f(x, y, z, t)] + N_{B21}[f(x, y, z, t)] = q_{B21}(x, z, t) \tag{3.119}$$

$$z = 0: \ L_{B30}[f(x, y, z, t)] + N_{B30}[f(x, y, z, t)] = q_{B30}(x, y, t) \tag{3.120}$$

$$z = 1: \ L_{B31}[f(x, y, z, t)] + N_{B31}[f(x, y, z, t)] = q_{B31}(x, y, t) \tag{3.121}$$

$$t = 0: f(x, y, z, t) = f_0(x, y, z), \frac{\partial}{\partial t} f(x, y, z, t) = f_{ot}^{'}(x, y, z), \tag{3.122}$$

where $L[\cdot]$ is the linear operator of unknown function $f(x, y, z, t)$ with the partial derivatives with respect to the spatial coordinates and the possible first-order partial derivative on time. In the boundary-value problem of partial differential equations, some boundary conditions in Eqs. (3.116)–(3.121) can also be replaced by other forms according to the measurable way of the problem. For example, in the N–S equation of fluid mechanics, the periodic boundary condition in a certain direction is sometimes used, e.g.,

$$f(x, 1, z, t) = f(x, 0, z, t). \tag{3.123}$$

In short words, such boundary conditions should be sufficient to make the boundary-value problem constitute a definite solution problem.

In addition, if an initial-value problem contains multi-unknown functions, we can use the column notation of the form

$$f(x, y, z, t) = [f_1(x, y, z, t), f_2(x, y, z, t), \ldots f_K(x, y, z, t)]. \tag{3.124}$$

After that, all notations for $f(x, y, z, t)$ can be replaced by this form and the governing equations associated with boundary and initial conditions formulated in Eqs. (3.115)–(3.122) are replaced by their corresponding matrix or operator form. It should be noted that in such a situation of multi-unknown functions, the number of the governing differential equations should be equal to the number of the unknown functions such that the problem is closed for the unknown functions. As follows, all deductions are implied in this foundation if the problem has multi-unknown functions.

When the time variable is considered as a parameter. At a specified each instant, it is evident that all Eqs. (3.101)–(3.112) can be used by replacing the all the decomposition coefficients varied with time, e.g.,

$$c_{n,jkl}(t) = \int_{R^3} f(x, y, z)\Phi_{n,jkl}(x, y, z)dxdydz = 2^{-3n/2} f_{n,(j+M_1)(k+M_1)(l+M_1)}(t).$$

After that, as the same is done in an one-dimensional performance, we can write the following formulae

$$\begin{aligned} f(x, y, z, t) &= P_n f(x, y, z, t) + \tilde{Q}_n f(x, y, z, t) \\ &= 2^{-3n/2} \sum_{j,k,l \in \Omega_Z} f_{n,jkl}(t)\Phi_{n,(j-M_1)(k-M_1)(l-M_1)}(x, y, z) \\ &\quad + \eta_{L[f]}(x, y, z, t) \end{aligned} \tag{3.125}$$

$$\begin{aligned} L[f(x, y, z, t)] &= P_n L[f(x, y, z, t)] + \tilde{Q}_n L[f(x, y, z, t)] \\ &= 2^{-3n/2} \sum_{j,k,l \in \Omega_Z} L[f_{n,jkl}(t)]\Phi_{n,(j-M_1)(k-M_1)(l-M_1)}(x, y, z) \\ &\quad + \eta_{L[f]}(x, y, z, t) \end{aligned} \tag{3.126}$$

$$\begin{aligned} \frac{\partial^2 f(x, y, z, t)}{\partial t^2} &= P_n \frac{\partial^2 f(x, y, z, t)}{\partial t^2} + \tilde{Q}_n \frac{\partial^2 f(x, y, z, t)}{\partial t^2} \\ &= 2^{-3n/2} \sum_{j,k,l \in \Omega_Z} \ddot{f}_{n,jkl}(t)\Phi_{n,(j-M_1)(k-M_1)(l-M_1)}(x, y, z) \\ &\quad + \eta_{\frac{\partial^2 f}{\partial t^2}}(x, y, z, t), \end{aligned} \tag{3.127}$$

where $\ddot{f}_{n,jkl}(t) = \frac{d^2 f_{n,jkl}(t)}{dt^2}$, and $f_{n,jkl}(t) = f(j2^n, k2^n, l2^n, t)$. Then

$$P_n f(x, y, z, t),\ P_n L[f(x, y, z, t)],\ P_n N[f(x, y, z, t)],\ P_n \frac{\partial^2 f(x, y, z, t)}{\partial t^2} \in V_n \tag{3.128}$$

$$\eta_f(x, y, z, t), \eta_{L[f]}(x, y, z, t), \eta_{N[f]}(x, y, z, t),\ \ \eta_{\frac{\partial^2 f}{\partial t^2}}(x, y, z, t) \in \tilde{W}_n \tag{3.129}$$

in which

$$\tilde{W}_n = W_n \cup W_{n+1} \cup \cdots = \sum_{i=n}^{\infty} \cup W_i \tag{3.130}$$

After Eqs. (3.125)–(3.127) are substituted into the differential equation of Eq. (3.115), then the Galerkin method is applied to the induced equation similar to what is done in one dimension, one can get a system of ordinary differential equations on variable time, i.e.,

$$\frac{d^2 f_{n,jkl}(t)}{dt^2} + L[f_{n,jkl}(t)] + N[f_{n,jkl}(t)] = q_{n,jkl}(t) \tag{3.131}$$

in which $0 \leq j, k, l \leq 2^n$ and.

$$\begin{aligned} q_{n,jkl}(t) &= 2^{3n/2} \int_{R^3} q(x, y, z, t)\Phi_{n,(j-M_1)(k-M_1)(l-M_1)}(x, y, z)dxdydz \\ &= q(j2^{-n}, k2^{-n}, l2^{-n}, t) \end{aligned}. \tag{3.132}$$

Since all truncation errors in Eqs. (3.125)–(3.127) have automatically disappeared by the orthogonal properties of the wavelet base functions in the calculations of the Galerkin method, hence, such induced equations in Eq. (3.131) are weakly closed.

To solve the ordinary differential equations in Eq. (3.131), one can employ a conventional numerical integral method, such as the fourth-order explicit Runge–Kutta method, to get the values of all decomposition coefficients $f_{n,jkl}(t)$ varying with time. After that, the numerical solution on the unknown function $f(x, y, z, t)$ is gained at the dyadic points with resolution level n in the defined domain.

3.6.4 Application Example—Closed Decomposition or Solution to the N–S Equations in Fluid Mechanics

Now, we apply the closed wavelet method mentioned above to the well-known Navier–Stokes (N–S) equations in the fluid mechanics.

Without losing generality for our purpose, we restrict our attention on the incompressible viscous flows in 3D space $\mathbf{x} \in \mathbb{R}^3$. Then, we can write the N–S equations as follows [11, 12]:

$$\begin{aligned} \rho\tfrac{\partial \mathbf{u}(\mathbf{x},t)}{\partial t} + \rho\mathbf{u}(\mathbf{x}, t) \cdot \nabla\mathbf{u}(\mathbf{x}, t) = \nu\Delta\mathbf{u}(\mathbf{x}, t) - \nabla p(\mathbf{x}, t) + \mathbf{g}(\mathbf{x}, t) \\ \nabla \cdot \mathbf{u}(\mathbf{x}, t) = 0, \qquad \mathbf{x} \in \mathbb{R}^3,\ t > 0 \end{aligned}. \tag{3.133}$$

Here, the velocity $\mathbf{u} = \mathbf{u}(\mathbf{x}, t) = u(x, y, z, t)\mathbf{i} + v(x, y, z, t)\mathbf{j} + w(x, y, z, t)\mathbf{k}$ and the pressure $p = p(\mathbf{x}, t)$ are the unknowns, $\mathbf{x} = x\mathbf{i} + y\mathbf{j} + z\mathbf{k} = (x, y, z)$; ρ is the mass density of the fluid; ν is the viscosity; $\mathbf{g}(\mathbf{x}, t)$ stands for a given, externally applied body force (e.g., gravity); $\nabla(= \mathbf{i}\frac{\partial}{\partial x} + \mathbf{j}\frac{\partial}{\partial y} + \mathbf{k}\frac{\partial}{\partial z})$ and $\Delta(= \nabla \cdot \nabla = \frac{\partial^2}{\partial x^2} + \frac{\partial^2}{\partial y^2} + \frac{\partial^2}{\partial z^2})$ are the vector gradient operator and the Laplacian operator in 3D space, respectively.

(**I**) *Raylods average—unclosed to the decomposition in mathematics*

The pioneering decomposition of the unknowns to the N–S equations into an average flow and a turbulent flow was proposed by Reynolds [2], i.e.,

$$\mathbf{u}(\mathbf{x}, t) = \overline{\mathbf{u}(\mathbf{x}, t)} + \mathbf{u}'(\mathbf{x}, t) \quad \text{and} \quad p(\mathbf{x}, t) = \overline{p(\mathbf{x}, t)} + p'(\mathbf{x}, t). \tag{3.134}$$

Here, $\overline{\mathbf{u}(\mathbf{x}, t)}$, $\overline{p(\mathbf{x}, t)}$ and $\mathbf{u}'(\mathbf{x}, t)$, $p'(\mathbf{x}, t)$ are the average and turbulent parts, respectively.

For saving space, here, we only introduce the time-averaging for an unknown function $g(\mathbf{x}, t)$ in a uniform form. In such method, the averaging calculation is defined by the following form:

$$\overline{g(\mathbf{x}, t)} = \lim_{T\to\infty} \frac{1}{T} \int_t^{t+T} g(\mathbf{x}, t)dt \approx \frac{1}{T} \int_t^{t+T} g(\mathbf{x}, t)dt \tag{3.135}$$

and

$$\overline{g'(\mathbf{x}, t)} = \lim_{T\to\infty} \frac{1}{T} \int_t^{t+T} g'(\mathbf{x}, t)dt \equiv 0 \tag{3.136}$$

$$\overline{\overline{g(\mathbf{x}, t)}} = \overline{g(\mathbf{x}, t)}, \quad \overline{\frac{\partial g'(\mathbf{x}, t)}{\partial t}} = \overline{g'(\mathbf{x}, t)} \equiv 0. \tag{3.137}$$

Under such average operation, substituting Eq. (3.134) into the N–S equations of Eq. (3.133), we get

$$\begin{aligned} &\rho\frac{\partial\{\overline{\mathbf{u}(\mathbf{x},t)}+\mathbf{u}'(\mathbf{x},t)\}}{\partial t} + \rho\{\overline{\mathbf{u}(\mathbf{x}, t)}+\mathbf{u}'(\mathbf{x}, t)\} \cdot \nabla\{\overline{\mathbf{u}(\mathbf{x}, t)}+\mathbf{u}'(\mathbf{x}, t)\} \\ &\quad = \nu\Delta\{\overline{\mathbf{u}(\mathbf{x}, t)}+\mathbf{u}'(\mathbf{x}, t)\} - \nabla\{\overline{p(\mathbf{x}, t)} + p'(\mathbf{x}, t)\} + \mathbf{g}(\mathbf{x}, t)\cdot \\ &\quad \nabla \cdot \{\overline{\mathbf{u}(\mathbf{x}, t)}+\mathbf{u}'(\mathbf{x}, t)\} = 0, \quad \mathbf{x} \in \mathbb{R}^3, \ t > 0 \end{aligned} \tag{3.138}$$

After the average operation and its characteristic displayed in Eqs. (3.135)–(3.137) are used to the induced Eq. (3.138), we have

$$\begin{aligned} &\rho\frac{\partial\overline{\mathbf{u}(\mathbf{x},t)}}{\partial t} + \rho\{\overline{\mathbf{u}(\mathbf{x}, t)}\nabla\overline{\mathbf{u}(\mathbf{x}, t)}+\overline{\mathbf{u}'(\mathbf{x}, t) \cdot \nabla\mathbf{u}'(\mathbf{x}, t)}\} = \nu\Delta\overline{\mathbf{u}(\mathbf{x}, t)} - \nabla\overline{p(\mathbf{x}, t)} + \overline{\mathbf{g}(\mathbf{x}, t)} \\ &\nabla \cdot \overline{\mathbf{u}(\mathbf{x}, t)} = 0, \quad \mathbf{x} \in \mathbb{R}^3, \ t > 0 \end{aligned} \tag{3.139}$$

or

$$\begin{aligned} &\rho\frac{\partial\overline{\mathbf{u}(\mathbf{x},t)}}{\partial t} + \rho\overline{\mathbf{u}(\mathbf{x}, t)}\nabla\overline{\mathbf{u}(\mathbf{x}, t)} = \nu\Delta\overline{\mathbf{u}(\mathbf{x}, t)} - \nabla\overline{p(\mathbf{x}, t)} - \rho\overline{\mathbf{u}'(\mathbf{x}, t) \cdot \nabla\mathbf{u}'(\mathbf{x}, t)} + \overline{\mathbf{g}(\mathbf{x}, t)} \\ &\nabla \cdot \overline{\mathbf{u}(\mathbf{x}, t)} = 0, \quad \mathbf{x} \in \mathbb{R}^3, \ t > 0 \end{aligned} \tag{3.140}$$

It is obvious that $\overline{\mathbf{u}'(\mathbf{x}, t) \cdot \nabla\mathbf{u}'(\mathbf{x}, t)} \neq 0$ which is induced by the nonlinearity term in the N–S equations, and Eq. (3.140) has the same formulae of the original N–S equations except for the induced nonzero term which is dependent upon the unknowns of turbulent flow. Thus, such average decomposition is not closed in mathematics. To close the induced Eq. (3.140) for the unknown average flow, many researchers have made efforts by means of some empirical formulae to formulate the nonzero term $\overline{\mathbf{u}'(\mathbf{x}, t) \cdot \nabla\mathbf{u}'(\mathbf{x}, t)}$ in terms of the unknown average flow $\overline{\mathbf{u}(\mathbf{x}, t)}$.

Each such connection is called as a turbulent model in fluid mechanics. Since such closed approaches are strongly dependent on the experience or empirical formula selected rather than in mathematics, the suitability of each of such obtained turbulent models for the unknown average flow is evidently restricted in some flows relevant to the condition of the empirical formula selected. Till today, no one such turbulent model has been found to satisfy whole flow characteristics what the N–S equations can behave for.

(**II**) *Closed basic equations for average flow based on the generalized Coiflets-based numerical method*

Now, let us apply the generalized Coiflets-based numerical method established above to the N–S equations such that the closed equations for characterizing the unknown average flow can be gained.

To the terms relevant to the unknowns in the N–S equations shown in Eq. (3.133), we introduce the following transition functions:

$$\boldsymbol{\theta}(\mathbf{x}, t) = \mathbf{u}(\mathbf{x}, t) \cdot \nabla \mathbf{u}(\mathbf{x}, t) \tag{3.141}$$

$$\boldsymbol{\xi}(\mathbf{x}, t) = \nu \Delta \mathbf{u}(\mathbf{x}, t) \tag{3.142}$$

$$\boldsymbol{\lambda}(\mathbf{x}, t) = \nabla p(\mathbf{x}, t) \tag{3.143}$$

$$\gamma(\mathbf{x}, t) = \nabla \cdot \mathbf{u}(\mathbf{x}, t). \tag{3.144}$$

Applying the expansion form of Eq. (3.125) to the unknown functions $\mathbf{u}(\mathbf{x}, t)$ and $p(\mathbf{x}, t)$ associated with the above transition functions, one gets

$$\begin{aligned} \mathbf{u}(\mathbf{x}, t) &= 2^{-3n/2} \sum_{i,j,k \in Z} \mathbf{u}_{n,ijk}(t) \Phi_{n,(i-M_1)(j-M_1)(k-M_1)}(\mathbf{x}) + \tilde{Q}_n \mathbf{u}(\mathbf{x}, t) \\ &= \overline{\mathbf{u}(\mathbf{x}, t)} + \mathbf{u}'(\mathbf{x}, t) \end{aligned} \tag{3.145}$$

$$\begin{aligned} p(\mathbf{x}, t) &= 2^{-3n/2} \sum_{i,j,k \in Z} p_{n,ijk}(t) \Phi_{n,(i-M_1)(j-M_1)(k-M_1)}(\mathbf{x}) + \tilde{Q}_n p(\mathbf{x}, t) \\ &= \overline{p(\mathbf{x}, t)} + p'(\mathbf{x}, t) \end{aligned} \tag{3.146}$$

$$\begin{aligned} \boldsymbol{\theta}(\mathbf{x}, t) &= 2^{-3n/2} \sum_{i,j,k \in Z} \boldsymbol{\theta}_{n,ijk}(t) \Phi_{n,(i-M_1)(j-M_1)(k-M_1)}(\mathbf{x}) + \tilde{Q}_n \boldsymbol{\theta}(\mathbf{x}, t) \\ &= \overline{\boldsymbol{\theta}(\mathbf{x}, t)} + \boldsymbol{\theta}'(\mathbf{x}, t) \end{aligned} \tag{3.147}$$

$$\boldsymbol{\xi}(\mathbf{x}, t) = 2^{-3n/2} \sum_{i,j,k \in Z} \boldsymbol{\xi}_{n,ijk}(t) \Phi_{n,(i-M_1)(j-M_1)(k-M_1)}(\mathbf{x}) + \tilde{Q}_n \boldsymbol{\xi}(\mathbf{x}, t)$$

$$= \overline{\boldsymbol{\xi}(\mathbf{x}, t)} + \boldsymbol{\xi}'(\mathbf{x}, t) \tag{3.148}$$

$$\begin{aligned}\boldsymbol{\lambda}(\mathbf{x}, t) &= 2^{-3n/2} \sum_{i,j,k\in Z} \boldsymbol{\lambda}_{n,ijk}(t)\Phi_{n,(i-M_1)(j-M_1)(k-M_1)}(\mathbf{x}) + \tilde{Q}_n\lambda(\mathbf{x}, t)\\ &= \overline{\boldsymbol{\lambda}(\mathbf{x}, t)} + \lambda'(\mathbf{x}, t)\end{aligned} \tag{3.149}$$

$$\begin{aligned}\gamma(\mathbf{x}, t) &= 2^{-3n/2} \sum_{i,j,k\in Z} \gamma_{n,ijk}(t)\Phi_{n,(i-M_1)(j-M_1)(k-M_1)}(\mathbf{x}) + \tilde{Q}_n\gamma(\mathbf{x}, t)\\ &= \overline{\gamma(\mathbf{x}, t)} + \gamma'(\mathbf{x}, t).\end{aligned} \tag{3.150}$$

Here, we use the notations of average and turbulent flows in Eqs. (3.145) and (3.146), for example,

$$\overline{\mathbf{u}(\mathbf{x}, t)} = 2^{-3n/2} \sum_{i,j,k\in Z} \mathbf{u}_{n,ijk}(t)\Phi_{n,(i-M_1)(j-M_1)(k-M_1)}(\mathbf{x}), \mathbf{u}'(\mathbf{x}, t) = \tilde{Q}_n\mathbf{u}(\mathbf{x}, t), \tag{3.151}$$

$$\overline{p(\mathbf{x}, t)} = 2^{-3n/2} \sum_{i,j,k\in Z} p_{n,ijk}(t)\Phi_{n,(i-M_1)(j-M_1)(k-M_1)}(\mathbf{x}), \quad p'(\mathbf{x}, t) = \tilde{Q}_n p(\mathbf{x}, t). \tag{3.152}$$

In such notations, all turbulent terms correspond to the truncation errors in the approximation of average flow. For the differential operator ∇ in the 3D Cartesian coordinate system acted to an unknown function, e.g., $p(x, y, z, t)$, its approximate expansion can be explicitly expressed by the form

$$P_n\nabla p(x, y, z, t) = 2^{-3n/2} \sum_{i,j,k\in Z} \nabla p_{n,ijk}(t)\Phi_{n,(i-M_1)(j-M_1)(k-M_1)}(\mathbf{x}), \tag{3.153}$$

where

$$\nabla p_{n,ijk}(t) = (\mathbf{i}\frac{\partial}{\partial x} + \mathbf{j}\frac{\partial}{\partial x} + \mathbf{k}\frac{\partial}{\partial x})p(x, y, z, t)|_{(x,y,x)=(i2^{-n},j2^{-n},k2^{-n})}. \tag{3.154}$$

Substituting Eqs. (3.145)–(3.150) into the N–S equations of Eq. (3.133), and taking the inner integration $< \cdot, \Phi_{n,(l-M_1)(m-M_1)(r-M_1)}(x) >= \int_{R^3} (\cdot)\Phi_{n,(l-M_1)(m-M_1)(r-M_1)}(\mathbf{x})\, d\mathbf{x}$ to each term in the induced equations, i.e., the Galerkin method is employed, then using the orthogonal properties of the form

$$< \Phi_{n,(i-M_1)(j-M_1)(k-M_1)}(\mathbf{x}), \Phi_{n,(l-M_1)(m-M_1)(r-M_1)}(x) >= \delta_{i,l}\delta_{j,m}\delta_{k,r}$$
$$< \Phi_{n,(l-M_1)(m-M_1)(r-M_1)}(\mathbf{x}), \tilde{Q}_n f(\mathbf{x}) >= \int_{\mathbb{R}^3} \Phi_{n,(l-M_1)(m-M_1)(r-M_1)}(\mathbf{x})\tilde{Q}_n f(\mathbf{x})d\mathbf{x} = 0$$

into all formulae relevant to Eqs. (3.145)–(3.150), we obtain

$$\rho\frac{d\mathbf{u}_{n,ijk}(t)}{dt}+\rho\mathbf{u}_{n,ijk}(t)\cdot\nabla\mathbf{u}_{n,ijk}(t)=\nu\Delta\mathbf{u}_{n,ijk}(t)-\nabla p_{n,ijk}(t)+\mathbf{g}_{n,ijk}(t)$$
$$\nabla\cdot\mathbf{u}_{n,ijk}(t)=0,\quad t>0,\quad i,j,k\in Z. \tag{3.155}$$

Here, $\mathbf{g}_{n,ijk}(t)=\mathbf{g}(x,y,z,t)|_{(x,y,x)=(i2^{-n},j2^{-n},k2^{-n})}$.

When the boundary conditions are either in a finite region or in a periodic infinite region or both, the N–S equations are in the bounded domain, e.g., $\mathbf{x}\in[0,1]^3\subset R^3$. After that, using Eq. (3.49) to the decomposition coefficients of partial derivative functions, one can replace them by the decomposition coefficients of original unknown function. For example,

$$\left[\frac{\partial^m f_{n,ijk}(t)}{\partial x^m},\frac{\partial^m f_{n,ijk}(t)}{\partial y^m},\frac{\partial^m f_{n,ijk}(t)}{\partial z^m}\right]$$
$$=\sum_{r\in Z_k}[f_{n,rjk}(t)\Gamma^{n,m}_{r-M_1,i-M_1},f_{n,irk}(t)\Gamma^{n,m}_{r-M_1,j-M_1},f_{n,ijr}(t)\Gamma^{n,m}_{r-M_1,k-M_1}]. \tag{3.156}$$

Substituting the transform of Eqs. (3.156) into (3.155), further, we finally obtain the following nonlinear ordinary differential equations on time t in the form

$$\rho\frac{d\mathbf{u}_{n,ijk}}{dt}+\rho\sum_{r\in Z_k}[u_{n,ijk}(t)\mathbf{u}_{n,rjk}(t)\Gamma^{n,1}_{r-M_1,i-M_1}+v_{n,ijk}(t)\mathbf{u}_{n,irk}(t)\Gamma^{n,1}_{r-M_1,j-M_1}$$
$$+w_{n,ijk}(t)\mathbf{u}_{n,ijr}(t)\Gamma^{n,1}_{r-M_1,k-M_1}]$$
$$=\nu\sum_{r\in Z_k}[\mathbf{u}_{n,rjk}(t)\Gamma^{n,2}_{r-M_1,i-M_1}+\mathbf{u}_{n,irk}(t)\Gamma^{n,2}_{r-M_1,j-M_1}+\mathbf{u}_{n,ijr}(t)\Gamma^{n,2}_{r-M_1,k-M_1}],$$
$$-\sum_{r\in Z_k}[p_{n,rjk}(t)\Gamma^{n,1}_{r-M_1,i-M_1}\mathbf{i}+p_{n,irk}(t)\Gamma^{n,1}_{r-M_1,j}\mathbf{j}+p_{n,ijr}(t)\Gamma^{n,1}_{r-M_1,k}\mathbf{k}]+\mathbf{g}_{n,ijk}(t) \tag{3.157}$$

$$\sum_{r\in Z_k}[u_{n,rjk}(t)\Gamma^{n,1}_{r-M_1,i-M_1}+v_{n,irk}(t)\Gamma^{n,1}_{r-M_1,j-M_1}+w_{n,ijr}(t)\Gamma^{n,1}_{r-M_1,k-M_1}]=0. \tag{3.158}$$

Here, $i,j,k\in Z_{[0,1]},\quad t>0$. It is obvious that all truncation errors or turbulent terms have automatically disappeared in the final ordinary differential equations of Eqs. (3.157) and (3.158) for the unknowns $\mathbf{u}_{n,ijk}$ and $p_{n,ijk}(t)$ at any resolution level n. In other words, when we take the generalized Coiflets-based solution approximations and their truncation errors of the unknown function as the parts of average flow and turbulent flow in the N–S equations, respectively, one finds that such a solution approach makes the differential governing equations for the average flow be independent on the truncated turbulent flow. Following Definitions 3.5 and 3.9, we know that such induced equations for the average flow are weakly closed.

References

1. Ascher UM, Mattheij RMM, Russell RD (1995) Numerical solution of boundary value problems for ordinary differential equations. Soci Industr Appl Math
2. Belytschko T, Lu YY, Gu L (1994) Element-free Galerkin methods. Int J Numer Meth Eng 37(2):229–256
3. Cockburn B, Karniadakis GE, Shu, CW (2012) Discontinuous Galerkin methods: theory, computation and applications. Springer Science and Business Media
4. Keller HB (2018) Numerical methods for two-point boundary-value problems. Courier Dover Publications
5. Coddington EA, Levinson N (1955) Theory of ordinary differential equations. Tata McGraw-Hill Education
6. Kubicek M, Hlavacek V (2008) Numerical solution of nonlinear boundary value problems with applications. Courier Corporation
7. Zheng XJ (1990) Theory and application of large deflection of circular thin plate. Jilin Science and Technology Press
8. Liu XJ, Zhou YH, Zhang L, Wang JZ (2014) Wavelet solutions of Burgers' equation with high Reynolds numbers. Science China 57(007):1285–1292
9. Hon YC, Mao XZ (1998) An efficient numerical sȘcheme for Burgers' equation. Appl Math Comput 95:37–50
10. Daubechies I (1992) Ten lectures on wavelets. Society for industrial and applied mathematics
11. Hinze JO (1975) Turbulence, 2nd edn. McGraw-Hill, New York
12. Wilcox DC (1995) Turbulence models for CFD, second ed., DCW Industries, Inc

Chapter 4
Error Analysis and Boundary Extension

Before we introduce the applications of the wavelet Galerkin method to solve the boundary-value problems, in this chapter, we introduce the error analysis and the boundary extension technology what we conducted such that we know when the accuracy of the applications is ensured. From Chap. 3, one finds that the errors in our method for solving nonlinear problems are mainly originated by the truncated error and the numerical integral of the generalized Gaussian integral scheme for the calculations of decomposition coefficients in the nonlinear terms, while the errors generated in the reconstruction of a function near the boundary ends can be substantially reduced by the boundary extension technology.

4.1 Error Estimation of Approximation of a Function

4.1.1 Truncation Error Analysis of the Wavelet-Based Approximation

For the simplicity of convenience, here, we rewrite the relevant expressions of approximation of a function $f(x) \in L^2(R)$ in a compact supported orthogonal wavelet as follows:

$$f(x) = \sum_{k \in Z} c_{n,k}\varphi_{n,k}(x) + \sum_{l \geq n}\sum_{k \in Z} d_{l,k}\psi_{l,k}(x), \tag{4.1}$$

in which

$$c_{n,k} = \int_R f(x)\varphi_{n,k}(x)dx, \tag{4.2a}$$

Y.-H. Zhou, *Wavelet Numerical Method and Its Applications in Nonlinear Problems*, Engineering Applications of Computational Methods 6,
https://doi.org/10.1007/978-981-33-6643-5_4

$$d_{l,k} = \int_R f(x)\psi_{l,k}(x)dx, \tag{4.2b}$$

are the scaling and wavelet decomposition coefficients, respectively. To the approximation of Eq. (4.1), the approximation is formulated by

$$f(x) \approx P_n f(x) = \sum_{k\in Z} c_{n,k}\varphi_{n,k}(x) = \sum_{k\in Z} \{\int_R f(x)\varphi_{n,k}(x)dx\}\varphi_{n,k}(x). \tag{4.3}$$

Hence, the truncated error of the approximation is

$$\eta_f(x) = \sum_{l\geq n}\sum_{k\in Z} d_{l,k}\psi_{l,k}(x) = f(x) - P_n f(x). \tag{4.4}$$

When the compactly supported orthogonal wavelet has up to $r-1$ order of vanishing moment, i.e.,

$$\int_R x^i \psi(x)dx = 0, \qquad i = 0, 1, 2, \ldots, r-1, \tag{4.5}$$

from Property 2.1, we know that the polynomial function of order $r-1$

$$P_r(x) = \sum_{i=0}^{r-1} \frac{f^{(i)}(x_0)}{i!}(x-x_0)^i \tag{4.6}$$

can be exactly presented by $P_n f(x)$, i.e.,

$$P_r(x) = \sum_{i=0}^{r-1} \frac{f^{(i)}(x_0)}{i!}(x-x_0)^i = P_n P_r(x). \tag{4.7}$$

Here, x_0 is a specified point in the definite region of $f(x)$ and $|x - x_0| < \delta$, in which δ is a small value. After the function $f(x)$ is formulated by Tayler's expansion at $x = x_0$, one gets

$$f(x) = P_r(x) + \frac{f^{(r)}(\xi)}{r!}(x-x_0)^r, \tag{4.8}$$

where $\xi \in (x_0, x)$ when $x > x_0$, and $\xi \in (x, x_0)$ when $x < x_0$. Denote

$$g_i(x) = (x-x_0)^i, \qquad i = 0, 1, \ldots, r-1. \tag{4.9}$$

Similar to Eq. (4.7), then, we have

$$g_i(x) = (x - x_0)^i = P_n(x - x_0)^i$$
$$= \sum_{k \in Z} [\int_R (x - x_0)^i \varphi_{n,k}(x) dx] \varphi_{n,k}(x), \qquad i = 0, 1, \ldots, r - 1. \tag{4.10}$$

Further,

$$g_i(x_0) = \delta_{0,i} = \sum_{k \in Z} [\int_R (x - x_0)^i \varphi_{n,k}(x) dx] \varphi_{n,k}(x_0), \qquad i = 0, 1, \ldots, r - 1. \tag{4.11}$$

Hence, from Eqs. (4.7) and (4.8), we have $f(x_0) = P_r(x_0)$ and

$$P_n f(x_0) = P_r(x_0) + \frac{f^{(r)}(\xi)}{r!} \sum_{k \in Z} [\int_R (x - x_0)^r \varphi_{n,k}(x) dx] \varphi_{n,k}(x_0)$$
$$= f(x_0) + \frac{f^{(r)}(\xi)}{r!} \sum_{k \in Z} [\int_R (x - x_0)^r \varphi_{n,k}(x) dx] \varphi_{n,k}(x_0). \tag{4.12}$$

Further,

$$|\eta_f(x_0)| = |f(x_0) - P_n f(x_0)|$$
$$= \left| \frac{f^{(r)}(\xi)}{r!} \right| \cdot \left| \sum_{k \in Z} [\int_R (x - x_0)^r \varphi_{n,k}(x) dx] \varphi_{n,k}(x_0) \right|. \tag{4.13}$$

Considering the support region of $\varphi_{n,k}(x)$, i.e., $\mathrm{Supp}[\varphi_{n,k}(x)] = [k2^{-n}, (k + \tilde{N})2^{-n}]$, and the summation times is not over $\tilde{N}$ in Eq. (4.13), one gets

$$|\eta_f(x_0)| = \left| \frac{f^{(r)}(\xi)}{r!} \right| \cdot \left| \sum_{k \in Z} [\int_{k2^{-n}}^{(k+\tilde{N})2^{-n}} (x - x_0)^r \varphi_{n,k}(x) dx] \varphi_{n,k}(x_0) \right|$$
$$\le \left| \frac{f^{(r)}(\xi)}{r!} \right| (\tilde{N}2^{-n})^r \left| \sum_{k \in Z} [\int_{k2^{-n}}^{(k+\tilde{N})2^{-n}} \varphi_{n,k}(x) dx] \varphi_{n,k}(x_0) \right|$$
$$= \left| \frac{f^{(r)}(\xi)}{r!} \right| (\tilde{N}2^{-n})^r 2^{n/2} \left| \sum_{k \in Z} [\int_{k2^{-n}}^{(k+\tilde{N})2^{-n}} \varphi_{\tilde{N}}(2^n x - k) dx] \varphi_{n,k}(x_0) \right|. \tag{4.14}$$

By using the variable transform in the integral of Eq. (4.14), we obtain

$$|\eta_f(x_0)| \le \left| \frac{f^{(r)}(\xi)}{r!} \right| (\tilde{N}2^{-n})^r 2^{-n/2} \left| \sum_{k \in Z} [\int_0^{\tilde{N}} \varphi_{\tilde{N}}(x) dx] \varphi_{n,k}(x_0) \right|. \tag{4.15}$$

Since $\int_0^{\tilde{N}} \varphi_{\tilde{N}}(x)dx]= 1$, Eq. (4.15) becomes into the form

$$\begin{aligned} |\eta_f(x_0)| &\le \left|\frac{f^{(r)}(\xi)}{r!}\right| (\tilde{N}2^{-n})^r 2^{-n/2} \left|\sum_{k\in Z} \varphi_{n,k}(x_0)\right| \\ &\le \left|\frac{f^{(r)}(\xi)}{r!}\right| (\tilde{N}2^{-n})^r \tilde{N} M \end{aligned}. \tag{4.16}$$

Here, $M = \max\{|\varphi_{\tilde{N}}(x)|; x \in [0, \tilde{N}]\}$. Denote $C = |\frac{f^{(r)}(\xi)}{r!}|\tilde{N}^{r+1}M(> 0)$. It is obvious that C is a finite bounded number. Therefore, Eq. (4.16) finally reduced by the form

$$|\eta_f(x_0)| = |f(x_0) - P_n f(x_0)| \le C2^{-nr}. \tag{4.17}$$

Since x_0 is an arbitrary point in the definite region of $f(x)$, thus, Eq. (4.17) is held for variable x. That is

$$|\eta_f(x)| = |f(x) - P_n f(x)| \le C2^{-nr}. \tag{4.18}$$

The error estimation of Eq. (4.18) tells us that when the approximation of Eq. (4.3) is used for a function $f(x) \in L^2(R)$, the accuracy of such approximation under a compactly supported orthogonal wavelet with resolution level n and the vanishing moments up to order of $r - 1$ is the order of precision $O(2^{-nr}) = C2^{-nr}$. For the generalized Coiflets with support region [0, 17] and $r = 6$ shown in Chap. 2, we have $2^{-nr} \approx 5.96\times10^{-8}, 9.31\times10^{-10}, 1.46\times10^{-11}$ when the resolution lever $n = 4, 5, 6$, respectively. In such situation, the accuracy of the approximation when $n = 6$ is enough for most problems. Of course, the calculations will intensively increase with increasing the resolution level although the accuracy can be further refined. From the generalized orthogonal wavelets introduced in Chap. 2, one finds that Daubechies wavelets has the highest vanishing moment under the same support region [1, 2]. This means that Daubechies wavelets could have the highest accuracy when the same support region and same resolution level are employed in the approximation. In the next subsection, however, one will find that such accuracy condition cannot be ensured when the approximation of Daubechies wavelets is used to expand a nonlinear term of $f(x)$. It is this reason to motivate us for finding another way based on the generalized Coiflets. Different from the Coiflets in the application of signal processing, some differently additional property requested to the generalized Coiflets in our researches for solving differential equations are considered as shown in Chap. 2, e.g., when M_1 and support region are taken by different values, one gets different generalized Coiflets.

4.1.2 Error Analysis to the Gaussian Integral for Decomposition Coefficients

The other main error emerged in the generalized Coiflets method for solving nonlinear problems are mainly originated from the numerical integral or the generalized Gaussian integral. From the numerical integral methods, we know that the generalized Gaussian integral is powerful, which has the advantages of not only high accuracy but also analytical expression of the integrated function. Here, we give a general discussion for it. As follows, we firstly introduce the error estimation of the generalized Gaussian integral for any compactly supported orthogonal wavelet in generality, then return to the case of the generalized Coiflets.

For simplicity, here, we rewrite the relevant equations which will be used. For a compactly supported orthogonal wavelet with scaling function $\varphi_{\tilde{N}}(x)$, its support region can be expressed by the form:

$$\operatorname{Supp}\varphi_N(x) = [0,\ \tilde{N}] = [0, 2N-1], \tag{4.19}$$

and its normalization condition is

$$M_0 = \int_0^{2N-1} \varphi_{\tilde{N}}(x)dx = 1. \tag{4.20}$$

When the vanishing moments of Eq. (4.5) are held, from Chap. 2, one can get the values of

$$M_i = \int_0^{2N-1} x^i \varphi_{\tilde{N}}(x)dx \quad , \quad i = 1,2,\ldots,r-1. \tag{4.21}$$

To calculate the decomposition coefficients $c_{n,k}$ in Eq. (4.2a), we select the generalized Gaussian integral method of the form [3–6]:

$$\int_0^{2N-1} \varphi_{\tilde{N}}(x)f(x)dx \approx \sum_{k=0}^{m} A_k f(x_k). \tag{4.22}$$

When Eq. (4.22) is accurately satisfied for arbitrary polynomials with orders up to $2m+1$, i.e., Eq. (4.22) has an algebraic precision of order $2m+1$, such integral scheme is called the generalized Gaussian integral for any $f(x) \in L^2(R)$ with the weight function $\varphi_{\tilde{N}}(x)$. Here, $x_k \in [0, 2N-1]$ and $A_k (k = 0, 1, 2, \ldots, m)$ are the unknowns and they are referred to Gaussian points and weight coefficients, respectively. Such unknowns are determined by the equations when Eq. (4.22) is accurately satisfied for $f(x) = x^i (i = 0, 1, \ldots, 2m+1)$. That is [7],

$$\int_0^{2N-1} x^i \varphi_{\tilde{N}}(x)dx = \sum_{k=0}^{m} A_k x_k^i \quad , \quad i = 0, 1, \ldots, 2m+1. \tag{4.23}$$

For such purpose to a compact supported orthogonal wavelet, the moments of the scaling function $\varphi_{\tilde{N}}(x)$ are considered in the establishment of a compactly supported orthogonal wavelet, which is fully corresponding to the condition of vanishing moments of its wavelet function $\psi_{\tilde{N}}(x)$. That is, Eq. (4.5) is held and the values of $M_i (i = 0, 1, \ldots, r-1)$ are obtained (see Chap. 2). Further, Eq. (4.23) becomes

$$\int_0^{2N-1} \varphi_{\tilde{N}}(x)x^i dx = \sum_{k=0}^{r/2--1} A_k x_k^i = M_i \quad , \quad i = 0, 1, \ldots, r-1. \tag{4.24}$$

Here, Eq. (4.21) has been taken into account. It is obvious that r should be an even integer greater than 2 and $2m+1 = r-1$ is employed.

The efficiency of the generalized Gaussian integral method is strongly dependent on the real solutions of unknowns A_k and $x_k (k = 0, 1, \ldots, r/2-1)$ in the system of nonlinear algebraic equations displayed in Eq. (4.24). If they exist and are found for a compactly supported orthogonal wavelet, the numerical integral can be further conducted. For example, Table 4.1 displays the real solutions of A_k and x_k for Daubechies wavelets. From it, one can find that when r is equal to 4 or over, the real solutions can be gained only at the third order of algebraic precision. Thus, Daubechies wavelets for the expansion of a nonlinear function has low accuracy no matter how to increase its support region or vanishing moments.

Now, let us return to the error generated by the generalized Gaussian integral. After a set of real solutions for A_k and x_k are obtained in the generalized Gaussian integral in Eq. (4.24), the decomposition coefficients in Eq. (4.2a) can be formulated by

$$c_{n,k} = \int_{k/2^n}^{(k+\tilde{N})/2^n} f(x)\varphi_{n,k}(x)dx \approx 2^{-n/2} \sum_{i=0}^{r/2-1} A_i f(\frac{k+x_i}{2^n}). \tag{4.25}$$

For such purpose, here, we rewrite Eq. (4.6) in the form $P_r(x) = \sum_{i=0}^{r-1} a_i x^i$ where a_i can be formulated by the coefficients in Eq. (4.6). Then, we have

$$f(x) = P_r(x) + \frac{f^{(r)}(\xi)}{r!}(x-x_0)^r. \tag{4.26}$$

It is evident that the error generated by the generalized Gaussian integral is original from the error part

$$\omega_r(x) = \frac{f^{(r)}(\xi)}{r!}(x-x_0)^r. \tag{4.27}$$

Table 4.1 Real solutions of Gaussian points and weight coefficientsfor Daubechies wavelets

Order of Vanishing moment	M_i	Gaussian points x_i	Weight coefficients A_i
$r = 2$	$M_0 = 1.000000000000000$	$x_0 = 0.500000000000000$	$A_0 = 1.000000000000000$
	$M_1 = 0.500000000000000$		
$r = 4$	$M_0 = 1.000000000000000$	$x_0 = 0.633974596215820$	$A_0 = 1.000000000000000$
	$M_1 = 0.633974596215820$	$x_1 = -3.23086494353463\text{E-}11$	$A_1 = 3.668395732555621\text{E-}36$
	$M_2 = 0.401923788647395$		
	$M_3 = 0.131091556791809$		
$r = 6$	$M_0 = 1.000000000000000$	$x_0 = 0.817401167800000$	$A_0 = 1.000000000000000$
	$M_1 = 0.817401167800000$	$x_1 = 6.661108984863969\text{E}9$	$A_1 = -3.406536343902618\text{E-}31$
	$M_2 = 0.668144669105689$	No real solutions of all x_i and A_i have been found for other cases	
	$M_3 = 0.445460044827100$		
	$M_4 = 0.117226346781396$		
	$M_5 = -0.046651091190452$		
$r = 8$	$M_0 = 1.000000000000000$	$x_0 = 1.005393213445220$	$A_0 = 1.000000000000000$
	$M_1 = 1.005393213445220$	$x_1 = -3.577244744394367\text{E}10$	$A_2 = 2.379077538892756\text{E-}33$
	$M_2 = 1.010815513644750$	No real solutions of all x_i and A_i have been found for other cases	
	$M_3 = 0.907360367223266$		
	$M_4 = 0.583771813560483$		
	$M_5 = 0.063077524235001$		
	$M_6 = -0.237749434751017$		
	$M_7 = -0.817841785128409$		

Since

$$P_n P_r(x) = \sum_{k \in Z} \left\{ \int_R P_r(x)\varphi_{n,k}(x)\varphi_{n,k}(x) = P_r(x) \right\} \tag{4.28}$$

is exactly held, one can get that the error appeared in the generalized Gaussian integral is formulated by

$$\eta_{c_{n,k}} = \int_R \{f(x) - P_r(x)\}\varphi_{n,k}(x)| = |\frac{f^{(r)}(\xi)}{r!} \int_{k/2^n}^{(k+\tilde{N})/2^n} (x - x_0)^r \varphi_{n,k}(x)dx|. \tag{4.29}$$

When $x, x_0 \in \text{Supp}[\varphi_{n,k}(x)]$ without losing generality, we have $|x - x_0| \le \tilde{N}2^{-n}$. Hence, Eq. (4.29) can be reduced into the form

$$\eta_{c_{n,k}} \le |\frac{f^{(r)}(\xi)}{r!}|(\tilde{N}2^{-n})^r \cdot 2^{-n/2}| \int_0^{\tilde{N}} \varphi_{\tilde{N}}(x)dx| \le |\frac{f^{(r)}(\xi)}{r!}|(\tilde{N}2^{-n})^r = D2^{-n(r+1/2)}. \tag{4.30}$$

Here, $D = |\frac{f^{(r)}(\xi)}{r!}|\tilde{N}^r$ is a finite real number. Comparing Eq. (4.30) with Eq. (4.18), one can find that the generalized Gaussian integral has the same order of precision as the truncated error once the Gaussian points and their weight coefficients can be obtained by real values. It should be noted that such deduction of Eq. (4.30) is based on that the real multi-Gaussian points and their corresponding real weight coefficients can be obtained, which is not easy for one to get them as displayed in Table 4.1 for Daubechies wavelets. When the generalized Gaussian integral method is employed in such situation, the accuracy of approximation under Daubechies wavelets cannot be ensured as the same order as that shown in Eq. (4.18). For the generalized Coiflets, however, a set of the real solutions for the multi-Gaussian points and corresponding weight coefficients are easily and always gained as shown in Chap. 2. From the solutions, one finds that Eq. (4.22) becomes the form:

$$c_{n,k} = \int_{k/2^n}^{(k+\tilde{N})/2^n} f(x)\varphi_{n,k}(x)dx \approx 2^{-n/2} f\left(\frac{k + M_1}{2^n}\right). \tag{4.31}$$

Further, the approximation can be formulated in the form:

$$f(x) \approx P_n f(x) = 2^{-n/2} \sum_{k \in Z} f\left(\frac{k + M_1}{2^n}\right)\varphi_{n,k}(x). \tag{4.32}$$

which has the same order of precision shown in both Eq. (4.30). Hence, the accuracy order formulated in Eq. (4.18) is ensured in the generalized Coiflets. It should be noted that another advantage using the generalized Coiflets is that the calculations in Eq. (4.32) is much down by corresponding to other cases of the multi-Gaussian points even if they are gained in real solutions. For the latter case, the approximation may be formulated by the form:

$$f(x) \approx P_n f(x) = 2^{-n/2} \sum_{k \in Z} \sum_{i=0}^{r/2-1} A_i f(\frac{k + x_i}{2^n}) \varphi_{n,k}(x), \qquad (4.33)$$

in which, r is an even integer. When $r = 4$ for the simplest case of Daubechies wavelets, for example, the reconstruction calculations in Eq. (4.33) are at least double to that of Eq. (4.32) except for the Gaussian points x_0 and x_1 non-integers. When the solutions of the Gaussian points are non-integers, the values of $f(\frac{k+x_i}{2^n})$ all are not located at the dyadic points in the definite region of $f(x)$. In such situation, the relationships displayed in Chap. 2 for the derivatives, integral and connection coefficients for the scaling function cannot be used since they are induced out on the basis of the dyadic points such that the relevant calculations, either pre- or post-data processing, cannot be exactly conducted. Once one uses the real Gaussian points in some approximate integers, as down in literature, such approximation obviously generates additionally implicit losing of accuracy in the expansion of Daubechies wavelets.

4.2 Error Estimations in Other Applications of the Generalized Coiflets

4.2.1 Error Analysis to the Decomposition Coefficient of Derivatives

To the derivatives of $f(x)$, denote $g_m(x) = d^m f(x)/dx^m (1 \le m \le r - 1)$. As done in Chap. 3, the derivative functions $g_m(x)$ can be expanded in the generalized Coiflets with the form:

$$g_m(x) \approx \sum_{k \in Z} g_{n,k} \varphi_{n,k}(x) = \sum_{k \in Z} f_{n,k}^{(m)} \varphi_{n,k}(x), \qquad (4.34)$$

where

$$g_{n,k}^{(m)} = \int_R g_m(x) \varphi_{n,k}(x) dx = g_m(\frac{k + M_1}{2^n}) = \frac{d^m f(x)}{dx^m}|_{x=(k+M_1)2^{-n}}. \qquad (4.35)$$

To take *mth*-order derivative of Eq. (4.3), we get

$$g_m(x) \approx \sum_{k \in Z} f_{n,k} \frac{d^m}{dx^m} \varphi_{n,k}(x) = \sum_{k \in Z} f_{n,k} \varphi_{n,k}^{(m)}(x). \tag{4.36}$$

Under the same precision order of approximations of Eqs. (4.35) and (4.36), thus, we get

$$\sum_{k \in Z} f_{n,k}^{(m)} \varphi_{n,k}(x) = \sum_{k \in Z} f_{n,k} \varphi_{n,k}^{(m)}(x). \tag{4.37}$$

Here, $f_{n,k} = c_{n,k}$. To taking calculation of $< \bullet, \varphi_{n,j}(x) >$ on Eq. (4.37), one gains

$$f_{n,j}^{(m)} = \sum_{k \in Z} \Gamma_{k,j}^{n,m} f_{n,k}. \tag{4.38}$$

Here, the connection coefficients $\Gamma_{k,j}^{n,m}$ shown in Chap. 2 is used. Then the approximation of $g_m(x) = d^m f(x)/dx^m$ can be gained by $P_n[d^m f(x)/dx^m]$. As displayed in Chap. 2, the connection coefficients can be obtained by an algebraic arithmetic with the accuracy of the generalized Coiflets used. Hence, the calculations of Eq. (4.38) and its reconstruction has the same precision of $f(x)$ when $f_{n,k}$ are exactly sampled. With the same reason, there are some errors in $f_{n,k}$ when $f(x)$ is unknown, which leads to the error of $f_{n,j}^{(m)}$ being much greater than that of $f_{n,k}$ because the derivatives amplify the error of $f_{n,j}^{(m)}$ as conventional numerical methods. To this case, we give an estimation of the amplified error as follows:

Denote the absolute errors of $f_{n,k}$ by $\varepsilon_{f_{n,k}} = |f_{n,k} - f_{n,k}^*|$ and $f_{n,k}^{(m)}$ by $\varepsilon_{f_{n,k}^{(m)}} = |f_{n,k}^{(m)} - \sum_{l \in Z_\Omega} \Gamma_{k,l}^{m,n} f_{n,l}^*|$. Then, to the induced error of $f_{n,j}^{(m)}$, we have

$$\begin{aligned}\varepsilon_{f_{n,k}^{(m)}} &= |f_{n,k}^{(m)} - \sum_{l \in Z_\Omega} \Gamma_{k,j}^{n,m} f_{n,l}^*| = |f_{n,k}^{(m)} - \sum_{l \in Z_\Omega} \Gamma_{k,l}^{n,m} (f_{n,l}^* - f_{n,l} + f_{n,l})| \\ &= |\sum_{l \in Z_\Omega} \Gamma_{k,l}^{n,m} (f_{n,l}^* - f_{n,l})|.\end{aligned} \tag{4.39}$$

Since

$$\begin{aligned}|\Gamma_{k,l}^{m,n}| &= |\int_R \frac{d^m \varphi_{n,k}(x)}{dx^m} \varphi_{n,l}(x) dx| = 2^n |\int_R \frac{d^m \varphi_{\tilde{N}}(2^n x - k)}{dx^m} \varphi_{\tilde{N}}(2^n x - l) dx| \\ &= 2^{nm} |\int_R \frac{d^m \varphi_{\tilde{N}}(x - k + l)}{dx^m} \varphi_{\tilde{N}}(x) dx| \le 2^{nm} C,\end{aligned}$$

where, $C = \max |\int_R \frac{d^m \varphi_{\tilde{N}}(x-k+l)}{dx^m} \varphi_{\tilde{N}}(x) dx|$, hence, Eq. (4.39) is further reduced in the form

$$\varepsilon_{f_{n,k}^{(m)}} \leq C(2\tilde{N}-1)2^{nm} \max \varepsilon_{f_{n,k}} \tag{4.40}$$

which reveals that the absolute error of $f_{n,k}^{(m)}$ in the mth-order derivative function is at least higher than the error of $f_{n,k}$ with order 2^{nm}.

4.2.2 Error Analysis to Decomposition Coefficients of Nonlinear Functions

From Chap. 3 , we know the fact that the generalized Gaussian integral method is mainly used for the expansion coefficients of the nonlinear terms in solving the boundary-value problems. To a nonlinear term of unknown function $f(x) \in L^2(R)$, denote it by a nonlinear operator $N[f(x)]$. Let $g(x) = N[f(x)]$. Applying Eq. (4.3) to $g(x) = N[f(x)]$, one immediately obtains

$$c_{n,k}^{g} = \int_R g(x)\varphi_{n,k}(x)dx \approx 2^{-n/2} \sum_{i=0}^{r/2-1} A_i g(\frac{k+x_i}{2^n}) = 2^{-n/2} \sum_{i=0}^{r/2-1} A_i N[f(\frac{k+x_i}{2^n})]. \tag{4.41}$$

To the generalized Coiflets as shown in Chap. 2, when the number of vanishing moments in Eq. (4.5) is equal to the number of relationships $M_j = M_1^j (j = 1, 2, \ldots, 2l)$, that is, $r = 2l$, it is so lucky for us that only one Gaussian point $x_1 = M_1$ is needed to be found for the case of multi-Gaussian points, and its corresponding weight coefficient with is equal to 1. Here, M_1 can be an integer preselected in the establishment of the wavelets. Further, Eq. (4.41) is reduced into the form:

$$c_{n,k}^{g} = \int_R g(x)\varphi_{n,k}(x)dx \approx 2^{-n/2} N[f(\frac{k+M_1}{2^n})]. \tag{4.42}$$

Further, we have

$$N[f(x)] \approx 2^{-n/2} \sum_{k \in Z} N[f(\frac{k+M_1}{2^n})]\varphi_{n,k}(x). \tag{4.43}$$

Then the approximation $N[f(x)]$ can be written as follows:

$$N[f(x)] \approx P_n N[f(x)] = 2^{-n/2} \sum_{k \in Z} N[f(\frac{k+M_1}{2^n})]\varphi_{n,k}(x) \tag{4.44}$$

To a $N[f(x)]$ with a known continuous function $f(x)$ in R, one can see that both $f(\frac{k+M_1}{2^n})$ and $N[f(\frac{k+M_1}{2^n})]$ are, respectively, the exact values of $f(x)$ and $N[f(x)]$

in the region. In such case, $P_n N[f(x)]$ and $P_n f(x)$ has the same order of accuracy. When $f(x)$ and $N[f(x)]$ is in a finite region, however, the decomposition and reconstruction have to be dependent on a boundary extension arithmetic. Once the natural extension is used to a known function, the accuracy of the approximations of $P_n N[f(x)]$ and $P_n f(x)$ are still same only controlled by the error in the generalized Coiflets. Since $f(x)$ is always unknown to be determined by a boundary-value problem in a finite domain, such natural extension is inefficient, and other boundary extension techniques should be used, e.g., Lagrange polynomial extension. In this case, there are some additional errors in such feasible boundary extension, which could be amplified by the nonlinear function. Denote the decomposition coefficients of $N[f(x), f^{(m)}(x)]$ by $N[f_{n,k}, f_{n,k}^{(m)}](= N[f_{n,k}, \sum\limits_{l\in Z_\Omega} \Gamma_{k,l}^{n,m} f_{n,l}])$ if $f^{(m)}(x)$ is contained. Then, we have

$$
\begin{aligned}
\varepsilon_N &= \left| N[f_{n,k}, \sum_{l\in Z_\Omega} \Gamma_{k,l}^{n,m} f_{n,l}] - N[f_{n,k}^*, \sum_{l\in Z_\Omega} \Gamma_{k,l}^{n,m} f_{n,l}^*] \right| \\
&\le \left| \frac{\partial}{f_{n,k}} N[f_{n,k}, \sum_{l\in Z_\Omega} \Gamma_{k,l}^{n,m} f_{n,l}]_{f_{n,k}=f_{n,k}^*} \varepsilon_{f_{n,k}} \right. \\
&\quad \left. + \frac{\partial}{f_{n,k}^{(m)}{}_{n,k}} N[f_{n,k}, \sum_{l\in Z_\Omega} \Gamma_{k,l}^{n,m} f_{n,l}]_{f_{n,k}=f_{n,k}^*} \sum_{l\in Z_\Omega} \Gamma_{k,l}^{n,m} \varepsilon_{f_{n,l}} \right| \\
&\le \{C_1 + C_2(2\tilde{N} - 1)\} 2^{nm} \max\left|\varepsilon_{f_{n,k}}\right|.
\end{aligned}
\tag{4.45}
$$

Here, $C_1 = \max|\frac{\partial}{\partial f(x)} \tilde{N}[f(x), f^{(m)}(x)]|$, $C_2 = \max|\frac{\partial}{\partial f^{(m)}(x)} \tilde{N}[f(x), f^{(m)}(x)]|$.

4.2.3 Error Analysis to the Approximation of Integral Functions

Here, we consider one multiply integration of $f(x)$ for an example, that is

$$
g_1(x) = \int_0^x f(y)dy. \tag{4.46}
$$

Substituting Eqs. (4.3) into (4.46), we get

$$
g_1(x) \approx \sum_{k\in Z} c_{n,k} \int_0^x \varphi_{n,k}(y)dy = g_1^*(x). \tag{4.47}
$$

According to Chap. 2, one finds that the values of function

$$\theta_1(x) = \int_0^x \varphi_{\tilde{N}}(y)dy \tag{4.48}$$

can be obtained at the dyadic points of resolution level n by an algebraic arithmetic, that is, $\theta_m(n, j) = \theta_m(x)|_{x=j2^{-n}}$ are gained. Since

$$\varphi_{n,k}(x)= 2^{n/2}\varphi_{\tilde{N}}(2^n x - k) \tag{4.49}$$

with $\mathrm{Supp}[\varphi_{n,k}(y_1)] = [k2^{-n}, (\tilde{N}+k)2^{-n}]$, the integral in the right side of Eq. (4.47) becomes the form:

$$\int_0^x \varphi_{n,k}(y)dy = \begin{cases} 0 & x \le k2^{-n} \\ 2^{n/2}\int_{k2^{-n}}^x \varphi_{\tilde{N}}(2^n y \quad k)dy & x \in (k2^{-n}, (\tilde{N}+k)2^{-n}) \\ 2^{-n/2} & x \ge (\tilde{N}+k)2^{-n} \end{cases}$$
$$= \begin{cases} 0 & x \le k2^{-n} \\ 2^{-n/2}\int_0^{x2^n-k} \varphi_{\tilde{N}}(y)dy & x \in (k2^{-n}, (\tilde{N}+k)2^{-n}) \\ 2^{-n/2} & x \ge (\tilde{N}+k)2^{-n} \end{cases}. \tag{4.50}$$

Denote

$$\Phi^1_{n,k}(x) - \int_0^x \varphi_{n,k}(y)dy. \tag{4.51}$$

Then, the values of $\Phi^1_{n,k}(x)$ at the dyadic points $j2^{-n}$ can be gained by

$$\Phi^1_{n,k}(j2^{-n}) = \begin{cases} 0 & j \le k \\ 2^{-n/2}\theta_1(j-k) & j \in (k, \tilde{N}+k) \\ 2^{-n/2} & j \ge \tilde{N}+k \end{cases}. \tag{4.52}$$

Since the values of $\theta_1(j)$ can be obtained at the integer points by an algebraic arithmetic, hence, the values of $\Phi^1_{n,k}(j2^{-n})$ for $\int_0^x \varphi_{n,k}(y)dy$ at the dyadic points can be gained too. In theory, this calculation for the integral has no loss of accuracy. Further, the approximation of Eq. (4.47) at the dyadic points can be got by the form:

$$g_1(j2^{-n}) = \int_0^{j2^{-n}} f(y)dy \approx \sum_{k\in Z} c_{n,k}\Phi^1_{n,k}(j2^{-n}). \tag{4.53}$$

Following the same procedure, one can further get those higher multiply integral of $f(x)$, e.g., $\int_0^x \int_0^{y_1} f(y)dydy_1$, if they are needed. For saving space, here, we omit their deductions. With the same reason, the data processing under Eq. (4.53) has the same order of precision as shown in Eq. (4.18). With the similar denotation used above, we have

$$\begin{aligned}\varepsilon_{g_1} &= |g_1(j2^{-n}) - g_1^*(j2^{-n})| \le |\sum_{k\in Z}(f_{n,k} - f_{n,k}^*)\Phi_{n,k}^1(j2^{-n})| + |\eta_f(x)| \\ &\le (2\tilde{N} - 1)C_3 \max|\varepsilon_{f_{n,k}}| + C2^{-nr}\end{aligned} \quad (4.2.54)$$

in which $C_3 = \max\{\int_0^x \varphi_{n,k}(y)dy\}$.

4.3 Boundary Extension Technology and Its Error Estimation

In the previous two sections, the error analyses are displayed for the approximation of a function $f(x)$ defined in the region $R = (-\infty, \infty)$ [8, 9]. In practice, almost all problems are considered in a finite region, e.g., $[a, b](b > a)$. As pointed in Chap. 3, such finite region for a function can be always reduced into the region of [0, 1] by using the variable transform of $y = (x - a)/(b - a)$. In such case, the following discussion is conducted in the region of [0, 1] without losing generality. According to the approximation of Eq. (4.3), one can find that the number of convolution summations at each dyadic point is finite and equal to $\tilde{N} - 1$, i.e.,

$$P_n f(j2^{-n}) = \sum_{k\in Z} c_{n,k}\varphi_{n,k}(j2^{-n}) = 2^{n/2} \sum_{k=j+1-\tilde{N}}^{j-1} c_{n,k}\varphi_{\tilde{N}}(j-k) \quad j = 0, 1, \ldots, 2^n, \quad (4.55)$$

where $\varphi_{\tilde{N}}(0) = \varphi_{\tilde{N}}(\tilde{N}) = 0$ has been considered. When the generalized Coiflets are used, we have

$$c_{n,k} = \int_{k2^{-n}}^{(k+\tilde{N})2^{-n}} f(x)\varphi_{n,k}(x)dx \approx 2^{-n/2} f(\frac{k + M_1}{2^n}) = 2^{-n/2} f_{n,k+M_1}. \quad (4.56)$$

From Eqs. (4.55) and (4.56), one can found that when x approaches the ends of region [0, 1], the calculations of both decomposition and reconstruction request the values of $f(x)$ outside [0, 1]. For example,

$$\begin{aligned}P_n f(0) &= \sum_{k=1-\tilde{N}}^{-1} f_{n,k+M_1}\varphi_{\tilde{N}}(-k) = \sum_{k=1-\tilde{N}+M_1}^{M_1-1} f_{n,k}\varphi_{\tilde{N}}(M_1 - k) \\ &= \sum_{k=1-\tilde{N}+M_1}^{-1} f_{n,k}\varphi_{\tilde{N}}(M_1 - k) + \sum_{k=0}^{M_1-1} f_{n,k}\varphi_{\tilde{N}}(M_1 - k)\end{aligned}, \quad (4.57)$$

$$
\begin{aligned}
P_n f(1) = \sum_{k=2^n+1-\tilde{N}}^{2^n-1} f_{n,k+M_1}\varphi_{\tilde{N}}(2^n-k) &= \sum_{k=2^n+1-\tilde{N}+M_1}^{2^n-1+M_1} f_{n,k}\varphi_{\tilde{N}}(2^n-k+M_1) \\
= \sum_{k=2^n+1-\tilde{N}+M_1}^{2^n} f_{n,k}\varphi_{\tilde{N}}(2^n-k+M_1) &+ \sum_{k=2^n+1}^{2^n-1+M_1} f_{n,k}\varphi_{\tilde{N}}(2^n-k+M_1)
\end{aligned}
\quad (4.58)
$$

For the generalized Coiflets, we always take $1 < M_1 < \tilde{N}$. In such situation, $f_{n,k} = f(k2^{-n})(k = 1 - \tilde{N} + M_1, 2 - \tilde{N} + M, \ldots, -1)$ and $k = 2^n + 1, 2^n + 2, \ldots, 2^n - 1 + M_1$ are outside [0, 1]. Denote

$$
\Omega_0 = [(1 - \tilde{N} + M_1)2^{-n}, 0) \quad , \quad \Omega_1 = (1, 1 + (M_1 - 1)2^{-n}] \quad (4.59)
$$

outside [0, 1] at $x = 0$ and $x = 1$, respectively. When the values of $f_{n,k}$ in both Ω_0 and Ω_1 are directly taken by zero, as pointed out in Chap. 3, the reconstruction in Eq. (4.55) near the ends may yield the jumping phenomena similar to Fourier's transform, which will directly lead to the loss of accuracy. In order to avoid this accuracy loss, we need to select a boundary extension method such that the value of $f(x)$ in $\Omega_e(= \Omega_0 \cup \Omega_1)$ can be expressed by the values of $f(x)$ in [0, 1], as done in other numerical methods.

4.3.1 General Criterion for Boundary Extension Based on Error Analysis

In generality, boundary extension function mentioned above can be denoted by $g(x)$ in $\Omega_e \cup [0, 1]$. After that, we have

$$
\overline{f}(x) = \begin{cases} f(x), x \in [0, 1] \\ g(x), x \in \Omega_e \end{cases} \quad (4.60)
$$

and the decomposition coefficients in Eq. (4.56) becomes

$$
\overline{c}_{n,k} = \int_{k2^{-n}}^{(k+\tilde{N})2^{-n}} \overline{f}(x)\varphi_{n,k}(x)dx \approx 2^{-n/2}\overline{f}(\frac{k+M_1}{2^n}) = \overline{f}_{n,k+M_1}. \quad (4.61)
$$

as well as its reconstruction is

$$
P_n\overline{f}(j2^{-n}) = 2^{-n/2} \sum_{k=j+1-\tilde{N}+M_1}^{j-1+M_1} \overline{f}_{n,k}\varphi_{\tilde{N}}(j-k) \quad , \quad x \in [0, 1]. \quad (4.62)
$$

For the extended function $g(x)$ in Ω_e, we have the following theorem:

Theorem 4.1 *If a continuous function* $f(x)$ *in* $[0, 1]$ *and* $g(x)$ *in* $\Omega_e \cup [0, 1]$ *are smooth enough, to the extended function* $\overline{f}(x)$ *defined by* Eq. (4.60), *the decomposition and reconstruction in* Eqs. (4.61) and (4.62) *satisfy the following error estimations*:

$$\left\| P_n \overline{f}(x) - f(x) \right\|_\infty \le D_1 \varepsilon_{g,f} + C_{\infty,1} 2^{-nr}, \tag{4.63}$$

$$\left\| P_n \overline{f}(x) - f(x) \right\|_{L^2[0,1]} \le 2^{-n/2} D_2 \varepsilon_{g,f} + C_{2,0} 2^{-nr} \tag{4.64}$$

in which D_1, D_2, $C_{\infty,1}$, $C_{2,0}$ are different bounded constants greater than zero, n and r are the resolution level and the number of vanishing moments in the generalized Coiflets. Here,

$$\varepsilon_{g,f} = \| g(x) - f(x) \|_\infty = \max\{|g(x)|\},\ x \in \Omega_e. \tag{4.65}$$

Proof According to Eqs. (4.56) and (4.61), when $1 - \tilde{N} + M_1 \le k \le -1$ to the left extension, one gets.

$$\begin{array}{l} |\overline{c}_{n,k} - c_{n,k}| = | \int_{k2^{-n}}^{0} [\overline{f}(x) - f(x)] \varphi_{n,k}(x) dx | \\ \quad \le \varepsilon_{g,f} | \int_{k2^{-n}}^{0} \varphi_{n,k}(x) dx | \le 2^{-n/2} \varepsilon_{g,f} B |k| \end{array}, \tag{4.66}$$

where Eq. (4.65) and $\overline{f}(x) = f(x)$ in $[0, 1]$ are used, and $B = \max\{\varphi_{\tilde{N}}(x)\} = 2^{-n/2} \max\{\varphi_{n,k}(x)\}$. When $2^n + 1 \le k \le 2^n + M_1 - 1$, with the similar progress above, we have

$$\begin{array}{l} |\overline{c}_{n,k} - c_{n,k}| = | \int_{1}^{1+k'2^{-n}} [g(x) - f(x)] \varphi_{n,k}(x) dx | \\ \quad \le 2^{-n/2} \varepsilon_{g,f} k' B \end{array}, \tag{4.67}$$

in which, the transform for the superscript $k = 2^n - \tilde{N} + k'$ is employed when $1 \le k' \le M_1 - 1$. Because the extension index of k is finite, hence, both Eqs. (4.66) and (4.67) tell us that

$$|\overline{c}_{n,k} - c_{n,k}| \le 2^{-n/2} \varepsilon_{g,f} \overline{B} \tag{4.68}$$

is held to the extension. Here, $\overline{B} = B \max\{|min\{k\}|, max\{k'\}\}$. When $0 \le k \le 2^n - \tilde{N}$, the decomposition coefficients $c_{n,k}$ are independent on the extension. Hence, we have $\overline{c}_{n,k} = c_{n,k}$.

Further, the left side of Eq. (4.63) can be expressed by the form:

$$\begin{aligned}\left\|P_n\overline{f}(x)-f(x)\right\|_\infty &= \max_{x\in[0,1]}\{|\sum_{k=1-\tilde{N}}^{2^n-1}\overline{c}_{n,k}\varphi_{n,k}(x)-f(x)|\}\\ &\le \max_{x\in[0,1]}\{|\sum_{k=1-\tilde{N}}^{2^n-1}(\overline{c}_{n,k}-c_{n,k})\varphi_{n,k}(x)|+||P^j f(x)-f(x)||_\infty\cdot \\ &\le \max_{x\in[0,1]}\{\sum_{k\in\Re}|\overline{c}_{j,k}-c_{j,k}|\times|\varphi_{n,k}(x)|\}+C_{\infty,1}2^{-nr}\end{aligned} \quad (4.69)$$

By means of results for $|\overline{c}_{n,k}-c_{n,k}|$, Eq. (4.69) becomes the form:

$$\begin{aligned}&||P_n\overline{f}(x)-f(x)||_\infty \le 2(\tilde{N}-1+M_1)2^{-n/2}\varepsilon_{g,f}2^{n/2}\overline{B}+C_{\infty,1}2^{-nr}\\ &= D_1\varepsilon_{g,f}+C_{\infty,1}2^{-nr}.\end{aligned}$$

Here, $D_1=2(\tilde{N}-1+M_1)\overline{B}$, integer set $\Re=\{k;k\subset[M_1-\tilde{N},-1]\cup[2^n-\tilde{N}+1,2^n]\}$, and Eq. (4.30) with $C_{\infty,1}=D$ is used. That is, Eq. (4.63) holds.

To the left side of Eq. (4.64), we have

$$\begin{aligned}&\left\|P_n\overline{f}(x)-f(x)\right\|_{L^2[0,1]}=\{\textstyle\int_0^1[P_n\overline{f}(x)-f(x)]^2dx\}^{1/2}\\ &=\{\textstyle\int_0^{(\tilde{N}-1)/2^n}[P_n\overline{f}(x)-f(x)]^2dx+\int_{(\tilde{N}-1)/2^n}^{1-(\tilde{N}-1)/2^n}[P_n\overline{f}(x)-f(x)]^2dx+.\\ &\quad+\textstyle\int_{1-(\tilde{N}-1)/2^n}^{1}[P_n\overline{f}(x)-f(x)]^2dx\}^{1/2}\end{aligned} \quad (4.70)$$

By using Eq. (4.68) and Eq. (4.30) in Eq. (4.70), we have

$$\begin{aligned}&\left\|P_n\overline{f}(x)-f(x)\right\|_{L^2[0,1]}\\ &\le\{2\times2^{-n}(\tilde{N}-1)(D_1\varepsilon_{g,f}+C_{\infty,1}2^{-nr})^2+[1-2(\tilde{N}-1)2^{-n}](C_{\infty,1}2^{-nr})^2\}^{1/2}\\ &\le\sqrt{2\times2^{-n}(\tilde{N}-1)}(D_1\varepsilon_{g,f}+C_{\infty,1}2^{-nr})+\sqrt{[1-2(\tilde{N}-1)2^{-n}]}C_{\infty,1}2^{-nr}\\ &=2^{-n/2}D_2\varepsilon_{g,f}+2^{-nr}C_{2,0}.\end{aligned} \quad (4.71)$$

Here, $D_2=\sqrt{2\times2^{-n}(\tilde{N}-1)}D_1$, and $C_{2,0}=(\sqrt{2\times2^{-n}(\tilde{N}-1)}+\sqrt{[1-2(\tilde{N}-1)2^{-n}]})C_{\infty,1}$. That is, the inequality of Eq. (4.64) is true. Thus, Theorem 4.1 is proved.

Theorem 4.1. provides us a criterion to evaluate the precision order of an extension function $g(x)$ employed, which is dependent on the preselected extension technology.

Example 4.1 zero or constant extension, that is $g(x)\equiv C$. This is the simplest but worst extension. For such extension, $\varepsilon_{g,f}=\max\{C-f(x)\}|_{x\in\Omega_o}$. Thus, both $\left\|P_n\overline{f}(x)-f(x)\right\|_\infty$ and $|P_n\overline{f}(x)-f(x)||_{L^2[0,1]}$ have only precision order of $O(2^0)$ and $O(2^{-n/2})$, respectively.

In order to ensure the precision order of original approximation, hence, the extension function should be chosen by a suitable way. For such purpose, we have the following theorem:

Theorem 4.2 *If an extension function $g(x)$ has C_r continuity in $[0, 1] \cup \Omega_e$ and the vanishing moments of the form $\int_R g^{(i)}(x)\psi(x)dx = 0 (i = 0, 1, \ldots, r-1)$, then the approximation of $\overline{f}(x)$ defined by Eq. (4.60) under the generalized Coiflets has the following inequalities:*

$$|\overline{c}_{n,k} - c_{n,k}| \leq C2^{-n(r+1/2)}, \tag{4.72}$$

$$\left\| P_n\overline{f}(x) - f(x) \right\|_{\infty} \leq M_{\infty}2^{-nr}, \tag{4.73}$$

$$\left\| P_n\overline{f}(x) - f(x) \right\|_{L^2[0,1]} \leq M_2 2^{-n(r+1/2)}. \tag{4.74}$$

Here, M_{∞} and M_2 are bounded constants, n and r are the resolution level and the number of vanishing moments in the generalized Coiflets.

Proof Since the conditions of $g(x)$, we know that the following function:

$$g(x) = \sum_{i=0}^{r-1} \frac{g^{(i)}(x_0)}{i!}(x - x_0)^i + \frac{g^{(r)}(\zeta)}{r!}(x - x_0)^r \quad , \quad x, x_0 \in [0, 1] \cup \Omega_0 \tag{4.75}$$

satisfy them. Here, $|\zeta - x_0| < |x - x_0|$. Denote $G_r(x) = \sum_{i=0}^{r-1} \frac{g^{(i)}(x_0)}{i!}(x - x_0)^i$. According to the property of the wavelet transform, then we have $P_n G_r(x) = G_r(x)$. By using Eqs. (4.30) to (4.75), further, we have

$$|\overline{c}_{n,k} - c_{n,k}| \leq (D+D_0)2^{-n(r+1/2)} = C_3 2^{-n(r+1/2)}, \tag{4.76}$$

in which, $C_3 = D+D_0$. That is, Eq. (4.72) is held. Similar to the process of proof deduction in Theorem 4.1, once Eq. (4.76) is employed in Eqs. (4.63) and (4.64), there is no difficulty for one to get Eqs. (4.73) and (4.74). Therefore, the proof for Theorem 4.2 is finished.

Theorem 4.2 tells us that the best extension is to take the extension function $g(x)$ being a polynomial function of order $r - 1$ when we expect that the precision order of approximation of the generalized Coiflets can be ensured after the extension. Once the order is lower than $r - 1$, it is evident that the accuracy will be down. When the extension function $g(x)$ is a polynomial function of order $r - 1$, we get $\varepsilon_{f_{n,k}} \leq C_3 2^{-n(r+1/2)} \leq C_3 2^{-nr}$ which is used in Sect. 4.2.

4.3.2 Boundary Extension Arithmetic Using Lagrange Polynomial Functions

Before we conduct the boundary extension using the base Lagrange polynomial functions, let us see the properties of them. Denote the p different interpolation or extrapolation points by $x_0, x_1, \ldots, x_p$ for the base Lagrange polynomial functions of order p [10, 11]

$$u_{p,i}(x) = \prod_{\substack{j=0 \\ \neq i}}^{p} \frac{x - x_j}{x_i - x_j} \quad , i = 0, 1, 2, \ldots, p \tag{4.77}$$

It is evident that $u_{p,i}(x_j) = \delta_{i,j}(i, j = 0, 1, \ldots, p)$. For any C_p continuous function $f(x)$, we have the following expansion on the base Lagrange functions:

$$g(x) = \sum_{i=0}^{p} g_i u_{p,i}(x) + \frac{g^{(p+1)}(\zeta)}{p!}\omega_p(x). \tag{4.78}$$

Here, $x, \xi \in [0, 1] \cup \Omega_e$, $\omega_p(x) = (x - x_0)(x - x_1)\cdots(x - x_p)$, and $g_i = g(x_i)(i, j = 0, 1, \ldots, p)$.

Theorem 4.3 *To any polynomial functions $g(x)$ of order up to p, we have the following exact expression of the form:*

$$g(x) = \sum_{i=0}^{p} g_i u_{p,i}(x) \tag{4.79}$$

Proof Without losing generality, we take the polynomial functions of the form $g(x) = x^k(k = 0, 1, \ldots, p)$. Due to $g^{(p+1)}(x) \equiv 0$ for such situation, hence, Eq. (4.77) becomes into Eq. (4.78), or explicitly,

$$x^k = \sum_{i=0}^{p} x_i^k u_{p,i}(x), \quad k = 0, 1, \ldots, p \tag{4.80}$$

is exactly held. Hence, the proof of Theorem 4.3 is finished.

According to Eq. (4.80), we can soon get the following property.

Property 4.1 To any base Lagrange polynomial functions of order j from 1 to p, we have.

$$\sum_{i=0}^{j} u_{j,i}(x) \equiv 1, \quad j = 1, 2, \ldots, p \tag{4.81}$$

Equation (4.81) tells us that the base Lagrange polynomial functions with order p can exactly express those polynomial functions of order without greater than the order of the base Lagrange polynomial functions, while the summation of all base Lagrange polynomial functions for each order j when $1 \leq j \leq p$ is identically equal to 1 as shown in Eq. (4.81).

Under the general request introduced in the previous subsection, here, we introduce the arithmetic of the boundary extension technology for the approximation of a function under the generalized Coiflets. Considering the dyadic points $j2^{-n}$ are used in the approximation at the resolution level n, $\mathrm{Supp}[\varphi_{\tilde{N}}(x)] = [0.\tilde{N}]$, and the number of vanishing moments $r(= \tilde{N}+1)/3)$, here, we take Lagrange extrapolation polynomial of order p such that all conditions in Theorem 4.2 are satisfied. That is,

$$g(x) = \sum_{i=0}^{p} g_i u_i(x), \quad 1 \leq p \leq r-1, \tag{4.82}$$

in which

$$u_i(x) = \prod_{\substack{j=0 \\ j \neq i}}^{p} \frac{x - x_j}{x_i - x_j}, \quad i = 0, 1, \ldots, p. \tag{4.83}$$

Here, $g_i = g(x_i)$ and $x_i = i2^{-n}$. It is obvious that $N_i(x_j) = \delta_{i,j}$ and

$$g(x_j) = \sum_{i=0}^{p} g_i u_i(x_j) = g_i \delta_{i,j}, \quad i, j = 0, 1, \ldots, p. \tag{4.84}$$

(I) Left extension outside [0, 1]: For such case, the dyadic points outer [0, 1] are $j2^{-n}$ $(j = 1-\tilde{N}+M_1, 2-\tilde{N}+M_1, \ldots, -1)$. We fixed the all points $x_i = i2^{-n}$ in [0, 1]. Then, for the extended points outer [0, 1], we have

$$g_{n,k} = g(k2^{-n}) = \sum_{i=0}^{p} g_i u_i(k2^{-n}), \quad -\tilde{N}+1+M_1 \leq k \leq -1. \tag{4.85}$$

Considering a natural extension for the values of $g_{n,k}$ outside [0, 1] formulated by those extrapolation values of function $f(x)$ in [0, 1], here, i.e.,

$$g_i = f(i2^{-n}) = f_{n,i}, \quad i = 0, 1, \ldots, p-1 \tag{4.86}$$

then, Eq. (4.85) becomes the form:

$$g_{n,k} = \sum_{i=0}^{p} f_{n,i} u_i(k2^{-n}), \quad -\tilde{N}+1 \leq k \leq -1 \tag{4.87}$$

in which

$$u_i(k2^{-n}) = \prod_{\substack{j=0 \\ j\neq i}}^{p} \frac{k-j}{i-j}, \quad -\tilde{N}+1 \leq k \leq -1, \quad i, j = 0, 1, \ldots, p. \tag{4.88}$$

(II) Right extension outside [0, 1]. For such case, the dyadic points outer [0, 1] are $j2^{-n}$ $(j = 2^n + 1, 2^n + 2, \ldots, 2^n + M_1 - 1)$. To the extension of Eq. (4.77), with the same way as the left extension, we take

$$x_i = 1 - i2^{-n}, \quad i = 0, 1, \ldots, p-1 \tag{4.89}$$

$$u_i(k2^{-n}) = \prod_{\substack{j=0 \\ j\neq i}}^{p} \frac{k+j-2^n}{i-j}, \quad k = 2^n + 1, 2^n + 2, \ldots, 2^n + M_1 - 1. \tag{4.90}$$

$$g_i = f(1 - i2^{-n}) = f_{n,2^n-i}, \quad i = 0, 1, \ldots, p. \tag{4.91}$$

Hence, the extension values outer [0, 1] at the end of $x = 1$ can be formulated by

$$g_{n,k} = \sum_{i=0}^{p} f_{n,2^n-i} u_i(k2^{-n}), \quad k = 2^n + 1, 2^n + 2, \ldots, 2^n + M_1 - 1. \tag{4.92}$$

According to Theorem 4.2, one finds that such continuous extrapolation extension has the precision of $O(2^{-np})$. When $p \geq r$, its precision is restricted by the order of $O(2^{-nr})$ in the approximate solution method using the generalized Coiflets. Therefore, the best continuous extension is to select $p = r - 1$ such that the precision in the numerical method is ensured with the best reform in the equations of unknowns $f_{n,k}$.

4.3.3 Numerical Test Examples of Approximations to a Known Function in a Finite Region

To show the ability of approximation method under the generalized Coiflets, here, we introduce some numerical test examples for some known functions in a finite region. For the simplicity and without losing generality, here, we take some known functions to numerically display the accuracy of the approximations when the generalized Coiflets are employed.

Example 4.2 Approximation to a known function $f(x) = x^6+20, x \in [0, 1]$.

To the 1D function in [0, 1], when the values of $f(x)$ outside [0, 1] are taken into account by different boundary extension programs, we get the decomposition coefficients under the generalized Coiflets with $\tilde{N} = 17$ and $M_1 = 7$ as follows:

$$c_{n,k} = \int_R (x^6 + 20)\varphi_{n,k}dx = \int_{k2^{-n}}^{(k+\tilde{N})2^{-n}} (x^6 + 20)\varphi_{n,k}(x)dx \approx (k + 7)2^{-n} + 20$$

and

$$f(x) \approx P_n f(x) = \sum_{k=1-\tilde{N}+M_1}^{2^n+M_1-1} \{ (k + 7)2^{-n} + 20\} \varphi_{n,k}(x).$$

In order to display the accuracy varying with different boundary extension programs, here, the zero extension, constant extension, periodic extension, natural extension, Lagrange polynomial extension with different order are applied in the numerical tests in this example.

(1) *Zero extension*: It means that all values outside [0, 1] needed in the reconstruction are chosen by zero.

(2) *Constant extension*: It means that all values outside [0, 1] at each end are taken by a constant which is equal to the value of the known function at the boundary.

(3) *Periodic extension*: This extension is usually used to the periodic boundary conditions, e.g., $f(0) = f(1)$ when the approximated function is unknown. This extension method implies that the values outside [0, 1] at each boundary are taken by a periodic condition about the ends, that is,

$$f(-i2^{-n}) = f(1 - i2^{-n}), \quad i = 1, 2, \ldots, 9$$

$$f(1+i2^{-n}) = f(i2^{-n}), \quad i = 1, 2, \ldots, 7$$

(4) *Natural extension*: This is conducted for a known function. In this method, the values outside [0, 1] are taken by the function values outside [0, 1]. It is obvious that such boundary extension is exact without loss of accuracy in the extension.

(5) *Lagrange polynomial extension with different order*: This boundary extension has been displayed in the previous subsection. Here, the order of Lagrange polynomial is taken by $p = 0, 1, 2, \ldots, 5$, respectively. Here, $p = 0$ is equivalent to the constant extension.

After that, the absolute errors are characterized by the form of both $E_\infty = \|P_n f(x) - f(x)\|_\infty$ in [0, 1] and $E_2 = \|P_n f(x) - f(x)\|_{L^2[0,1]}$. Since the dyadic points outside [0, 1] at $x = 0$ and $x = 1$ has 9 and 6 numbers, respectively, due to the transition of $M_1 = 7$ dyadic points, hence, to the leftmost and rightmost cases,

8 dyadic points in [0, 1] is the least request. In other words, the resolution level can be taken in the region of $n \geq 3$. Figure 4.1 displays the absolute errors varying with the resolution level n employed in the generalized Coiflets, where Fig. 4.1a, b are, respectively, drawn by E_∞ and E_2 to the above extensions.

From Fig. 4.1, one sees that the worst extensions are the periodic and zero extensions, in which the absolute errors are in the order of $O(1)$ and they are almost no decreasing with increasing the resolution level *n*, and the best one is the natural extension. When Lagrange polynomials are used in the boundary extension, the results reveal that absolute error values when $n = 3$, i.e., $E^{(3)}$, and the slope ratio *k*, are down with increasing the order *p* of Lagrange polynomial function from zero to 5. When we display the values of $E^{(3)}$ varying with *p*, as shown in Fig. 4.2, the data can be fitted by a suitably selected fitting function, e.g., here, $E^{(3)}(p) = \vartheta(p) = \alpha_0 + \alpha_1 p + \alpha_2 p^2$. Once the method of least squares is used to fit the obtained data of $E^{(3)}$ in this example, we get $\alpha_0 = -3.386$, $\alpha_1 = -1.078$, $\alpha_2 = -0.206$, and $R^2 = 0.9995$ for the case of E_∞, and to $\alpha_0 = -5.370$, $\alpha_1 = -1.248$, $\alpha_2 = -0.148$, and $R^2 = 0.9999$ for the case of E_2. The values of $R^2 = 0.9995$ and $R^2 = 0.9999$ for evaluating the fittings tells us that the fitting results are so excellent in these two cases. When one transfers the logarithm scaling into the decimal system, one gets $E(n, p) = C(p)2^{-k(p)(n-3)}$. Here, the symbol *p* is employed for a variable, and $C(p) = 2^{E^{(3)}(p)}$. It is obvious that the greatest $C(p)$ is appeared when $n = 3$ in each order of Lagrange polynomial extensions. Table 4.2 displays their values varying with *p*. According to Figs. 4.1 and 4.2 as well as Table 4.2, the absolute errors of E_2 is always lower than those of E_∞, while the slope ratios $k = k(p)$ or the down orders $2^{-k(p)n}$ of absolute error for E_2 is always lower than that of E_∞ about 0.5 or $2^{-0.5n}$ except for E_2 with $n = 5$ where the accuracy of generalized Coiflets employed should be in the order change of 2^{-6n}. These test results demonstrate that the previous error analysis is true. When $p = 5$

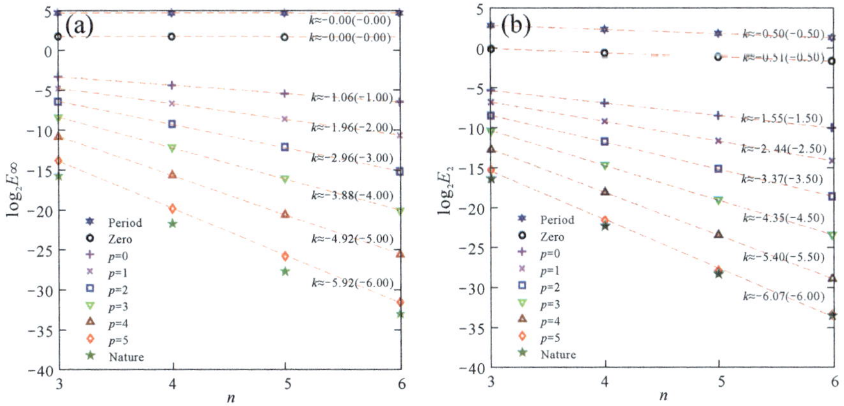

Fig. 4.1 Absolute errors E_∞ (Left) and E_2 (Right) of approximations $P_n f(x)$ of $f(x) = x^6 + 20$ in [0, 1] when the generalized Coiflets with $\tilde{N} = 17$ and different boundary extension arithmetic are used. Here, the slope ratio k to each line implies $\log_2 E = E^{(3)} + k(n-3)$ in which $E^{(3)}$ is the value of E when $n = 3$

Fig. 4.2 Changing of $E^{(3)}$ with order index p which is fitted by the function $E^{(3)}(p) = \vartheta(p) = \alpha_0 + \alpha_1 p + \alpha_2 p^2$. To the case E_∞, we get $\alpha_0 = -3.386$, $\alpha_1 = -1.078$, $\alpha_2 = -0.206$, and $R^2 = 0.9995$; and to the case E_2, we get $\alpha_0 = -5.370$, $\alpha_1 = -1.248$, $\alpha_2 = -0.148$, and $R^2 = 0.9999$

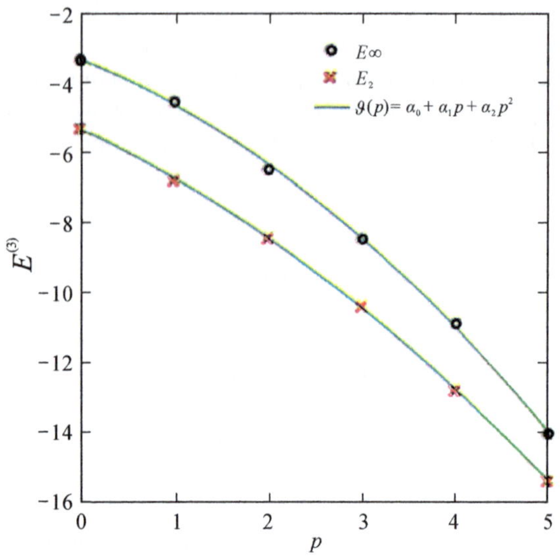

Table 4.2 Values of $C(p)$ varying with the order p of Lagrange polynomial extension for the cases of E_∞ and E_2

p	0	1	2	3	4	5
E_∞	9.49×10^{-2}	4.18×10^{-2}	1.10×10^{-2}	2.82×10^{-3}	5.34×10^{-4}	6.23×10^{-5}
E_2	2.46×10^{-2}	8.48×10^{-3}	2.85×10^{-3}	7.40×10^{-4}	1.44×10^{-4}	2.44×10^{-5}

and $n = 4, 5, 6$, for example, we have the respective errors in the orders of the form $E_\infty = 6.23 \times 10^{-5} \times 2^{-5.92}(<1.38 \times 10^{-6})$, $6.23 \times 10^{-5} \times 2^{-11.84}(<2.15 \times 10^{-8})$, and $6.23 \times 10^{-5} \times 2^{-14.76}(<2.69 \times 10^{-9})$.

Example 4.3 Approximation to the derivative functions of a known function. Here, we give two known functions of $f(x) = \sin(\pi x)$ and $f(x) = x^6$ in [0, 1] as examples.

To the former one of a periodic function, it is obvious that the periodic extension is equivalent to the natural extension. Denote the derivative functions of the specific functions by

$$f^{(m)}(x) = \frac{d^m f(x)}{dx^m}, \quad m = 1, 2, 3, 4 \tag{4.93}$$

One can soon give their explicit functions for both $f(x) = \sin(\pi x)$ and $f(x) = x^6$, which are the standard to evaluate the accuracy of their corresponding approximations when the generalized Coiflets-based approximation are used to the derivative functions in Eq. (4.93). If we directly use the approximation method to the explicit formulation of Eq. (4.93), the results of both approximations and their accuracy all are the same as those displayed in the original function $f(x)$ in [0, 1].

Here, we use the approximation method to them as same as done in the solving differential equations. That is, when the decomposition coefficients $f_{n,k+M_1} = f(\frac{k+M_1}{2^n})$ for the known function both in and outside [0, 1] are obtained by the generalized Coiflets with $\tilde{N} = 17$ and $n = 5$, i.e., the natural extension is employed, one can find that all values of $f_{n,k+M_1}$ obtained are the values of the original function. After that, the decomposition coefficients $f_{n,j}^{(m)}$ for the $(m \geq 1)$ can be gained by means of the formulation of in terms $f_{n,j}^{(m)}$ of $f_{n,k}$, see Eq. (4.38), i.e.,

$$f_{n,j}^{(m)}(j2^{-n}) = \sum_{k \in Z} \Gamma_{k-M_1,j-M_1}^{n,m} f_{n,k} \tag{4.94}$$

in the region $[0, 1] \cup \Omega_e$. Then, the approximate to the derivative functions is the reconstruction of each derivate function of the form:

$$f^{(m)}(x) \approx P_n f^{(m)}(x) = \sum_{k=1-\tilde{N}+M_1}^{2^n+M_1-1} \sum_{k \in Z} \Gamma_{k,j}^{n,m} f_{n,k} \varphi_{n,k-M_1}(x), \quad m \geq 1. \tag{4.95}$$

In order to give a comparison of the accuracy between the reconstruction of $f(x)$ and its derivative functions, here, there are two ways in the approximations of the derivative functions. One is from the direct calculations of Eq. (4.94) without reconstruction, and another is from the reconstruction of Eq. (4.95). In fact, these two ways can be used to the original function $f(x)$. To the former one, it is obvious that such decomposition coefficients $f_{n,k} = f(k2^{-n})(k = 0, 1, 2, \ldots, 2^n)$ are exactly equal to the values of $f(x)$ at the dyadic points in [0, 1] at the resolution level n. In other words, there is no accuracy lose when the decomposition coefficients $f_{n,k} = f(k2^{-n})$ are used to numerically express the function $f(x)$. Similar to this case, the obtained $f_{n,j}^{(m)}(j2^{-n})$ by Eq. (4.94) are also exact to the derivatives functions in [0, 1] when $f(x)$ is natural extension.

Figures 4.3 and 4.4 exhibit the absolute error distribution of the approximations of derivative functions for $f(x) = \sin(\pi x)$ and $f(x) = x^6$ in [0, 1], respectively, in which a comparison of them with that of the absolute errors of approximation for their original functions is also given. Here, the absolute errors are calculated by $E(x) = |f(x) - P_n f(x)|$ or a similar form. The fact of existing errors displayed in Figs. 4.3a and 4.4a tells us that there is some error in the calculations of connection coefficients in Eq. (4.94) since the wavelet is used for the calculations. If the connection coefficients are exactly calculated, otherwise, there should be no errors in the calculations of $f_{n,j}^{(m)}(j2^{-n})$ by means of Eq. (4.94) in theory because no errors appeared in the calculations of $f_{n,k+M_1}$ in the natural extension. At the same time, the absolute errors of $f_{n,j}^{(m)}(j2^{-n})$ in [0, 1] all are within ones of $P_n f(x)$, and most errors for the derivative functions are lower than the errors of their $P_n f(x)$.

After the approximations for the derivative functions are gained by the reconstruction of Eq. (4.95), their absolute error distributions are drawn in the Figs. 4.3b and

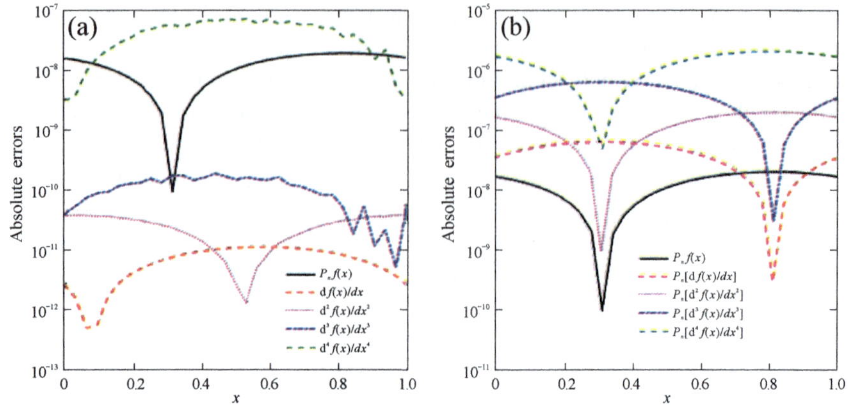

Fig. 4.3 A comparison of the distribution of absolute errors of approximations to the derivative functions $f^{(m)}(x)$ and to a known function $f(x) = \sin(\pi x)$ in [0, 1] using the generalized Coiflet with $\tilde{N} = 17$ and $n = 5$ as well as the natural extension. Here, the symbol with and without P_n are used to identify the approximations from Eqs. (4.95) and (4.94), respectively

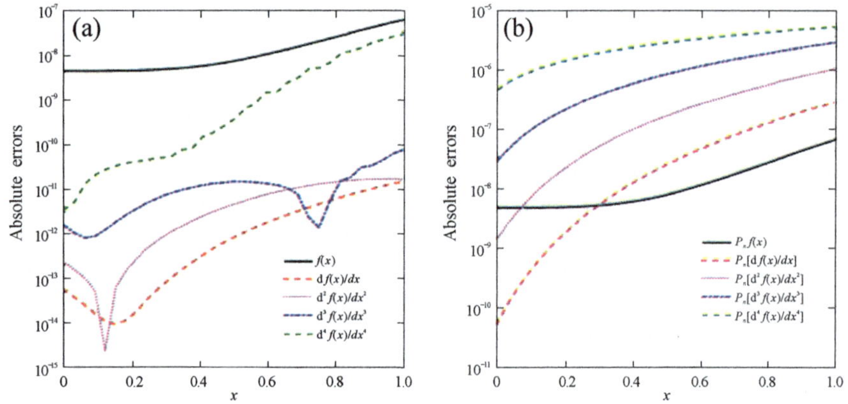

Fig. 4.4 A comparison of the distribution of absolute errors of approximations to the derivative functions $f^{(m)}(x)$ and to a known function $f(x) = x^6$ in [0, 1] using the generalized Coiflet with $\tilde{N} = 17$ and $n = 5$ as well as the natural extension. Here, the symbol with and without P_n are used to identify the approximations from Eq. (4.95) and Eq. (4.94), respectively

4.4b for the specific functions of $f(x) = \sin(\pi x)$ and $f(x) = x^6$ in [0, 1], respectively. Comparing them with their corresponding Figs. 4.3a and 4.4a, one finds that the error of each derivative function approximation based on the reconstruction is always higher than that without using the reconstruction, and that of $P_n f(x)$. As the derivative order increases, the error increases in the magnitude of order two due to the amplifying effect inherently existed in the derivative calculations if there are some errors in the calculations. Comparing to the error of $P_n f(x)$, for example, the absolute error of the reconstructed approximation to the fourth-order derivative function is within 10^{-5}, while the error of $P_n f(x)$ is within 10^{-7}.

From Figs. 4.5 and 4.6, one finds that the approximations of $P_n f(x)$ has almost no loss of accuracy under Lagrange polynomial extension, but the accuracy of the approximations for their derivative functions all are substantively increased by artificial boundary extension. As shown in the previous section, such Lagrange polynomial extension is the best one within the accuracy of the generalized Coiflets

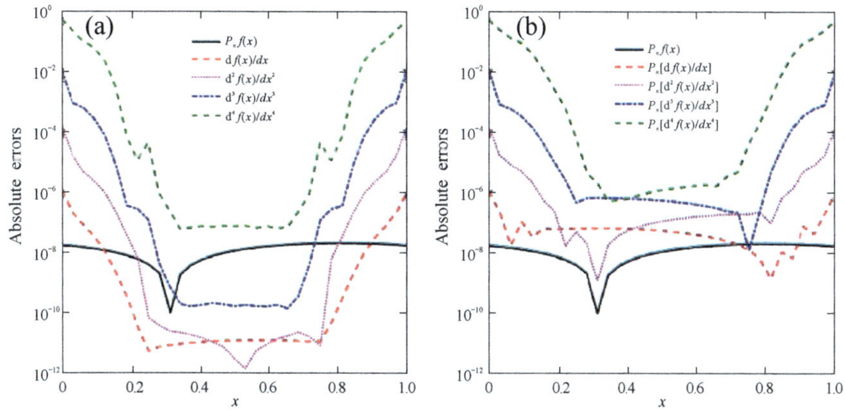

Fig. 4.5 A comparison of the distribution of absolute errors of approximations to the derivative functions $f^{(m)}(x)$ and to a known function $f(x) = \sin(\pi x)$ in [0, 1] using the generalized Coiflet with $\tilde{N} = 17$ and $n = 5$ as well as Lagrange polynomial extension. Here, the symbol with and without P_n are used to identify the approximations from Eq. (4.95) and Eq. (4.94), respectively

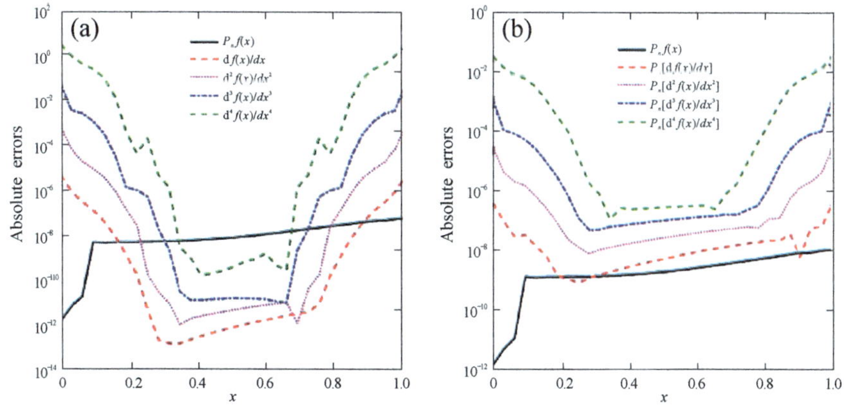

Fig. 4.6 A comparison of the distribution of absolute errors of approximations to the derivative functions $f^{(m)}(x)$ and to a known function $f(x) = x^6$ in [0, 1] using the generalized Coiflet with $\tilde{N} = 17$ and $n = 5$ as well as Lagrange polynomial extension. Here, the symbol with and without P_n are used to identify the approximations from Eq. (4.95) and Eq. (4.94), respectively

employed in the numerical tests, however, there are unavoidable errors to $f(k2^{-n})$ in Lagrange polynomial extension, and such errors are much amplified by Eq. (4.94) in the approximate calculations of the derivative functions. As shown in the previous section, the amplified factor to errors in the connection coefficients for derivatives of the base scaling functions is at least order of 2^{mn}, where m and n are the derivative order and the resolution level, respectively. When $m = 4$ and $n = 5$, we get $2^{nm} = 2^{20} \sim 10^7$, which implies that the accuracy order drops at least seven orders for the approximation of fourth-order derivative function. Although such an effect is mainly appeared in the regions near the ends of [0, 1], it can be decreased to the middle region. At present, such similar errors amplified by the derivative function calculations also appeared in all other numerical methods.

Example 4.4 Approximation to the nonlinear functions of a known function $f(x) = \sin(\pi x)$ in [0, 1]. Here, we consider the nonlinear functions of the forms:

$$f^2(x) = sin^2(\pi x),\ f^3(x) = sin^3(\pi x),\ f(x)f^{(1)}(x) = \pi \sin(\pi x)\cos(\pi x).$$

From the calculations of the generalized Coiflets, the calculations of the decomposition coefficients to the above nonlinear functions are, respectively, of the form:

$$f^2_{n,k} = sin^2(\pi k2^{-n}),\ f^3_{n,k} = sin^3(\pi k2^{-n}),\ f_{n,k}f^{(1)}_{n,k} = \sin(\pi k2^{-n})f^{(1)}_{n,k},$$

in which, $f^{(1)}_{n,k} = f^{(1)}(k2^{-n})$ and $f_{n,k} = f(k2^{-n}) = sin(\pi k2^{-n})$. Here, $f^{(1)}_{n,k}$ should be calculated by Eq. (4.94) rather than by its explicit formula when the generalized Coiflets is applied with Lagrange polynomial extension of order 5. The absolute errors of them are displayed in Fig. 4.7a, b for $n = 5$ and $n = 8$, respectively. From it, one can see that the absolute error of the nonlinear function has some increasing with the order of nonlinear function, and the difference between the approximations of nonlinear functions and $P_n f(x)$ is down with the resolution level n increasing. When $n = 5$, for example, the accuracy of $P_n f(x)$ is near the order of 10^{-8}, while the errors of $P_n f^3(x)$ and $P_n[f(x)f^{(1)}(x)]$ are within 10^{-6}. When $n = 8$, however, both errors of $P_n f^3(x)$ and $P_n[f(x)f^{(1)}(x)]$ are within 10^{-8} and the accuracy of $P_n f(x)$ has a litter increasing.

Example 4.5 Approximation to one order integral, i.e., $I(x) = \int_0^x f(y)dy,\ x \in [0, 1]$, of a known function $f(x) = \sin(\pi x)$ in [0, 1].

To such integral, the calculation equations are displayed in subsection 4.2.3. When Lagrange polynomial extension of order 5 is used, Fig. 4.8 shows its absolute error distribution and the change ratio of the absolute error at $x = 0.5$ varying with resolution level n. According to Fig. 4.8a, b, one sees that the accuracy of the integral is higher than that of the approximation of the integrated function, and the change ratio approaches to the highest accuracy for the integrated function.

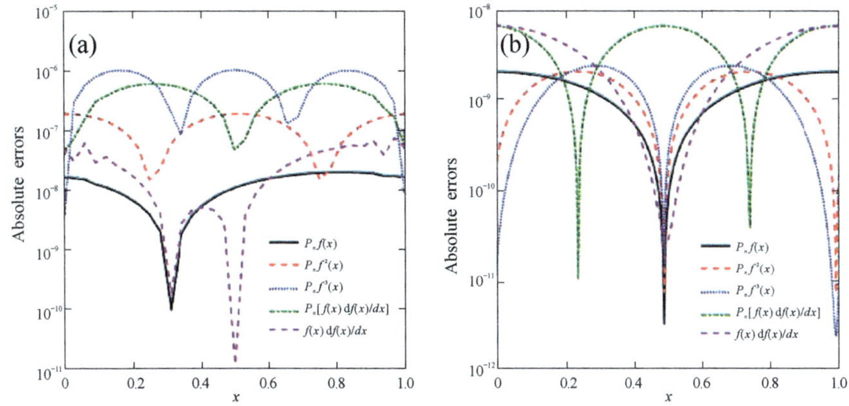

Fig. 4.7 A comparison of distribution of absolute errors of approximations with $n = 5$ (Left) and $n = 8$ (Right) to the nonlinear functions of $f^2(x)$, $f^3(x)$, $f(x)df(x)/dx$ and to a known function $f(x) = \sin(\pi x)$ in [0, 1] using the generalized Coiflet with $\tilde{N} = 17$ and Lagrange polynomial extension of order 5. Here, the symbol with and without P_n are used to identify the approximations in Eq. (4.95) and Eq. (4.94), respectively

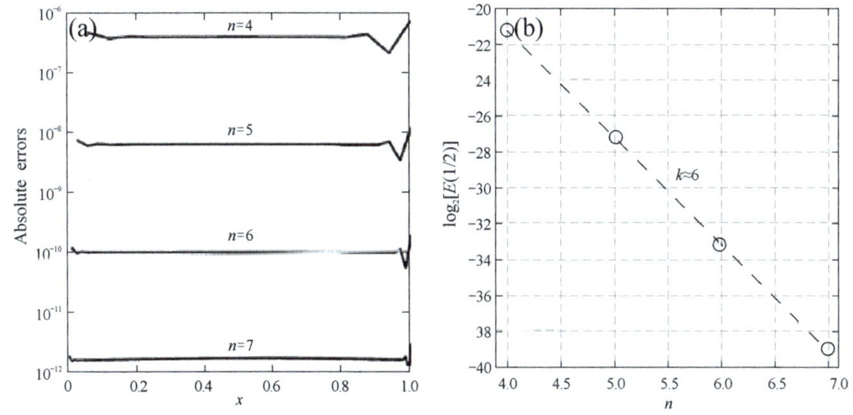

Fig. 4.8 Absolute error distribution of $I(x) = \int_0^x f(y)dy$, $x \in [0, 1]$ with $f(x) = \sin(\pi x)$ using the generalized Coiflets with $\tilde{N} = 17$ and different resolution level n (Left), and change ratio of absolute error at $x = 0.5$ varying with the resolution level (Right)

References

1. Ten DI (1999) lectures on wavelets. Society for Industrial and Applied Mathematics, Philadelphia
2. Meyer Y (1992) Wavelets and operators. Cambridge University Press, New York
3. Zhou YH, Wang JZ (1997) Generalized wavelet Gaussian integral method and its application in differential equations (in Chinese). Shanghai University Press, Modern Mathematics and Mechanics MMM-VII, Shanghai, pp 464–467

4. Zhou YH, Wang JZ (1999) Generalized Gaussian integral method for wavelet scaling function calculation and its application (in Chinese). Acta Mathematicav Scientia 19:293–300
5. Wang JZ, Zhou YH (1998) Error estimation of the generalized wavelet Gaussian integral method (in Chinese). J Lanzhou Univ 34:26–30
6. Wang JZ (2001) Generalized theory and arithmetic of orthogonal wavelets and applications to researches of mechanics including piezoelectric smart structures (in Chinese). PhD dissertation, Lanzhou University, China
7. Wang C (1994) A concise tutorial on numerical analysis (in Chinese). Higher Education Press, Beijing
8. Mallat SG (1989) A theory for multiresolution signal decomposition: the wavelet representation. IEEE Trans Pattern Anal Mach Intell 11:674–691
9. Cohen A (2003) Numerical analysis of wavelet methods. Elsevier, New York
10. Werner W (1984) Polynomial interpolation: lagrange versus Newton. Mathematics of Computation. 43:205–217
11. Rashed MT (2004) Lagrange interpolation to compute the numerical solutions of differential, integral and integro-differential equations. Appl Math Comput 151:869–878

Chapter 5
Wavelet-Based Solutions for Linear Boundary-Value Problems

The Galerkin method is one of the most popular weighted residual methods, as whose performance shows a good balance among accuracy, computation, and stability [1]. In the solution of mechanics problems, the weighted residual method can be strictly linked to the variational principle, further guaranteeing the reliability and universality of the method in the calculation and analysis of various engineering problems. During the implementation of the method, when selecting a wavelet series to represent the approximate solution, the corresponding Galerkin method becomes the so-called wavelet Galerkin method. Considering the unique position of the Galerkin method in numerical calculations, the wavelet Galerkin method almost becomes the earliest attempt of solving mechanics problems among all wavelet-based numerical methods, which continues to attract the interest of researchers so far.

As early as 1990, Latto and Tenenbaum [2] studied the Burgers equation using the wavelet Galerkin method. After that, various wavelet-based numerical techniques have attracted widespread attention from researchers. For example, Amaratunga et al. [3] solved the simple 1D second-order boundary-value problem by using the Daubechies scaling function as a trial function. However, as pointed out by Xu and Shann [4], there exist two difficult problems to be solved for the wavelet Galerkin method, which directly uses the wavelet function or the scale function as a trial function. The first one is how to eliminate the jitter instability of the solution near the boundary, which arose by the truncation of the series and the implementation of the essential boundary conditions. The second one is how to accurately calculate various integrals (i.e., the so-called connection coefficients) of the products of wavelet/scaling basis functions and their derivatives. To apply boundary conditions, early researchers proposed the compatibility matrix method [3, 5]. However, this method can only make the solution approximately satisfy the boundary conditions. To remedy this drawback, Lu et al. [6] proposed a technical approach that can implement the boundary conditions by appending additional equations to the boundary nodes. Diaz et al. [7] successfully introduced the Lagrange multiplier method to deal with the boundary conditions associated with the wavelet Galerkin method. Nakagoshi and Noguchi

Y.-H. Zhou, *Wavelet Numerical Method and Its Applications in Nonlinear Problems*, Engineering Applications of Computational Methods 6,
https://doi.org/10.1007/978-981-33-6643-5_5

[8] also studied the bending of Mindlin plates by using the same technique. Applications have shown that the aforementioned methods of boundary condition treatment are effective. Particularly, the Lagrange multiplier method appears extremely universal. However, such an approach increases the dimension of the algebraic equations about the unknown solution, which eventually increases the computational cost. In consideration of this issue, Zhou et al. [9, 10] established the so-called generalized orthogonal Coiflets, and applied them to the wavelet Galerkin method. Given that the Coiflets exhibit quasi-interpolation, then essential boundary conditions can be realized by directly replacing relevant equations, resolving such a difficulty in the wavelet Galerkin method. For the calculation of connection coefficients, Chen et al. [11] proposed an accurate approach for that of Daubechies wavelets in a finite domain. However, their method contains some errors on specific derivations [9, 12]. In 1998, Zhou et al. [13] independently proposed a similar method on the exact calculation of high-order derivatives of the scaling functions of Daubechies wavelets. Later on, Zhou and Wang noticed the work by Chen [11] and corrected the errors in [11]. In 2001, detailed procedures for exact calculation of connection coefficients of general orthogonal wavelets were systematically discussed in the dissertation [9]. Subsequently, in 2007, Zhang et al. [12] also noticed and corrected the errors in [11], and published the results as a comment.

After solving the two aforementioned basic problems, researches on the wavelet Galerkin method have achieved considerable progress. The method is then widely used in the study of mechanical problems. For example, Reginska et al. [14] solved the linear heat conduction equation by using the Galerkin method associated with Meyer wavelet and discussed the stability and convergence of the method [15]. Kim and Jang [16] used an interpolated wavelet to establish an adaptive wavelet Galerkin method and successfully used it in the loading analysis of thin-walled box beams. There are also many other studies by using the wavelet Galerkin method, such as Kim et al. [17] optimized the topology of elastic structures and Qassab and Nair [18] solved the free vibration problem of elastic ropes. Zhou and Wang successfully solved the static, dynamic, and buckling problems of beam structures, and established a vibration control model on the piezoelectric intelligent structures based on the generalized Coiflets associated Galerkin method [9, 19–29]. The proposed methods have been applied in different engineering problems [30–35] including the vibration control of large spatial structures [32]. Other notable applications of the wavelet Galerkin method have included the elastoplastic problems, dynamic stress concentration problems, linear elastic fracture problems, and fluid mechanics [36–44].

We can see that the wavelet Galerkin method has achieved certain success in its applications to various fields in engineering, especially in linear problems, and exhibited better precision than conventional Galerkin method. Most recently, as the resolution of the two basic problems on treatment of boundary conditions and calculation of connection coefficients, the wavelet Galerkin method has gradually become a powerful tool for solving differential equations [9, 45–47].

5.1 One-Dimensional Boundary-Value Problems

5.1.1 The Wavelet Approximation Incorporating Boundary Extension

In this section, we will discuss the approximation of a function defined on a finite interval. Technique of boundary extension based on Lagrange polynomials will be adopted.

In view of the fact that all 1D boundary-value problems can be studied by transforming them into the interval [0,1] by means of coordinate translation and expansion, here we consider the function $f(x) \in L^2[0, 1]$ without loss of generality. According to Eqs. (4.85) and (4.92), when nodal points are considered in equally spacing by $1/2^n$, we can construct a new function $\tilde{f}(x)$, by boundary extension from the original one $f(x)$, as follows:

$$\tilde{f}(x) \approx \begin{cases} \sum_{i=0}^{r-1} u_{0,i}(x) f(x_i), & x < 0, \\ f(x), & 0 \le x \le 1, \\ \sum_{i=0}^{r-1} u_{1,i}(x) f(x_{2^n - i}), & x > 1 \end{cases} \tag{5.1}$$

where

$$\begin{cases} u_{0,i}(x) = \prod_{\substack{j=0 \\ j \neq i}}^{r-1} \frac{2^n x - j}{i - j}, \\ u_{1,i}(x) = \prod_{\substack{j=0 \\ j \neq i}}^{r-1} \frac{2^n (x-1) + j}{i - j}. \end{cases} \tag{5.2}$$

Considering the scaling function with compact support $[0, \widetilde{N}]$, the function $\tilde{f}(x)$ on [0,1] can be approximated as

$$\tilde{f}(x) \approx \sum_{k=1-\tilde{N}}^{2^n - 1} \tilde{f}\left(\frac{M_1 + k}{2^n}\right) \varphi(2^n x - k), \quad x \in [0, 1]. \tag{5.3}$$

After the linear transformation of the summation index is used, we have

$$\tilde{f}(x) \approx \sum_{k=1-\widetilde{N}+M_1}^{2^n - 1 + M_1} \tilde{f}\left(\frac{k}{2^n}\right) \varphi(2^n x - k + M_1), \quad x \in [0, 1]. \tag{5.4}$$

Substitution of Eq. (5.1) into Eq. (5.4) yields

$$\tilde{f}(x) \approx \sum_{k=0}^{r-1} f_{n,k}\left[\sum_{l=1-\tilde{N}+M_1}^{-1} u_{0,k}\left(\frac{l}{2^n}\right)\varphi(2^n x - l + M_1) + \varphi(2^n x - k + M_1)\right]$$
$$+ \sum_{k=r}^{2^n - r} f_{n,k}\varphi(2^n x - k + M_1)$$
$$+ \sum_{k=2^n-r+1}^{2^n} f_{n,k}\left[\sum_{l=2^n+1}^{2^n-1+M_1} u_{1,2^n-k}\left(\frac{l}{2^n}\right)\varphi(2^n x - l + M_1) + \varphi(2^n x - k + M_1)\right]. \tag{5.5}$$

We define a modified scaling function as

$$\phi_{n,k}(x) = \begin{cases} \sum\limits_{l=1-\tilde{N}+M_1}^{-1} u_{0,k}\left(\frac{l}{2^n}\right)\varphi(2^n x - l + M_1) + \varphi(2^n x - k + M_1), & 0 \le k \le r-1, \\ \varphi(2^n x - k + M_1), & r \le k \le 2^n - r, \\ \sum\limits_{l=2^n+1}^{2^n-1+M_1} u_{1,2^n-k}\left(\frac{l}{2^n}\right)\varphi(2^n x - l + M_1) + \varphi(2^n x - k + M_1), & 2^n - r + 1 \le k \le 2^n, \end{cases} \tag{5.6}$$

where $\varphi(x)$ is the generalized Coiflet-type orthogonal scaling function with the first-order moment $M_1 = 7$, and the number of vanishing moments of the corresponding wavelet function $r = 6$, which was developed in Chap. 2. Then in terms of Eqs. (5.1) and (5.6), we can obtain

$$f(x) \approx P^n f(x) = \sum_{k=0}^{2^n} f_{n,k}\phi_{n,k}(x), \quad x \in [0,\ 1]. \tag{5.7}$$

According to the analysis in Sect. 4.3, the approximation error can be expressed as

$$\left\| d^m f(x)/dx^m - d^m P^n f(x)/dx^m \right\|_{L^2[0,1]} = C2^{-n(r-m)}, \tag{5.8}$$

where non-negative integers $m < r = 6$, and the constant C only depends on the smoothness of the function $f(x)$.

For a function $f(x)$ with period $P = 1$, i.e., $f(x) = f(x+1)$, following a similar way we can have

$$f(x) \approx \sum_{k=0}^{2^n-1} f(k/2^n)\phi^P_{n,k}(x), \quad x \in [0, 1] \tag{5.9}$$

in which the modified wavelet basis can be written as

$$\phi_{n,k}^{P}(x) = \begin{cases} \varphi(2^n x - 2^n - k + 7) + \varphi(2^n x - k + 7), & k \in [0, 6], \\ \varphi(2^n x - k + 7), & k \in [7, 2^n - 10], \\ \varphi(2^n x + 2^n - k + 7) + \varphi(2^n x - k + 7), & k \in [2^n - 9, 2^n - 1]. \end{cases} \tag{5.10}$$

5.1.2 Galerkin-Wavelet Solution Program

To demonstrate how the wavelet Galerkin method can be used, we consider the following linear boundary-value problem

$$\begin{cases} \mathbf{L}u(x) = f(x), \ 0 \leq x \leq 1 \\ \left.\frac{d^i u}{dx^{i_x}}\right|_{x=0} = 0, \qquad \left.\frac{d^{j_s} u}{dx^{j_s}}\right|_{x=1} = 0 \end{cases} \tag{5.11}$$

where L denotes the differential operator, i_s, j_s, and s are non-negative integers, whose value determines the solvability of Eq. (5.11).

According to Eq. (5.7), functions $u(x)$ and $f(x)$ can be expanded as

$$u(x) = P^n u(x) + \widetilde{Q}^n u(x) = \sum_{k=0}^{2^n} u(k/2^n)\phi_{n,k}(x) + \widetilde{Q}^n u(x), \tag{5.12}$$

$$f(x) = \sum_{k=0}^{2^n} f\left(\frac{k}{2^n}\right)\phi_{n,k}(x_1) + \widetilde{Q}^n f(x). \tag{5.13}$$

And boundary conditions can be imbedded in Eq. (5.6) and applied in Eq. (5.12). From Eq. (5.9), we known that the truncation errors in Eqs. (5.12) and (5.13) are $Q^n u(x) = O(2^{-nr})$ and $Q^n f(x) = O(2^{-nr})$, respectively.

When Eqs. (5.12) and (5.13) are substituted into Eq. (5.11), we obtain

$$\sum_{k=0}^{2^n} u(k/2^n)\mathbf{L}\phi_{n,k}(x) - \sum_{k=0}^{2^n} f\left(\frac{k}{2^n}\right)\phi_{n,k}(x_1) = Q(x). \tag{5.14}$$

The remainder of the equation is

$$Q(x) = \widetilde{Q}^n f(x) - \mathbf{L}\widetilde{Q}^n u(x). \tag{5.15}$$

Following the basic procedure of the Galerkin method, after the weight function $\phi_{n,l}(x)$, $l = 0, 1, 2, \ldots, 2^n$, are multiplied on both sides of Eq. (5.14), and the integrals are taken to the induced equation on x from 0 to 1, then we have

$$\mathbf{AU} - \mathbf{BF} \approx \mathbf{0}, \tag{5.16}$$

where matrix $\mathbf{A} = \left\{a_{lk} = \int_0^1 \mathbf{L}\phi_{n,k}(x)\phi_{n,l}(x)dx\right\}$, $\mathbf{B} = \left\{b_{lk} = \int_0^1 \mathbf{L}^1\phi_{n,k}(x)\phi_{n,l}(x)dx\right\}$, vector $\mathbf{U} = \{u_k = u(k/2^n)\}$, and $\mathbf{F} = \{f_k = f(k/2^n)]\}$, the subscripts $k, l = 0, 1, 2, \ldots, 2^n$, in which, the generalized connection coefficients a_{lk} and b_{lk} can be accurately obtained using the method described in Appendix 5.1.

Solving Eq. (5.16), we can obtain the function values $u(k/2^n), k = 0, 1, 2, \ldots, 2^n$. When these values are substituted into Eq. (5.12), the approximate solution $P^n u(x)$ for Eq. (5.11) can be obtained.

As mentioned above, we have obtained the wavelet-based solution for linear differential equations as shown in Eq. (5.11). From the entire solution process, one can see that the solution format is highly uniform for any form of linear problems. During the discretization process, this solution format maintains a high degree of standardization, making the current wavelet solution easier to use than conventional methods.

5.1.3 Numerical Solution Results of 1D Poisson Equation

In this subsection, the efficiency and accuracy of the proposed wavelet method will be demonstrated by numerically solving the Poisson equations with various boundary conditions. To effectively evaluate the performance of the proposed method, thereafter, the maximum absolute error L_∞, the mean absolute error L_1, the relative error norm L_2, and the order of convergence $R_{\infty,1}$ will be considered, which are, respectively, defined as follows:

$$L_\infty = \max_k\left\{\left|u_k^{num} - u_k^{exact}\right|\right\}, \tag{5.17}$$

$$L_1 = \sum_k \left|u_k^{mum} - u_k^{exact}\right|/N, \tag{5.18}$$

$$L_2 = \int_\Omega \left(u^{mum} - u^{exact}\right)^2 d\Omega \Big/ \int_\Omega \left(u^{exact}\right)^2 d\Omega, \tag{5.19}$$

$$R_{\infty,1} = \frac{\log\left[L_{\infty,1}(N_2)/L_{\infty,1}(N_1)\right]}{\log\left[N_1^{1/D}/N_2^{1/D}\right]}, \tag{5.20}$$

where N and D stand for the number of grid points and the spatial dimension of the problem [48, 49].

Table 5.1 Mean absolute error L_1 and order of convergence R_1 for Example 5.1

Present			FDM [50]		
Grid N	L_1-error	Order R_1	Grid N	L_1-error	Order R_1
16	1.351E-04	–	40	4.422E-04	–
32	6.756E-06	4.32	80	1.132E-04	1.97
64	2.986E-07	4.50	160	2.736E-05	2.04

Table 5.2 Maximum absolute error L_∞ and order of convergence R_∞ for the problem

Present			FDM [50]		
Grid N	L_∞-error	Order R_∞	Grid N	L_∞-error	Order R_∞
16	7.801E-04	–	40	9.236E-04	–
32	6.578E-05	3.57	80	2.654E-04	1.79
64	4.532E-06	3.86	160	7.306E-05	1.86

Example 5.1 The 1D Poisson equation with Dirichlet boundary conditions is considered as follows:

$$\begin{cases} \frac{\partial^2 p}{\partial x^2} = 4\left[(1-2x)^2\pi^2 - 2\right]\sin(2\pi x) + 16\pi(1-2x)\cos(2\pi x) \\ p(0) = p(1) = 0, \quad x \in [0, 1] \end{cases}, \tag{5.21}$$

whose exact solution is $p(x) = -(1-2x)^2 \sin(2\pi x)$ [50].

Tables 5.1 and 5.2 represent the errors of the numerical solutions of the 1D Poisson problem in Eq. (5.21), acquired by using the proposed wavelet method with various values of grid points N. As shown in these tables, under the condition of less nodal points, the results obtained by using wavelet method exhibit considerably better numerical accuracy than those by using the FDM [50].

5.2 2D and 3D Boundary-Value Problems

5.2.1 Galerkin-Wavelet Solution Program

In the previous section, the closed wavelet method for solving linear boundary-value problems has been described in detail. Numerical properties, such as solution accuracy and convergence speed, have also been systematically discussed. However, in many fields of natural science and engineering, many problems should be described by using multidimensional boundary-value problems. As an example, we consider a linear 3D problem as follows:

$$\begin{cases} \mathbf{L}u(x,y,z) + f(x,y,z) = 0 \\ \left.\frac{\partial^{i_{x,s}+i_{y,s}+i_{z,s}} u(x,y,z)}{\partial^{i_{x,s}} x \partial^{i_{y,s}} y \partial^{i_{z,s}} z}\right|_{x,y,z=0,1} = 0 \end{cases}, \quad x, y, z \in [0, 1], \tag{5.22}$$

where L is the linear differential operators, values of integers i_x, i_y,i_z, and s constitute the solution condition of the problem.

In accordance with the approximation format of multidimensional wavelets, the unknown function $u(x, y, z)$ can be characterized as

$$u(x,y,z) \approx \sum_{k=0}^{2^{n_x}} \sum_{l=0}^{2^{n_y}} \sum_{m=0}^{2^{n_z}} u(k/2^{n_x}, l/2^{n_y}, m/2^{n_z}) \phi_{n_x,k}^x(x) \phi_{n_y,l}^y(y) \phi_{n_z,m}^z(z). \tag{5.23}$$

Simultaneously, on the basis of Sect. 3.6.2, the term $f(x, y, z)$ in the equation is regarded as a new function, which can be expanded to

$$f(x,y,z) \approx \sum_{k=0}^{2^{n_x}} \sum_{l=0}^{2^{n_y}} \sum_{m=0}^{2^{n_z}} f(k/2^{n_x}, l/2^{n_y}, m/2^{n_z})] \phi_{n_x,k}^x(x) \phi_{n_y,l}^y(y) \phi_{n_z,m}^z(z). \tag{5.24}$$

When Eqs. (5.23) and (5.24) are substituted into Eq. (5.22), we can derive

$$\begin{aligned} &\sum_{k=0}^{2^{j_x}} \sum_{l=0}^{2^{j_y}} \sum_{m=0}^{2^{j_z}} u(k/2^{n_x}, l/2^{n_y}, m/2^{n_z}) \mathbf{L}[\phi_{n_x,k}^x(x) \phi_{n_y,l}^y(y) \phi_{n_z,m}^z(z)] \\ &+ \sum_{k=0}^{2^{j_x}} \sum_{l=0}^{2^{j_y}} \sum_{m=0}^{2^{j_z}} f(k/2^{n_x}, l/2^{n_y}, m/2^{n_z}) (\phi_{n_x,k}^x(x) \phi_{n_y,l}^y(y) \phi_{n_z,m}^z(z)) \approx 0. \end{aligned} \tag{5.25}$$

Following the standard procedure of the Galerkin method, the weight function $\phi_{n_x,k_0}^x(x)\phi_{n_y,l_0}^y(x)\phi_{n_z,m_0}^z(z)$ is multiplied on both sides of Eq. (5.34), with subscripts $k_0 = 0, 1, \ldots, 2^{n_x}$, $l_0 = 0, 1, \ldots, 2^{n_y}$, and $m_0 = 0, 1, \ldots, 2^{n_z}$. Taking integration in the region $[0, 1]^3$ to the induced equations, we get

$$\mathbf{AU} + \mathbf{BF} \approx \mathbf{0}, \tag{5.26}$$

where matrices

$$\mathbf{A} = \left\{ a_{op} = \int_0^1 \int_0^1 \int_0^1 \mathbf{L}[\phi_{n_x,k}^x(x) \phi_{n_y,l}^y(y) \phi_{n_z,m}^z(z)] \phi_{n_x,k_0}^x(x) \phi_{n_y,l_0}^y(x) \phi_{n_z,m_0}^z(z) dx dy dz \right\}, \tag{5.27}$$

$$\mathbf{B} = \left\{ b_{op} = \int_0^1 \int_0^1 \int_0^1 \phi_{n_x,k}^x(x) \phi_{n_y,l}^y(y) \phi_{n_z,m}^z(z) \phi_{n_x,k_0}^x(x) \phi_{n_y,l_0}^y(x) \phi_{n_z,m_0}^z(z) dx dy dz \right\}, \tag{5.28}$$

and vectors $\mathbf{U} = \{u_p = u(k/2^{n_x}, l/2^{n_y}, m/2^{n_z})\}$, $\mathbf{F} = \{f_p = f(k/2^{n_x}, l/2^{n_y}, m/2^{n_z})\}$; and subscripts $o = (2^{n_y}+1)[(2^{n_x}+1)k+l]+m$, $p = (2^{n_y}+1)[(2^{n_x}+1)k_0+l_0]+m_0$, $k, k_0 = 0, 1, \ldots, 2^{n_x}$, $l, l_0 = 0, 1, \ldots, 2^{n_y}$, and $m, m_0 = 0, 1, \ldots, 2^{n_z}$. The generalized connection coefficients a_{op} and b_{op} in Eqs. (5.27) and (5.28) can be finally transformed into the product of the 1D connection coefficients, which can be accurately obtained using the method described in Appendix 5.1.

Similar to the one-dimensional case, we directly discretize the boundary conditions and replace the corresponding equations in Eq. (5.26). Then the function values $u(k/2^{n_x}, l/2^{n_y}, m/2^{n_z})$, $k_0 = 0, 1, \ldots, 2^{n_x}$, $l_0 = 0, 1, \ldots, 2^{n_y}$, and $m_0 = 0, 1, \ldots, 2^{n_z}$ can be obtained by solving the algebraic Eq. (5.26). Substitution of $u(k/2^{n_x}, l/2^{n_y}, m/2^{n_z})$ into Eq. (5.23) gives an approximate solution for Eq. (5.22).

Considering that the approximation format on multidimensional linear operators is similar to that of the 1D case, the aforementioned wavelet method for solving multidimensional linear boundary-value problems also exhibits the closed features as discussed in the previous section. Thus, the error estimate of the linear algebraic Eq. (5.26) can be easily obtained. And the current wavelet closed method can also solve the multidimensional problems in a very uniform way.

5.2.2 Numerical Solution Results of 2D Laplace and Poisson Problem

For the numerical solutions of multidimensional problems, we first take two-dimensional problems as examples.

Example 5.2 Consider the following 2D Laplace equation

$$\begin{cases} \frac{\partial^2 p}{\partial x^2} + \frac{\partial^2 p}{\partial y^2} = 0, \qquad x, y \subset (0, 2), \\ p(0, y) = -y^3,\ p(2, y) = -8 + 12y + 6y^2 - y^3, \\ p(x, 0) = -x^3,\ p(x, 2) = -8 + 12x + 6x^2 - x^3 \end{cases} \tag{5.29}$$

whose exact solution is $p(x, y) = 3x^2y + 3xy^2 - x^3 - y^3$ [48, 51].

Figure 5.1 shows the relative error norm L_2 of the numerical solutions for Eq. (5.29), which is obtained by using the local boundary integral equation method (LBIE) [51], the meshless local Petrov–Galerkin method (MLPG) [48], and the proposed wavelet method here. As shown in this figure, the proposed wavelet solution is much more accurate than other methods. And the accuracy is nearly independent of the number of grid points N, which differs from those obtained by using the LBIE and MLPG. It can be seen from Fig. 5.1 that the order of convergence of the adopted wavelet method is about 7.5 [48, 51]. Such a high order of accuracy may be attributed to that the wavelet expansion of the unknown function is exact, as solution of Eq. (5.29) is a polynomial with order less than the vanishing moment $r = 6$, of the

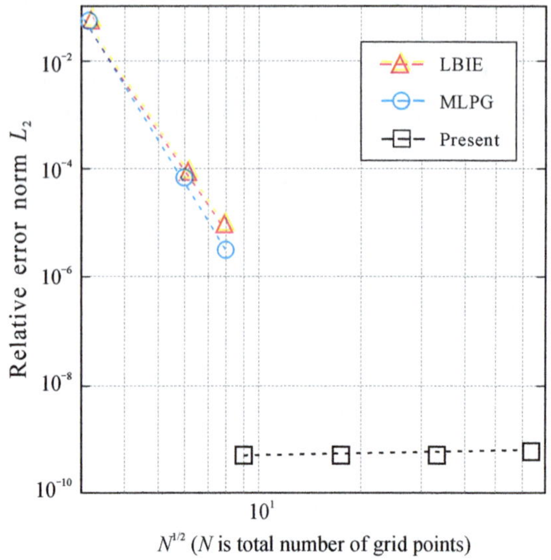

Fig. 5.1 Relative error norm L_2 varying with the grid number in one direction for the numerical solutions of Example 5.2. Here, a comparison of the proposed wavelet-based method (marked by "present") with others using LBIE and MLPG

wavelet function we used [9, 52, 53]. The very slight error of the proposed solution shown in Fig. 5.1 may be caused by rounding errors that are not specially addressed in this study.

Example 5.3 Consider the 2D Poisson equation

$$\frac{\partial^2 p}{\partial x^2} + \frac{\partial^2 p}{\partial y^2} = -8\pi^2 \sin(2\pi x)\sin(2\pi y), \quad x, y \in [0, 1]. \tag{5.30}$$

The equation is, respectively, subjected to the Dirichlet boundary conditions [49]

$$p(0, y) = p(1, y) = p(x, 0) = p(x, 1) = 0 \tag{5.31}$$

and the mixed boundary conditions

$$\begin{cases} \frac{\partial p}{\partial x} = 2\pi \sin(2\pi y), \quad x = 0, \\ p + \frac{\partial p}{\partial x} = 2\pi \sin(2\pi y), \quad x = 1, \\ p(x, 0) = p(x, 1) = 0. \end{cases} \tag{5.32}$$

The exact solution for this problem is $p(x, y) = \sin(2\pi x)\sin(2\pi y)$.

Tables 5.3 and 5.4 show the comparisons of the mean absolute error L_1 and the maximum absolute error L_∞ between solutions obtained by using different numerical methods for the 2D Poisson equation (5.30) with the Dirichlet boundary conditions (5.31). As shown in Tables 5.3 and 5.4, the present wavelet solutions are more accurate than those obtained by using the second-order finite volume method (TFVD) and the fourth-order finite volume method (FFVD) [49]. From Tables 5.3 and 5.4, we

Table 5.3 Error L_1 and order of convergence R_1 for Example 5.3

Grid N	Present wavelet		TFVD [49]		FFVD [49]	
	L_1-error	Order R_1	L_1-error	Order R_1	L_1-error	Order R_1
16 × 16	3.795E-05	–	–	–	–	–
32 × 32	7.566E-07	5.65	–	–	–	–
64 × 64	1.394E-08	5.76	1.075E-04	–	1.361E-07	–
128 × 128	2.710E-10	5.68	2.644E-05	2.02	8.490E-09	4.00
256 × 256	6.920E-12	5.29	6.562E-06	2.01	5.302E-10	4.00
512 × 512	2.589E-12	1.42	1.635E-06	2.00	3.312E-11	4.00

Table 5.4 Error L_∞ and order of convergence R_∞ for Example 5.3

Grid N	Present wavelet		TFVD [49]		FFVD [49]	
	L_∞-error	Order R_∞	L_∞-error	Order R_∞	L_∞-error	Order R_∞
16 × 16	1.959E-04	–	–	–	–	–
32 × 32	6.868E-06	4.83	–	–	–	–
64 × 64	2.217E-07	4.95	2.306E-04	–	3.182E-07	–
128 × 128	7.002E-09	4.98	5.457E-05	2.08	1.970E-08	4.01
256 × 256	2.199E-10	4.99	1.330E-05	2.04	1.228E-09	4.00
512 × 512	7.396E-12	4.89	3.286E-06	2.02	7.652E-11	4.00

can also see that the order of convergence $R_{\infty,1}$ of the proposed wavelet method is approximately 5, which clearly exceeds that of the finite volume methods [49]. Figure 5.2 shows the relation between the errors and the number of grid points N for the 2D Poisson equation (5.30) with the mixed boundary conditions (5.32). As shown in Fig. 5.2, the proposed wavelet method for solving the Poisson problems with complex boundary conditions is also highly accurate and efficient, with an order of convergence of $R_{\infty,1} \approx 4$ and a mean absolute error of $L_1 \approx 2.27 \times 10^{-7}$ for the number of grid points $N = 128 \times 128$.

5.2.3 *Numerical Solution Results of 3D Poisson Equation*

Example 5.4 Consider the 3D Poisson equation with Dirichlet boundary conditions

$$\begin{cases} \frac{\partial^2 p}{\partial x^2} + \frac{\partial^2 p}{\partial y^2} + \frac{\partial^2 p}{\partial z^2} = -3\pi^2 \sin(\pi x)\sin(\pi y)\sin(\pi z) \\ p(x, y, z) = 0, \quad x, y, z = 0, 1, \quad x, y, z \in [0, 1] \end{cases}. \tag{5.33}$$

It can be easily verified that the exact solution is $p(x, y, z) = \sin(\pi x)\sin(\pi y)\sin(\pi z)$ [54].

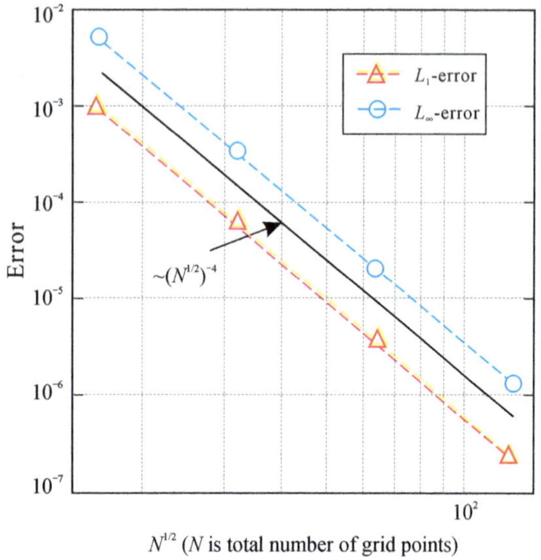

Fig. 5.2 Mean absolute error L_1 and maximum absolute error L_∞ varying with the grid points in one direction of the wavelet solutions for Example 5.3 with mixed boundary conditions

Table 5.5 Maximum absolute error L_∞ and order of convergence R_∞ for Example 5.4

Grid N	Present wavelet		FOS [54]		CDS [54]		HWM [55]	
	L_∞	R_∞	L_∞	R_∞	L_∞	R_∞	L_∞	R_∞
$8 \times 8 \times 8$	1.89E-04	–	2.35E-04	–	1.30E-02	–	1.35E-04	–
$16 \times 16 \times 16$	6.71E-06	4.82	1.43E-05	4.04	3.22E-03	2.01	3.55E-05	1.92
$32 \times 32 \times 32$	2.20E-07	4.93	9.04E-07	3.98	8.04E-04	2.00	–	–

Table 5.5 lists the maximum absolute error L_∞ and the order of convergence R_∞ of the numerical solutions for this 3D Poisson problem of Eq. (5.33), which are obtained by using the proposed wavelet method and other conventional methods. The results listed in Table 5.5 clearly show that the proposed wavelet solutions are more accurate than those provided by the fourth-order compact scheme (FOS) [54], the central difference scheme (CDS) [54], and the Haar wavelet method (HWM) [55].

Example 5.5 Consider the 3D Poisson equation

$$\frac{\partial^2 p}{\partial x^2} + \frac{\partial^2 p}{\partial y^2} + \frac{\partial^2 p}{\partial z^2} = e^x + e^y + e^z, \quad x, y, z \in [0, 1], \tag{5.34}$$

which is subjected to the Dirichlet boundary conditions that are extracted from the exact solution $p(x, y, z) = e^x + e^y + e^z$.

Relative to the exact solution, Fig. 5.3 shows the scaling curves of mean absolute error L_1 and the maximum absolute error L_∞ of the proposed wavelet-based solutions varying with the number of grid points in one-dimensional direction. As the number

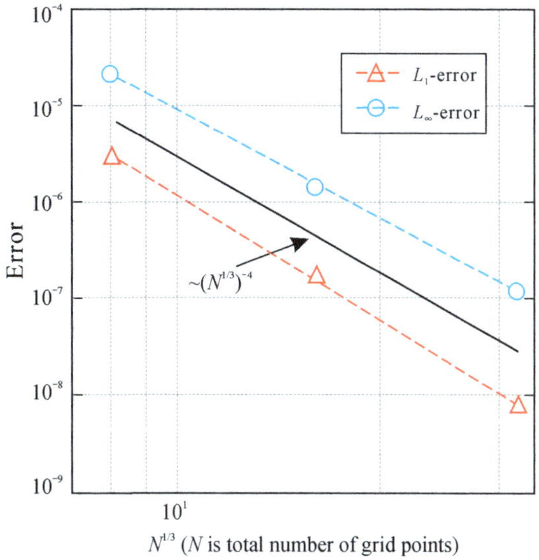

Fig. 5.3 Scaling curves of mean absolute error L_1 and maximum absolute error L_∞ of the wavelet-based solutions for Example 5.4 varying with the number of grid points in one direction or $N^{1/3}$

of grid points increase, the absolute errors decrease with $\sim (N^{1/3})^{-4}$, where N stands for the total meshes in the 3D problem. Hence, the proposed wavelet method achieves good accuracy and efficiency in solving the 3D Poisson problem of Eq. (5.34), e.g., the mean absolute error L_1 can reach 7.830×10^{-9} when the number of grid points $N = 32 \times 32 \times 32$.

In this section, an approximation scheme for an L^2-function defined on a 3D bounded space by combining techniques of boundary extension and Coiflet-type wavelet expansion is introduced. On the basis of such approximation scheme, a modified wavelet Galerkin method is proposed for the solution of Poisson equations with various boundary conditions. By numerically solving 1D, 2D, and 3D Poisson problems, the results demonstrate that the proposed wavelet achieves considerably better accuracy and convergence rate than many existing methods. It also demonstrates that the wavelet method has good capability in dealing with problems with mixed boundary conditions.

5.3 Deflection of Thin Rectangular Plate with Variable Thickness

5.3.1 Differential Equation with Variable Coefficients

In accordance with small deflection theory of thin plates, the following equilibrium equation for the variable thickness rectangular plate structure is derived

$$\nabla^2\left(D\nabla^2 w\right) - (1-\mu)\left(D_{,xx}w_{,yy} - 2D_{,xy}w_{,xy} + D_{,yy}w_{,xx}\right) = q,\ x, y \in (a, b)., \tag{5.35}$$

where $D = D(x, y)$ is the bending stiffness, $w = w(x, y)$ is the deflection of the plate, μ is Poisson's ratio, and $q = q(x, y)$ is the transversal distribution loads on the plate. By defining $D = D_0 D_1$, $D_0 = E/12\left(1-\mu^2\right)$, $D_1 = t(x, y) = h^3(x, y)$, where $h(x, y)$ is the thickness of the plate, and introducing $\bar{x} = x/a$, $\bar{y} = y/a$, $\bar{w} = w/a$, and $\bar{q} = qa^3/D_0$, we can write

$$\nabla^2\left(\overline{D}_1\nabla^2\bar{w}\right) - (1-\mu)\left(\overline{D}_{1,\bar{x}\bar{x}}\bar{w}_{,\bar{y}\bar{y}} - 2\overline{D}_{,\bar{x}\bar{y}}\bar{w}_{,xy} + \overline{D}_{,\bar{y}\bar{y}}\bar{w}_{,\bar{x}\bar{x}}\right) = \bar{q},\quad \bar{x}, \bar{y} \in (0, 1). \tag{5.36}$$

In the subsequent analysis, the "-" above each variable is disregarded. For example, w will still be used to define $\bar{w}$. Here, $a = b$ is chosen for a square thin plate.

5.3.2 Wavelet-Based Solution Program

When the conventional wavelet Galerkin method is applied to solve Eq. (5.36), one finds that such method will inevitably lead to the calculation of the integral of the form $\int_0^1 f(x)\phi_{m,k}(x)\phi_{m,j}^{(n)}(x)dx$, and such a calculation is extremely difficult. As shown in the follows, such difficulty can be overcome on the basis of the generalized wavelet Gaussian integral formula. Considering sufficiently smooth functions $f(x)$ and $u(x)$, we have the following relations

$$fu_{,x} = (fu)_{,x} - uf_{,x}, \tag{5.37}$$

$$fu_{,xx} = (fu)_{,xx} - 2\left(uf_{,x}\right)_{,x} + uf_{,xx}. \tag{5.38}$$

Then, Eq. (5.36) can be transformed into

$$\begin{aligned}&\nabla^4(wD_1) + \nabla^2\left(w\nabla^2 D_1\right) - 2\left[\nabla^2\left(wD_{1,x}\right)_{,x} + \nabla^2\left(wD_{1,y}\right)_{,y}\right]\\&- (1-\mu)\left[\left(wD_{1,xx}\right)_{,yy} + \left(wD_{1,yy}\right)_{,xx} - 2\left(wD_{1,xy}\right)_{,xy}\right] = q.\end{aligned} \tag{5.39}$$

As an example, here, the boundary conditions of a clamped plate are taken into account:

$$\begin{aligned}&w|_{x=0} = w|_{x=1} = 0,\ w|_{y=0} = w|_{y=b/a} = 0,\ w_{,x}|_{x=0}\\&= w_{,x}|_{x=1} = 0,\ w_{,y}|_{y=0} = w_{,y}|_{y=b/a} = 0.\end{aligned} \tag{5.40}$$

For simplicity, we set $a = b$. It can be easily verified that for sufficiently smooth functions $s(x, y)$, the function $w(x, y)s(x, y)$ also satisfies the boundary conditions in Eq. (5.40). We apply the generalized wavelet Gaussian integral method to the wavelet approximation of wD_1, $w\nabla^2 D_1$, $wD_{1,y}$, etc. Then, substitution of the approximation formulae into Eq. (5.39) yields

$$\begin{aligned}
&\sum_{i=1}^{2^n-1}\sum_{j=1}^{2^n-1}(wD_1)_{ij}\nabla^4\left[\phi_{n,i}\phi_{n,j}\right]+\sum_{i=1}^{2^n-1}\sum_{j=1}^{2^n-1}\left(w\nabla^2 D_1\right)_{ij}\nabla^2\left[\phi_{n,i}\phi_{n,j}\right]\\
&\quad-2\sum_{i=1}^{2^n-1}\sum_{j=1}^{2^n-1}\left(wD_{1,x}\right)_{ij}\left\{\nabla^2\left[\phi_{n,i}(x)\phi_{n,j}(x)\right]\right\}_{,x}\\
&\quad-2\sum_{i=1}^{2^n-1}\sum_{j=1}^{2^n-1}\left(wD_{1,y}\right)_{ij}\left\{\nabla^2\left[\phi_{n,i}(x)\phi_{n,j}(x)\right]\right\}_{,y}\\
&\quad-(1-\mu)\sum_{i=1}^{2^n-1}\sum_{j=1}^{2^n-1}\left(wD_{1,xx}\right)_{ij}\left[\phi_{n,i}(x)\phi_{n,j}(x)\right]_{,yy}\\
&\quad-(1-\mu)\sum_{i=1}^{2^n-1}\sum_{j=1}^{2^n-1}\left(wD_{1,yy}\right)_{ij}\left[\phi_{n,i}(x)\phi_{n,j}(x)\right]_{,xx}\\
&\quad+2(1-\mu)\sum_{i=1}^{2^n-1}\sum_{j=1}^{2^n-1}\left(wD_{1,xy}\right)_{ij}\left[\phi_{n,i}(x)\phi_{n,j}(x)\right]_{,yy}\approx q(x,y).
\end{aligned}$$

Selecting $\phi_{n,i'}(x)\phi_{n,j'}(y)$ as the weight function and multiplying both sides of the preceding formula by the weight function, we can derive the following algebraic equation in matrix form by considering the inner product

$$\mathbf{FW}=\mathbf{Q}, \tag{5.41}$$

where the coefficient matrix $\mathbf{F}$ and loading vector $\mathbf{Q}$ are

$$\begin{aligned}
\mathbf{F}=&\left\{D_{1,ij}\int_0^1\int_0^1\phi_{n,i'}(x)\phi_{n,j'}(y)\nabla^4\left[\phi_{n,i}(x)\phi_{n,j}(y)\right]dxdy\right.\\
&+\left(\nabla^2 D_1\right)_{ij}\int_0^1\int_0^1\phi_{n,i'}(x)\phi_{n,j'}(y)\times\nabla^2\left[\phi_{n,i}(x)\phi_{n,j}(y)\right]dxdy\\
&-2\left(D_{1,x}\right)_{ij}\int_0^1\int_0^1\phi_{n,i'}(x)\phi_{n,j'}(y)\left\{\nabla^2\left[\phi_{n,i}(x)\phi_{n,j}(y)\right]\right\}_{,x}dxdy\\
&-2\left(D_{1,y}\right)_{ij}\int_0^1\int_0^1\phi_{n,i'}(x)\phi_{n,j'}(y)\left\{\nabla^2\left[\phi_{n,i}(x)\phi_{n,j}(x)\right]\right\}_{,y}dxdy\\
&-(1-\mu)\left(D_{1,xx}\right)_{ij}\times\int_0^1\int_0^1\phi_{n,i'}(x)\phi_{n,j'}(y)\left[\phi_{n,i}(x)\phi_{n,j}(x)\right]_{,yy}dxdy
\end{aligned}$$

$$- (1-\mu)\left(D_{1,yy}\right)_{ij}\int_0^1\int_0^1 \phi_{n,i'}(x)\phi_{n,j'}(y)\times\left[\phi_{n,i}(x)\phi_{n,j}(x)\right]_{,xx}dxdy$$

$$- 2(1-\mu)\left(D_{1,xy}\right)_{ij}\int_0^1\int_0^1 \phi_{n,i'}(x)\phi_{n,j'}(y)\left[\phi_{n,i}(x)\phi_{n,j}(x)\right]_{,xy}dxdy \quad (5.42a)$$

$$\mathbf{Q} = \left\{\int_0^1\int_0^1 \phi_{n,i'}\phi_{n,j'}q(x,y)dxdy\right\}^T, \quad (5.42b)$$

where $i', j', i, j = 1, 2, \ldots, 2^n - 1$, the integration calculation of Eq. (5.42) can be realized by using the method introduced previously. Then unknown function values w_{ij} can be determined by solving Eq. (5.41).

5.3.3 Numerical Solution Results

Example 5.6 As an example, here, we consider the plate thickness with the form $h(x, y) = 2y/5 + 9/5$ and the external distribution load of the form

$$\begin{aligned} q(x,y) = {} & 16\pi^4 D_1(x,y)[4\cos(2\pi x)\cos(2\pi y) - \cos(2\pi x) - \cos(2\pi y)] \\ & + 4\pi^4 D_{1,yy}(x,y)[0.3\cos(2\pi x) + \cos(2\pi y) - 1.3\cos(2\pi x)\cos(2\pi y)] \\ & + 16\pi^3 D_{1,y}(x,y)[2\cos(2\pi x)\sin(2\pi y) - \sin(2\pi y)]. \end{aligned}$$

For such case, Eq. (5.39) has an analytical solution of the form

$$w(x,y) = [1 - \cos(2\pi x)][1 - \cos(2\pi y)].$$

Following the wavelet-based solution program introduced above, the numerical solution results are obtained by the Coiflet-based method with $\widetilde{N} = 17$ and $M_1 = 7$ when the resolution level n is taken by 3 and 4, respectively. Comparing with its exact solution of the problem, Fig. 5.4 displays the numerical results obtained and the exact solution as well as the absolute error between them. When the resolution level $n = 4$, for example, one knows that the meshes in the domain are 16×16. From Fig. 5.4, we find that the numerical solution results for the case are closely approaching to the exact solution, which shows that this wavelet-based method has high accuracy and low calculation. When $n = 4$, for example, Fig. 5.4b, d imply that the relative error for the maximum deflection at the center of the plate is within 0.3%.

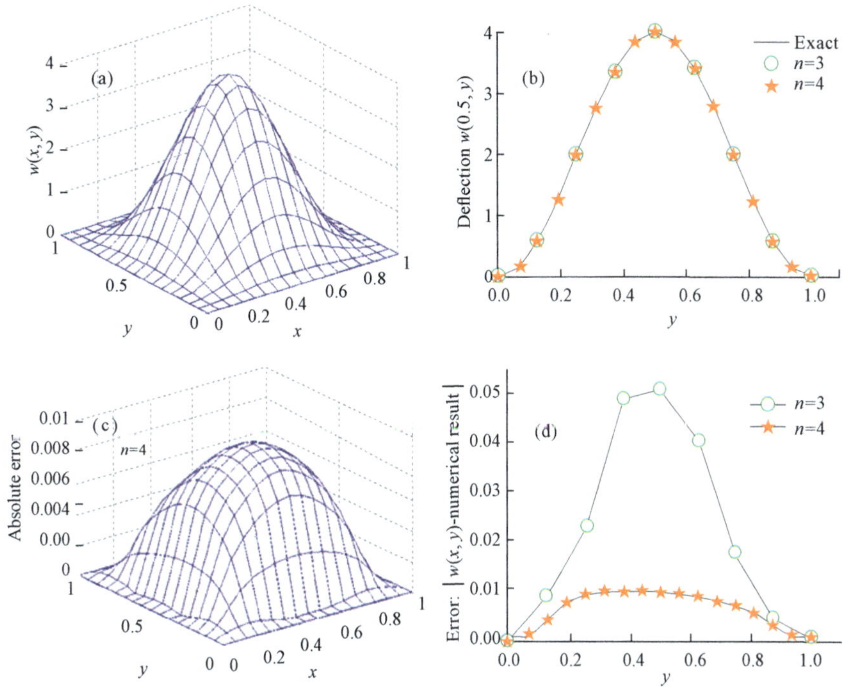

Fig. 5.4 Approximate deflection solutions of the coiflet-based method and their absolute errors between the exact solution and the approximation: **a** Deflection surface of approximation solution when $n = 4$; **b** Deflection curves of $w(0.5, y)$ from approximations and exact solution; **c** Absolute error distribution of approximation solution for $n = 4$ relative to exact solution; **d** Absolute error curves of approximation solutions relative to exact solution of $w(0.5, y)$

Appendix 5.1 Calculation of Connection Coefficients of Modified Basis Function

According to the definition of the modified scaling basis given in Eq. (5.6), the generalized connection coefficients can be expressed as

$$\Gamma_{k.l}^{n,m} = \int_0^1 \frac{d^m \phi_{n,k}(x)}{dx^m} \phi_{n,l}(x) dx = \begin{cases} \overset{\frown}{\Gamma}{}_{k.l}^{n,m} & \text{for } k \in [0, r-1], \\ \tilde{\Gamma}_{k.l}^{n,m} & \text{for } k \in [r, 2^n - r], \\ \breve{\Gamma}3_{k.l}^{n,m} & \text{for } k \in [2^n - r + 1, 2^n] \end{cases}, \tag{5.43}$$

where

$$
\widehat{\Gamma}^{n,m}_{k,l} = \begin{cases}
\sum\limits_{l_1=-9}^{-1}\sum\limits_{l_2=-9}^{-1} u_{0,k}\left(\frac{l_1}{2^n}\right)u_{0,l}\left(\frac{l_2}{2^n}\right)\overline{\Gamma}^{n,m}_{l_1-7,l_2-7} + \overline{\Gamma}^{n,m}_{k-7,l-7} & \\
+ \sum\limits_{l_1=-9}^{-1} u_{0,k}\left(\frac{l_1}{2^n}\right)\overline{\Gamma}^{n,m}_{l_1-7,l-7} + \sum\limits_{l_2=-9}^{-1} u_{0,l}\left(\frac{l_2}{2^n}\right)\overline{\Gamma}^{n,m}_{k-7,l_2-7}, & \text{for } l \in [0, r-1], \\
\sum\limits_{l_1=-9}^{-1} u_{0,k}\left(\frac{l_1}{2^n}\right)\overline{\Gamma}^{n,m}_{l_1-7,l-7} + \overline{\Gamma}^{n,m}_{k-7,l-7}, & \text{for } l \in \left[r, 2^n - r\right], \\
\sum\limits_{l_1=-9}^{-1}\sum\limits_{l_2=2^n+1}^{2^n+6} u_{0,k}\left(\frac{l_1}{2^n}\right)u_{1,2^n-l}\left(\frac{l_2}{2^n}\right)\overline{\Gamma}^{n,m}_{l_1-7,l_2-7} + \overline{\Gamma}^{n,m}_{k-7,l-7} & \\
+ \sum\limits_{l_1=-9}^{-1} u_{0,k}\left(\frac{l_1}{2^n}\right)\overline{\Gamma}^{n,m}_{l_1-7,l-7} + \sum\limits_{l_2=2^n+1}^{2^n+6} u_{1,2^n-l}\left(\frac{l_2}{2^n}\right)\overline{\Gamma}^{n,m}_{k-7,l_2-7}, & \text{for } l \in \left[2^n - r + 1, 2^n\right],
\end{cases} \tag{5.44}
$$

$$
\widetilde{\Gamma}^{n,m}_{k,l} = \begin{cases}
\sum\limits_{l_2=-9}^{-1} u_{0,l}\left(\frac{l_2}{2^n}\right)\overline{\Gamma}^{n,m}_{k-7,l_2-7} + \overline{\Gamma}^{n,m}_{k-7,l-7}, & \text{for } l \in [0, r-1], \\
\overline{\Gamma}^{n,m}_{k-7,l-7}, & \text{for } l \in [r, 2^n - r], \\
\sum\limits_{l_2=2^n+1}^{2^n+6} u_{1,2^n-l}\left(\frac{l_2}{2^n}\right)\overline{\Gamma}^{n,m}_{k-7,l_2-7} + \overline{\Gamma}^{n,m}_{k-7,l-7}, & \text{for } l \in [2^n - r + 1, 2^n],
\end{cases} \tag{5.45}
$$

$$
\breve{\Gamma}^{n,m}_{k,l} = \begin{cases}
\sum\limits_{l_1=2^n+1}^{2^n+6}\sum\limits_{l_2=-9}^{-1} u_{1,2^n-k}\left(\frac{l_1}{2^n}\right)u_{0,l}\left(\frac{l_2}{2^n}\right)\overline{\Gamma}^{n,m}_{l_1-7,l_2-7} + \overline{\Gamma}^{n,m}_{k-7,l-7} & \\
+ \sum\limits_{l_1=2^n+1}^{2^n+6} u_{1,2^n-k}\left(\frac{l_1}{2^n}\right)\overline{\Gamma}^{n,m}_{l_1-7,l-7} + \sum\limits_{l_2=-9}^{-1} u_{0,l}\left(\frac{l_2}{2^n}\right)\overline{\Gamma}^{n,m}_{k-7,l_2-7}, & \text{for } l \in [0, r-1], \\
\sum\limits_{l_1=2^n+1}^{2^n+6} u_{1,2^n-k}\left(\frac{l_1}{2^n}\right)\overline{\Gamma}^{n,m}_{l_1-7,l-7} + \overline{\Gamma}^{n,m}_{k-7,l-7}, & \text{for } l \in \left[r, 2^n - r\right], \\
\sum\limits_{l_1=2^n+1}^{2^n+6}\sum\limits_{l_2=2^n+1}^{2^n+6} u_{1,2^n-k}\left(\frac{l_1}{2^n}\right)u_{1,2^n-l}\left(\frac{l_2}{2^n}\right)\overline{\Gamma}^{n,m}_{l_1-7,l_2-7} + \overline{\Gamma}^{n,m}_{k-7,l-7} & \\
+ \sum\limits_{l_1=2^n+1}^{2^n+6} u_{1,2^n-k}\left(\frac{l_1}{2^n}\right)\overline{\Gamma}^{n,m}_{l_1-7,l-7} + \sum\limits_{l_2=2^n+1}^{2^n+6} u_{1,2^n-l}\left(\frac{l_2}{2^n}\right)\overline{\Gamma}^{n,m}_{k-7,l_2-7}, & \text{for } l \in \left[2^n - r + 1, 2^n\right].
\end{cases} \tag{5.46}
$$

According to Eq. (5.43), the calculation of the generalized connection coefficients $\Gamma^{n,m}_{k,l}$ can be reduced to the calculation of the integral

$$
\overline{\Gamma}^{n,m}_{k,l} = \int_0^1 \frac{d^m\varphi(2^n x - k)}{dx^m}\varphi(2^n x - l)dx. \tag{5.47}
$$

Let $y = 2^n x - l$, Eq. (5.47) can be transformed into

$$
\overline{\Gamma}^{n,m}_{k,l} = 2^{n(m-1)}\int_{-l}^{2^n - l} \frac{d^m\varphi(y - k + l)}{dy^m}\varphi(y)dy. \tag{5.48}
$$

Then Eq. (5.47) can be further expressed as

$$
\overline{\Gamma}^{n,m}_{k,l} = 2^{n(m-1)}[\Gamma^m_{k-l}(2^j - l) - \Gamma^m_{k-l}(-l)]. \tag{5.49}
$$

Since the compact support of the scaling function is $\left[0, \widetilde{N}\right]$, the relationship can be obtained according to Sect. 2.5.3

$$\Gamma_k^m(x) = 0, \quad |k| \geq \widetilde{N} \| x \leq 0 \| \; x \leq k, \tag{5.50}$$

$$\Gamma_k^m(x) = \overline{\Gamma}_k^m = \int_{-\infty}^{+\infty} \frac{d^m \varphi(x-k)}{dx^m} \varphi(x) dx, \quad x \geq \widetilde{N}, \tag{5.51}$$

$$\overline{\Gamma}_k^m = 0, \quad |k| \geq \widetilde{N}. \tag{5.52}$$

It can be seen from Eqs. (5.50) to (5.52) that the integral $\Gamma_k^m(x)$ needs to be calculated separately. The detailed calculation process of these integrals can be found in Sect. 2.5.3.

References

1. Fletcher CAJ (1984) Computational Galerkin methods. Springer, New York
2. Latto A, Tenenbaum E (1990) Compactly supported wavelets and the numerical solution of Burgers' equation. Comptes Rendus de Académie des Sci—Ser I—Math 311(13):903–909
3. Amaratunga K, Williams JR (1993) Wavelet based green's function approach to 2D PDEs. Eng Comput 10(4):349–367
4. Xu JC, Shann WC (1992) Galerkin-wavelet methods for two-point boundary value problems. Numer Math 63(1):123–144
5. Qian S, Weiss J (1993) Wavelets and the numerical solution of partial differential equations. J Comput Phys 106(1):155–175
6. Lu D, Ohyoshi T, Miura K (1997) Treatment of boundary conditions in one-dimensional wavelet-Galerkin method. JSME Int J Ser A Solid Mech Mater Eng 40(4):382–388
7. Díaz AR (1999) A wavelet–Galerkin scheme for analysis of large-scale problems on simple domains. Int J Numer Meth Eng 44(11):1599–1616
8. Nakagoshi S, Noguchi H (2001) A modified wavelet Galerkin method for analysis of Mindlin plates. JSME Int J Ser A Solid Mech Mater Eng 44(4):610–615
9. Wang JZ (2001) Generalized theory and arithmetic of orthogonal wavelets and applications to researches of mechanics including piezoelectric smart structures. PhD dissertation. Gansu: Lanzhou University (in Chinese)
10. Wang JZ, Wang XM, Zhou YH (2010) Numerical solutions to differential equations with strong nonlinearities based on series expansion of orthogonal scaling functions. J Lanzhou Univ (Natural Sciences) 46(5):96–101 (in Chinese)
11. Chen MQ, Hwang C, Shih YP (1996) The computation of wavelet-Galerkin approximation on a bounded interval. Int J Numer Meth Eng 39:2921–2924
12. Zhang T, Tian YC, Tadé MO, Utomo J (2007) Comments on 'the computation of wavelet-Galerkin approximation on a bounded interval. Int J Numer Meth Eng 72(2):244–251
13. Zhou YH, Wang JZ, Zheng XJ (1998) Applications of wavelet Galerkin FEM to bending of beam and plate structures. Appl Math Mech (English Edition) 19(8):745–755
14. Regińska T, Lars E (1997) Solving the sideways heat equation by a wavelet-Galerkin method. Inverse Prob 13(4):1093
15. Regińska T, Eldén L (2000) Stability and convergence of the wavelet-Galerkin method for the sideways heat equation. J Inverse Ill-posed Prob 8(1):31–49

16. Kim YY, Jang GW (2002) Hat interpolation wavelet-based multi-scale Galerkin method for thin-walled box beam analysis. Int J Numer Meth Eng 53(7):1575–1592
17. Kim JE, Jang GW, Kim YY (2003) Adaptive multiscale wavelet-Galerkin analysis for plane elasticity problems and its applications to multiscale topology design optimization. Int J Solids Struct 40(23):6473–6496
18. Al-Qassab M, Nair S (2004) Wavelet-Galerkin method for the free vibrations of an elastic cable carrying an attached mass. J Sound Vib 270(1–2):191–206
19. Liu XJ (2014) A wavelet method for uniformly solving nonlinear problems and its application to quantitative research on flexible structures with large deformation. PhD dissertation. Gansu: Lanzhou University. 2014 (in Chinese)
20. Wang X M (2012) Closed solution to nonlinear problem of beams/plates based on wavelet. PhD dissertation. Gansu: Lanzhou University (in Chinese)
21. Zhang L (2016) High-precision wavelet numerical methods and applications to nonlinear structural analysis. PhD dissertation. Gansu: Lanzhou University (in Chinese)
22. Zhou J, Zhou YH, Liew KM (2011) Wavelet-based method for stability analysis of vibration control systems with multiple delays. Comput Mech 47(2):161–170
23. Zhang L, Wang JZ, Zhou YH (2014) Wavelet method applied to specific adhesion of elastic solids via molecular bonds. Theor Appl Mech Lett 4(4):041009
24. Liu XJ, Wang JZ, Zhou YH (2011) A modified wavelet-Galerkin method for computation in structural mechanics with strong nonlinearity. Chin J Solid Mech 32(3):249–257 (in Chinese)
25. Liu XJ, Wang JZ, Zhou YH (2013) A wavelet method for studying large deflection bending problems of flexible beams. Chin J Solid Mech 33(S1):159–165 (in Chinese)
26. Wang XM, Liu XJ, Wang JZ, Zhou YH (2015) A wavelet method for bending of circular plate with large deflection. Acta Mech Solida Sin 28(1):83–90
27. Zhang L, Wang JZ, Zhou YH (2015) Wavelet solution for large deflection bending problems of thin rectangular plates. Arch Appl Mech 85(3):355–365
28. Zhang L, Wang JZ, Zhou YH (2016) Large deflection and post-buckling analysis of non-linearly elastic rods by wavelet method. Int J Non-Linear Mech 78:45–52
29. Wang JZ, Wang XM, Zhou YH (2012) A wavelet approach for active-passive vibration control of laminated plates. Acta Mech Sin 28(2):520–531
30. Yang Z, Liao S (2017) A HAM-based wavelet approach for nonlinear ordinary differential equations. Commun Nonlinear Sci Numer Simul 48:439–453
31. Fenik S, Starek L (2010) Optimal PI controller with position feedback for vibration suppression. J Vib Control 16(13):2023–2034
32. Si HW, Li DX (2003) Researches on wavelet-based vibration control of large space smart structures. J Nat Univ Defense Technol 25(3):14–18 (in Chinese)
33. Koziol P, Hryniewicz Z (2006) Analysis of bending waves in beam on viscoelastic random foundation using wavelet technique. Int J Solids Struct 43(22–23):6965–6977
34. Koziol P, Mares C, Esat I (2008) Wavelet approach to vibratory analysis of surface due to a load moving in the layer. Int J Solids Struct 45(7–8):2140–2159
35. Hryniewicz Z (2011) Dynamics of rayleigh beam on nonlinear foundation due to moving load using adomian decomposition and coiflet expansion. Soil Dyn Earthquake Eng 31(8):1123–1131
36. Tanaka S, Okada H, Okazawa S (2012) A wavelet Galerkin method employing B-spline bases for solid mechanics problems without the use of a fictitious domain. Comput Mech 50(1):35–48
37. Sannomaru S, Tanaka S, Yoshida K-i, Bui T Q, Okazawa S, Hagihara S (2017) Treatment of Dirichlet-type boundary conditions in the spline-based wavelet Galerkin method employing multiple point constraints. Appl Math Model 43:592–610
38. Tanaka S, Sannomaru S, Imachi M, Hagihara S, Okazawa S, Okada H (2015) Analysis of dynamic stress concentration problems employing spline-based wavelet Galerkin method. Eng Anal Bound Elem 58:129–139
39. Tanaka S, Okada H, Okazawa S, Fujikubo M (2013) Fracture mechanics analysis using the wavelet Galerkin method and extended finite element method. Int J Numer Meth Eng 93(10):1082–1108

40. Tanaka S, Suzuki H, Ueda S, Sannomaru S (2015) An extended wavelet Galerkin method with a high-order B-spline for 2D crack problems. Acta Mech 226(7):2159–2175
41. Dahmen W, Kunoth A, Urban K (1996) A wavelet Galerkin method for the stokes equations. Computing 56(3):259–301
42. Rathish Kumar BV, Mehra M (2005) Wavelet-taylor Galerkin method for the Burgers equation. BIT Numer Math 45(3):543–560
43. Mehra M, Kumar V (2007) Fast wavelet-taylor Galerkin method for linear and non-linear wave problems. Appl Math Comput 189(2):1292–1299
44. Mittal RC, Kumar S (2010) A numerical study of stationary solution of viscous Burgers' equation using wavelet. Int J Comput Math 87(6):1326–1337
45. Zhou YH, Wang JZ, Zheng XJ (1998) Applications of wavelet Galerkin FEM to bending of beam and plate structures. Appl Math Mech 19(8):697–706 (in Chinese)
46. Zhou YH, Wang JZ (1999) Generalized Gaussian integral method for calculations of scaling function transform of wavelets and its applications. Acta Mathematica Scientia 19(3):293–300 (in Chinese)
47. Liu XJ, Zhou YH, Wang XM, Wang JZ (2013) A wavelet method for solving a class of nonlinear boundary value problems. Commun Nonlinear Sci Numer Simul 18:1939–1948
48. Atluri SN, Zhu T (1998) A new meshless local Petrov-Galerkin (MLPG) approach in computational mechanics. Comput Mech 22(2):117–127
49. Barad M, Colella P (2005) A fourth-order accurate local refinement method for Poisson's equation. J Comput Phys 209(1):1–18
50. Gibou F, Fedkiw RP, Cheng LT, Kang M (2002) A second-order-accurate symmetric discretization of the Poisson equation on irregular domains. J Comput Phys 176(1):205–227
51. Zhu T, Zhang JD, Atluri SN (1998) A local boundary integral equation (LBIE) method in computational mechanics, and a meshless discretization approach. Comput Mech 21(3):223–235
52. Meyer Y (1992) Wavelets and operators. Cambridge University Press
53. Liu XJ, Zhou YH, Zhang L, Wang JZ (2014) Wavelet solutions of Burgers' equation with high Reynolds numbers. Sci China: Technol Sci 57(7):1285–1292
54. Zhang J (1998) Fast and high accuracy multigrid solution of the three dimensional Poisson equation. J Comput Phys 143(2):449–461
55. Shi Z, Cao Y, Chen Q (2012) Solving 2D and 3D Poisson equations and biharmonic equations by the Haar wavelet method. Appl Math Model 36(11):5143–5161

Chapter 6
Wavelet-Based Laplace Transformation for Initial- and Boundary-Value Problems

6.1 Essentials of Laplace Transformation

6.1.1 Laplace Transform of a Function and its Inverse Transform

Definition 6.1 If the function $f(t)$ is defined on $t \geq 0$, and the integral $\int_0^{+\infty} f(t)e^{-st}dt$ (s is a complex parameter) converges in a domain of s, such an integral is referred to as the Laplace transform of $f(t)$, denoted by

$$F(s) = \mathcal{L}[f(t)] = \int_0^{+\infty} f(t)e^{-st}dt. \tag{6.1}$$

When the Laplace transform $F(s)$ is obtained to $f(t)$, then its inverse transform can be expressed as

$$f(t) = \mathcal{L}^{-1}[F(s)] = \frac{1}{2\pi i}\int_{\beta-i\infty}^{\beta+i\infty} F(s)e^{st}ds,\ t > 0,$$

where $s = \beta + i\omega$, ω is a real variable, and β is a real constant satisfying the relation $f(t)e^{-\beta t} \in L^2(R)$.

Theorem 6.1 (The existence theorem of the Laplace transform) *If a function $f(t)$ satisfies the following conditions*:

(1) *When $t < 0$, $f(t) = 0$;*

Y.-H. Zhou, *Wavelet Numerical Method and Its Applications in Nonlinear Problems*,
Engineering Applications of Computational Methods 6,
https://doi.org/10.1007/978-981-33-6643-5_6

Table 6.1 Laplace transform of some common functions

$f(t)$	$F(s)$	$f(t)$	$F(s)$
1	$\frac{1}{s}$	$t \sin at$	$\frac{2as}{(s^2+a^2)^2}$
e^{at}	$\frac{1}{s-a}$	$t \cos at$	$\frac{s^2-a^2}{(s^2+a^2)^2}$
$t^m (m > -1)$	$\frac{\Gamma(m+1)}{s^{m+1}}$	$t \sinh at$	$\frac{2as}{(s^2-a^2)^2}$
$\sin at$	$\frac{a}{s^2+a^2}$	$t \cosh at$	$\frac{s^2+a^2}{(s^2-a^2)^2}$
$\cos at$	$\frac{s}{s^2+a^2}$	$\sin^2 t$	$\frac{1}{2s} - \frac{s}{s^2+4}$
$\sinh at$	$\frac{a}{s^2-a^2}$	$\cos^2 t$	$\frac{1}{2s} + \frac{s}{s^2+4}$
$\cosh at$	$\frac{s}{s^2-a^2}$	$\sin at \sin bt$	$\frac{2abs}{[s^2+(a+b)^2][s^2+(a-b)^2]}$

(2) *When $t \to +\infty$, the growth rate of $f(t)$ does not exceed a certain exponential function, that is, there are constants $M > 0$ and $c \geq 0$ such that $f(t)$ satisfies*

$$|f(t)| \leq Me^{ct}, \quad 0 \leq t < +\infty.$$

A function that satisfies the above conditions is called its increase not exceeding exponential, and c is its growth index. Then the Laplace transform of $f(t)$ must exist on the half-plane $\mathrm{Re}(s) > c$, and the integral at the right end converges absolutely and uniformly on $\mathrm{Re}(s) \geq c_1 > c$, as well as $F(s)$ is an analytic function in the half-plane of $\mathrm{Re}(s) > c$.

Table 6.1 displays the Laplace transform of some common functions. Next, we will introduce several basic properties of the Laplace transform, which are very useful in the computation of the Laplace transform.

Property 6.1 *If α and β are constants, and $\mathcal{L}[f_1(t)] = F_1(s)$, $\mathcal{L}[f_2(t)] = F_2(s)$, then, we have*

$$\begin{aligned}\mathcal{L}[\alpha f_1(t) + \beta f_2(t)] &= \alpha\mathcal{L}[f_1(t)] + \beta\mathcal{L}[f_2(t)],\\ \mathcal{L}^{-1}[\alpha F_1(s) + \beta F_2(s)] &= \alpha\mathcal{L}^{-1}[F_1(s)] + \beta\mathcal{L}^{-1}[F_2(s)].\end{aligned} \tag{6.2}$$

There is no difficulty for one to prove this property. In fact, Property 6.1 displays that the Laplace transform is linear.

Property 6.2 *If $\mathcal{L}[f(t)] = F(s)$ and a is a real number, then, we have*

$$\mathcal{L}[e^{at} f(t)] = F(s-a). \tag{6.3}$$

Proof According to Eq. (6.1), we have

$$\mathcal{L}[e^{at}f(t)] = \int_0^{+\infty} e^{at}f(t)e^{-st}dt = \int_0^{+\infty} f(t)e^{-(s-a)t}dt.$$

It can be seen from this relation that the right side of the above formula just replaces s with $s-a$ in $F(s)$, so one gets

$$\mathcal{L}[e^{at}f(t)] = F(s-a) \quad (\mathrm{Re}(s-a) > c).$$

Thus, the proof of Property 6.2 is finished.

Property 6.3 *If $\mathcal{L}[f(t)] = F(s)$, and when $t < 0$, $f(t) = 0$, then for any non-negative real number τ, we have*

$$\mathcal{L}[f(t-\tau)] = \mathrm{e}^{-s\tau}F(s). \tag{6.4}$$

Proof According to Eq. (6.1), we have

$$\mathcal{L}[f(t-\tau)] = \int_0^{+\infty} f(t-\tau)e^{-st}dt = \int_0^{\tau} f(t-\tau)e^{-st}dt + \int_{\tau}^{+\infty} f(t-\tau)e^{-st}dt.$$

Because of $f(t-\tau) = 0$ for $t < \tau$, the first integral at the right end of the above formula is zero. And for the second integral, letting $t - \tau = u$, then, we have

$$\mathcal{L}[f(t-\tau)] = \int_0^{+\infty} f(u)e^{-s(u+\tau)}du = \mathrm{e}^{-s\tau}\int_{\tau}^{+\infty} f(u)e^{-su}du = \mathrm{e}^{-s\tau}F(s) \quad (\mathrm{Re}(s) > c).$$

The above properties can make finding the Laplace transform of a more complex function easy once $F(s)$ is gained, e.g., those listed in Table 6.1.

Example 6.1 Consider the function.

$$f(t) = 3 + t^2 + 2e^{-t} + \sin 2t + e^{2t}t^3 + g(t-2)$$

in which $g(t) = 0$ for $t < 0$ and $g(t) = 1$, otherwise.

For such $f(t)$, the first three items are the common functions and their Laplace transforms are listed in Table 6.1. The latter two can be dealt with by using Properties 6.2 and 6.3, respectively. Eventually, we have

$$\begin{aligned}
&\mathcal{L}[3 + t^2 + 2e^{-t} + \sin 2t + e^{2t}t^3 + g(t-2)] \\
&= \mathcal{L}[3] + \mathcal{L}[t^2] + \mathcal{L}[2e^{-t}] + \mathcal{L}[\sin 2t] + \mathcal{L}[e^{2t}t^3] + \mathcal{L}[g(t-2)] \\
&= \frac{3}{s} + \frac{2}{s^3} + \frac{2}{s+1} + \frac{2}{s^2+4} + \frac{6}{(s-2)^4} + \frac{1}{s}e^{-2s}.
\end{aligned}$$

6.1.2 *Laplace Transforms of Function Derivative and Integral*

As Laplace transforms of the *n*-th order derivative function, $f^{(n)}(t)$, and the integral $\int_0^t f(u)du$ are sometimes necessary for solving differential or integral equations, and therefore we review some of their essential relations of Laplace transform in the following.

Property 6.4 *If a function $f(t)$ is differentiable except at a finite number of points, and continuous on the whole interval, there exists the relation*

$$\mathcal{L}[f^{(1)}(t)] = sF(s) - f(0) \tag{6.5}$$

in which $F(s)$ is the Laplace transform of $f(t)$.

Proof Assume that $f(t)$ is differentiable except at points $t_1, t_2, \ldots, t_n$. Then, we have

$$\mathcal{L}[f^{(1)}(t)] = \int_0^{+\infty} e^{-st} f^{(1)}(t)dt = \int_0^{t_1} e^{-st} f^{(1)}(t)dt + \int_{t_1}^{t_2} e^{-st} f^{(1)}(t)dt + \cdots + \int_{t_n}^{+\infty} e^{-st} f^{(1)}(t)dt. \tag{6.6}$$

To a typical term on the right side by part integration as follows, it yields

$$\begin{aligned}\int_{t_r}^{t_{r+1}} e^{-st} f^{(1)}(t)dt &= e^{-st} f(t)\Big|_{t_r}^{t_{r+1}} + s\int_{t_r}^{t_{r+1}} e^{-st} f(t)dt \\ &= e^{-st_{r+1}} f(t_{r+1}-) - e^{-st_r} f(t_r+) + s\int_{t_r}^{t_{r+1}} e^{-st} f(t)dt.\end{aligned}$$

in which $f(t_{r+1}-) = \lim_{t\to t_{r+1}-0} f(t)$ and $f(t_r+) = \lim_{t\to t_r+0} f(t)$ are the left and right values of $f(t)$ at the points of t_{r+1} and t_r, respectively. Since $f(t)$ is continuous at each point t_i, there is $f(t_i-) = f(t_i+) = f(t_i)$. Thus, we have

$$\int_{t_r}^{t_{r+1}} e^{-st} f^{(1)}(t)dt = e^{-st_{r+1}} f(t_{r+1}) - e^{-st_r} f(t_r) + s\int_{t_r}^{t_{r+1}} e^{-st} f(t)dt.$$

Submitting the above relation into Eq. (6.6), one can obtain the Laplace transform of Eq. (6.5), which means that this property is held.

By the mathematical induction method and Property 6.4, we can have the following general algorithm.

Property 6.5 *If a function $f(t)$ is smooth enough, e.g., to its* n-*th order derivative function, $f^{(n)}(t)$, there is the relation*

$$\mathcal{L}[f^{(n)}(t)] = sF(s) - \sum_{m=0}^{n-1} s^{n-m} f^{(m)}(0) \tag{6.7}$$

in which $F(s)$ is the Laplace transform of $f(t)$.

For the Laplace transform of the integral $\int_0^t f(u)du$, there is the following algorithm.

Property 6.6 *For a function $f(t)$, there is the relation*

$$\mathcal{L}[\int_0^t f(u)du] = F(s)/s \tag{6.8}$$

in which $F(s)$ is the Laplace transform of $f(t)$.

Proof From the definition, we have

$$\begin{aligned}\mathcal{L}[\int_0^t f(u)du] &= \int_0^{\infty} e^{-st}\left(\int_0^t f(u)du\right)dt \\ &= -\frac{e^{-st}}{s}\left(\int_0^t f(u)du\right)\Bigg|_0^{\infty} + \int_0^{\infty}\frac{e^{-st}}{s} f(t)dt\end{aligned}$$

through integration by parts. As e^{-st} tends to zero with $t \to \infty$, then $-\frac{e^{-st}}{s}\left(\int_0^t f(u)du\right)\Big|_0^{\infty}$ should be zero at both of the lower and upper limits. Further, Eq. (6.8) is achieved, which means that this property is held.

Example 6.2 Using the Laplace transform to find the solution of equation $y'' + 2y' - 3y = e^{-t}$ with the initial condition $y|_{t=0} = 0$ and $y'|_{t=0} = 1$.

Let the solution of the equation be $y = y(t)(t \geq 0)$ and $\mathcal{L}[y(t)] = Y(s)$. Taking the Laplace transform on both sides of the differential equation, and considering the initial-value conditions, we have

$$s^2Y(s) - 1 + 2sY(s) - 3Y(s) = \frac{1}{s+1}.$$

This is an algebraic equation with an unknown Laplace transform $Y(s)$. Solving such algebraic equation, one can gain

$$Y(s) = \frac{s+2}{(s+1)(s-1)(s+3)}.$$

This is the Laplace transform of initial-value problem in this example.

In order to find the inverse transformation of $Y(s)$, it can be converted by the following form of partial fractions, that is

$$Y(s) = \frac{s+2}{(s+1)(s-1)(s+3)} = \frac{-\frac{1}{4}}{s+1} + \frac{\frac{3}{8}}{s-1} + \frac{-\frac{1}{8}}{s+3}.$$

It is obvious that all terms above can be found in Table 6.1. Hence, their inverse Laplace transforms or original functions can be analytically obtained, and the solution of the problem is finally gained by the form

$$y(t) = -\frac{1}{4}e^{-t} + \frac{3}{8}e^{t} - \frac{1}{8}e^{-3t} = \frac{1}{8}(3e^{t} - 2e^{-t} - e^{-3t}).$$

One can substitute the gained solution into the differential equation and its initial conditions such that these equations all are satisfied.

Example 6.3 Using the Laplace transform to find the solution of equation $ty'' + (1 - 2t)y' - 2y = 0$ satisfying the initial condition $y|_{t=0} = 0$, $y'|_{t=0} = 2$.

Denote $Y(s)$ be the Laplace transform of the function $y(t)$. Then to the term of $ty(t)$, its Laplace transform can be written as

$$\mathcal{L}[ty(t)] = \int\limits_{0}^{+\infty} ty(t)e^{-st}dt = -\frac{d}{ds}\int\limits_{0}^{+\infty} y(t)e^{-st}dt = -\frac{d}{ds}Y(s).$$

Further, $\mathcal{L}[ty'(t)] = \int_0^{+\infty} ty'(t)e^{-st}dt = -\frac{d}{ds}\int_0^{+\infty} y'(t)e^{-st}dt = -\frac{d}{ds}[sY(s) - y(0)]$,

$$\mathcal{L}[ty''(t)] = \int\limits_{0}^{+\infty} ty''(t)e^{-st}dt = -\frac{d}{ds}\int\limits_{0}^{+\infty} y''(t)e^{-st}dt = -\frac{d}{ds}[s^2Y(s) - sy(0) - y'(0)].$$

Using the Laplace transform to the differential equation and considering the initial conditions, we get

$$(2 - s)Y'(s) - Y(s) = 0.$$

Here, $s \neq 0$ is considered. To the above-induced equation using the separable variables, one gets

$$\frac{dY}{Y} = -\frac{ds}{s - 2}.$$

Then integrating the above equation, we get the analytical solution of the form

$$\ln Y(s) = -\ln(s - 2) + \ln C,$$

or

$$Y(s) = \frac{C}{s - 2}.$$

Here, C is an integral constant. By means of Table 6.1, finally, the general solution $y(t) = Ce^{2t}$ to this example is gained. In order to determine the constant C, substitution of the solution into the initial conditions yields

$$C = y(0) = 1.$$

Thus, the determinant solution is obtained by $y(t) = e^{2t}$.

According to the above introduction, one has found that a differential equation can be converted to either an algebraic equation or the ordinary differential equation on the Laplace transform of the unknown function after the Laplace transform is applied. Once the Laplace transform function for a problem is gained, the solution for the problem is also strongly dependent on whether or not its Laplace inverse transform can be gained. For the simplest cases listed in Table 6.1, such Laplace inverse transform can be analytically gained when the Laplace transform functions are only consisted of the terms in Table 6.1. To many problems in engineering, however, their Laplace transform functions are always outer those in Table 6.1. In theory, the original solution of the problems can be obtained by solving such equations of the Laplace transform. When the form of the Laplace transform function or equation is relatively complicated, however, the original unknown function cannot be always obtained through such analytic method displayed above. Under such situation, it is necessary that a numerical method should be used to obtain the inversion Laplace transform. At present, it has been known that such numerical inverse Laplace transform is generally an ill-posed problem [1]. Due to no a uniform numerical approach being suitable for different cases, recently, there are various numerical approaches in the literature attempting to conquer this difficulty [2], e.g., the Gaver-Stehfest method, the Schapery's method, the Möbius transformation methods, the Talbot method, and the Fourier series method. The Gaver–Stehfest method is popular due to its simplicity and adequacy for exponentially decaying functions. When this method is used, one sees that each vector element should be constructed by using a numerical simulation, which makes the method to be awfully expensive in computation. To the Schapery method, the implementation is relatively easy when a matrix decomposition in the method is available. However, two obvious restrictions are displayed in the Schapery method. The first one is that it requires the solutions being a steady state. Another is that there is no reliable approach for selecting the relevant parameters in the method. In general, each numerical method has its own advantages and disadvantages [2]. As following in this chapter, we mainly introduce a wavelet-based method proposed by the author's group to conquer this difficulty [18, 19].

6.2 Wavelet-Based Laplace Transforms

Hereafter, we introduce an efficient and robust Laplace inversion method established based on the generalized Coiflets expansion of functions in the transformed or frequency domain.

6.2.1 Quantitative Spectrum Feature of Scaling Function Employed

For a function $f(t) \in L^2(R)$, from Sect. 2.4.1, the Fourier transform and its inversion, respectively, are defined as

$$\widehat{f}(\omega) = \int_{-\infty}^{+\infty} f(t)\mathrm{e}^{-i\omega t} dt,$$

$$f(t) = \frac{1}{2\pi} \int_{-\infty}^{+\infty} \widehat{f}(\omega)\mathrm{e}^{i\omega t} d\omega.$$

By applying the Fourier transform to Eqs. (2.2) and (2.6), we derive

$$\widehat{\varphi}(\omega) = P\left(e^{-i\omega/2}\right)\widehat{\varphi}(\omega/2), \tag{6.9}$$

$$\widehat{\psi}(\omega) = Q\left(e^{-i\omega/2}\right)\widehat{\varphi}(\omega/2), \tag{6.10}$$

where P and Q are defined by

$$P(z) = \frac{1}{2}\sum_{k=0}^{\widetilde{N}} p_k z^k, \tag{6.11}$$

$$Q(z) = \frac{1}{2}\sum_{k=0}^{\widetilde{N}} q_k z^k. \tag{6.12}$$

Here, z stands for a complex variable in the frequency space, and $i = \sqrt{-1}$. Based on the normalization property, we have $\widehat{\varphi}(0) = \int_R \varphi(t)dt = 1$. Therefore, Eq. (6.9) can be applied recursively, then a product formula can be obtained [3, 4]

$$\widehat{\varphi}(\omega) = \prod_{k=1}^{\infty} P\left(e^{-i\omega/2^k}\right). \tag{6.13}$$

The convergence of this product was discussed in [3, 4]. Further, the combination of Eqs. (6.10) and (6.13) yields

$$\widehat{\psi}(\omega) = Q\left(e^{-i\omega/2}\right)\prod_{k=1}^{\infty} P\left(e^{-i\omega/2^{k+1}}\right). \tag{6.14}$$

From Chap. 2, we know that the function $\widehat{\varphi}(\omega)$ is a low-pass filter, while $\widehat{\psi}(\omega)$ a band-pass filter. When the generalized Coiflets is employed, as the same as Fig. 2.5 displayed, Fig. 6.1a redraws the energy spectra $\left|\widehat{\varphi}(\omega)\right|^2$ for $r = 2, 4, 6$, where the support region is $\widehat{N} = 3r - 1$. As r increases, $\left|\widehat{\varphi}(\omega)\right|^2$ approaches to the energy spectrum of the ideal low-pass filter. Comparing the low-pass filter with the ideal case, Fig. 6.1b gives the relative errors between $\left|\widehat{\varphi}(\omega)\right|^2$ and the energy spectrum of the ideal low-pass filter. From it, one sees that this relative error can be reduced to less than 0.3% within the frequency domain $\left[0, \pi/2\right]$ when $r = 6$, which indicates

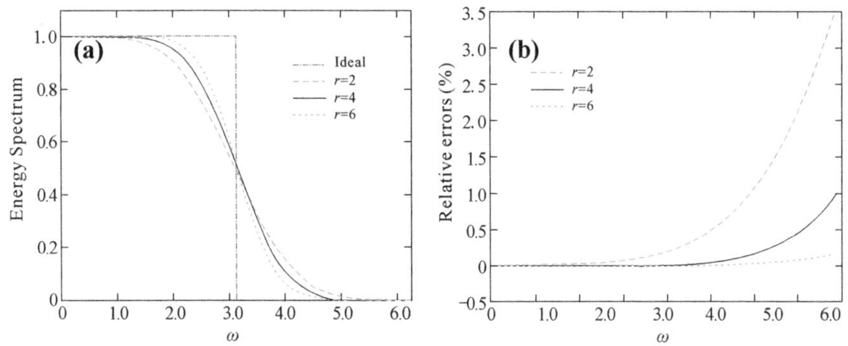

Fig. 6.1 Comparison between the energy spectrum of the ideal low-pass filter and $\left|\widehat{\varphi}(\omega)\right|^2$ for $r = 2, 4, 6$ in $[0, \pi/2]$: **a** comparison of the energy spectrum function (Left); **b** relative errors (Right)

that the generalized Coiflets scaling function exhibits a good low-pass characteristic for sufficiently large r. At the same time, Fig. 2.6 told us that such perfect domain of the low-pass filter is also dependent on the resolution level n. When n increases, the perfect domain of the generalized Coiflets low-pass filter increases fast as displayed in Chap. 2.

6.2.2 Fourier Transform and Inverse Transform Based on Wavelet

When Fourier transform is applied to Eq. (4.1), the following equation can be obtained:

$$\begin{aligned}\widehat{f}(\omega) &= \sum_{k=-\infty}^{+\infty} c_{n,k}\int_{-\infty}^{\infty}\varphi_{n,k}(t)e^{-i\omega t}dt + \sum_{j=n}^{+\infty}\sum_{k=-\infty}^{+\infty} d_{j,k}\int_{-\infty}^{\infty}\psi_{j,k}(t)e^{-i\omega t}dt \\ &= 2^{-n/2}\widehat{\varphi}\left(\frac{\omega}{2^n}\right)\sum_{k=-\infty}^{+\infty} c_{n,k}e^{-i\omega k/2^n} + \sum_{j=n}^{+\infty} 2^{-j/2}\widehat{\psi}\left(\frac{\omega}{2^j}\right)\sum_{k=-\infty}^{+\infty} d_{j,k}e^{-i\omega k/2^j}.\end{aligned} \tag{6.15}$$

Here, $f(x)$ in Eq. (4.1) is replaced by $f(t)$. As pointed out in Chap. 2, such function $\widehat{f}(\omega)$ is also in $L^2(R)$. In such case, using Eq. (4.1) to expand $\widehat{f}(\omega)$, i.e.,

$$\widehat{f}(\omega) = \sum_{k=-\infty}^{+\infty} \hat{c}_{n,k}\varphi_{n,k}(\omega) + \sum_{j=n}^{+\infty}\sum_{k=-\infty}^{+\infty} \hat{d}_{j,k}\psi_{j,k}(\omega) \tag{6.16a}$$

in which

$$\hat{c}_{n,i} = \int_{-\infty}^{+\infty}\widehat{f}(\omega)\varphi_{n,i}(\omega)d\omega, \quad \hat{d}_{j,k} = \int_{-\infty}^{+\infty}\widehat{f}(\omega)\psi_{j,k}(\omega)d\omega. \tag{6.16b}$$

Applying the above Fourier inversion formula to Eq. (6.16a), one gets

$$\begin{aligned}f(t) &= \frac{1}{2\pi}\sum_{k=-\infty}^{+\infty} \hat{c}_{n,k}\int_{-\infty}^{\infty}\varphi_{n,k}(\omega)e^{i\omega t}d\omega + \frac{1}{2\pi}\sum_{j=n}^{+\infty}\sum_{k=-\infty}^{+\infty} \hat{d}_{j,k}\int_{-\infty}^{\infty}\psi_{j,k}(\omega)e^{i\omega t}d\omega \\ &= \frac{1}{2\pi\, 2^{n/2}}\varphi_*\left(-\frac{t}{2^n}\right)\sum_{k=-\infty}^{+\infty} \hat{c}_{n,k}e^{itk/2^n} + \frac{1}{2\pi}\sum_{j=n}^{+\infty} 2^{-j/2}\psi_*\left(-\frac{t}{2^j}\right)\sum_{k=-\infty}^{+\infty} \hat{d}_{j,k}e^{itk/2^j}.\end{aligned} \tag{6.17}$$

Here, $\varphi_*(-t/2^n) = \int_{-\infty}^{+\infty} \varphi_{n,k}(\omega)e^{i\omega t/2^n}d\omega$ and $\psi_*(-t/2^n) = \int_{-\infty}^{+\infty} \psi_{n,k}(\omega)e^{i\omega t/2^j}d\omega$ are similar to the Fourier transforms for $\varphi_{n,k}(\omega)$ and $\psi_{n,k}(\omega)$, respectively. According to the definitions of them, we know that both $\varphi_*(-t/2^n)$ and $\psi_*(-t/2^n)$ are independent on the index k. Using the generalized wavelet-Gaussian integral method [3–5] or the generalized Coiflets displayed in Chap. 2, we have the following relations

$$\hat{c}_{n,k} = \int_{-\infty}^{+\infty} \hat{f}(\omega)\varphi_{n,k}(\omega)d\omega \approx 2^{-n/2}\hat{f}\left(\frac{k+M_1}{2^n}\right).$$

$$\hat{d}_{j,k} = \int_{-\infty}^{+\infty} \hat{f}(\omega)\psi_{j,k}(\omega)d\omega = \sum_{m=0}^{\widetilde{N}}(-1)^m p_{3N-1-m}\int_{-\infty}^{+\infty} 2^{j/2}\hat{f}(\omega)\varphi\left(2^{(j+1)}\omega - 2k - m\right)d\omega.$$

$$\approx 2^{-j/2-1}\sum_{m=0}^{\widetilde{N}}(-1)^m p_{\widetilde{N}-1}\hat{f}\left(\frac{M_1+m+2k}{2^{j+1}}\right).$$

After that, Eq. (6.17) can be used to invert $\widehat{f}(\omega)$ once the coefficients $\hat{c}_{n,k}$ and $\hat{d}_{j,k}$ are known. If $\widehat{f}(\omega)$ is a smooth function, as pointed out previously, then the approximations of $\hat{c}_{n,k}$ and $\hat{d}_{j,k}$ have high accuracy when the resolution level n is large enough such that the expression of Eq. (6.16a) converges. In accordance with Eqs. (6.15)–(6.17), the formulae for Fourier transform and its inversion can be rewritten as follows:

$$\widehat{f}(\omega) = \lim_{n\to\infty}\left\{\frac{1}{2^n}\widehat{\varphi}\left(\frac{\omega}{2^n}\right)\sum_k f\left(\frac{M_1+k}{2^n}\right)e^{-i\omega k/2^n}\right\}, \tag{6.18}$$

$$f(t) = \lim_{n\to\infty}\left\{\frac{1}{2^{n+1}\pi}\varphi_*\left(-\frac{t}{2^n}\right)\sum_k \widehat{f}\left(\frac{M_1+k}{2^n}\right)e^{itk/2^n}\right\}. \tag{6.19}$$

If we defined

$$f_n(t) = \frac{1}{2^{n+1}\pi}\varphi_*\left(-\frac{t}{2^n}\right)\sum_k \widehat{f}\left(\frac{M_1+k}{2^n}\right)e^{itk/2^n}, \tag{6.20}$$

then Eq. (6.19) can be simply written as

$$f(t) = \lim_{n\to\infty} f_n(t). \tag{6.21}$$

It should be noted that $\widehat{\varphi}(\omega/2^n)$ is a low-pass filter whose frequency band is concentrated in interval $[-2^n\pi, 2^n\pi]$, as shown in Fig. 2.6. To the inversion of Eq. (6.19), if we consider $\widehat{f}(\omega)$ as a signal and $f(t)$ as its frequency spectrum, then $f_n(t)$ will be the frequency spectrum after the signal $\widehat{f}(\omega)$ has been filtered by a low-pass filter. That is, $f_n(t)$ should coincide (approximately) with $f(t)$ within the frequency interval $[-2^n\pi, 2^n\pi]$. When n tends toward infinity, the domain of coincidence will become $(-\infty, \infty)$.

To numerically compute the inversion series $f_n(t)$ using Eq. (6.20), the finest frequency bandwidth of $f(t)$ should be previously selected, i.e., an appropriate range of variable ω of $\widehat{f}(\omega)$ should cover the main domain of $\widehat{f}(\omega)$. Denote the main domain by $[\omega_{\min}, \omega_{\max}]$. Then, the maximum and minimum values of k in the summation of Eq. (6.20) can be determined by

$$k_{\max} = \omega_{\max}2^n - 1, \quad k_{\min} = \omega_{\min}2^n - \widehat{N} + 1. \tag{6.22}$$

Example 6.4 Using the wavelet-based Fourier inverse transform to the known function

$$f(t) = \begin{cases} \sin(10\pi t), & \mathrm{t} \in [0, 5] \\ 0 & \text{other} \end{cases}. \tag{6.23}$$

For this case, one can gets its exact Fourier transform in the form

$$\widehat{f}(\omega) = 10\pi\left(1 - e^{-5\omega i}\right)/\left(10^2\pi^2 - \omega^2\right). \tag{6.24}$$

After that, we use the above wavelet-based method to give its inversion results. When the generalized Coiflets are taken by $r = 6$ and $M_1 = 7$, the obtained results are plotted in Fig. 6.2, where $\omega_{\min} = -100$, and $\omega_{\max} = 100$. When the resolution level $n = 0$, Fig. 6.2a gives the inversed function $f(t)$ in [0, 5] and its energy spectrum $||\widehat{\varphi}(\omega)||$. From it, one sees that the approximation of Fourier inverse transform is

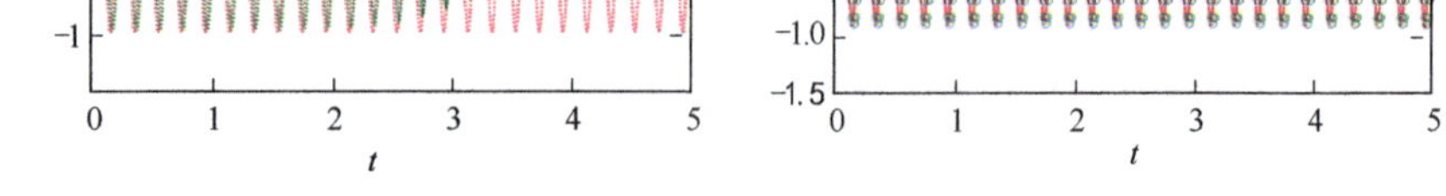

Fig. 6.2 Numerical results of the wavelet-based Fourier inverse transform to Example 6.4 when $\omega \in [-100, 100]$ and $r = 6$:**a** comparison of the numerical results with exact values associated with the energy spectrum $||\widehat{\varphi}(\omega)||$ when $n = 0$ for the original function $f(x) = \sin(10\pi t), \quad t \in [0, 5]$ (Left); **b** the results of wavelet-based Fourier inversion with $n = 4$ compared with its exact values (Right)

well in agreement with its exact or original function formulated in Eq. (6.23) in the region of [0, 2.3], where the energy spectrum $||\widehat{\varphi}(\omega)||$ in the domain approaches to the case of the ideal low-pass filter. When the resolution level is taken as $n = 4$, as displayed in Fig. 6.2b, the numerical inverse results all are coincident with their exact values well in [0, 5]. This fact tells us that the wavelet-based Fourier inverse transform has high accuracy and low calculations.

6.2.3 Laplace Transform and Inverse Transform Based on Wavelet

Assume that $f(t) = 0$ when $t < 0$, Eq. (6.1) can be rewritten as

$$F(\beta + i\omega) = \int_{-\infty}^{+\infty} f(t)e^{-\beta t}e^{-i\omega t}dt = \int_{0}^{+\infty} f(t)e^{-\beta t}e^{-i\omega t}dt. \tag{6.25}$$

Denote

$$h(t) = f(t)e^{-\beta t}, \tag{6.26}$$

then we have

$$\widehat{h}(\omega) = F(\beta + i\omega) = \int_{-\infty}^{+\infty} h(t)e^{-i\omega t}dt = \int_{0}^{+\infty} h(t)e^{-i\omega t}dt. \tag{6.27}$$

By applying the previous Fourier inversion formulae, i.e., Eqs. (6.20) and (6.21), to Eq. (6.27), the following expressions are obtained:

$$h_n(t) = \frac{1}{2^{n+1}\pi}\varphi_*\left(-\frac{t}{2^n}\right)\sum_{k=-\infty}^{+\infty}\widehat{h}\left(\frac{k + M_1}{2^n}\right)e^{ikt/2^n} \tag{6.28}$$

and

$$h(t) = \lim_{n\to\infty} h_n(t). \tag{6.29}$$

Finally, the wavelet-based formulae for the Laplace inversion is gained

$$f_n(t) = \frac{e^{\beta t}}{2^{n+1}\pi}\varphi_*\left(-\frac{t}{2^n}\right)\sum_{k=-\infty}^{+\infty} F\left(\beta + i\frac{M_1 + k}{2^n}\right)e^{ikt/2^n}, \tag{6.30}$$

$$f(t) = \lim_{n\to\infty} f_n(t) \tag{6.31}$$

where $f(t)e^{-\beta t} \in L^2(R)$ is used. From Eq. (6.30), one sees that such wavelet-based Laplace inversion can be evidently employed by almost independently selecting the parameter β. When $\beta = 0$, for example, the Laplace transform and its inversion can be, respectively, degraded to the Fourier transform and its inversion except for $f(t) = 0$ when $t < 0$.

Example 6.5 Wavelet-based Laplace transform and its inversion to the known real function $u(t) = (t/10\pi)\sin(10\pi t)$ and complex function $v(t) = \sin(it)\sin(5\pi t)$, where $i = \sqrt{-1}$. Here, the region of time is [0,5].

Here, we take $\beta = 1$ for $u(t)$ and $\beta = 2$ for $v(t)$, respectively, in the wavelet-based Laplace inversions. After Eq. (6.30) is applied to these two cases, one can easily gain their inversion form

$$u_n(t) = \frac{e^t}{2^{n+1}10\pi^2}\varphi_*\left(-\frac{t}{2^n}\right)\sum_{k=-\infty}^{\infty}\frac{\sin\left(2\arctan\left(10\pi/\left(1+i(M_1+k)/2^n\right)\right)\right)}{\left(1+i(M_1+k)/2^n\right)^2+100\pi^2}e^{ikt/2^n}, \tag{6.32}$$

$$v_n(t) = \frac{e^t}{2^{n+1}\pi}\varphi_*\left(-\frac{t}{2^n}\right)\sum_{k=-\infty}^{\infty}\frac{5\pi i}{2}\left(\frac{1}{\left((M_1+k)/2^n-1\right)^2+25\pi^2} - \frac{1}{\left((M_1+k)/2^n+1\right)^2+25\pi^2}\right)e^{ikt/2^n}. \tag{6.33}$$

Based on Eqs. (6.32) and (6.33), the numerical inverse results are displayed in Figs. 6.3 and 6.4, respectively. Here, $r = 6$, $M_1 = 7$, $\omega_{\min} = -100$, and $\omega_{\max} = 100$ are taken in these two cases. From them, one sees that when the resolution level in the employed wavelet is taken as zero, i.e., $n = 0$; the obtained numerical results

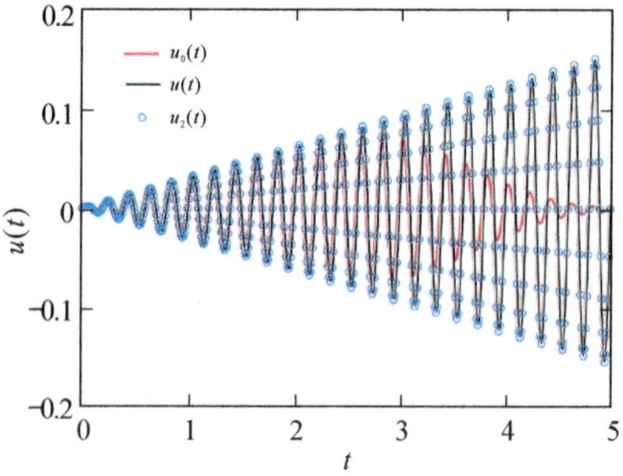

Fig. 6.3 Numerical results of the generalized Coiflets-based Laplace inversion method to the known real function $u(t) = (t/10\pi)\sin(10\pi t)$ when $n = 0$ and 2, which are identified by the subscript of $u(t)$ in the figure

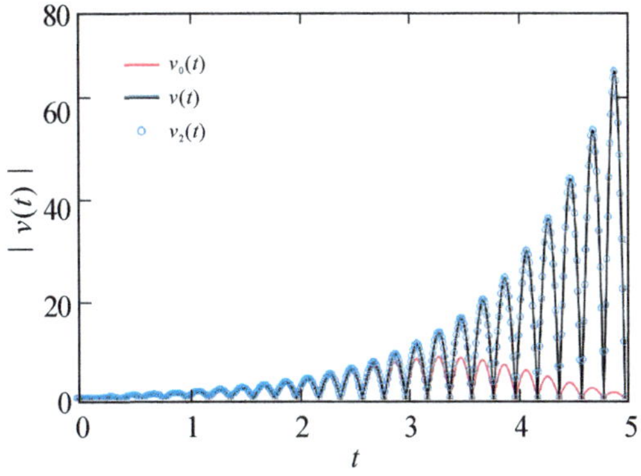

Fig. 6.4 The numerical results of the generalized Coiflet-based Laplace inversion method to the known complex function $v(t) = \sin(ti)\sin(5\pi t)$ when $n = 0$ and 2, which are identified by the subscript of $v(t)$ in the figure. Here, $|v(t)|$ stands for the magnitude of the complex function

are well in agreement with their exact values in the region of about [0, 2.5]. When $n = 2$, all numerical inverse results are fully coincident with their exact values in the given region of [0, 5].

6.3 Application Examples: Numerical Solution of A Fractionally Damped Dynamic System

Fractional derivative constitutive models are mainly used to characterize the viscoelastic behavior of materials, which has recently received considerable attention [6–11, 14, 15]. The use of such "fractionally damped" dynamic models is motivated to a large extent by the fact that fractional derivative models can well describe the frequency dependence of dynamic properties where fewer parameters are typically required to represent material behavior compared with, for example, the traditional Kelvin and Maxwell-based models.

6.3.1 Dynamic Equations with Fractional Damping

In an effort to generalize the constitutive relation of viscoelasticity, a number of researchers have applied viscoelastic constitutive relations with fractional derivatives:

$$\sum_{i=0}^{A} p_i D^{\alpha_i} \sigma = \sum_{j=0}^{B} q_j D^{\beta_j} \varepsilon, \tag{6.34}$$

where σ and ε represent the stress and strain, respectively, D^{α} and D^{β} are the operator of the fractional derivative, i.e., $0 \leq \alpha_0, \alpha_1, \cdots, \alpha_A \leq 1$, and $0 \leq \beta_0, \beta_1, \cdots, \beta_B \leq$

1, and p_i and q_i are the material parameters to be determined from a suitable curve fit with test data [10].

In accordance with the extended Riemann–Liouville definition [11–13], the fractional α-order derivative of the function $f(t)$ is defined as

$$D^{\alpha} f(t) = \frac{1}{\Gamma(n-\alpha)} \frac{d^n}{dt^n} \int_0^t (t-\tau)^{n-\alpha-1} f(\tau) d\tau, \quad 0 < n - \alpha < 1, \tag{6.35}$$

where n is an integer, and $\Gamma(\bullet)$ is the Gamma function. In the real-time domain, the presence of integral–differential operators makes the solution more computationally complex than ordinary operators. Once it is transferred into the frequency domain, however, the model becomes easier to handle in the calculations. For such purpose, we still use the definition of Laplace transform in Eq. (6.1), i.e.,

$$Y(s) = \mathcal{L}[y(t)] = \int_0^{+\infty} y(t) e^{-st} dt.$$

When $n - 1 < \alpha \leq n$ in the fractional derivatives, Eq. (6.7) for the Laplace transform of a derivative function can be generalized by the form [12]

$$\mathcal{L}\left[D^{\alpha} y(t)\right] = s^{\alpha} Y(s) - \sum_{k=0}^{n-1} s^k D^{\alpha-1-k} y(0). \tag{6.36}$$

When $n = 1$, Eq. (6.36) becomes

$$\mathcal{L}\left[D^{\alpha} y(t)\right] = s^{\alpha} Y(s) - D^{\alpha-1} y(0). \tag{6.37}$$

Here, the term $D^{\alpha-1} y(0)$ is a fractional derivative of function $y(t)$ at $t = 0$. For a reasonable $y(t)$, such $D^{\alpha-1} y(0)$ is always equal to zero. Thus, we have

$$\mathcal{L}\left[D^{\alpha} y(t)\right] = s^{\alpha} Y(s). \tag{6.38}$$

Example 6.6 Consider the viscoelastic behavior of a single degree-of-freedom oscillator that consists of a discrete mass and a viscoelastic spring governed by a fractional calculus law in conjunction with Newton's second law. The equation of motion can be written in the following form [11]:

$$m \frac{d^2}{dt^2} y(t) + r(t) = f(t), \; y(0) = h_0, \; \dot{y}(0) = h_1, \tag{6.39}$$

where m represents the mass coefficient, $f(t)$ is the externally applied force, $r(t)$ is the force in the viscoelastic spring, $y(t)$ is the displacement of the mass, and $\dot{y} = dy/dt$. Similar to Eq. (6.34), the constitutive equation between the force $r(t)$ and the displacement $y(t)$ for the spring can be formulated as

$$\sum_{i=0}^{A} p_i D^{\alpha_i} r(t) = \sum_{j=0}^{B} q_j D^{\beta_j} y(t), \tag{6.40}$$

where $0 \le \alpha_i, \beta_j \le 1$.

Applying the above Laplace transform formulae to Eqs. (6.39) and (6.40), we get

$$ms^2 Y(s) + R(s) = F(s) + smh_0 + mh_1 \tag{6.41}$$

and

$$\left(\sum_{i=0}^{A} p_i s^{\alpha_i}\right) R(s) = \left(\sum_{j=0}^{B} q_j s^{\beta_j}\right) Y(s), \tag{6.42}$$

i.e.,

$$R(s) = \frac{\sum_{j=0}^{B} q_j s^{\beta_j}}{\sum_{i=0}^{A} p_i s^{\alpha_i}} Y(s). \tag{6.43}$$

Here, h_0 and h_1 are the constants to be determined by the initial conditions. Substitution of Eq. (6.43) into Eq. (6.41) yields

$$Y(s) = G(s)[F(s) + mh_0 s + mh_1], \tag{6.44}$$

where $G(s) = \sum_{i=0}^{A} p_i s^{\alpha_i} / [ms^2 \sum_{i=0}^{A} p_i s^{\alpha_i} + \sum_{j=0}^{B} q_j s^{\beta_j}]$, or,

$$G(s) = \frac{\sum_{i=0}^{A} p_i s^{\alpha_i}}{\sum_{k=0}^{A+B+1} c_k s^{\gamma_k}} \tag{6.45}$$

in which $c_k = mp_k$, $\gamma_k = 2 + \alpha_k$ for $0 \le k \le A$; and $c_k = q_{k-A-1}$, $\gamma_k = \beta_{k-A-1}$ for $A + 1 \le k \le A + B + 1$.

Taking the Laplace inversion to Eq. (6.44), we have the following general solution

$$y(t) = mh_0\dot{g}(t) + mh_0 g(0)\delta(t) + mh_1 g(t) + \int\limits_0^t g(t-\tau)f(\tau)d\tau, \qquad (6.46)$$

where the Green function $g(t)$ is the inverse Laplace transform of $G(s)$, and $\delta(t)$ is the δ-function, i.e., $\delta(t) = 1$ for $t = 0$ and $\delta(t) = 0$, otherwise.

It is obvious that such general solution expressed by Eq. (6.46) is strongly dependent on the inversion $g(t)$ of the fractional Green function $G(s)$. Here, the wavelet-based Laplace inversion is employed to get $g(t)$. Then $\beta = 1$ to such problems because the Green function must be finite due to damping, one gets

$$g_n(t) = \frac{e^t}{2^{n+1}\pi}\varphi_*\left(-\frac{t}{2^n}\right)\sum_{k=-\infty}^{\infty}\frac{\sum_{i=0}^{A} p_i\left(1+i(M_1+k)/2^n\right)^{\alpha_i}}{\sum_{k=0}^{A+B+1} c_k\left(1+i(M_1+k)/2^n\right)^{\gamma_k}}e^{ikt/2^n}. \qquad (6.47)$$

Under the above wavelet-based Laplace transform solution, we give its numerical predictions of three cases as follows.

Case 1: Free vibration when $f(t) \equiv 0$. For simplicity, we take the values of the parameters in the system with $m = 1$, $A = 0$, $p_0 = 1$, $\alpha_0 = 0$, $B = 1$, $q_0 = 100$, $\beta_0 = 0$ and, $q_1 = 10^{2-\beta_1}$, and the different values of $\beta_1 = 0, 0.1, 0.5, \pi/5, 0.9$. That is, the constitutive equation of Eq. (6.34) is of the form $\sigma = q_0\varepsilon + q_1 D^{\beta_1}\varepsilon$.

To such dynamic problem selected, the function of $G(s)$ can be analytically gained by the form

$$G(s) = \frac{1}{s^2 + c_2 s^{\gamma_2} + 100}, \qquad (6.48)$$

where $c_2 = q_1$ and $\gamma_2 = \beta_1$. Then, following the wavelet-based Laplace inverse transform, one can get the analytic solution $g_n(t)$ similar to Eq. (6.47). Further, the dynamic solution of $y(t)$ can be gained by substituting the gained $g_n(t)$ into Eq. (6.46).

Figure 6.5 presents the dynamic responses obtained by using Eq. (6.47) for the case 1 when $r = 6$, $M_1 = 7$, $n = 1$, $\omega_{\max} = -\omega_{\min} = 250$ and the different values of β_1 selected. The results show that the damping effect decreases with an increase in $\gamma_2 = \beta_1$ when the other parameters have been fixed. When $\gamma_2 = \beta_1 = 0$, such dynamic system is a full free vibration without damping. According to the selected parameters for this special case, one can get its period being 0.44435. From Fig. 6.5, one finds that the corresponding curve marked by $\gamma_2 = 0$ has 6.75 cycles in time region [0,3]. Hence, the period of such free vibration is numerically given by 0.44444. Thus, the relative error between the exact and the numerical periods is within 0.1%, which exhibits a high accuracy of the numerical solution of the wavelet-based Laplace method with very low meshes coincident with the resolution level $n = 1$. When $\beta_1 = 0.5$, the response results gained here for the free damping vibration are also in a good agreement with the previous one by other methods in [7, 12].

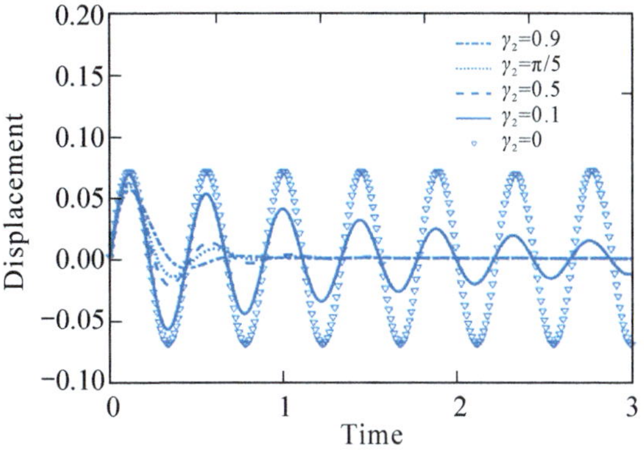

Fig. 6.5 Quantitative results of a fractionally damped single degree-of-freedom system in Case 1 obtained by the wavelet-based Laplace inverse method for different $\gamma_2 = \beta_2 = 0, 0.1, 0.5, \pi/5, 0.9$

Case 2: Example in Ref. [15]. If $m = 1$, $A = 0$, $B = 1$, $\alpha_1 = \beta_1 = 0$, $\alpha_0 = \beta_0$, then the Laplace transform of the Green function in Eq. (6.45) becomes

$$G(s) = \frac{c_0 s^{\alpha_0} + c_1 s^{\alpha_1}}{c_0 s^{\gamma_0} + c_1 s^{\gamma_1} + c_2 s^{\gamma_2} + c_3 s^{\gamma_3}}, \tag{6.49}$$

where $c_0 = p_0$, $c_1 = p_1 = 1$, $c_2 = q_0$, $c_3 = q_1$, $\gamma_0 = 2 + \alpha_0$, $\gamma_1 = 2 + \alpha_1 = 2$, $\gamma_2 = \beta_0$, $\gamma_3 = \beta_1 = 0$. In [15] for this case, further, $p_0 = b^{\alpha_0}$, $q_0 = 2b^{\alpha_0}$, $q_1 = 1$, $\alpha_0 = 1/2$, $b = 1$. Figure 6.6 gives the quantitative results of the analytic solution of Eq. (6.46) on the basis of Eq. (6.49) and the above parameters selected, and a comparison of them and the numerical results obtained in [15]. From the comparison, one sees that the predicted response from the wavelet-based Laplace transform method is quantitatively in good agreement with the existing results in [15], when $\alpha_0 = 1/2$, $b = 1$.

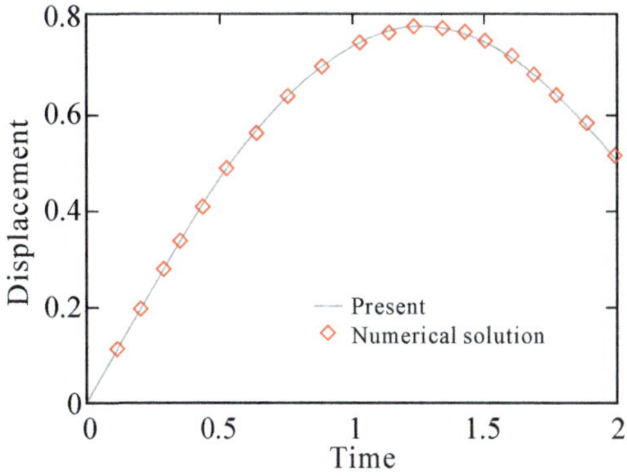

Fig. 6.6 Dynamic response of a fractionally damped single degree-of-freedom system given in Case 2 gained by the wavelet-based Laplace inverse transform

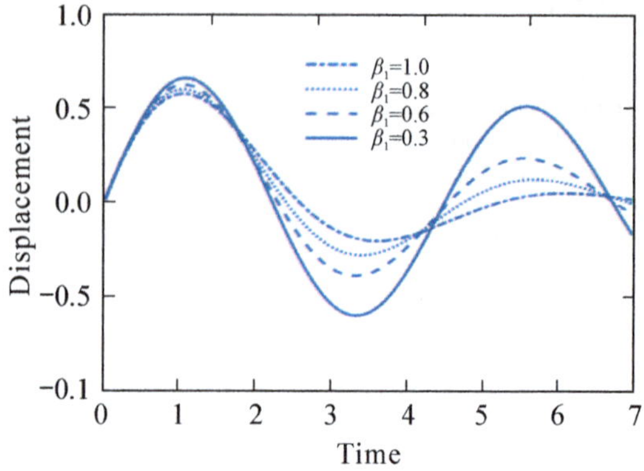

Fig. 6.7 Dynamic responses of a fractionally damped single degree-of-freedom system given in Case 3 with different β_1 (=0.3, 0.6, 0.8, and 1.0) when the wavelet-based Laplace inverse transform is employed

Case 3: Free damping vibration varying with β_1. Here, $m = 1$, $A = 2$, $B = 2$, $p_0 = 0.2$, $p_1 = 0.2$, $q_1 = 1$, $p_2 = q_2 = 1$, $q_0 = 1$, $\alpha_0 = \beta_0 = 0$, $\alpha_1 = 0.1$, $\alpha_2 = \pi/7$, and $\beta_2 = 0.8$.

For such case, the Laplace transform of the Green function in Eq. (6.45) becomes

$$G(s) = \frac{0.2 + 0.2s^{0.1} + s^{\pi/7}}{c_0 s^{\gamma_0} + c_1 s^{\gamma_1} + c_2 s^{\gamma_2} + c_3 s^{\gamma_3} + c_4 s^{\gamma_4} + c_5 s^{\gamma_5}}, \tag{6.50}$$

where $c_0 = mp_0 = 0.2$, $c_1 = mp_1 = 0.2$, $c_2 = mp_2 = 0.2$, $c_3 = q_0 = 1$, $c_4 = q_1 = 1$, $c_5 = q_2 = 1$, $\gamma_0 = 2 + \alpha_0 = 2$, $\gamma_1 = 2 + \alpha_1 = 2.1$, $\gamma_2 = 2 + \alpha_2 = 2 + \pi/7$, $\gamma_3 = \beta_0 = 0$, $\gamma_4 = \beta_1$, $\gamma_5 = \beta_2 = 0.8$. Further,

$$G(s) = \frac{0.2 + 0.2s^{0.1} + s^{\pi/7}}{s^{2+\pi/7} + 0.2s^{2.1} + 0.2s^2 + s^{0.8} + s^{\beta_1} + 1}. \tag{6.51}$$

Following the previous program, one can get the analytical solution of such a dynamic system on the wavelet-based Laplace inverse method. Figure 6.7 illustrates the dynamic response curves of the system in case 3 with different β_1 when $r = 6$, $M_1 = 7, n = 2, \omega_{\max} = \omega_{\min} = 250$. From Fig. 6.7, we find that amplitude decreases with increasing β_1 but the frequency is almost no change.

6.3.2 *Numerical Solution for Nonlinear Fractional Dynamic System with Single–Degree-of-Freedom*

For simplicity and without loss of generality, here, we consider a single–degree-of-freedom vibration system with fractional derivative type damping and a nonlinear restoring force. To such a dynamic system displayed in [15], the governing equation and initial conditions can be given by the form

$$c_1\ddot{y}(t) + c_2 D_t^{\alpha} y(t) + g(y(t)) = f(t), \quad t > 0, \tag{6.52}$$

$$y(0) = y_0, \ \dot{y}(0) = y_1, \tag{6.53}$$

where $c_1\ddot{y}(t)$ represents the inertia force, $c_2 D_t^{\alpha} y(t)$ $(0 < \alpha < 2)$ is the damping force, $f(t)$ is the external excitation, $g(y(t))$ is the nonlinear restoring force. Here, $D_t^{\alpha} y(t)$ is chosen as the Caputo fractional derivative of $y(t)$ [16], i.e.,

$$D_t^{\alpha} y(t) = \begin{cases} y^{(n)}(t) & \alpha = n \in \mathbf{N}, \\ \frac{1}{\Gamma(n-\alpha)} \int\limits_0^t \frac{y^{(n)}(\tau)}{(t-\tau)^{\alpha+1-n}} d\tau, & n-1 < \alpha < n \end{cases} \tag{6.54}$$

in which $y^{(n)}(t) = d^n y(t)/dt^n$, and $\Gamma(\cdot)$ is the Gamma function. Denote the Laplace transform of $y(t)$ by $Y(s)$, i.e., $\mathcal{L}[y(t)] = Y(s)$. Then the Laplace transform of $D_t^{\alpha} y(t)$ can be expressed as [16]

$$\mathcal{L}[D_t^{\alpha} y(t)] = s^{\alpha}\mathcal{L}[y(t)] - \sum_{m=0}^{n-1} s^{\alpha-1-m} y^{(m)}(0^+), \quad n-1 < \alpha \le n, \ n \in \mathbf{N}. \tag{6.55}$$

Applying the Laplace transform to the governing Eq. (6.52) associated with the initial conditions of Eq. (6.53) and considering Eq. (6.55), we have

$$(c_1 s^2 + c_2 s^{\alpha})Y(s) + \mathcal{L}[g(y(t)) - f(t)] = \sum_{n=0}^{1} c_1 s^n y_n + \sum_{n=0}^{[\alpha]} c_2 s^{\alpha-1-n} y_n \tag{6.56}$$

or

$$Y(s) + O(s)\{\mathcal{L}[g(y(t)) - f(t)]\} = Q(s), \tag{6.57}$$

in which $O(s) = 1/(c_1 s^2 + c_2 s^{\alpha})$, $Q(s) = O(s)\left(\sum\limits_{n=0}^{1} c_1 s^n y_n + \sum\limits_{n=0}^{[\alpha]} c_2 s^{\alpha-1-n} y_n\right)$, $[\alpha]$ stands for the integer part of α.

Using the inverse Laplace transform to Eq. (6.57), we get

$$y(t) + \int\limits_0^t o(t-\tau)[g(y(\tau)) - f(\tau)]d\tau = q(t). \tag{6.58}$$

Here,

$$o(t) = \mathcal{L}^{-1}[R(s)] = \frac{1}{c_1} t E_{2-\alpha,2}\left(-\frac{c_2}{c_1} t^{2-\alpha}\right), \; q(t) = \mathcal{L}^{-1}[Q(s)] \tag{6.59}$$

in which $E_{2-\alpha,2}(-c_2 t^{2-\alpha}/c_1)$ is the generalized Mittag–Leffler type functions defined in the power series

$$E_{a,b}(z) = \sum_{k=0}^{\infty} \frac{z^k}{\Gamma(ak+b)}. \tag{6.60}$$

From Eq. (6.59), one can see that the integral kernel $o(t)$ is a non-singular smooth function with a property of $o(0) = 0$ when $0 < \alpha \leq 2$.

Now, let us use the wavelet-based method to solve Eq. (6.58). For the integral variable τ in Eq. (6.58), the base scaling function containing the boundary extension showing in Sect. 5.1 can be rewritten by the form

$$\phi_{n,k}^{[0,t]}(\tau) = \begin{cases} \sum\limits_{l=1-N+M_1}^{-1} u_{0,k}\left(\frac{l}{2^n}\right)\varphi(2^n\tau - l + M_1) + \varphi(2^n\tau - k + M_1), & 0 \leq k \leq r-1, \\ \varphi(2^n\tau - k + M_1), & r \leq k \leq t2^n - r, \\ \sum\limits_{l=t2^n+1}^{t2^n-1+M_1} u_{1,t2^n-k}\left(\frac{l}{2^n}\right)\varphi(2^n\tau - l + M_1) + \varphi(2^n\tau - k + M_1), & t2^n - r + 1 \leq k \leq t2^n. \end{cases} \tag{6.61}$$

Using Eq. (6.61) to approximate the term $o(t-\tau)[g(y(\tau)) - f(\tau)]$ with $\tau \in [0, t]$ in Eq. (6.58), we have

$$o(t-\tau)[g(y(\tau)) - f(\tau)] \approx \sum_{k=0}^{t2^n} o(t - t_k)[g(y_k) - f_k]\phi_{n,k}^{[0,t]}(\tau), \tag{6.62}$$

where $t_k = k/2^n$, $g(y_k) = g(y(t_k))$, and $f_k = f(t_k)$. Integrating both sides of Eq. (6.62), we obtain

$$\int_0^t o(t-\tau)[g(y(\tau)) - f(\tau)]d\tau \approx \sum_{k=0}^{t2^n} o(t - t_k)[g(y_k) - f_k]\Phi_{n,k}(t) \tag{6.63}$$

in which $\Phi_{n,k}(t) = \int_0^t \phi_{n,k}^{[0,t]}(\tau)d\tau$ can be obtained by the definition of Eqs. (6.61) and (4.51). Substituting Eq. (6.63) into Eq. (6.58), and then set $t = t_i$, we get

$$y_i + \sum_{k=0}^{i} o_{i-k}[g(y_k) - f_k]\Phi_{n,k}(t_i) \approx q_i, \tag{6.64}$$

where $y_i = y(t_i)$, $t_i = i/2^n$ $(i = 1, 2, 3, \ldots)$, $o_{i-k} = o(t_i - t_k)$, and $q_i = q(t_i)$. According to Eq. (6.59), there should be $o(0) = 0$ for $0 < \alpha \le 2$, which implies that $o(i-k)[g(y_k) - f_k]\Phi_{n,k}(t_i) = 0$ when $k = i$. Thus, Eq. (6.64) can be further simplified into the form

$$y_i \approx \sum_{k=0}^{i-1} o_{i-k}[f_k - g(y_k)]\Phi_{n,k}(t_i) + q_i. \tag{6.65}$$

Equation (6.65) tells us that the solution y_i can be directly obtained step-by-step as the index i increases. In this process, no matrix inversion is needed.

Example 6.7 We consider the following fractional equation given by Li [17],

$$D^2y(t) + D^{1.234}y(t) + D^{0.333}y(t) + y^3(t) = f(t), \quad t > 0, \tag{6.66}$$

$$y(0) = 0, \, \dot{y}(0) = 0. \tag{6.67}$$

When the external force is taken as

$$f(t) = \frac{2}{\Gamma(2)}t + \frac{2}{\Gamma(2.766)}t^{1.766} + \frac{2}{\Gamma(3.667)}t^{2.667} + \left(\frac{1}{3}t^3\right)^3,$$

such governing equations in this example have the exact solution of the form $y(t) = t^3/3$.

Following the above program to such a system, after the Laplace transform is used in Eq. (6.66), one can get the equation similar to Eq. (6.57), while the corresponding convolution kernel $o(t)$ can be written in the form

$$o(t) = \mathcal{L}^{-1}[O(s)], \, O(s) = \frac{1}{s^2 + s^{1.234} + s^{0.333}}. \tag{6.68}$$

Applying the wavelet-based Laplace inversion Eqs. (6.30) and (6.31) to Eq. (6.68), we have

$$o(t) = \lim_{q\to\infty}\left\{\frac{e^t}{2^{q+1}\pi}\varphi_*\left(-\frac{t}{2^q}\right)\sum_{k=-\infty}^{\infty} O\left(1 + i\frac{M_1 + k}{2^q}\right)e^{itk/2^q}\right\}, \tag{6.69}$$

where $\widehat{\varphi}(\omega)$ is the Fourier transform of the scaling function $\varphi(t)$ and $i = \sqrt{-1}$.

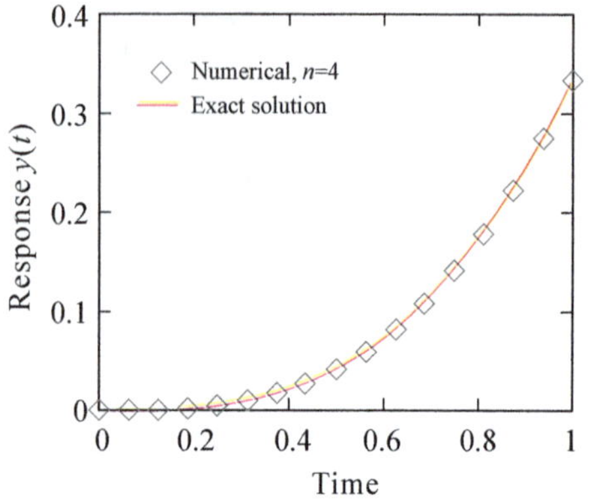

Fig. 6.8 Comparison between the numerical and exact solutions of Example 6.7. Here, "Numerical" means the results obtained by the wavelet-based Laplace inverse transform method when the resolution level $n = 4$

After that, the solution similar to Eq. (6.65) can be obtained by

$$y_i \approx \sum_{k=0}^{i-1} o_{i-k}(f_k - y_k^3)\Phi_{n,k}(t_i), \tag{6.70}$$

where $o(t)$ is given by Eqs. (6.68) and (6.69).

Figure 6.8 shows the comparison between the exact result of this problem and its numerical approximation of Eq. (6.70) from the wavelet-based Laplace inverse transform method when $n = 4$ and time step of 1/16. From this figure, one finds that the approximate solution from the generalized Coiflets wavelet-based method is well coincided with its exact one. Table 6.2 gives the absolute errors of the approximations for different resolution level, i.e., $n = 4, 5, 6$. From this, one sees that the maximum absolute errors of ~1.0E-5 have appeared when $n = 4$, and as n increases, the errors decrease in order. From Ref. [17], one can also find that Li solved the same equation by using a method based on the so-called Chebyshev wavelets, where there are a set of 48 nonlinear algebraic equations needed to be solved and the absolute error of its approximate solution at $t = 0.5$ is 1.269211E-04, which is at least one order larger than the absolute error by using Eq. (6.70) when $n \geq 4$, see Table 6.2. It should be noted that the approximate solution Eq. (6.70) does not involve any calculations of inversion of matrixes and without solving nonlinear algebraic equations. Thus, the solution based on the generalized Coiflets-based Laplace inverse transform method

Table 6.2 Absolute error of numerical solutions under different resolution level n

t	$n = 4$	$n = 5$	$n = 6$
0.25	6.1702945e-05	1.0564871e-06	1.9098283e-07
0.50	5.2252928e-06	2.1596250e-06	9.0243531e-08
0.75	8. 2326125e-06	4.2263352e-06	8.2589488e-08
1.00	1.0274991e-05	7.6263066e-06	1.1919017e-07

has not only high accuracy but also easy and low computations compared with those methods employed in literature.

Example 6.8 Nonlinear vibration system with fractional damping is considered by the form

$$D^2y(t) + \frac{8}{10}D^{1/2}y(t) + y^3(t) = f(t), \quad t > 0, \tag{6.71}$$

$$y(0) = 0,\ \dot{y}(0) = 0. \tag{6.72}$$

When the external force is selected by the form

$$f(t) = 2\left(t - \frac{9}{10}\right)\left(t - \frac{7}{10}\right) + 4t\left(t - \frac{7}{10}\right) + 4t\left(t - \frac{9}{10}\right) + 2t^2 + \frac{8}{10\Gamma(1/2)}\left(\frac{128}{35}\sqrt{t^7} - \frac{128}{25}\sqrt{t^5} + \frac{42}{25}\sqrt{t^3}\right) + \left[t^2\left(t - \frac{9}{10}\right)\left(t - \frac{7}{10}\right)\right]^3,$$

one can verify that the following exact solution of the dynamic system is

$$y(t) = t^2(t - 9/10)(t - 7/10).$$

Following the solution procedure suggested above, we convert Eq. (6.71) into the form of Eq. (6.65). The corresponding convolution kernel $o(t)$ in Eq. (6.65) can be given by Eq. (6.59) with parameters $c_1 = 1$, $c_2 = 8/10$, $\alpha = 1/2$, which is

$$o(t) = tE_{1.5,2}\left(-\frac{8}{10}t^{1.5}\right), \tag{6.73}$$

where $E_{a,b}(z)$ is the generalized Mittag–Leffler type functions defined by Eq. (6.60).

After that, the approximate solution of this example can be gained by

$$y_i \approx \sum_{k=0}^{i-1} o_{i-k}(f_k - y_k^3)\Phi_{n,k}(t_i). \tag{6.74}$$

As shown in Fig. 6.9, when $n = 7$, the approximate result of Eq. (6.74) is well in agreement with its exact solution quantitatively. This example problem was also solved by Ray et al. [20] by using the modified Adomain Decomposition Method (ADM), and by Atanackovic et al. [21] using a different numerical scheme with an expansion formula for fractional derivative. Table 6.3 displays a quantitative comparison between the exact results and some existing approximate results. From Table 6.3, we can see that the approximate solutions given by our Coiflets-based Laplace inverse transform method have an accuracy of at least five effective numbers,

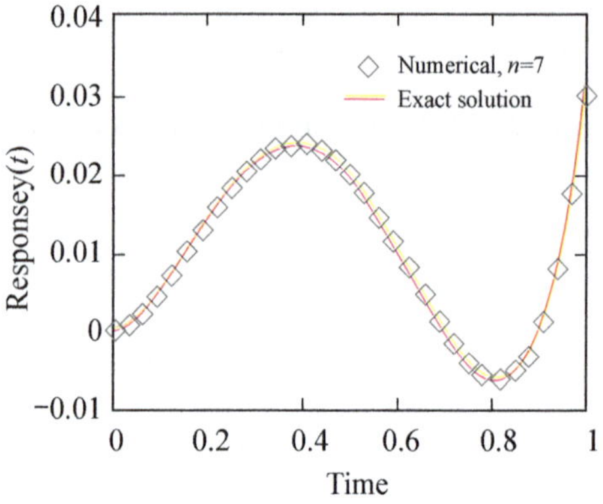

Fig. 6.9 Comparison between the wavelet numerical and exact solutions for Example 6.8

Table 6.3 Comparison of numerical solutions of Example 6.8 with its exact ones

t	Exact	Coiflet-based	Ray [20]	Atanackovic [21]
0.25	0.01828130	0.01828127	0.0182813	0.0182531908
0.50	0.02000000	0.02000003	0.020026	0.0198515236
0.75	−0.00421875	−0.00421871	−0.0419593	−0.004492587
1.00	0.03000000	0.0300000	0.0300995	0.0293800105

while other results obtained by, e.g., Ray et al. [20] and Atanackovic et al. [21] have only two effective numbers except for $t = 0.25$.

6.3.3 Numerical Solution for the Multi-term Time-Fractional Diffusion-Wave Equation

The time-fractional, diffusion-wave equation [16, 22–26, 29, 30], which describes the anomalous diffusion-wave phenomenon arising from different circumstances [16, 22–26, 29, 30], is obtained by replacing the first- or second-order time derivative in a standard diffusion or wave equation with the sum of a fractional derivative of the order $\alpha_m \in (0, 2]$. That is,

$$\begin{cases} \sum_{m=1}^{M} \mu_m D_t^{\alpha_m} u(x,t) = k\dfrac{\partial^2 u(x,t)}{\partial x^2} + q(x,t), 0 < x < l, t > 0, 0 < \alpha_m \le N, \\ D_t^n u(x,0) = g_n(x), n = 0, 1, \ldots, N-1, \\ a_0 u(0,t) + b_0 u_{,x}(0,t) = u_0(t), \\ a_l u(l,t) + b_l u_{,x}(l,t) = u_l(t) \end{cases} \tag{6.75}$$

where μ_m, k, a_0, b_0, a_l, and b_l are constants, $q(x, t)$, g_n, $u_0(t)$, and $u_l(t)$ are known functions, and the partial differential is used by

$$f_{,x}(a, y) = \partial f(x, y)/\partial x|_{x=a}.$$

In this problem, the unknown function $u(x, t)$ can be a concentration, a probability distribution, or displacement for diffusion, a general transport, or a wave motion, respectively.

It should note that the generalized boundary conditions contain nearly all classic types of boundary conditions. For example, when $a_0 = a_l = 0$, they become into the Dirichlet boundary conditions; if $b_0 = b_l = 0$, then they are the Neumann boundary conditions; and when they are arbitrary of nonzero, the boundary conditions are called as the general Robin boundary conditions.

To obtain the analytical solution for Eq. (6.75), the Laplace transform is applied to Eq. (6.75), then we get

$$\sum_{m=1}^{M} \mu_m s^{\alpha_m} U(x, s) - k\frac{\partial^2 U(x, s)}{\partial x^2} = Q(x, s) + \sum_{m=1}^{M}\sum_{n=0}^{[\alpha_m]} \mu_m s^{\alpha_m - 1 - n} g_n(x), \quad (6.76)$$

where $[\alpha]$ represents the maximal integer that is less than α. $U(x, s)$, $Q(x, s)$ are the Laplace transform of $u(x, t)$ and $q(x, t)$, respectively.

Multiplying both sides of Eq. (6.76) by e^{-px}, and integrating the induced equation with respect to x from 0 to x_0 $(0 \le x_0 \le l)$, one gets

$$\begin{aligned}\left(\sum_{m=1}^{M} \mu_m s^{\alpha_m} - kp^2\right)\widetilde{U}(p, s, x_0) &= \widetilde{Q}(p, s, x_0) + \sum_{m-1}^{M}\sum_{n-0}^{[\alpha_m]} \mu_m s^{\alpha_m - 1 - n}\widetilde{G}_n(p, x_0) \\ &+ k\left[U_{,x}(x_0, s)e^{-px_0} - U_x(0, s) + pU(x_0, s)e^{-px_0} - pU(0, s)\right]\end{aligned}, \quad (6.77)$$

where

$$\widetilde{U}(p, s, x) \equiv \int_0^x U(\tau, s)e^{-p\tau}d\tau = \int_0^x\int_0^{+\infty} u(\tau, t)e^{-st-p\tau}dtd\tau, \quad (6.78)$$

$$\widetilde{Q}(p, s, x) \equiv \int_0^x Q(\tau, s)e^{-p\tau}d\tau = \int_0^x\int_0^{+\infty} q(\tau, t)e^{-st-p\tau}dtd\tau, \quad (6.79)$$

$$\widetilde{G}_n(p, x) \equiv \int_0^x g_n(\tau)e^{-p\tau}d\tau. \quad (6.80)$$

In the Laplace transform space, the homogeneous equation of Eq. (6.77) is

$$\left(\sum_{m=1}^{M}\mu_m s^{\alpha_m} - kp^2\right)\widetilde{U}(p, s, x_0) = 0.$$

To the non-trivial solution of $\widetilde{U}(p, s, x_0)$, its necessary and sufficient condition is

$$\sum_{m=1}^{M}\mu_m s^{\alpha_m} - kp^2 = 0, \tag{6.81}$$

which constitutes an eigenvalue equation. To the eigenvalue p in Eq. (6.81), two solutions are obtained by

$$p_1 = -p_2 = \sqrt{\sum_{m=1}^{M}\mu_m s^{\alpha_m}/k}. \tag{6.82}$$

Due to $s = \beta + i\omega$, Eq. (6.82) tells us that real component β can be arbitrarily taken such that $p_1 \neq p_2 \neq 0$. Substituting the different roots p_1 and p_2 into Eq. (6.77), respectively, we have

$$\begin{cases} \widetilde{Q}(p_1, s, x_0) + \sum\limits_{m=1}^{M}\sum\limits_{n=0}^{[\alpha_m]}\mu_m s^{\alpha_m-1-n}\widetilde{G}_n(p_1, x_0) \\ +k\left[U_{,x}(x_0, s)e^{-p_1x_0} - U_{,x}(0, s) + p_1U(x_0, s)e^{-p_1x_0} - p_1U(0, s)\right] = 0 \\ \widetilde{Q}(p_2, s, x_0) + \sum\limits_{m=1}^{M}\sum\limits_{n=0}^{[\alpha_m]}\mu_m s^{\alpha_m-1-n}\widetilde{G}_n(p_2, x_0) \\ +k\left[U_x(x_0, s)e^{-p_2x_0} - U_x(0, s) + p_2U(x_0, s)e^{-p_2x_0} - p_2U(0, s)\right] = 0 \end{cases} \tag{6.83}$$

From Eq. (6.83), further, the functions of $U(x_0, s)$ and $U_x(x_0, s)$ can be obtained by

$$U(x, s) = \frac{\Phi_1(x, s) + \Psi_1(x, s)}{k(p_1 - p_2)} + \frac{(e^{p_1x} - e^{p_2x})U_x(0, s) + (p_1e^{p_2x} - p_2e^{p_2x})U(0, s)}{p_1 - p_2}, \tag{6.84}$$

$$U_x(x, s) = \frac{\Phi_2(x, s) + \Psi_2(x, s)}{k(p_1 - p_2)} + \frac{(p_1e^{p_2x} - p_2e^{p_x x})U_x(0, s) + p_1p_2(e^{p_2x} - e^{p_x x})U(0, s)}{p_1 - p_2}, \tag{6.85}$$

where x_0 is replaced with x, and

$$\Phi_1(x, s) = \widetilde{Q}(p_2, s, x)e^{p_2x} - \widetilde{Q}(p_1, s, x)e^{p_1x}, \tag{6.86}$$

$$\Phi_2(x,s) = p_2\widetilde{Q}(p_1,s,x)e^{p_1 x} - p_1\widetilde{Q}(p_2,s,x)e^{p_2 x}, \tag{6.87}$$

$$\Psi_1(x,s) = \sum_{m=1}^{M}\sum_{n=0}^{[\alpha_m]} \mu_m s^{\alpha_m-1-n}\left[\widetilde{G}_n(p_2,x)e^{p_2 x} - \widetilde{G}_n(p_1,x)e^{p_1 x}\right], \tag{6.88}$$

$$\Psi_2(x,s) = \sum_{m=1}^{M}\sum_{n=0}^{[\alpha_m]} \mu_m s^{\alpha_m-1-n}\left[p_2\widetilde{G}_n(p_1,x)e^{p_1 x} - p_1\widetilde{G}_n(p_2,x)e^{p_2 x}\right]. \tag{6.89}$$

Applying the Laplace transform to the boundary conditions in Eq. (6.75), we get

$$a_0 U(0,s) + b_0 U_x(0,s) = U_0(s) = \int_0^{+\infty} u_0(t)e^{-st}dt, \tag{6.90}$$

$$a_l U(l,s) + b_l U_x(l,s) = U_l(s) = \int_0^{+\infty} u_l(t)e^{-st}dt. \tag{6.91}$$

Let variable x in Eqs. (6.84) and (6.85) be l. Then substitution of them into Eq. (6.91) yields

$$AU(0,s) + BU_{,x}(0,s) = T, \tag{6.92}$$

where

$$\begin{cases} A = a_l k(p_1 e^{p_1 l} - p_2 e^{p_2 l}) + b_l k p_1 p_2 (e^{p_2 l} - e^{p_1 l}) \\ B = a_l k(e^{p_1 l} - e^{p_2 l}) + b_l k(p_1 e^{p_2 l} - p_2 e^{p_1 l}) \\ T - k(p_1 - p_2)U_l(s) - a_l[\Phi_1(l,s) + \Psi_1(l,s)] - b_l[\Phi_2(l,s) + \Psi_2(l,s)] \end{cases}. \tag{6.93}$$

From Eqs. (6.90) and (6.92), one can obtain

$$\begin{aligned} U(0,s) &= \frac{Tb_0 - BU_0(s)}{Ab_0 - Ba_0} \\ U_{,x}(0,s) &= \frac{AU_0(s) - Ta_0}{Ab_0 - Ba_0}. \end{aligned} \tag{6.94}$$

Substituting Eq. (6.94) into Eq. (6.84), we obtain the following exact solution of the form in the Laplace domain

$$U(x,s) = \Gamma_0(x,s) + \frac{\Gamma_1(x,s) + \Gamma_2(x,s) + \Gamma_3(x,s) + \Gamma_4(x,s) + \Gamma_5(x,s) + \Gamma_6(x,s)}{\Theta(s)}. \tag{6.95}$$

Here, Eq. (6.82) is considered, and

$$\Gamma_0(x,s) = \frac{\Phi_1(x,l) + \Psi_1(x,l)}{2kp_1}, \tag{6.96}$$

$$\Theta(s) = 2kp_1[(a_0a_l - b_0b_lp_1^2)(e^{-p_1l} - e^{p_1l}) + p_1(a_lb_0 - a_0b_l)(e^{-p_1l} + e^{p_1l})], \tag{6.97}$$

$$\Gamma_1(x,s) = [a_0a_l(e^{p_1x} - e^{-p_1x}) - a_lb_0p_1(e^{p_1x} + e^{-p_1x})]\Phi_1(l,s), \tag{6.98}$$

$$\Gamma_2(x,s) = [a_0b_l(e^{p_1x} - e^{-p_1x}) - b_0b_lp_1(e^{p_1x} + e^{-p_1x})]\Phi_2(l,s), \tag{6.99}$$

$$\Gamma_3(x,s) = [a_0a_l(e^{p_1x} - e^{-p_1x}) - a_lb_0p_1(e^{p_1x} + e^{-p_1x})]\Psi_1(l,s), \tag{6.100}$$

$$\Gamma_4(x,s) = [a_0b_l(e^{p_1x} - e^{-p_1x}) - b_0b_lp_1(e^{p_1x} + e^{-p_1x})]\Psi_2(l,s), \tag{6.101}$$

$$\Gamma_5(x,s) = 2kp_1[(a_l - b_lp_1)e^{-p_1(l-x)} - (a_l + b_lp_1)e^{p_1(l-x)}]U_0(s), \tag{6.102}$$

$$\Gamma_6(x,s) = 2kp_1[(a_0 + b_0p_1)e^{-p_1x} - (a_0 - b_0p_1)e^{p_1x}]U_l(s). \tag{6.103}$$

Finally, substituting Eq. (6.95) into Eq. (6.30), we gain the solution of Eq. (6.75) in the time–space domain as follows

$$u(x,t) \approx u_j(x,t) = \frac{e^{\beta t}}{2^{j+1}\pi}\varphi_*\left(-\frac{t}{2^j}\right)\sum_{k\in Z} U\left(x, \beta + i\frac{k+M_1}{2^j}\right)e^{ikt/2^j}. \tag{6.104}$$

Example 6.9 Exact solution of a nonhomogeneous fractional differential equation with the Robin boundary conditions. Such governing equations are of the form

$$\begin{cases} D_t^{1.5}u(x,t) - \frac{1}{2}D_t^{0.5}u(x,t) = \frac{3}{2}\frac{\partial^2 u(x,t)}{\partial x^2} + f(x,t),\ 0 < x < \pi, t > 0, \\ f(x,t) = \frac{45\sqrt{\pi}t^3 - 48t^{5/2} + 40t^{3/2}}{30\sqrt{\pi}}\sin(x) - 3\sqrt{\frac{t}{\pi}}, \\ u(x,0) = 2x, \\ u_{,t}(x,0) = 3, \\ 2u(0,t) + 3u_{,x}(0,t) = 3t^3 + 6t + 6, \\ -u(\pi,t) - 4u_{,x}(\pi,t) = 4t^3 - 3t - 8 - 2\pi. \end{cases} \tag{6.105}$$

Following our proposed wavelet-based solution program, i.e., using Eqs. (6.95) and (6.96)–(6.103), we get the Laplace transform of the unknown function $u(x,t)$ in the form

$$U(x,s) = \frac{3}{s^2} + \frac{2x}{s} + \frac{6}{s^4}\sin(x). \tag{6.106}$$

For such analytical function in Eq. (6.106), when the Laplace inverse transform is applied, the exact solution of Eq. (6.105) governing this problem is analytically obtained as follows:

$$u(x,t) = t^3 \sin(x) + 3t + 2x. \tag{6.107}$$

Daftardar–Gejji and Bhalekar [27] applied the separation of variables method to solve the boundary-value problems of fractional diffusion-wave equations with Dirichlet and Neumann boundary conditions. However, the solution for such equations with Robin boundary conditions is not considered in their work [27]. Following their solution procedure, if the separation of variables method is used to solve the equation with Robin boundary conditions, then the following transcendental or nonlinear equation must numerically solve to obtain the eigenvalue λ:

$$\tan(\sqrt{2\lambda/3}\pi) = -\frac{5\sqrt{2\lambda/3}}{2+8\lambda}. \tag{6.108}$$

It is obvious that to find all the solutions of the transcendental equations from Eq. (6.108) is normally hard since such nonlinear algebraic equation has infinite roots. In general, only a few low-order eigenvalues can be obtained, which makes the resulting solution insufficiently accurate. However, the exact solution formulated by Eq. (6.107) of this problem governed by the fractional differential equation with the Robin boundary conditions can be analytically obtained by our proposed method.

Example 6.10 Comparison of the wavelet-based solution and others to a homogeneous fractional differential equation considered in [27]

$$\begin{cases} D_t^{0.9}u(x,t) = \partial^2 u(x,t)/\partial x^2, 0 < x < \pi, t > 0 \\ u(0,t) = u(\pi,t) = 0 \\ u(x,0) = \sin(x) \end{cases}. \tag{6.109}$$

Again, the exact solution of the example in the Laplace domain can be gained in the form

$$U(x,s) = \frac{\sin(x)}{s^{0.1}+s}. \tag{6.110}$$

By substituting Eq. (6.110) into Eq. (6.104), the wavelet-based solution of Eq. (6.109) in the time–space domain can be further obtained by

$$u_j(x,t) = \frac{\sin x e^{\beta t}}{2^{j+1}\pi}\widehat{\varphi}\left(-\frac{t}{2^j}\right)\sum_{k\in Z}\frac{e^{ikt/2^j}}{(\beta+i(k+M_1)2^{-j})^{0.1}+\beta+i(k+M_1)2^{-j}}. \tag{6.111}$$

As a comparison, we briefly give these solution approaches and results in literature, e.g., Mittag-Leffler-type functions [28] have been widely applied to the solution of fractional differential equations. Such type of functions satisfies the following relation

$$\mathcal{L}^{-1}\left[\frac{s^{a-b}}{s^a+\lambda}\right]=t^{b-1}E_{a,b}(-\lambda t^a), \tag{6.112}$$

where $E_{a,b}(z)$ is the generalized Mittag–Leffler-type functions defined by Eq. (6.60).

Using Eqs. (6.60) and (6.112), Daftardar–Gejji and Bhalekar [27] also obtained the solution of Eq. (6.109) in the form

$$u(x,t)=\sin(x)E_{0.9,1}(-t^{0.9})=\sin(x)\lim_{N\to\infty}\sum_{k=0}^{N}\frac{(-t^{0.9})^k}{\Gamma(0.9k+1)}. \tag{6.113}$$

Figure 6.10 gives the comparison between solutions $u(\pi/2,t)$ of approximations from Eq. (6.111) when resolution level $n=6$ and that from Eq. (6.113) for different truncated number N in the infinite summation. As shown in this figure, the wavelet-based solution result from Eq. (6.111) smoothly changes in the time domain of $[0,35]$ as time t increases. However, the solution result from Eq. (6.113) is strongly dependent on the truncation number N. For each truncation number N, the approximate solution of Eq. (6.113) in the low interval of t is well coincident with the wavelet-based solution of Eq. (6.111), but when time approaches a critical instant and over it, there is a jump increase in the approximate solution of Eq. (6.113), which means that the convergent interval of such approximate solution is strongly dependent on the truncation number N and time variable.

In theory, the jumping phenomenon appeared in the approximate solution of Eq. (6.113) mainly originated from the term $(-t^{0.9})^k$ when both t and N (or k) increase large enough. As pointed out in [31], the convergence rate reduces rapidly

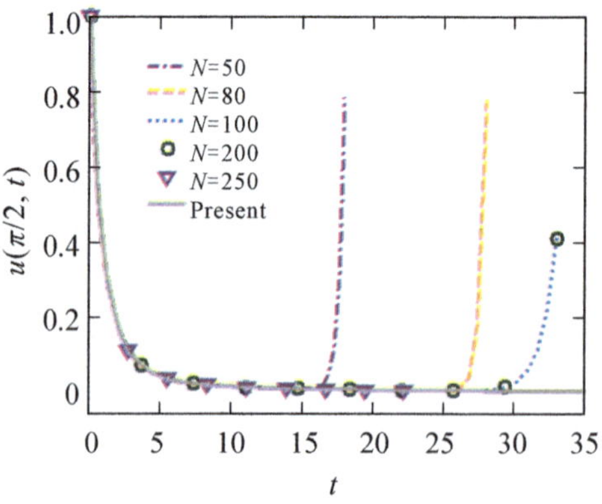

Fig. 6.10 Comparison of numerical solutions between the wavelet-based method marked by "present" and Daftardar–Gejji and Bhalekar method identified by the number "N" truncated in summation

as the truncation number N and time increase, resulting in an extremely slow rate of convergence for a larger N. For example, as shown in Fig. 6.10, with N increasing from 50 to 80, the convergence interval is extended from about [0, 15] to [0, 25], enlarging the time interval of approximation about 10 units. However, the convergence interval is extended only about 2.5 units (from about [0, 25] to [0, 27.5]) when N increases from 80 to 100. Further, there is even no visible extension in the approximate solution of Eq. (6.113) when the truncation number N increases from 100 to 200, see Fig. 6.10. In fact, the absolute value of each term $(-t^{0.9})^k$ becomes a huge number when both the time t and the power exponent k are larger enough. Such a situation may make the convergent interval of the approximation of Eq. (6.113) smaller. Also as shown in Fig. 6.10, for example, only the approxiamte solution in [0, 22] can be directly obtained by Eq. (6.113) when $N = 250$, while the interval of an efficient solution is about [0, 27.5] when $N = 100$. This fact tells us that the approximate solution in Eq. (6.113) does not converge as N approaches to infinite. It is this fact that makes an estimation for the approximate solution of Eq. (6.113) in a relative large time interval becomes impossible if no a or more specialized algorithms are added. Once some specific techniques are added, the solution method usually becomes very complicated and time-consuming [31]. In contrast, the suitable interval of the wavelet-based solution of Eq. (6.111) will extend linearly as the resolution level j increases, and there is no other technique requested in the computation. As displayed in Fig. 6.10, a convergent numerical solution in the interval [0, 35] can be directly obtained through Eq. (6.111) by taking $j = 6$. When the resolution level increases, the suitable interval of the wavelet-based solution increases too.

Example 6.11 Wave propagation with no Markovian diffusion process for a particle trapped in finite space with absorbing-reflecting boundaries and a sharp initial condition [32]. The governing equations of this example are of the form

$$\begin{cases} D_t^{\alpha} p(x,t) = \dfrac{\partial^2 p(x,t)}{\partial^2 x}, \quad 0 < x < 1, t > 0, 0 < \alpha \le 1 \\ p(x,t) = \delta(x - 1/2), \\ p(0,t) = 0, \dfrac{\partial p(1,t)}{\partial x} = 0, \ t > 0 \end{cases}. \tag{6.114}$$

To Eq. (6.114), the Laplace transform is of the form

$$P(x,s) = \begin{cases} \frac{\gamma e^{\gamma(1-2x)/2}(1+e^{\gamma})(e^{2\gamma x}-1)}{2(1+e^{2\gamma})s}, & if \quad x \le 0.5 \\ \frac{\gamma e^{-\gamma(1-2x)/2}(e^{\gamma}-1)(e^{2\gamma}+e^{2\gamma x})}{2(1+e^{2\gamma})s}, & if \quad x > 0.5 \end{cases}, \tag{6.115}$$

where $\gamma = \sqrt{s^{\alpha}}$. Further, the integrated survival probability $w(t)$ in the Laplace transform domain can be obtained by integrating Eq. (6.115) as follows

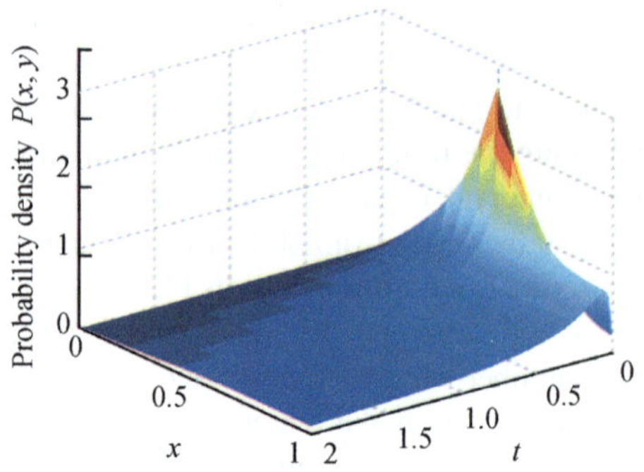

Fig. 6.11 Wavelet-based solution results for the probability density function $p(x, t)$ varying with time and space coordinates when $\alpha = 0.75$

$$W(s) = \int_0^1 P(x, s)dx = \frac{1 - \cosh(\gamma - 2)\text{sech}(\gamma)}{s}. \tag{6.116}$$

After that, the wavelet-based solution can be obtained by substituting Eqs. (6.115) and (6.116) into Eq. (6.104), respectively. To save space, here, we omit their detail formulae.

When $\alpha = 0.75$, e.g., the quantitative results of solution $p(x, t)$ based on our proposed method in this section are plotted in Fig. 6.11 varying with time and space coordinates. From Fig. 6.11, one sees that the short-time behavior of the diffusion process is dominated by the initial value. That is to say, when an external application with an initial concentration jump occurs only at the point $x = 1/2$, such external action in the subsequent propagation is fast deceased in a short-time domain and in the region containing $x = 1/2$. When $t \geq 0.5$, e.g., $p(x, t)$ becomes almost uniform and approaches to zero. When we take different values of α, for example, 0.5, 0.75, and 1, their solutions of $p(1, t)$, i.e., the evolution of the probability density $p(x, t)$ at the reflecting boundary $x = 1$ are displayed in Fig. 6.12, while their corresponding integrated survival probability $w(t)$ are shown in Fig. 6.13. According

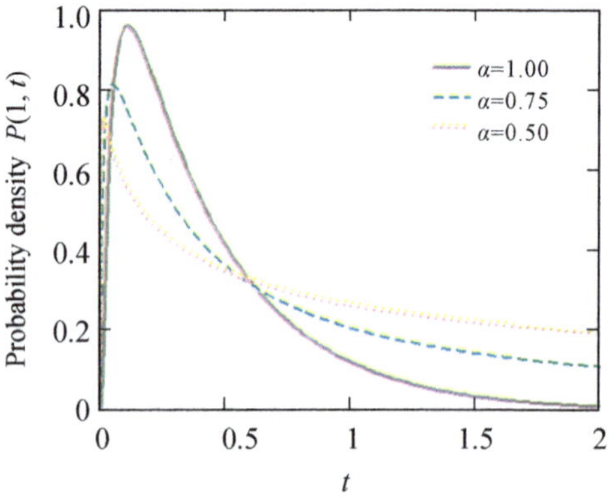

Fig. 6.12 Wavelet-based solution results of the probability density function $p(1, t)$ at the reflecting boundary $x = 1$ varying with time when $\alpha = 0.5, 0.75, 1$, respectively

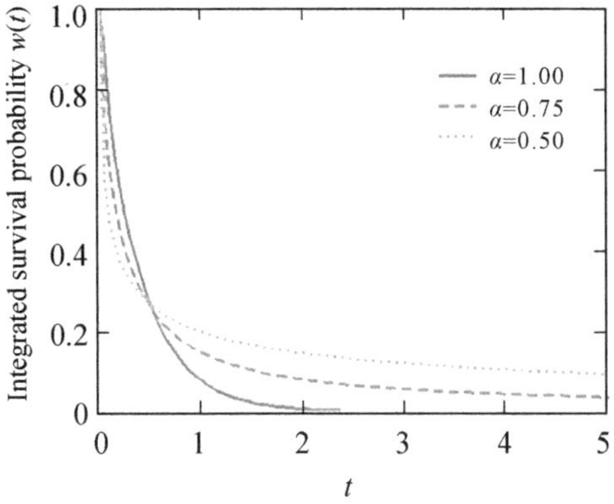

Fig. 6.13 Wavelet-based solution results of the integrated survival probability $w(t)$ varying with time when $\alpha = 0.5, 0.75, 1$, respectively

to Figs. 6.12 and 6.13, we know that the slow decay of the sub-diffusion solution for long periods displays the Brownian behavior, and that a small order of fractional derivative corresponds to a long survival time which is expected to a sub-diffusion phenomenon. Such a detailed discussion can be found in Metzler and Klafter [32].

6.3.4 Numerical Solution for Nonlinear Fractional Diffusion-Wave Equation

Consider a fractional diffusion-wave equation with a nonlinear term and inhomogeneous source terms as follows [33, 34]

$$c_1 \frac{\partial^2}{\partial t^2} u(x,t) + c_2 D_t^{\alpha} u(x,t) - \frac{\partial^2 u}{\partial x^2} + \chi(x,t,u) = \psi(x,t), \quad 0 < x < b, t > 0, \tag{6.117}$$

with the initial and boundary conditions,

$$u(x,0) = g_0(x), \quad u_t(x,0) = g_1(x), \quad 0 \leq x \leq b, \tag{6.118}$$

$$u(0,t) = 0, \quad u(b,t) = 0, \qquad t \geq 0 \tag{6.119}$$

where $0 < \alpha \leq 2$, c_1 and c_2 are constants, $\chi(x,t,u)$ stands for a nonlinear term of unknown function $u(x,t)$, and $\psi(x,t)$ is an inhomogeneous source term.

Similar to our previous solution program for the initial-value problem given in Sect. 6.3.2, we denote the Laplace transform of $u(x,t)$ by $U(x,s)$, i.e., $\mathcal{L}[u(x,t)] = U(x,s)$. Applying the Laplace transform to the fractional diffusion-wave Eq. (6.117) and taking into account the initial and boundary conditions in Eqs. (6.118) and (6.119), we have

$$(c_1s^2 + c_2s^\alpha)U(x,s) - \mathcal{L}[H(x,t)] = \sum_{n=0}^{1} c_1 s^n g_n(x) + \sum_{n=0}^{[\alpha]} c_2 s^{\alpha-1-n} g_n(x), \tag{6.120}$$

where

$$H(x,t) = \frac{\partial^2}{\partial x^2}u(x,t) - \chi(x,t,u) + \psi(x,t). \tag{6.121}$$

Further, Eq. (6.120) can be rewritten as

$$U(x,s) - O(s)\{\mathcal{L}[H(x,t)]\} = Q(x,s). \tag{6.122}$$

Here, $Q(x,s) = O(s)\left(\sum_{n=0}^{1} c_1 s^n g_n(x) + \sum_{n=0}^{[\alpha]} c_2 s^{\alpha-1-n} g_n(x)\right)$, and $O(s) = 1/(c_1s^2 + c_2s^\alpha)$.

Applying the inverse Laplace transform to Eq. (6.122), we get

$$u(x,t) - \int_0^t o(t-\tau)H(x,\tau)d\tau = q(x,t), \tag{6.123}$$

where $o(t)$ is given by Eq. (6.59), which is a non-singular smooth function with property of $o(0) = 0$ due to $0 < \alpha \leq 2$ and $q(x,t) = \mathcal{L}^{-1}[Q(x,s)]$. Obviously, function $q(x,t)$ can be easily obtained once the initial conditions in Eq. (6.119) are specified.

Using Eq. (6.59) to approximate the function $o(t-\tau)H(x,\tau)$ with $\tau \in [0,t]$, we have

$$o(t-\tau)H(x,\tau) \approx \sum_{k=0}^{t2^n} o(t-t_k)H(x,t_k)\phi_{n,k}^{[0,t]}(\tau). \tag{6.124}$$

Substituting Eq. (6.124) into Eq. (6.123), and setting $t = t_i$, we obtain

$$u(x,t_i) \approx \sum_{k=0}^{i-1} o_{i-k}H(x,t_k)\Phi_{n,k}(t_i) + q(x,t_i) \tag{6.125}$$

or

$$u(x,t_k) \approx \sum_{l=1}^{b2^m-1} u(x_l,t_k)\phi_{m,l}^{[0,b]}(x), \tag{6.126}$$

where $x_l = l/2^m$. Denote

$$H_{j,k} \equiv H(x_j, t_k) \approx \sum_{l=1}^{b2^m-1} u_{l,k} \frac{d^2\phi_{m,l}^{[0,b]}(x_j)}{dx^2} + \psi(x_j, t_k) - \chi(x_j, t_k, u_{j,k}) \quad (6.127)$$

in which $u_{l,k} = u(x_l, t_k)$. We further obtain

$$u_{j,i} \approx \sum_{k=0}^{i-1} o_{i-k} H_{j,k} \Phi_{n,k}(t_i) + q_{j,i}. \quad (6.128)$$

To clearly show the method applied in the time–space problem, let us summarize the procedure of the above method as follows

(a) When $t = 0$ or $i = 0$, we have $u_{j,0} = g_0(x_j)$.
(b) When $t = 1/2^n$ or $i = 1$, we have

$$H_{j,0} \approx \sum_{l=1}^{b2^m-1} u_{l,0} \frac{d^2\phi_{m,l}^{[0,b]}(x_j)}{dx^2} + \psi(x_j, t_0) - \chi(x_j, t_0, u_{j,0}), \quad (6.129a)$$

$$u_{j,1} \approx r_1 H_{j,0} \Phi_{n,0}\left(\frac{1}{2^n}\right) + q_{j,1}. \quad (6.129b)$$

(c) When $t = i_0/2^n$ or $i = i_0$, we have already known $u_{j,i<i_0}$, $H_{j,i<i_0-1}$, and

$$H_{j,i_0-1} \approx \sum_{l=1}^{b2^m-1} u_{l,i_0-1} \frac{d^2\phi_{m,l}^{[0,b]}(x_j)}{dx^2} + \psi(x_j, t_{i_0-1}) - \chi(x_j, t_{i_0-1}, u_{j,i_0-1}),$$

thus we can obtain

$$u_{j,i_0} \approx \sum_{k=0}^{i_0-1} r_{i_0-k} H_{j,k} \Phi_{n,k}(t_{i_0}) + q_{j,i_0}. \quad (6.130)$$

(d) The process (a), (b), (c) can be repetitively proceeded to any time t.

It can be seen that such a solution procedure decouples the complicated spatial–temporal dependences, making the solution of the nonlinear fractional differential Eqs. (6.117)–(6.119) become very simple. And no inversion of matrix and other special solution techniques are needed to deal with the nonlinear spatial and temporal operators.

Example 6.12 One had used in Rashidinia et al. [35]. For such case, the nonlinear fractional diffusion-wave equation is of the form

$$\frac{\partial^\alpha u}{\partial t^\alpha} - \mu \frac{\partial^2 u}{\partial x^2} + \chi(x,t,u) = \psi(x,t), \quad 0 \le x \le 1, 1 < \alpha \le 2, t > 0, \tag{6.131}$$

which subjects to the following initial and boundary conditions

$$u(x,0) = 0, u_t(x,0) = 0, \quad 0 \le x \le 1, \tag{6.132}$$

$$u(0,t) = 0, u(1,t) = t^3, \quad t \ge 0, \tag{6.133}$$

where $\chi(x,t,u) = u^2$, $\psi(x,t) = 6tx^3 + (tx)^6 - 6t^3x$, $\mu = 1$, when $\alpha = 2$. To such a case, the exact solution of Eqs. (6.131)–(6.133) in a closed form is $u(x,t) = x^3t^3$.

In order to take the above solution program, we first introduce a function transform, i.e., $u(x,t) = v(x,t) + x^5t^3$, such that the inhomogeneous boundary condition in this example is reduced into homogeneous on the unknown auxiliary function $v(x,t)$, while the source term becomes into $\tilde{\psi}(x,t) = 6x^3t + (xt)^6 - 6xt^3 - 6x^5t + 20x^3t^3$. After that, the above solution steps are employed for such a reduced equation of $v(x,t)$, that is,

$$r(t) = \mathcal{L}^{-1}\left[\frac{1}{s^\alpha}\right] = \frac{t^{-1+\alpha}}{\Gamma(\alpha)} \tag{6.134}$$

and

$$v_{j,i} \approx \sum_{k=0}^{i-1} o_{i-k} H_{j,k} \Phi_{n,k}(t_i) \tag{6.135}$$

in which

$$H_{j,k} \approx \sum_{l=1}^{2^m-1} v_{l,k} \frac{d^2\phi_{m,l}^{[0,1]}(x_j)}{dx^2} + \tilde{\psi}_{j,k} - (v_{j,k} + x_j^5 t_k^3)^2, \tilde{\psi}_{j,k} = \tilde{\psi}(x_j, t_k).$$

Substituting this equation into Eq. (6.130), the approximate solution is finally gained by the wavelet-based method.

When the wavelet resolution levels for time and space variables are chosen as $n = 8$ and $m = 4$, respectively, the quantitative results of the approximate solution of Eqs. (6.131)–(6.133) are, respectively, displayed in Fig. 6.14 for the time response at the space point $x = 0.25$, and for the displacement configurations at the instant $t = 0.25$ under different fractional order α. From the figure, one sees that when $\alpha = 2$, all approximate solutions by means of the wavelet-based method are well in

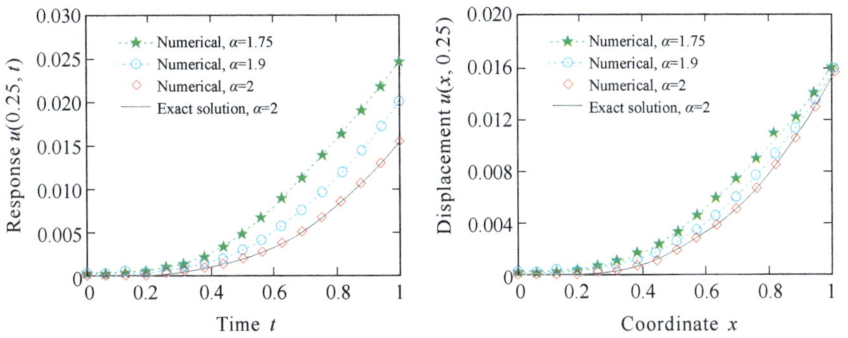

Fig. 6.14 Wavelet-based numerical solutions of Example 6.12 for $\alpha = 1.75, \alpha = 1.9, \alpha = 2$ under resolution level $n = 8$ time variable and $m = 4$ for space variable: Time-responses at space point $x = 0.25$ (Left); Displacement configurations at instant $t = 0.25$ (Right)

agreement with its exact solution $u(x, t) = x^3 t^3$ quantitatively. With decreasing the factor α from $\alpha = 2$ through $\alpha = 1.9$ to $\alpha = 1.75$, we find that the magnitude of approximate solution increases except for the boundary points.

Example 6.13 A nonlinear fractional Sine-Gordon equation with a non-polynomial nonlinear term. Such nonlinear problem can be expressed by

$$D_t^{\alpha} u - u_{xx} + \sin u = f(x, t), \quad 0 \le x \le 1, t > 0. \tag{6.136}$$

and the initial and the boundary conditions

$$u(x, 0) = 0, \quad u_t(x, 0) = 0, \qquad 0 < x \le 1, \tag{6.137}$$

$$u(0, t) = 0, \quad u(1, t) = 0, \qquad t \ge 0. \tag{6.138}$$

When $f(x, t) = 10t^{0.2} \sin(\pi x)/\Gamma(0.2) + \pi^2 t^2 \sin(\pi x) + \sin(t^2 \sin(\pi x))$ and $\alpha = 1.8$ are taken into account, we can obtain an analytical solution $u(x, t) = t^2 \sin(\pi x)$ that satisfies Eqs. (6.136)–(6.138). Following a similar procedure as that has been applied to Eqs. (6.131)–(6.133), we can obtain the numerical solution of Eqs. (6.136)–(6.138) as follows

$$u_{j,i} \approx \sum_{k=0}^{i-1} o_{i-k} H_{j,k} \Phi_{n,k}(t_i) \tag{6.139}$$

in which $o(t) = t^{0.8} / \Gamma(1.8)$, $f_{j,k} = f(x_j, t_k)$, and

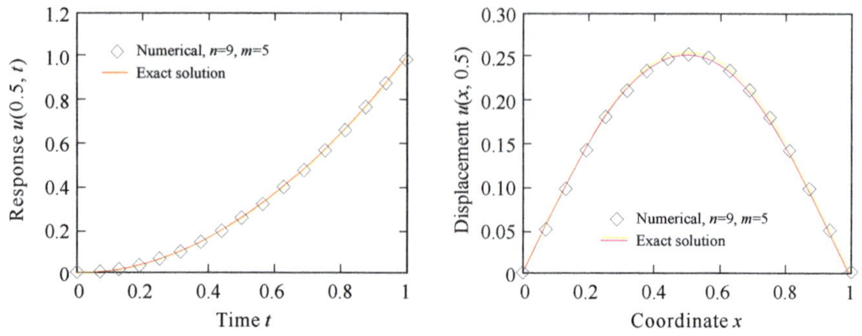

Fig. 6.15 Comparison between the numerical and exact solutions of the nonlinear fractional Sine-Gordon equation in Example 6.13: Time-response at point $x = 0.5$ (Left); Displacement configuration at instant $t = 0.5$ (Right)

$$H_{j,k} \approx \sum_{l=1}^{2^m-1} u_{l,k} \frac{d^2\phi_{m,l}^{[0,1]}(x_j)}{dx^2} + f_{j,k} - \sin(u_{j,k}).$$

When the resolution levels $n = 9$ for time variable and $m = 5$ for space variable are chosen, Fig. 6.15 gives the numerical results of the nonlinear fractional Sine-Gordon equation obtained by the wavelet-based method for the cases at space point $x = 0.5$ and instant $t = 0.5$. To correspond its exact solution $u(x, t) = t^2 \sin(\pi x)$, such approximation solution has the relative error within 1%.

References

1. Al-Shuaibi A (2001) Inversion of the Laplace transform via post-Widder formula. Integr Transform Spec Funct 11(3):225–232
2. Cohen AM (2007) Numerical methods for Laplace transform inversion. Springer, New York
3. Daubechies I (1988) Orthonormal bases of compactly supported wavelets. Commun Pure Appl Math 41(7):909–996
4. Daubechies I (1993) Orthonormal bases of compactly supported wavelets II: Variations on a theme. SIAM J Math Anal 24(2):499–519
5. Beylkin G, Coifman R, Rokhlin V (1991) Fast wavelet transforms and numerical algorithms. Commun Pure Appl Math 44(2):141–183
6. Rossikhin YA, Shitikova MV (1997) Applications of fractional calculus to dynamic problems of linear and nonlinear hereditary mechanics of solids. Appl Mech Rev 50(1):15–67
7. Ingman D, Suzdalnitsky J (2001) Iteration method for equation of viscoelastic motion with fractional differential operator of damping. Comput Methods Appl Mech Eng 190(37–38):5027–5036
8. Miller KS (1993) The Mittag-Leffler and Related Functions. Integral transforms and special functions. Integr Transform Spec Funct 1(1):41–49
9. Agrawal OP (2001) Stochastic analysis of dynamic system containing fractional derivatives. J Sound Vib 247(5):927–938

10. Jones DIG (2001) Handbook of viscoelastic vibration damping. Wiley, Arizona
11. Enelund M, Mähler L, Runesson K, Josefson BL (1999) Formulation and integration of the standard linear viscoelastic solid with fractional order rate laws. Int J Solid Struct 36(16):2417–2442
12. Oldham KB, Spanier J (1974) The Fractional calculus: integrations and differentiations of arbitrary order. Academic Press, New York
13. Samko SG, Kilbas AA, Marichev OI (1993) Fractional integrals and derivatives. Gordon and Breach Science Publisher, Yverdon, Yverdon-les-Bains, Switzerland
14. Suarez LE, Shokooh A (1997) An eigenvector expansion method for the solution of motion containing derivatives. J Appl Mech 64(3):629–635
15. Enelund M, Olsson P (1999) Damping described by fading memory-analysis and application to fractional derivative models. Int J Solids Struct 36(7):939–970
16. Caputo M, Mainardi F (1971) Linear models of dissipation in anelastic solids. La Rivista Del Nuovo Cimento 1(2):161–198
17. Yuanlu LI (2010) Solving a nonlinear fractional differential equation using Chebyshev wavelets. Commun Nonlinear Sci Numer Simul 15(9):2284–2292
18. Wang JZ, Zhou YH, Gao HJ (2003) Computation of the Laplace inverse transform by application of the wavelet theory. Commun Numer Methods Eng 19:959–975
19. Liu XJ, Wang JZ, Wang XM, Zhou YH (2014) Exact solutions of multi-term fractional diffusion-wave equations with Robin type boundary conditions. Appl Math Mech 35(001):49–62
20. Ray SS, Chaudhuri KS, Bera RK (2006) Analytical approximate solution of nonlinear dynamic system containing fractional derivative by modified decomposition method. Appl Math Comput 182(1):544–552
21. Atanackovic TM, Stankovic B (2008) On a numerical scheme for solving differential equations of fractional order. Mech Res Commun 35(7):429–438
22. Huang F, Guo B (2010) General solutions to a class of time fractional partial differential equations. Appl Math Mech (English Edition) 31(7):815–826
23. Mainardi F (1996) Fractional relaxation-oscillation and fractional diffusion-wave phenomena. Chaos Solitons Fractals 7(9):1461–1477
24. Caputo M (1967) Linear models of dissipation whose Q is almost frequency independent-II. Geophys J Int 13(5):529–539
25. Nigmatullin RR (1986) The realization of the generalized transfer equation in a medium with fractal geometry. Physica Status Solidi (B). 133(1):425–430
26. Nigmatullin RR (1984) To the theoretical explanation of the universal response. Physica Status Solidi (B). 123(2):739–745
27. Daftardar-Gejji V, Bhalekar S (2008) Boundary value problems for multi-term fractional differential equations. J Math Anal Appl 345(2):754–765
28. Wang JZ (2011) Fractional stochastic description of hinge motions in single protein molecules. Chin Sci Bull 56:495–501
29. Metzler R, Klafter J (2000) The random walk's guide to anomalous diffusion: a fractional dynamics approach. Phys Rep 339:1–77
30. Tarasov VE (2013) Review of some promising fractional physical models. Int J Mod Phys B 27(09):1330005
31. Welch SWJ, Ropper RAL, Duren RG (1999) Application of time-based fractional calculus methods to viscoelastic creep and stress relaxation of materials. Mech Time-Depend Mater 3:279–303
32. Metzler R, Klafter J (2000) Boundary value problems for fractional diffusion equations. Physica A 278(1–2):107–125
33. Zhou YH, Wang XM, Wang JZ, Liu XJ (2011) A wavelet numerical method for solving nonlinear fractional vibration, diffusion and wave equations. Comput Model Eng Sci 7(2):137–160

34. Zhou YH, Wang JZ, Zheng XJ (2001) A numerical inversion of the Laplace transform by use of the scaling function transform of wavelet theory. Acta Mathematica Scientia 21A(1):86–93 (in Chinese)
35. Rashidinia J, Ghasemi M, Jalilian R (2010) Numerical solution of the nonlinear Klein-Gordon equation. J Comput Appl Math 233(8):1866–1878

Chapter 7
Wavelet-Based Solutions for Boundary-Value Problems

Nonlinear boundary-value problems (NBVPs) arise from almost every scientific and engineering field, especially mechanical theory of beam and plate structures. Commonly, one-dimensional NBVPs can be written into a general form

$$\mathbf{T}\left(x, u, \frac{du}{dx}, \ldots, \frac{d^n u}{dx^n}\right) = 0, x \in [a, b] \tag{7.1}$$

with corresponding boundary conditions, where $\mathbf{T}$ is a nonlinear operator and u is an unknown function of variable x. Since the solutions of NBVPs are critically important for analyzing scientific and engineering problems, to find accurate and efficient methods for solving Eq. (7.1) has attracted considerable research attention. In Chap. 3, we have established a wavelet approximation scheme of bounded functions based on the techniques of boundary extension and generalized Coiflet-type wavelet expansion, which can effectively eliminate the undesired oscillating error near boundary points. In addition, when applying this modified wavelet approximation to the Galerkin method, a popular and powerful solution method to the NBVPs, the interpolating property of this approximation can avoid numerically calculating the integral of multifold-products of scaling functions and their derivatives, and make every nonlinear term of unknown functions in NBVPs to be expressed in a linear-liking combination of single-point sampling of a composite of the unknown function at corresponding nodal points, which is crucially important for the successful employment of the wavelet Galerkin-type methods.

In this chapter, we will introduce the basic idea and procedure of the proposed wavelet algorithm through the solutions of general nonlinear differential equations. Various numerical examples will be considered to further demonstrate the high accuracy and efficiency of the wavelet-based methods with relative low calculations.

Y.-H. Zhou, *Wavelet Numerical Method and Its Applications in Nonlinear Problems*, Engineering Applications of Computational Methods 6,
https://doi.org/10.1007/978-981-33-6643-5_7

7.1 Expansion of Nonlinear Operator Equation in One Dimension and Error Estimations

As a specific example, we consider a reduced form of Eq. (7.1)

$$\mathbf{L}[u(x)] + \mathbf{F}[u(x)] + p(x) = 0, \tag{7.2}$$

where **L** is a differential operator and **F** is a linear or nonlinear operator. As discussed in Chap. 4, when $u(x)$ is approximated by a truncated series and **F** is a linear operator, the residual error of the solution is independent of the approximate solution and depends only on the truncation error. In such a case, methods based on this expansion are closed. However, for a nonlinear operator, **N**, almost all existing conventional approximation schemes do not exhibit closure property, i.e., the corresponding approximate results are usually coupled with their truncation errors. In Sect. 3.4.1, we only analyzed the applications of three expansion schemes to nonlinear equations with the square-type nonlinearity. Nevertheless, the nonlinear operator can couple high-order and low-order terms, where Theorem 3.1 is invalid for nonlinear operators, regardless of which expansion scheme is used. Therefore, if the expansion scheme, Eq. (3.24) in Chap. 3, of the approximate solution is directly substituted into the nonlinear terms in the calculation as conventional methods, then the approximate part of the nonlinear term cannot be decoupled from its own truncation error. Accordingly, the methods based on this calculation are not closed. In practice, however, nearly all existing methods adopt this approach to obtain the approximate expression schemes for nonlinear terms [1–40]. Therefore, these existing methods do not exhibit a closure property.

However, when the function is expanded by the bases of generalized Coiflets-type scaling functions, this problem is apparently solved as displayed in Chap. 3. When the boundary extension shown in Chap. 5 is employed for the expansion via the scaling function, we can get the following theorems similar to that shown in Chap. 4.

Theorem 7.1 For any nonlinear operator, **N**, i.e., $\mathbf{N}[f(x)] \in L^2[0, 1]$, when the generalized Coiflets with resolution level j is employed, we have.

$$\mathbf{N}[f(x)] = P_j\mathbf{N}[f(x)] + \widetilde{Q}_j\mathbf{N}[f(x)], \tag{7.3}$$

where

$$\left\| \widetilde{Q}_j\mathbf{N}[f(x)] \right\|_{L^2[0,1]} = O(2^{-jN})$$

and

$$\mathbf{N}[f(x)] \approx P_j\mathbf{N}[f(x)] = \sum_{k=0}^{2^j} \mathbf{N}[f(k/2^j)]\phi_{j,k}(x), \quad x \in [0, 1]. \tag{7.4}$$

In Eq. (7.4), the expansion coefficients are explicitly represented, no matter what general nonlinear operator is applied. The form of expansion decouples the retained approximation from the truncation error. Therefore, we say that the wavelet algorithm based on this operation scheme is closed.

Not only one-dimensional problems, but also multidimensional problems can exhibit similar operations and conclusions. In the following, a 3D function is used as an example to illustrate the expansion scheme.

Theorem 7.2 For any nonlinear operator **N** that satisfies $\mathbf{N}[f(x, y, z)] \in L^2[0, 1]^3$, when the 3D generalized Coiflets with resolution level j is employed, we have

$$\begin{aligned} \mathbf{N}[f(x, y, z)] &\approx P_{j_{\min}}\mathbf{N}[f(x, y, z)] \\ &= \sum_{k=0}^{2^{j_x}}\sum_{l=0}^{2^{j_y}}\sum_{m=0}^{2^{j_z}} \mathbf{N}[f(k/2^{j_x}, l/2^{j_y}, m/2^{j_z})]\phi^x_{j_x,k}(x)\phi^y_{j_y,l}(y)\phi^z_{j_z,m}(z) \end{aligned} \tag{7.5}$$

and

$$\left\|\mathbf{N}[f(x, y, z)] - P_{j_{\min}}\mathbf{N}[f(x, y, z)]\right\|_{L^2[0,1]^3} \le C_0 2^{j_{\min} N}, \tag{7.6}$$

where the integer $j_{\min} = \min\{j_x, j_y, j_z\}$, and C_0 is a positive constant.

We can regard $\mathbf{N}[f(x, y, z)]$ as a new function. Then, Theorem 7.2 can be proven in accordance with Theorem 7.1.

Like the 1D case, the expansion scheme as shown in Sect. 5.2, for multidimensional functions provides an explicit representation of the expansion coefficient in the approximation scheme of relevant nonlinear terms. Consequently, the approximate part and the truncation error are decoupled. Therefore, the wavelet algorithm based on this operation is closed.

From the preceding discussion, a unified expansion form for unknown function and the linear or nonlinear operator can be explicitly formulated by the form:

$$u(x) = \sum_{k=0}^{2^j} u(k/2^j)\phi_{j,k}(x) + \sum_{n \ge j}\sum_{k \in Z} \bar{d}_{n,k}\psi_{n,k}(x), \tag{7.7}$$

$$\mathbf{F}[u(x)] = \sum_{k=0}^{2^j} \mathbf{F}[u(k/2^j)]\phi_{j,k}(x) + \sum_{n \ge j}\sum_{k \in Z} \tilde{d}_{n,k}\psi_{n,k}(x). \tag{7.8}$$

Substituting Eqs. (7.7) and (7.8) into Eq. (7.2) yields

$$\sum_{k=0}^{2^j} u(k/2^j)\mathbf{L}[\phi_{j,k}(x)] + \sum_{k=0}^{2^j} \mathbf{F}[u(k/2^j)]\phi_{j,k}(x) + p(x) = Q(x), \tag{7.9}$$

where the residual error has the form

$$Q(x) = -\sum_{n\geq j}\sum_{k\in Z}\overline{d}_{n,k}\psi_{n,k}(x) - \sum_{n\geq j}\sum_{k\in Z}\tilde{d}_{n,k}\psi_{n,k}(x). \tag{7.10}$$

Such residual error in Eq. (7.10) is similar to the residual error of the linear problem analyzed in Sect. 5.1, which is independent of the expansion coefficient $u(k/2^j)$ and the scaling function $\phi_{j,k}(x)$ employed. This finding indicates that the approximate solution and the high-order truncation are decoupled, i.e., the solution is closed. The corresponding wavelet expansion coefficients $\overline{d}_{n,k}$ and $\tilde{d}_{n,k}$ in the residual error are small quantities with the same order as $O(2^{-jr})$ and are independent of the approximate solution, where j and r are the resolution level and the number of vanishing moment in the generalized Coiflets employed. It is obvious that when the resolution level, j, is sufficiently large, such error function is always a higher order small quantity relative to the main approximations in Eqs. (7.7) and (7.8) no matter how magnitude of the approximation solution is large.

Comparing the solution process from Eqs. (7.7) and (7.8) to the conventional expansion of nonlinear term shown in Chap. 3, it can be easily found that the fundamental difference between the present and the conventional treatments is that the former directly expands the nonlinear term according to the wavelet bases to obtain its approximate representation, while the latter substitutes the approximation of the unknown function into the nonlinear term to obtain its approximation. In addition, it can be seen from the expansions in Eqs. (7.7) and (7.8) that the expansion coefficients for the nonlinear term, $\mathbf{F}[u(x)]$, are explicitly related to the function values of $u(x)$ at discrete nodal points. This interesting property of approximation allows Eq. (7.9) only and explicitly contains unknown quantities of $u(k/2^j)$, and no new unknown quantities are introduced. Therefore, many conventional numerical methods for the solutions of NBVPs, such as the Galerkin method, can be directly used to further obtain the approximate solutions of the problems. In the following, we display how the wavelet approximation can be applied to the Galerkin method for solving general nonlinear NBVPs.

7.2 Galerkin-Wavelet Solution Program

In order to make the description on the solution procedures be clear enough, we first consider the so-called semi-linear problems, that is, the nonlinear operator only acts on the unknown function itself and the terms related to the derivatives of the unknown function is linear [41],

$$\begin{cases} \mathbf{L}^0 u(x) + \mathbf{L}^1\mathbf{N}[u(x)] = 0, \quad 0 \leq x \leq 1 \\ \left.\frac{d^{i_s}u}{dx^{i_s}}\right|_{x=0} = 0, \quad \left.\frac{d^{j_s}u}{dx^{j_s}}\right|_{x=1} = 0 \end{cases}, \tag{7.11}$$

where $\mathbf{L}^0$ and $\mathbf{L}^1$ are differential operators, $\mathbf{N}$ is a nonlinear operator, which can be in a form of integer power, transcendental or sum of multiples, i_s, j_s, and s are non-negative integers. Many problems in the fields of natural science and engineering technology can usually be transformed into the form of Eq. (7.11).

The unknown function, $u(x)$, can be expressed as

$$u(x) = P_j u(x) + \widetilde{Q}_j u(x) = \sum_{k=0}^{2^j} u(k/2^j)\phi_{j,k}(x) + \widetilde{Q}_j u(x). \tag{7.12}$$

According to Sect. 4.1, we have the truncation error $\widetilde{Q}_j u(x) = O(2^{-jN})$. At the same time, according to Eq. (3.31), the nonlinear term, $\mathbf{N}[u(x)]$, in the equation can be expanded to

$$\begin{aligned}\mathbf{N}[u(x)] &= P_j\mathbf{N}[u(x)] + \widetilde{Q}_j\mathbf{N}[u(x)] \\ &= \sum_{k=0}^{2^j} \mathbf{N}[u(k/2^j)]\phi_{j,k}(x) + \widetilde{Q}_j\mathbf{N}[u(x)].\end{aligned} \tag{7.13}$$

Once again, according to Sect. 4.2, the truncation error is $\widetilde{Q}_j\mathbf{N}[u(x)] = O(2^{-jr})$. Substituting Eqs. (7.12) and (7.13) into Eq. (7.11), we obtain

$$\sum_{k=0}^{2^j} u(k/2^j)\mathbf{L}^0\phi_{j,k}(x) + \sum_{k=0}^{2^j} \mathbf{N}[u(k/2^j)]\mathbf{L}^1\phi_{j,k}(x) = Q(x) \approx 0. \tag{7.14}$$

The residual error of the equation is

$$Q(x) = -\mathbf{L}^0\widetilde{Q}_j u(x) - \mathbf{L}^1\widetilde{Q}_j\mathbf{N}[u(x)]. \tag{7.15}$$

Following the basic procedure of the conventional Galerkin method, after both sides of the Eq. (7.14) are multiplied with the weight functions, $\phi_{j,l}(x)$, $l = 0, 1, 2, \ldots, 2^j$, then integrate it from 0 to 1, a set of nonlinear algebraic equations can be gained by the matrix form

$$\mathbf{AU} + \mathbf{BD} \approx \mathbf{0}. \tag{7.16}$$

In the above equation, square matrices

$$\mathbf{A} = \left\{ a_{lk} = \int_0^1 \mathbf{L}^0\phi_{j,k}(x)\phi_{j,l}(x)dx \right\}, \mathbf{B} = \left\{ b_{lk} = \int_0^1 \mathbf{L}^1\phi_{j,k}(x)\phi_{j,l}(x)dx \right\},$$

and vectors

$$\mathbf{U} = \left\{u_k = u(k/2^j)\right\}, \mathbf{D} = \left\{d_k = \mathbf{N}[u(k/2^j)]\right\}$$

where subscript $k, l = 0, 1, 2, \ldots, 2^j$, and the generalized connection coefficients, a_{lk} and b_{lk}, can be accurately obtained by the method described in Appendix 5.1. According to Sect. 5.1, then directly discretize the boundary conditions in Eq. (7.11) and replace the corresponding equations in the equation group of Eq. (7.16). Finally, using the Newton iteration method or other methods with high accuracy and stability to solve the nonlinear algebraic equations of Eq. (7.16), the nodal values $u(k/2^j)$, $k = 0, 1, 2, \ldots, 2^j$ can be obtained accordingly. Substituting them into Eq. (7.12), then we get the approximate solution $P_j u(x)$ of Eq. (7.11).

In the above, we mainly introduced the wavelet-based solution method of the semi-linear nonlinear differential equations of form (7.11). From it, one can see that the solution format is highly uniform for general NBVPs. For different nonlinear forms, only the vector **D** and matrix **B**, in the nonlinear algebraic equations of Eq. (7.16) are different, which only affects the specific process of the subsequent Newton iteration method. However, in the whole process of discretization, whether it is an integer power nonlinear problem or a non-integer power nonlinear problem that is generally difficult to deal with by existing methods [41], the current wavelet-based method maintains a high degree of standardization. This makes the current wavelet method easier to use than existing methods, and even easier to develop standardized solvers.

We further consider two other classes of nonlinear operators, including quasi-linear and fully nonlinear differential equations that nonlinear terms contain the highest order derivative of the unknown function [42],

$$\begin{cases} \mathbf{L}^0 u(x) + \mathbf{L}^1\mathbf{N}[\mathbf{L}^2 u(x)] = 0, \quad 0 \le x \le 1 \\ \left.\dfrac{d^{i_s} u}{dx^{i_s}}\right|_{x=0} = 0, \quad \left.\dfrac{d^{j_s} u}{dx^{j_s}}\right|_{x=1} = 0 \end{cases}, \tag{7.17}$$

where $\mathbf{L}^0$, $\mathbf{L}^1$, and $\mathbf{L}^2$ denote differential operators; and **N** is a nonlinear operator. The boundary conditions are similar to the definite conditions described in Eq. (7.11) that satisfies. By introducing the function transform, $v(x) = \mathbf{L}^2[u(x)]$, Eq. (7.17) can be reduced into

$$\begin{cases} \mathbf{L}^0 u(x) + \mathbf{L}^1\mathbf{N}[v(x)] = 0, \quad 0 \le x \le 1 \\ v(x) = \mathbf{L}^2 u(x) \end{cases}. \tag{7.18}$$

The nonlinear operator in Eq. (7.18) no longer involves the derivative of the function and returns to the form of Eq. (7.11), which can be solved using the afore-mentioned wavelet-based method. The solutions for quasi-linear and fully nonlinear problems are also unified into the normalized solution for semi-linear problems.

Further analysis shows that the residual error, Eq. (7.15), of the approximation equation, Eq. (7.14), is completely independent of the undetermined coefficients $u(k/2^j), k = 0, 1, 2, \ldots, 2^j$ that determine the approximate solution. By contrast, it is completely dependent on the resolution level j. From Chap. 4, the following error estimation can be obtained:

$$Q(x) = -\mathbf{L}^0 \widetilde{Q}_j u(x) - \mathbf{L}^1 \widetilde{Q}_j \mathbf{N}[u(x)] = O(2^{-j(r-N_{\max})}), \tag{7.19}$$

where $N_{\max}$ is the order of the highest derivative in the differential operators $\mathbf{L}^0$ and $\mathbf{L}^1$. Given that the weight function satisfies the relationship $|\phi_{j,k}(x)| < C$, where C is a constant, the error induced in the nonlinear algebraic equation, Eq. (7.16), is also $O(2^{-j(r-N_{\max})})$. From Eq. (7.19), the residual error equation $Q(x)$ is completely independent of the approximate solution for almost any form of nonlinear term $\mathbf{N}[u(x)]$. As long as the resolution level j is sufficiently large, small $Q(x)$ is ensured as a high-order small value relative to all the reservations in the approximation solution in Eq. (7.14). No matter what nonlinear problem is weak or strong, the approximation equation, Eq. (7.14), used in the current wavelet closed method is a reliable and excellent approximation of the original equation, Eq. (7.11), ensuring that the solution obtained using the approximation equation based on Eq. (7.14) is a valid approximate solution for the original equation, Eq. (7.11).

Similar to the solution scheme for Eq. (7.11), the approximation equation for Eq. (7.17) is

$$\begin{cases} \mathbf{L}^0 P_j u(x) + \mathbf{L}^1 P_j \mathbf{N}[v(x)] = Q_1(x) = -\mathbf{L}^0 \widetilde{Q}_j u(x) + \mathbf{L}^1 \widetilde{Q}_j \mathbf{N}[v(x)] \\ P_j v(x) - \mathbf{L}^2 P_j u(x) = Q_2(x) = \mathbf{L}^2 \widetilde{Q}_j u(x) - \widetilde{Q}_j v(x). \end{cases} \tag{7.20}$$

In accordance with Sect. 4.2, the residual errors $Q_1(x) = O(2^{-j(r-N_{\max})})$ and $Q_2(x) = O(2^{-j(r-N_2)})$ of Eq. (7.20) can also be obtained, where N_2 is the highest derivative in the differential operator $\mathbf{L}_2$. Thus, similar to the preceding analysis process, the approximate equation, Eq. (7.20), can be guaranteed to provide an effective approximation of Eq. (7.18), i.e., Eq. (7.17), regardless of whether for a strong or weak nonlinear problem, as long as the resolution level j is sufficiently large. Furthermore, the validity of the approximate solution based on the approximation equation is guaranteed.

As pointed out in Chap. 3, the solution scheme of most conventional methods [41] are not closed. Following them, the unknown function $u(x)$ to be solved can be expanded by an approximation and its truncation, i.e.,

$$u(x) = \tilde{u}(x) + g(x). \tag{7.21}$$

Substitution the preceding equation into Eq. (7.11) yields

$$\mathbf{L}^0 \tilde{u}(x) + \mathbf{L}^1 \mathbf{N}[\tilde{u}(x)] = -\mathbf{L}^0 g(x) - \mathbf{L}^1 \{\mathbf{N}[\tilde{u}(x) + g(x)] - \mathbf{N}[\tilde{u}(x)]\} = Q_3(x). \tag{7.22}$$

On the basis of this kind of expanding method, the residual error $Q_3(x)$ of the approximate equation, Eq. (7.22), is usually dependent on the approximate solution $\tilde{u}(x)$, or vice versa. In such situation, the conventional solution methods are suitable to the case only when Eq. (7.11) is a weak nonlinear problem. In other words, only when the change in the nonlinear part caused by the variation of the approximate solution is considerably smaller than that of the linear part, the residual error $Q_3(x)$ of the calculation be ensured by a high-order small quantity relative to the main term reserved in the solution at the left side of Eq. (7.22). For a strong nonlinear problem, however, such high-order small quantity for the residual error $Q_3(x)$ cannot be ensured when the magnitude of the approximation solution increases large over a critical value. Thus, the solution obtained by solving this approximation equation is no longer a valid approximate solution for the original equation, Eq. (7.11), when it becomes in strong nonlinearity. This fundamental reason also explains why the existing methods are generally effective in solving weak nonlinear problems, but frequently ineffective in solving strong nonlinear problems. Comparing the conventional solution methods, the wavelet closed method presented in this book can ensure that the residual error of the approximation equation is a high-order small amount relative to the retained approximations of a nonlinear problem with either weak or strong nonlinearity.

In addition, Eq. (7.22) shows that since the base function series used in the conventional approximation scheme appears inside the nonlinear operator, as pointed out in Chap. 3, such methods cannot directly solve the problems with non-integer nonlinear terms. However, the wavelet-based method can do them without addition of steps of losing accuracy due to the expansion of the nonlinear term liking a linear case as shown above.

7.3 Numerical Solution Examples of Application to 1D Nonlinear Problems

To test the validity of the presented wavelet closed method and to quantitatively analyze its numerical accuracy, such as the accuracy of the solution and convergence speed, two cases of nonlinear problems are specifically studied in this section.

7.3.1 Solution Results of 1D Bratu Equation with Exponential Nonlinearity

Consider the 1D Bratu equation that is widely used to describe reaction diffusion problems, the cosmic expansion process, and many other physical problems, which is frequently used to test numerical algorithms [43–46]. The 1D Bratu equation is of the form

$$\begin{cases} \frac{d^2u(x)}{dx^2} + \lambda e^{u(x)} = 0 \\ u(0) = u(1) = 0 \end{cases}, \tag{7.23}$$

where parameter $\lambda > 0$. For such nonlinear problem with non-integer power nonlinearity, it is found that its exact solution can be gained by

$$u(x) = -2\ln\left\{\frac{\cosh[(x-1/2)\theta/2)]}{\cosh(\theta/4)}\right\}, \tag{7.24}$$

where parameter θ is determined by using the equation $\theta = \sqrt{2\lambda}\cosh(\theta/4)$. When parameter λ is less than, equal to, or greater than the critical value $\lambda_c \approx 3.513830719$, Eq. (7.23) has either two solutions, or one solution, or no solution [47–52], respectively. This means that there is a bifurcation in the solution of Eq. (7.23).

Evidently, the traditional Galerkin method cannot be directly used for such exponential nonlinear term in Eq. (7.24), but the wavelet Galerkin method introduced in this book can be directly processed as shown below.

Under the generalized Coiflets, the unknown function or approximation solution, $u(x)$, can be expressed as

$$u(x) \approx \sum_{k=1}^{2^j-1} u(k/2^j)\phi_{j,k}(x). \tag{7.25}$$

By applying it to the nonlinear operator in Eq. (7.23), $\mathbf{N}[u(x)] = \lambda e^{u(x)}$, similar to Eqs. (7.13), (7.14), and (7.16), the following expansions are obtained:

$$e^{u(x)} \approx \sum_{k=0}^{2^j} e^{u(k/2^j)}\phi_{j,k}(x), \tag{7.26}$$

$$\sum_{k=1}^{2^j-1} u(k/2^j)\frac{d^2\phi_{j,k}(x)}{dx^2} + \lambda\sum_{k=1}^{2^j-1} e^{u(k/2^j)}\phi_{j,k}(x) + \lambda[\phi_{j,0}(x) + \phi_{j,2^j}(x)] \approx 0. \tag{7.27}$$

Further, the system of nonlinear algebraic equations can be gained by using the Garlekin's method, i.e.,

$$\mathbf{AU} + \lambda\mathbf{BD} + \lambda\mathbf{C} \approx \mathbf{0}, \tag{7.28}$$

where matrices

$$\mathbf{A} = \left\{a_{lk} = \int_0^1 d^2\phi_{j,k}(x)/dx^2\phi_{j,l}(x)dx\right\} \text{ and } \mathbf{B} = \left\{b_{lk} = \int_0^1 \phi_{j,k}(x)\phi_{j,l}(x)dx\right\};$$

vectors

$$\mathbf{U} = \left\{u_k = u(k/2^j)\right\}, \mathbf{D} = \left\{d_k = e^{u(k/2^j)}\right\}, \text{ and, } \mathbf{C} = \{c_l = \lambda(b_{l0} + b_{l2^j})\};$$

and subscript k, $l = 1, 2, \ldots, 2^j - 1$. Here, the generalized connection coefficients a_{lk} and b_{lk} can be accurately obtained using the method described in Appendix 5.1.

By solving the nonlinear algebraic equation, Eq. (7.28), the unknown node values $u(k/2^j), k = 1, 2, \ldots, 2^j - 1$ can be obtained. Then, the solution $u(x)$ for Eq. (7.23) can be reconstructed from Eq. (7.25). Here, the classical Newton iteration method and the step-by-step iteration are used to solve Eq. (7.28) as λ increases. When λ is very small, the initial values in the Newton iterations are taken to zero. Once a solution is gained for a given λ, its solution is used as an initial input of the Newton iterations for an increased λ large than previous step λ. When in the solution occurs a bifurcation, we need to search for the directions of bifurcated solutions at the bifurcation point. Once the directions are known, the solution along with one bifurcated solution can be tracked out.

Figures 7.1, 7.2, and 7.3 show the wavelet approximation solutions for Eq. (7.23) and their error distribution when parameter $\lambda = 1$, 2, and 3, respectively, where the resolution level is taken by $j = 4$, 5, and 6, respectively. From these figures, one sees that the wavelet closed method exhibits extremely high precision. For example, when $j = 6$, the absolute value of the relative error is basically maintained within the order of 10^{-6} in the entire solution region, and the boundary jump phenomenon is also effectively suppressed.

In addition, the error order decreases rapidly as the resolution level j increases. When $\lambda = 1$, for example, the relative error of the wavelet-based solution decreases from 10^{-4} to 10^{-6} when $j = 4$ to 10^{-7}–10^{-8} when $j = 6$. In order to characterize this scaling low of the errors varying with the resolution level or the node number $n = 2^j - 1$, Fig. 7.4 displays such scaling relationships between the absolute errors appeared in the approximations and the resolution level j or the number of nodes

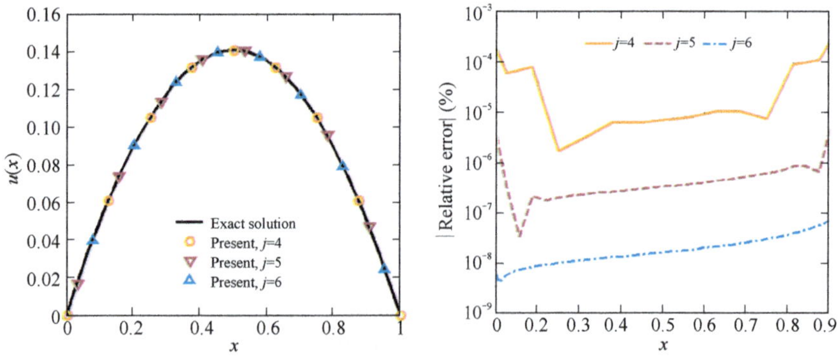

Fig. 7.1 Comparison between the wavelet-based approximate solution obtained by different resolution levels $j = 4, 5, 6$ and the exact solution when parameter $\lambda = 1$: approximate solution $u(x)$ (Left); relative errors of the approximations relative to the exact solution (Right)

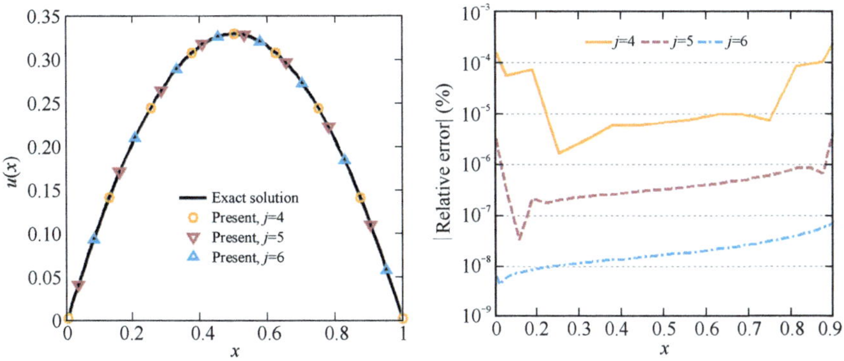

Fig. 7.2 Comparison between the wavelet-based approximate solution obtained by different resolution levels $j = 4, 5, 6$ and the exact solution when parameter $\lambda = 2$: approximate solution $u(x)$ (Left); relative error of the approximations relative to the exact solution (Right)

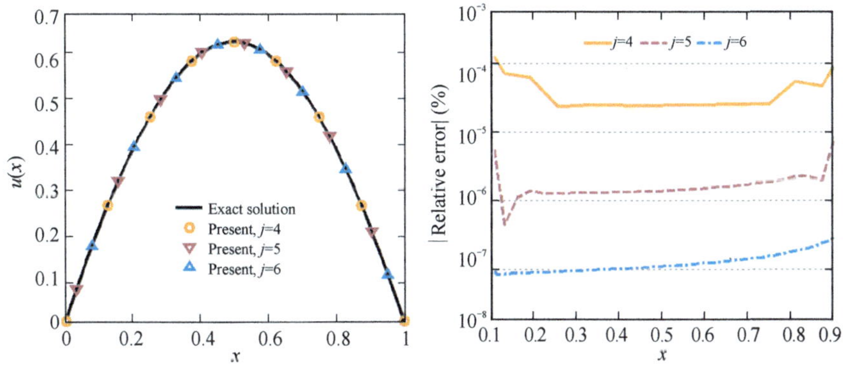

Fig. 7.3 Comparison between the wavelet-based approximate solution obtained by different resolution levels $j = 4, 5, 6$ and the exact solution when parameter $\lambda = 3$: approximate solution $u(x)$ (Left); relative errors of the approximations relative to the exact solution (Right)

$n = 2^j - 1$. As shown in Fig. 7.4, the absolute errors of the approximation solutions all decreased in the scaling form of $n^{-4}C_0(2^{-4j}C_1)$ as the number of computation nodes n (or resolution level j) increases, where $C_0(C_1)$ is a positive constant. That is, the wavelet method introduced in this study achieves a fourth-order convergence rate when such equation is solved by the wavelet-based method. In addition, Fig. 7.4 tells us that the absolute error of the approximate solution has a fewer increasing but their decreasing scales have the same with increasing the parameter λ in the nonlinearity of the system. This property can be further verified in Fig. 7.5 where the dependence of the relative error of $u(1/2)$ on parameter λ is presented. According to Fig. 7.5, we know that when parameter λ increases from 0 to 3, the relative error of the approximate solution is basically maintained at the same magnitude for a given resolution level. Such relative error is significantly increased only when λ is close to

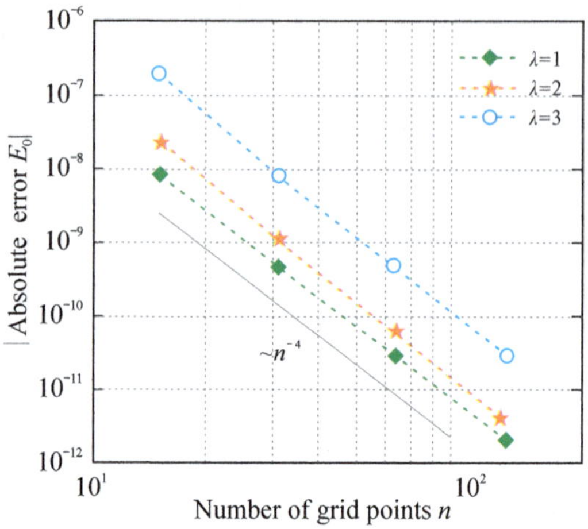

Fig. 7.4 Scaling dependence of the absolute error E_0 of $u(1/2)$ on the number of nodes $n = 2^j - 1$ used in the wavelet-based calculations when parameters $\lambda = 1, 2,$ and 3, respectively. Reference scaling line: $E_0 \sim n^{-4}$; j stands for the resolution level used in the wavelet

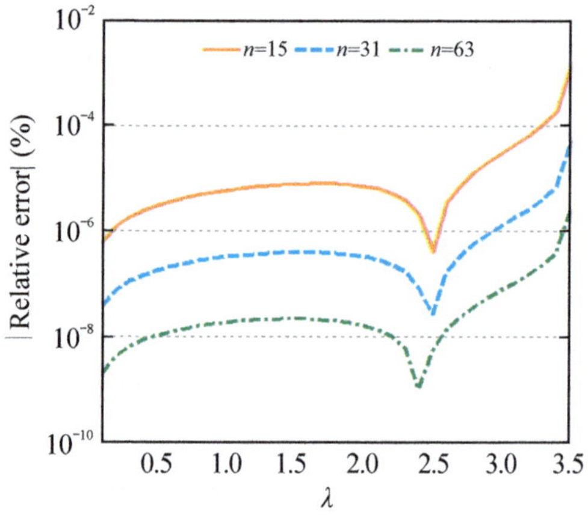

Fig. 7.5 Variation curves of the relative error of wavelet-based approximation $u(1/2)$ changed with parameter λ in the nonlinear term of 1D Bratu equation with different nodes

and over the critical value $\lambda_c \approx 3.513830719$ in the solution of the problem due to the decrease in the solvability of equations [47–50]; however, the magnitude of such increase is almost maintained within one order. This result shows that the accuracy of the wavelet closed method introduced in this study is insensitive to the nonlinearity of the system, and high-precision approximate solutions can be obtained for both weak and strong nonlinear systems.

By contrast, the absolute errors of those typical conventional methods without closure property and the wavelet-based method for solving this nonlinear problem are shown in Tables 7.1, 7.2, and 7.3, for example, the spline interpolation method [47], B-spline interpolation method [48], Lie-group shooting method (LGSM) [49], differential transformation method (DTM) [50], Laplace transform method [51], and polynomial decomposition method [52]. According to these tables, one sees that the accuracy of the wavelet-based method is significantly higher than those of

Table 7.1 Absolute value of the absolute error when $\lambda = 1$

x	Wavelet $j = 6$	Spline [47]	B-Spline [48]	LGSM [49]	DTM [50]	Laplace [51]	Decomposition [52]
0.1	4.22E−12	5.77E−10	2.98E−06	7.51E−07	–	1.98E−06	2.69E−03
0.2	8.85E−12	2.47E−10	5.46E−06	1.02E−06	1.00E−10	3.94E−06	2.02E−03
0.3	1.50E−11	4.56E−11	7.33E−06	7.05E−07	2.00E−10	5.85E−06	1.52E−04
0.4	2.09E−11	7.64E−11	8.50E−06	5.24E−07	2.00E−10	7.70E−06	2.20E−03
0.5	2.67E−11	1.46E−10	8.89E−06	5.07E−09	2.00E−10	7.47E−06	3.02E−03
0.6	3.21E−11	7.64E−11	8.50E−06	5.14E−07	2.00E−10	1.11E−05	2.20E−03
0.7	3.71E−11	4.56E−11	7.33E−06	8.95E−07	3.00E−10	1.26E−05	1.52E−04
0.8	4.17E−11	2.47E−10	5.46E−06	1.01E-06	2.00E−10	1.35E−05	2.02E−03
0.9	4.71E−11	5.77E−10	2.98E−06	7.42E-07	2.00E−10	1.20E−05	2.69E−03

Table 7.2 Absolute value of the absolute error when $\lambda = 2$

x	Wavelet $j = 6$	Spline [47]	B-Spline [48]	LGSM [49]	DTM [50]	Laplace [51]	Decomposition [52]
0.1	7.56E−12	7.71E−09	1.72E−05	4.03E−06	1.30E−06	2.13E−03	1.52E−02
0.2	1.84E−11	1.41E−08	3.26E−05	5.70E−06	2.56E−06	4.21E−03	1.47E−02
0.3	3.15E−11	1.98E−08	4.49E−05	5.22E−06	3.77E−06	7.19E−03	5.89E−03
0.4	4.37E−11	2.42E−08	5.28E−05	3.07E−06	4.87E−06	8.00E−03	3.25E−03
0.5	5.48E−11	2.60E−08	5.56E−05	1.46E−08	5.85E−06	7.60E−03	7.99E−03
0.6	7.43E−11	2.42E−08	5.28E−05	3.05E−06	7.66E−06	1.09E−02	3.25E−03
0.7	7.20E−11	1.98E−08	4.49E−05	5.19E−06	7.29E−06	1.19E−02	5.89E−03
0.8	7.77E−11	1.41E−08	3.26E−05	5.68E−06	7.71E−06	1.24E−02	1.47E−02
0.9	8.38E−11	7.71E−09	1.72E−05	4.01E−06	7.41E−06	1.09E−02	1.52E−02

Table 7.3 Absolute value of the absolute error when $\lambda = 3.51$

x	Wavelet $j = 6$	Spline [47]	B-Spline [48]	LGSM [49]
0.1	7.34E−09	7.61E−06	3.84E−02	4.45E−05
0.2	1.81E−08	5.83E−06	7.48E−02	7.12E−05
0.3	2.56E−08	7.19E−06	1.06E−01	7.30E−05
0.4	3.09E−08	7.89E−06	1.27E−01	4.47E−05
0.5	3.30E−08	7.31E−06	1.35E−01	7.76E−07
0.6	3.16E−08	7.89E−06	1.27E−01	4.56E−05
0.7	2.70E−08	7.19E−06	1.06E−01	7.20E−05
0.8	2.00E−08	5.83E−06	7.48E−02	7.05E−05
0.9	1.17E−08	7.61E−06	3.84E−02	4.41E−05

the six typical methods, where the best one in the conventional six methods is the spline interpolation method [47], and the worst one is the polynomial decomposition method [52]. When the nonlinear problem is weak, e.g., λ from 1 to 2, the error of the spline interpolation method [47] increases at least one order, but the accuracy of the wavelet-based method remains unchanged. When the nonlinear problem approaches to strong, e.g., $\lambda = 3.51$ which is near the bifurcation point, the wavelet-based method has still accuracy of order 10^{-8}, which is at least 2 orders higher than the accuracy of the polynomial decomposition method [52].

It should be pointed out that no result of DTM [50], Laplace transform method [51], and polynomial decomposition method [52] is contained in Table 7.3 due to the accuracy of their methods decrease sharply for the case $\lambda = 3.51$ near the bifurcation point.

Now, let us introduce the tracking two solutions of this problem on the basis of the wavelet-based method. As pointed out previously, there are two solutions in this problem when $\lambda < \lambda_c \approx 3.513830719$, and there is one or no solution when $\lambda > \lambda_c$. If there is no solution in this nonlinear problem when $\lambda > \lambda_c$, such bifurcation point belongs to a type of limit point. Here, the solution is tracked to the case of limit point. As λ increases from zero to λ_c, one piece of approximate solution can be tracked out as displayed above. Then, other piece of approximate solution can be obtained by decreasing λ from λ_c to zero. When the resolution level $j = 4$, such tracked two solutions are displayed in Fig. 7.6 for $u\,(1/2)$ varying with λ. From Fig. 7.6, one sees that when λ increases from zero to λ_c, the value of $u\,(1/2)$ in the wavelet-based solution increases from zero to about 1.5. After that, the value of $u\,(1/2)$ monotonously increases up to about 7.3 when λ decreases from λ_c to zero. For simplicity, we denote the solution in increasing λ by the lower part and

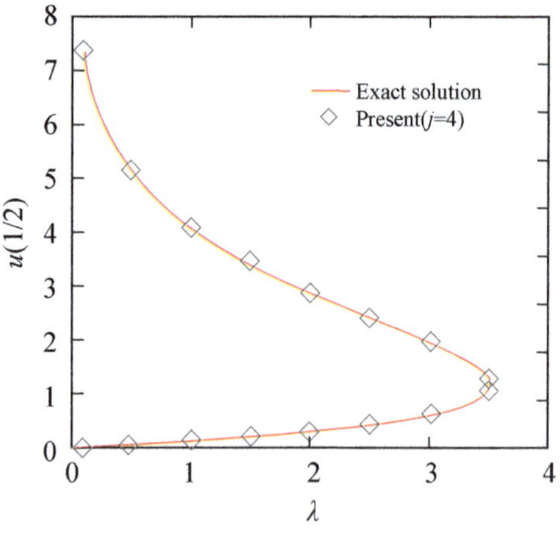

Fig. 7.6 Numerical results of $u\,(1/2)$ in the wavelet-based solution, when the resolution level is taken by 4, of the 1D nonlinear Bratu equation varying with parameter λ in the nonlinear term. The lower part of the solution is gained by increasing λ from zero to λ_c, i.e., the maximum value of λ in the figure, and the upper part is corresponding to the decreasing of λ from λ_c to zero

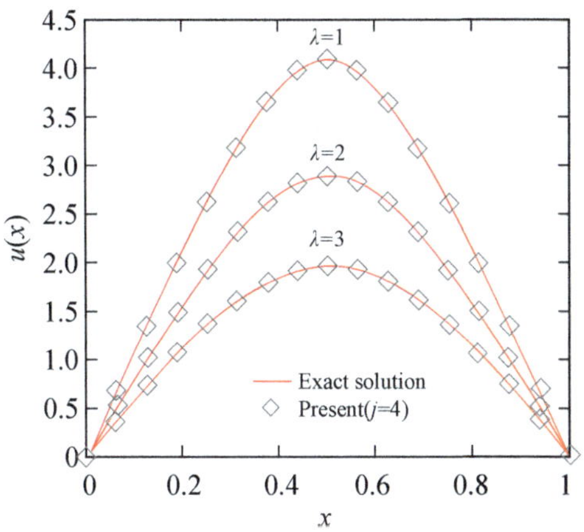

Fig. 7.7 Space configuration comparison of the wavelet-based solution and the exact one to the 1D nonlinear Bratu equation when the parameter λ decreases from 3 to 1, i.e., the upper part of the solution in the Bratu equation

the solution in decreasing λ by upper part. Figure 7.7 displays the upper solution configuration in space when $\lambda = 3, 2, 1$, respectively. Both in Figs. 7.6 and 7.7, their corresponding exact solutions are also plotted out. From them, one sees that the wavelet-based solutions all are well agreement with their exact values quantitatively only when $j = 4$.

The quantitative results of the wavelet-based solution to the 1D nonlinear Bratu equation clearly reveals that such solution method has high accuracy and low calculations, and the same program can be effectively used to solve both weak and strong nonlinear problems in one nonlinear differential equation.

As Neilan et al. pointed out in [42], to solve high-order or fully nonlinear problems is considerably more difficult than dealing with low-order or semi-nonlinear problems. In other words, many solution methods that are effective in dealing with low-order or semi-nonlinear problems fail to solve high-order or completely nonlinear problems [42]. At follows, such more complex nonlinear problems are selected to further test the available application scope of the wavelet closed method.

7.3.2 Solution Results of 1D Boundary-Value Problem with Sine Nonlinearity

The following nonlinear differential equation is considered by

$$\begin{cases} \lambda_2 \frac{d^2u(x)}{dx^2} + \gamma \sin\left[\frac{d^2u(x)}{dx^2}\right] + u(x) = q(x) \\ = -\lambda_2 \pi^2 \sin(\pi x) - \gamma \sin[\pi^2 \sin(\pi x)] + \sin(\pi x) \\ u(0) = u(1) = 0 \end{cases}, \tag{7.29}$$

where λ_2 and γ are constants. The readers can evaluate that such nonlinear problem has the exact solution of $u(x) = \sin(\pi x)$.

Before we apply the standard wavelet-based solution scheme described in Sect. 7.2, let us introduce a function transform of the form

$$v(x) = d^2u(x)/dx^2. \tag{7.30}$$

Then, the differential equation in Eq. (7.29) can be rewritten as

$$\lambda_2 \frac{d^2u(x)}{dx^2} + \gamma \sin[v(x)] + u(x) = q(x). \tag{7.31}$$

Applying the expansion schemes of the wavelet-based method in Sect. 7.1 to each term in Eq. (7.31), one gets

$$u(x) \approx \sum_{k=1}^{2^j-1} u\left(\frac{k}{2^j}\right)\phi_{j,k}(x), \tag{7.32}$$

$$v(x) \approx \sum_{k=0}^{2^j} v\left(\frac{k}{2^j}\right)\phi_{j,k}(x), \tag{7.33}$$

$$\sin[v(x)] \approx \sum_{k=0}^{2^j} \sin[v\left(\frac{k}{2^j}\right)]\phi_{j,k}(x), \tag{7.34}$$

$$q(x) \approx \sum_{k=0}^{2^j} q\left(\frac{k}{2^j}\right)\phi_{j,k}(x). \tag{7.35}$$

Substituting Eqs. (7.32)–(7.35) into Eqs. (7.30) and (7.31), we have

$$\sum_{k=0}^{2^j} v(k/2^j)\phi_{j,k}(x) \approx \sum_{k=1}^{2^j-1} u(k/2^j)\frac{d^2\phi_{j,k}(x)}{dx^2}, \tag{7.36}$$

$$\sum_{k=1}^{2^j-1} u(k/2^j)\left[\lambda_2 \frac{d^2\phi_{j,k}(x)}{dx^2} + \phi_{j,k}(x)\right]$$
$$+ \gamma \sum_{k=0}^{2^j} \sin[v(k/2^j)]\phi_{j,k}(x) \approx \sum_{k=0}^{2^j} q(k/2^j)\phi_{j,k}(x). \tag{7.37}$$

Multiplying both sides of Eq. (7.36) by weight functions $\phi_{j,l}(x)$, $l = 0, 1, 2, \ldots, 2^j$ and integrating the induced equation respect to variable x from 0 to 1, we get

$$\mathbf{V} = \mathbf{A}^{-1}\mathbf{B}\mathbf{U}, \tag{7.38}$$

where matrices

$$\mathbf{A} = \left\{ a_{lk} = \int_0^1 \phi_{j,k}(x)\phi_{j,l}(x)dx \right\} \text{ and } \mathbf{B} = \left\{ b_{lm} = \int_0^1 d^2\phi_{j,m}(x)/dx^2\phi_{j,l}(x)dx \right\}.$$

As well as vectors

$$\mathbf{V} = \{v_k = v(k/2^j)\} \text{ and } \mathbf{U} = \{u_m = u(m/2^j)\}$$

in which subscripts k, $l = 0, 1, \ldots, 2^j$ and $m = 1, 2, \ldots, 2^j - 1$.

Moreover, taking the same steps as above, i.e., both sides of Eq. (7.37) are multiplied by the weight functions $\phi_{j,l}(x)$, $l = 1, 2, \ldots, 2^j - 1$, and are integrated from 0 to 1, then we gain the following algebraic equations in matrix form

$$\mathbf{CU} + \gamma\mathbf{DE} = \mathbf{DQ}, \tag{7.39}$$

where

$$\mathbf{C} = \left\{ c_{lk} = \Gamma_{l,k}^{0,0} + \lambda_2\Gamma_{l,k}^{0,2} \right\} \text{ and } \mathbf{D} = \left\{ d_{lm} = \int_0^1 \phi_{j,m}(x)\phi_{j,l}(x)dx \right\};$$

vectors

$$\mathbf{E} = \left\{ e_m = \sin[\{\mathbf{A}^{-1}\mathbf{BU}\}_m] \right\} \text{ and } \mathbf{Q} = \{q_m = q(m/2^j)\};$$

and subscripts $k, l = 1, 2, \ldots, 2^j - 1$ and $m = 1, 2, \ldots, 2^j - 1$. The generalized connection coefficients $\Gamma_{l,k}^{0,n} = \int_0^1 d^n\phi_{j,k}(x)/dx^n\phi_{j,l}(x)dx$ can be accurately obtained by using the method described in Appendix 5.1.

By solving the nonlinear algebraic equation, Eq. (7.39), the node values $u(k/2^j)$, $k = 1, 2, \ldots, 2^j - 1$, can be obtained by the same program of solving the nonlinear algebraic equations as introduced previously. Then, the solution $u(x)$ for Eq. (7.29) can be reconstructed through Eq. (7.32). For simplicity in this test, the parameters in Eq. (7.29) are taken by $\lambda_2 = 1$, and different $\gamma \neq 0$ which behaves for the level of nonlinear term in the second-order fully nonlinear differential equation. That is, the nonlinearity becomes stronger with increasing γ.

Figures 7.8, 7.9, and 7.10 show the relative error distribution in space of the wavelet-based solution with different resolution level when parameter γ is chosen by 10, 100, and 1000, respectively, where the scaling law of the absolute errors varying with the resolution level are also displayed. From these figures, one sees that the absolute error order is not dependent on the parameter γ when a resolution

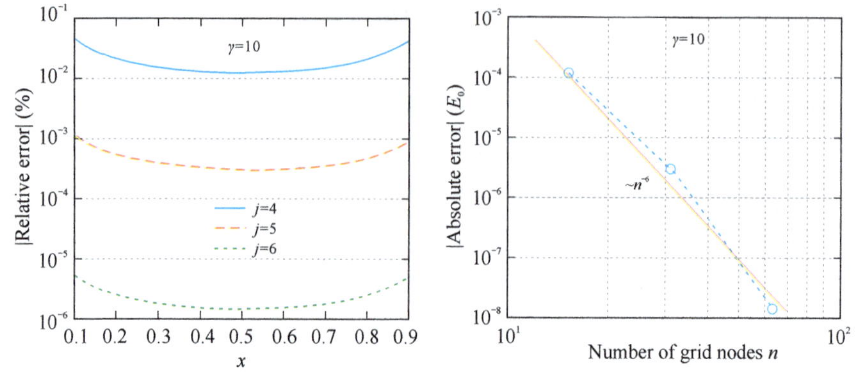

Fig. 7.8 Error curves of the wavelet-based solution of 1D boundary-value problem with nonlinearity term $\gamma\ \sin\{d^2u(x)/dx^2\}$ with different resolution level j when $\gamma = 10$: space distribution of the relative error along x (Left); scaling dependence of the absolute error E_0 of $u(1/2)$ on and the node number $n = 2^j–1$ (Right)

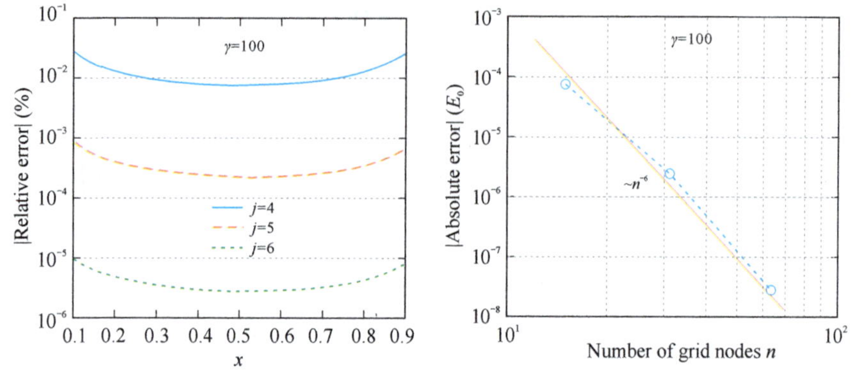

Fig. 7.9 Error curves of the wavelet-based solution of 1D boundary-value problem with nonlinearity term $\gamma\ \sin(d^2u(x)/dx^2)$ with different resolution level j when $\gamma = 100$: space distribution of the relative error along x (Left); scaling dependence of the absolute error E_0 of $u(1/2)$ on and the node number $n = 2^j–1$ (Right)

level is chosen. When the resolution level increases from 4 to 6, however, the relative errors all decrease from the order 10^{-1}–10^{-2} to 10^{-5}–10^{-6}, while the absolute errors change from order ~10^{-4} to ~10^{-8}. According to 7.8, 7.9, and 7.10, one can also see that the scaling dependence of the absolute errors on the node number n is ~n^{-6}.

To explicitly reveal the accuracy of the wavelet-based solution independent on the parameter γ, Fig. 7.11 displays the curves of the relative errors of the solution $u(1/2)$ varying with γ for different resolution level.

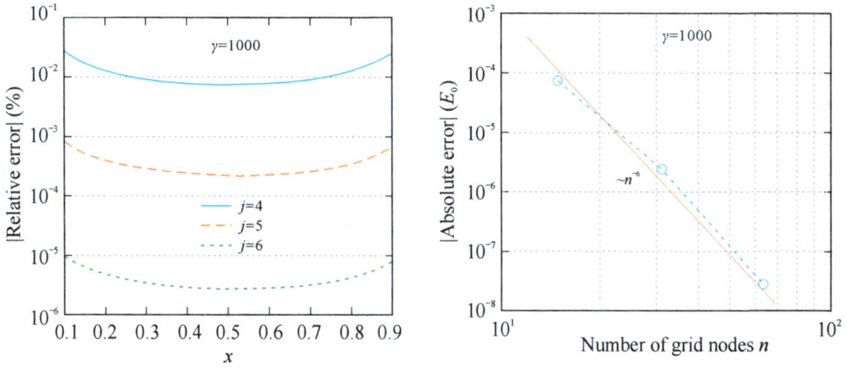

Fig. 7.10 Error curves of the wavelet-based solution of 1D boundary-value problem with nonlinearity term $\gamma \sin\{d^2u(x)/dx^2\}$ with different resolution level j when $\gamma = 1000$: distribution of the relative error along x (Left); scaling dependence of the absolute error E_0 of $u(1/2)$ and the number of nodes $n = 2^j - 1$ (Right)

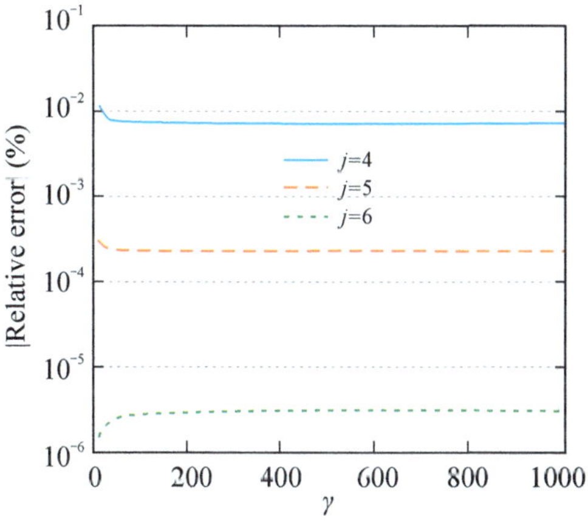

Fig. 7.11 Relative error curves of the wavelet-based solution $u(1/2)$ varying with parameter γ in nonlinearity term $\gamma \sin\{d^2u(x)/dx^2\}$ of the 1D boundary-value problem

7.3.3 Numerical Solution of 2D Bratu Equation

In this subsection, we apply the wavelet-based solution method to solve the 2D nonlinear Bratu equation of the form [53, 54]

$$\begin{cases} \frac{\partial^2 u(x,y)}{\partial x^2} + \frac{\partial^2 u(x,y)}{\partial y^2} + \lambda e^{u(x,y)} = f(x, y) \\ u(x, y)|_{x,y=0,1} = 0, \quad 0 < x, y < 1. \end{cases} \tag{7.40}$$

Following the generalized procedure introduced in Sect. 7.2, $u(x, y)$ can be expanded as

$$u(x,y) \approx \sum_{k=1}^{2^j-1}\sum_{l=1}^{2^j-1} u(k/2^j, l/2^j)\phi_{j,k}(x)\phi_{j,l}(y). \tag{7.41}$$

Similar to Eq. (7.34), the nonlinear term in Eq. (7.40) can be directly expanded by

$$e^{u(x,y)} \approx \sum_{k=0}^{2^j}\sum_{l=0}^{2^j} e^{u(k/2^j, l/2^j)}\phi_{j,k}(x)\phi_{j,l}(y). \tag{7.42}$$

In addition, we further have

$$f(x,y) \approx \sum_{k=0}^{2^j}\sum_{l=0}^{2^j} f(k/2^j, l/2^j)\phi_{j,k}(x)\phi_{j,l}(y). \tag{7.43}$$

Substitution of Eqs. (7.41)–(7.43) into Eq. (7.40) yields

$$\begin{aligned}&\sum_{k=1}^{2^j-1}\sum_{l=1}^{2^j-1} u(k/2^j, l/2^j)\left[\frac{d^2\phi_{j,k}(x)}{dx^2}\phi_{j,l}(y) + \phi_{j,k}(x)\frac{d^2\phi_{j,l}(y)}{dy^2}\right]\\ &+\sum_{k=0}^{2^j}\sum_{l=0}^{2^j}\left[\lambda e^{u(k/2^j, l/2^j)} - f\left(\frac{k}{2^j}, \frac{l}{2^j}\right)\right]\phi_{j,k}(x)\phi_{j,l}(y) \approx 0.\end{aligned} \tag{7.44}$$

By using the Garlekin method to Eq. (7.44) with the weight functions, $\phi_{j,n}(x)\phi_{j,m}(y)$, $n, m = 0, 1, 2, \ldots, 2^j$, and integrating the induced equations with respect to (x, y) in the region $[0, 1]^2$, we get a set of nonlinear algebraic equations of the matrix form

$$\mathbf{AU} + \lambda\mathbf{BD} + \lambda\mathbf{C} \approx \mathbf{Q}, \tag{7.45}$$

where matrices

$$\mathbf{A} = \left\{a_{op} = \overline{a}_{nk}\tilde{a}_{ml} + \tilde{a}_{nk}\overline{a}_{ml}\right\} \text{ and } \mathbf{B} = \left\{b_{op} = \tilde{a}_{nk}\tilde{a}_{ml}\right\}, \tag{7.46}$$

vectors

$$\mathbf{U} = \left\{u_p = u(k/2^j, l/2^j)\right\}, \mathbf{D} = \left\{d_p = e^{u(k/2^j, l/2^j)}\right\}, \tag{7.47}$$

$$\mathbf{C} = \left\{c_o = (\tilde{a}_{n0} + \tilde{a}_{n2^j})\sum_{l=0}^{2^j}\tilde{a}_{ml} + (\tilde{a}_{m0} + \tilde{a}_{m2^j})\sum_{k=1}^{2^j-1}\tilde{a}_{nk}\right\}, \tag{7.48}$$

and

$$\mathbf{Q} = \left\{ q_o = \sum_{k=0}^{2^j} \sum_{l=0}^{2^j} f(k/2^j, l/2^j) \tilde{a}_{nk} \tilde{a}_{ml} \right\}. \tag{7.49}$$

Among these matrices and vectors, the generalized connection coefficients

$$\overline{a}_{lk} = \int_0^1 d^2\phi_{j,k}(x)/dx^2 \phi_{j,l}(x)dx \text{ and } \tilde{a}_{lk} = \int_0^1 \phi_{j,k}(x)\phi_{j,l}(x)dx;$$

and subscripts $o = (2^j - 1)(n - 1) + m$; $p = (2^j - 1)(k - 1) + l$; and k, l, n, $m = 1,2,\ldots,2^j - 1$.

The unknown node value $u(k/2^j, l/2^j)$, $k, l = 1, 2, \ldots, 2^j - 1$ can be obtained by solving the above nonlinear algebraic equation, Eq. (7.45), using a mature method such as Newton iteration method. After that, the approximate solution of Eq. (7.40) can be gained by the reconstruction of Eq. (7.41).

To test the effectiveness of the aforementioned wavelet closed method, firstly, the accuracy and convergence rate of this method are quantitatively investigated when the external application in Eq. (7.40) has the form

$$f(x, y) = \lambda e^{\sin(\pi x)\sin(\pi y)} - 2\pi^2 \sin(\pi x)\sin(\pi y). \tag{7.50}$$

In such case, the 2D Bratu equation has the following exact solution

$$u(x, y) = \sin(\pi x)\sin(\pi y).$$

When the resolution level $j = 4$, 5, and 6, respectively, is taken in the wavelet-based solution method to solve the 2D nonlinear Bratu equation for the application expressed in Eq. (7.50), the distribution curves of relative errors of the approximate solutions on the exact solution with different λ are shown in Fig. 7.12. From this figure, one sees that the relative errors are only dependent on the resolution level rather than on the parameter. When the resolution level j increases from 4 to 6, the relative errors are within the order from 10^{-3}–10^{-4} to 10^{-6}–10^{-7}. Considering the relationship between the node number and the resolution level, i.e., $n = (2^j - 1) \times (2^j - 1)$, we have that the node number is corresponding to 15×15, 31×31, 63×63 in Fig. 7.12, respectively. Hence, the accuracy and efficiency of the wavelet-based solution method are also confirmed by applying it in the 2D nonlinear Bratu equation for the case when it has exact solution in the example.

Further, let us see the scaling dependence of absolute error of the wavelet-based solution on the node number employed in this numerical example. Figure 7.13 clearly shows such scaling dependence where the absolute error of the current wavelet approximation decreases rapidly with the node number n in the form of $n^{-2.5}$. That

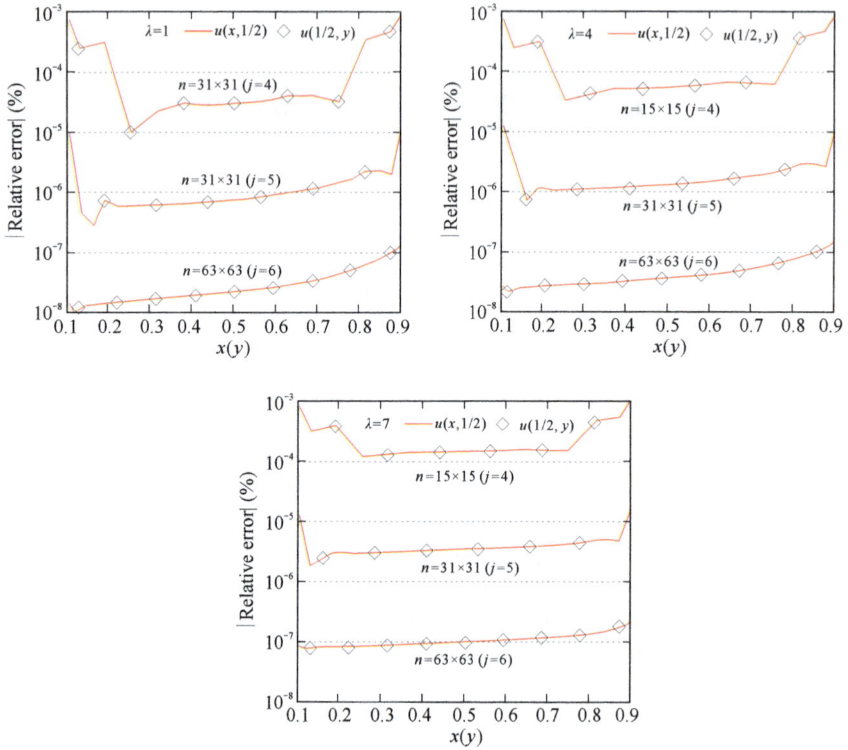

Fig. 7.12 Relative error curves of the wavelet-based solution to the 2D nonlinear Bratu equation with the application of Eq. (7.50) when the resolution levels $j = 4$, 5, and 6, respectively, for $\lambda = 1$, $\lambda = 4$ and $\lambda = 7$

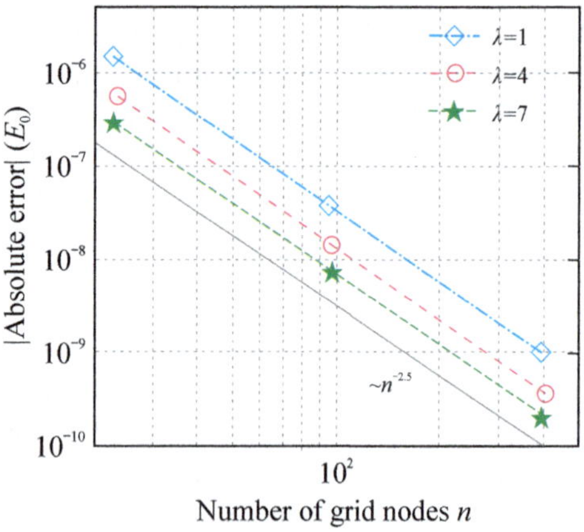

Fig. 7.13 Scaling dependence of absolute errors E_0 of $u(1/2, 1/2)$ of the wavelet-based solution to the 2D nonlinear Bratu equation with the application of Eq. (7.50) varying with the resolution level. Here, $\lambda = 1$, 4, and 7, and the node number $n = (2^j - 1) \times (2^j - 1)$

is, the wavelet closed method has a 2.5-order convergence speed for this example of the 2D Bratu equation.

Next, let us consider the case that the 2D nonlinear Bratu equation has no exact solution. For simplicity and without losing generality, here, we take $f(x, y) \equiv 0$ as an example, which is the case of classical 2D Bratu equation. Similar to the 1D Bratu equation, when parameter $\lambda > \lambda_c$, $\lambda = \lambda_c$, and $\lambda < \lambda_c$, the 2D Bratu equation also has zero, one, and two solutions, respectively [55–58]. In such situation, the critical value $\lambda = \lambda_c$ is a type of limit bifurcation point. To track such solution, here, we select the resolution level $j = 4$ in the wavelet-based solution method to numerically show the approximate solution results. By using the wavelet-based method to solve Eq. (7.40) when $f(x, y) \equiv 0$, the critical value λ_c is searched and it is located in the region between 6.80 and 6.81. This obtained critical value is highly consistent with the currently accepted value $\lambda_c = 6.808124423$ [55–58], which is significantly better than $\lambda_c = 7.12222$ obtained using the FDM and $\lambda_c = 6.93957823$ obtained using FEM [59]. In the numerical tracking to this critical value, moreover, when $\lambda = 6.80$, the wavelet closed method can find two solutions of Eq. (7.40). When $\lambda = 6.81$, however, the iterative process is divergent in solving the nonlinear algebraic equation, Eq. (7.45), regardless of the initial value of the iteration, which reveals that Eq. (7.40) has no solution when $\lambda = 6.81$. Therefore, the critical value of λ should be located in the region of (6.80, 6.81).

From Eq. (7.40), we know that the parameter λ characterizes the context of nonlinear term in the classical 2D Bratu equation. As the parameter increases, in other words, the effect of the nonlinear term on the solution increases too. When the parameter enters a relatively large value, the nonlinear effect becomes strong, or the governing equation becomes a strong nonlinear problem. In order to get the quantitative solution of such strong nonlinear problem with high accuracy and low calculation, here, the resolution level is taken by $j = 6$. From Fig. 7.12, we know that the accuracy of such wavelet-based solution has order of 10^{-7}. To explicitly display the solution results, Table 7.4 lists the central values of two solutions existing in the classical 2D Bratu equation with parameter λ from 0.1 to 6.8, obtained by the wavelet-based method at resolution level $j = 6$. In Table 7.4, $u_{\text{low}}(1/2, 1/2)$ and $u_{\text{up}}(1/2, 1/2)$ represent the lower piece solution and the upper piece solution, respectively. The lower piece solution is tracked by increasing the parameter from 0.1 to 6.8, while the upper piece solution is obtained by decreasing the parameter from 6.8 to 0.1. These

Table 7.4 Wavelet approximate solution $u(1/2, 1/2)$ for the classical 2D Bratu equation at different λ values

λ	$u_{\text{low}}(1/2, 1/2)$	$u_{\text{up}}(1/2, 1/2)$	λ	$u_{\text{low}}(1/2, 1/2)$	$u_{\text{up}}(1/2, 1/2)$
0.1	7.408093229E−03	1.138218777E+01	4.0	3.955259988E−01	3.454936050E+00
1.0	7.810101876E−02	7.548494384E+00	5.0	5.569597638E−01	2.845945007E+00
2.0	1.668957564E−01	5.072459550E+00	6.0	7.971089810E−01	2.239910842E+00
3.0	2.703663546E−01	4.156756192E+00	6.8	1.323450119E + 00	1.462270493E+00

results are close to the solution obtained by using FEM [59]. Thus, the validity of the wavelet-method solution is confirmed, and all properties in the classical 2D Bratu equation are ensured by the wavelet-based solution method.

References

1. Cole JD (1968) Perturbation method in applied mathematics. Blaisdell Publishing Company, Waltham
2. Nayfeh AH (2000) Perturbation method. Wiley, New York
3. Von Dyke M (1975) Perturbation methods in fluid mechanics. The Parabolic Press, Stanford
4. Ali AH, Al-Saif ASJ (2008) Adomian decomposition method for solving some models of nonlinear partial differential equations. Basrah J Sci A 26:1-11
5. Liao S (2012) Homotopy analysis method in nonlinear differential equations. Springer, Berlin
6. Von Kármán T (1910) Festigkeits problem in maschinenbau. Encyklopädie Der Mathematischen Wissenschaften. 4:311–385
7. Way S (1934) Bending of circular plate with large deflection. J Appl Mech 56:627–637
8. Vincent JJ (1931) The bending of a thin circular plate. Phil Mag 12:185–197
9. Chien W (1947) Large deflection of a circular clamped plate under uniform pressure. Chinese J Phys 7:102–113
10. Chien W, Yeh K (1954) On the large deflection of circular plates. China Sci 3:405–437
11. Qian WC (1987) A selection of scientific papers by Qian Weichang. Fujian Education Press, Fuzhou
12. Qian WC, Lin HS, Hu HC et al (1954) Large deflection of elastic circular thin plate. Science Press, Beijing
13. Causon DM, Mingham CG (2010) Introductory finite difference method for PDEs. Ventus Publishing, London
14. Delfour M, Fortin M, Payr G (1981) Finite-difference solutions of a non-linear Schrödinger equation. J Comput Phys 44:277–288
15. Chen YM (1975) Numerical computation of dynamic stress intensity factors by a Lagrangian finite-difference method. Eng Fract Mech 7:653–660
16. Brian PLT (1961) A finite-difference method of high-order accuracy for the solution of three-dimensional transient heat conduction problems. AIChE J 7:367–370
17. Beam RM, Warming RF (1978) An implicit factored scheme for the compressible Navier-Stokes equations. Am Inst Aeron Astron J 16:393–402
18. Narasimhan TN, Witherspoon PA (1976) An integrated finite difference method for analyzing fluid flow in porous media. Water Resour Res 12:57–64
19. Yuste SB, Quintana-Murillo J (2012) A finite difference method with non-uniform timesteps for fractional diffusion equations. Comput Phys Commun 183:2594–2600
20. Buckmire R (2004) Application of a Mickens finite-difference scheme to the cylindrical Bratu-Gelfand problem. Numer Methods Part Diff Equ: Int J 20:327–337
21. Qiu JB (1991) Theory and application of weighted residual value method. Aerospace Press, Beijing
22. Wang XZ (2007) Computational mechanics. Lanzhou University Press, Lanzhou
23. Bramble JH, Schatz AH (1971) Least squares methods for 2mth order elliptic boundary-value problems. Math Comput 25:1–32
24. Xiu D, Hesthaven JS (2006) High-order collocation methods for differential equations with random inputs. Soc Indus Appl Math J Sci Comput 27:1118–1137
25. Cockburn B, Shu CW (2006) The local discontinuous Galerkin method for time-dependent convection-diffusion systems. Soc Indus Appl Math J Numer Anal 35:2440–2463
26. Wang P, Yao Y, Tulin MP (1995) An efficient numerical tank for non-linear water waves, based on the multi-subdomain approach with BEM. Int J Numer Meth Fluids 20:1315–1337

27. Feng X, Neilan M (2009) Vanishing moment method and moment solutions for fully nonlinear second order partial differential equations. J Sci Comput 38:74–98
28. Atluri SN, Zhu T (1998) A new Meshless Local Petrov-Galerkin approach in computational mechanics. Comput Mech 22:117–127
29. Zhu T, Zhang J, Atluri SN (1998) A meshless local boundary integral equation method for solving nonlinear problems. Comput Mech 22:174–187
30. Zhang X, Liu Y (2004) Meshless method. Tsinghua University Press, Beijing
31. Sauter SA, Schwab C (2011) Boundary element methods. Springer, Berlin
32. Sutradhar A, Paulino GH, Gray LJ (2008) Symmetric Galerkin boundary element method. Springer, Berlin
33. Caglar H, Caglar N, Özer M (2008) Fifth-degree B-spline solution for nonlinear fourth-order problems with separated boundary conditions. J Phys: Conf Ser 96:012031
34. Wang XC (2004) Finite element method. Tsinghua University Press, Beijing
35. Desai YM, Eldho TI, Shah AH (2011) Finite element method with applications in engineering. Dorling Kindersley Pvt. Ltd., Noida
36. Wang XC, Shao M (1997) Basic principles and numerical methods of finite element method. Tsinghua University Press, Beijing
37. Chen Y, Lee J, Eskandarian A (2007) Meshless methods in solid mechanics. Springer Science & Business Media, Inc., New York
38. Gu ZQ, Kou GM (1997) Active vibration control. National Defense Industry Press, Beijing
39. Ja YH (1983) Using the shooting method to solve boundary-value problems involving nonlinear coupled-wave equations. Opt Quant Electron 15:529–538
40. Bert CW, Malik M (1996) Differential quadrature method in computational mechanics: a review. Appl Mech Rev 49:1–28
41. Tadmor E (2012) A review of numerical methods for nonlinear partial differential equations. Bull Am Math Soc 49:507–554
42. Neilan M, Glowinski R, Feng X (2013) Recent developments in numerical methods for fully nonlinear second order partial differential equations. SIAM Rev 55:205–267
43. Guo Y (2008) Nonlinear partial differential equations. Tsinghua University Press, Beijin
44. Bebernes J, Eberly D (1987) Mathematical problems from combustion theory. Springer, New York
45. Abbott JP (1978) An efficient algorithm for the determination of certain bifurcation points. J Comput Appl Math 4:19–27
46. Frank-Kamenetskii DA (1967) Diffusion and heat transfer in chemical kinetics, 2nd edn. Plenum Press, New York
47. Jalilian R (2010) Non-polynomial spline method for solving Bratu's problem. Comput Phys Commun 181:1868–1872
48. Caglar H, Caglar N, Özer M et al (2010) B-spline method for solving Bratu's problem. Int J Comput Math 87: 885–1891
49. Abbasbandy S, Hashemi MS, Liu CS (2011) The Lie-group shooting method for solving the Bratu equation. Commun Nonlinear Sci Numer Simul 16:4238–4247
50. Hassan IHAH, Ertürk VS (2007) Applying differential transformation method to the one-dimensional planar Bratu problem. Int J Contemp Math Sci 2:1493–1504
51. Khuri SA (2004) A new approach to Bratu's problem. Appl Math Comput 147:131–137
52. Deeba E, Khuri SA, Xie S (2000) An algorithm for solving boundary value problems. J Comput Phys 159:125–138
53. Liu XJ, Zhou YH, Wang XM, Wang JZ (2013) A wavelet method for solving a class of nonlinear boundary value problems. Commun Nonlinear Sci Numer Simul 18:1939–1948
54. Liu XJ, Wang JZ, Zhou YH (2013) Wavelet solution of a class of two-dimensional nonlinear boundary value problems. CMES-Comput Model Eng Sci 92:493–505
55. Chang SL, Chien CS (2003) A multigrid-lanczos algorithm for the numerical solutions of nonlinear eigenvalue problems. Int J Bifurcation Chaos 13:1217–1228
56. Mohsen A, Sedeek LF, Mohamed SA (2008) New smoother to enhance multigrid-based methods for Bratu problem. Appl Math Comput 204:325–337

57. Doedel EJ, Sharifi H (2000) Collocation methods for continuation problems in nonlinear elliptic PDEs. Notes Numer Fluid Mech 74:105–118
58. Fedoseyev AI, Friedman MJ, Kansa EJ (2000) Continuation for nonlinear elliptic partial [differential equations discretized by the multiquadratic method. Int J Bifurcation Chaos 10:481–492
59. Odejide SA, Aregbesola YAS (2006) A note on two dimensional Bratu problem. Kragujevac J Math 29:49–57

Chapter 8
Space–Time Fully Decoupled Wavelet-Based Solution to Nonlinear Problems

The nonlinear initial-boundary-value problems (NIBVPs), i.e., the time-dependent NPDEs, provide a quantitative description for many nonlinear phenomena which play crucial roles in various scientific and engineering applications, including fluid mechanics, electromagnetic waves, nonlinear optics, population dynamics, solid-state physics, chemical kinetics, and plasma physics [1–9]. Thus, finding a solution with high accuracy to NIBVPs is critically important to understand and resolve various problems in science and engineering. However, until now, it is still exceedingly difficult to obtain a satisfactory solution to many nonlinear time-dependent PDEs, either analytically or numerically [1–9].

In the past decades, some analytical and semi-analytical methods have been developed to solve the nonlinear time-dependent PDEs, such as the Tanh method [10, 11], the Adomian's decomposition method [12, 13], the homotopy perturbation method [14], the variational iteration method [15], and the differential transform methods [16]. However, the scope of applications of these analytical techniques seems to be confined to a very few special nonlinear time-dependent PDF problems [10–17]. Therefore, a considerable number of studies have been conducted to numerically solve various nonlinear time-dependent PDEs [18–42]. These numerical methods can be roughly classified into two groups.

One is the space–time coupled formulation [43, 44] in which the space–time domain is discretized simultaneously by treating the temporal dimension as an additional spatial dimension [18–21]. For example, Shao and Wu [18] studied the nonlinear Klein–Gordon and sine-Gordon equations by applying the Chebyshev tau meshless method to simultaneously discretize space and time variables. Pekmen and Tezer-Sezgin [19] proposed a space–time coupled discretizing procedure based on the differential quadrature method to solve the nonlinear Klein–Gordon and sine-Gordon equations. Bulbul and Sezer [20] discretized the nonlinear Klein–Gordon equation in space and time by using the Taylor polynomial collocation method. These space–time coupled methods normally have no time step stability restriction and can maintain good accuracy in both space and time even for a relatively large

Y.-H. Zhou, *Wavelet Numerical Method and Its Applications in Nonlinear Problems*, Engineering Applications of Computational Methods 6,
https://doi.org/10.1007/978-981-33-6643-5_8

time step [18–21, 43, 44]. However, when a space–time coupled formulation is used to study the long time evolutions of nonlinear problems, a large number of grid points in time are required, making the size of algebraic equations prohibitively large and the corresponding computational cost considerably expensive, even to the point of being unacceptable for available computing resources [18–21, 43, 44]. Although several technologies have been developed to alleviate these drawbacks, such as time-stepping procedures and matrix-free techniques, the space–time coupled approach remains prohibitively expensive, particularly for 3D unsteady problems [18, 43].

The other one is that the most popular methods for solving nonlinear space–time coupled equations may be the formulations [43, 44] in which space and time is performed independently [23–42]. For example, Kuo and Luis [23] applied the Galerkin method to the spatial discretization of the nonlinear Klein–Gordon equation and obtained a system of nonlinear ordinary differential equations that can be solved using the FDM. A similar space–time performed method was proposed by Argyris and Haase to solve the nonlinear Schrodinger and sine-Gordon equations [24]. Dehghan and Shokri [25] used a θ-weighted scheme for the temporal discretization of the nonlinear sine-Gordon equation, resulting in a set of boundary-value problems solved by using the radial basis function collocation method. Other similar space–time formulations that combined various numerical techniques, such as the collocation method, FEM, FPM, the compact difference scheme, FDM, and the Runge–Kutta method, have also been developed to solve nonlinear time-dependent PDE problems [26–42].

These independent space–time performed formulations are commonly associated with relatively low computational effort compared with space–time coupled formulations [14–35]. However, when most of these space–time decoupled formulations are employed for solving nonlinear time-dependent PDEs, especially for solving fully nonlinear and quasi-linear wave equations, some matrices generated in the spatial discretization of nonlinear terms are dependent on the time-dependent unknown vector [14–16, 18–21, 23, 24, 26–29, 31–33, 36, 37]. Thus, these matrixes must be recalculated at each time step, thereby consuming considerable computing resources. For example, Mohebbi and Dehghan [33] applied the element-free reproducing kernel particle Ritz method to transform the Klein–Gordon equation

$$\partial^2 u/\partial t^2 - a\partial^2 u/\partial x^2 + bu + cu^k + f = 0$$

into a system of nonlinear ordinary differential equations

$$\mathbf{M}d^2\mathbf{U}/dt^2 + [-a\mathbf{K}_0 + b\mathbf{M} + c\mathbf{K}_1(\mathbf{U})]\mathbf{U} + \mathbf{F} = 0 \tag{8.1}$$

with

$$\mathbf{M} = \int_{\Omega} \mathbf{\Psi}^T \mathbf{\Psi} d\Omega, \quad \mathbf{K}_0 = \int_{\Omega} d\mathbf{\Psi}^T / dx d\mathbf{\Psi} / dx d\Omega,$$

$$\mathbf{K}_1(\mathbf{U}) = \int_{\Omega} \mathbf{\Psi}^T (\Psi \mathbf{U})^{k-1} \mathbf{\Psi} d\Omega, \quad \mathbf{F} = \int_{\Omega} \mathbf{\Psi}^T f(x,t) d\Omega \tag{8.2}$$

in which $\mathbf{U}(t)$ is the time-dependent unknown column vector, and transversal vector function, $\mathbf{\Psi}(x)$, represents the shape functions in spatial domain Ω [33]. Equation (8.1) shows that the matrix, $\mathbf{K}_1(\mathbf{U})$, from the spatial discretization of nonlinear term, $u^k(x,t)$, relies explicitly on the unknown vector, $\mathbf{U}(t)$, and should be updated at each time step in the subsequent Newmark time integration [33]. The above problems are also encountered in many existing collocation techniques [16–21, 26, 29] when used for spatial discretization of fully nonlinear or quasi-linear time-dependent PDEs [36, 37]. Such repeated recalculations of matrices from the spatial discretization of nonlinear terms can be considered as re-performing the spatial discretization at each time step [14, 15, 31–37]. Therefore, the decoupling between spatial and temporal discretization in these existing space–time decoupled formulations is incomplete because they cannot divide the procedure for solving nonlinear wave problems into two completely separate steps [34–37]. In such situation, additional computational cost is needed to update the matrices at each time step, which are generated in the spatial discretization of nonlinear terms in time-dependent PDEs [34–37]. Except for such considerable computations which appeared in the mentioned methods by renewing the matrix corresponding to the nonlinear term(s) in the governing equation(s), in fact, the time–space coupled nonlinear PDEs cannot be fully decoupled by the above solution methods and due to that no method contains a closed solution program with respect to either time variable or space variable or both, which usually cannot ensure the accuracy of the numerical solutions, especially for strong nonlinear problems.

In this chapter, a space–time fully decoupled wavelet-based method is introduced for solving the nonlinear time-dependent PDEs. By applying a modified wavelet Galerkin method for the spatial discretization, the nonlinear time-dependent PDEs are reduced into a system of nonlinear ordinary differential equations, in which the approximation part is independent on those of truncation in space, and all matrices are constant over time and don't need to be updated in the subsequent time integration. After the essential program is introduced in the next section, some typical nonlinear time-dependent PDEs are quantitatively solved to demonstrate the validity and efficiency of the wavelet-based solution method.

8.1 Galerkin-Wavelet Solution Program

8.1.1 Spatial Discretization by Wavelet

As discussed in Chap. 7, by introducing some intermediate variables at most, the nonlinear initial-boundary-value problem always can be written into the general form

$$\begin{cases} \partial u_i(x,t)/\partial t = \sum_{l=1}^{M_1} \mathbf{L}_{i,l}^0 u_l(x,t) + \sum_{l=1}^{M_2} \mathbf{L}_{i,l}^1 \mathbf{N}_{i,l}[u_1(x,t),\ldots,u_M(x,t),x,t] + f_i(x,t),\ 0 < x < 1 \\ \partial^{j_{i,s}^0} u_i(x,t)/\partial^{j_{i,s}^0} x|_{x=0} = \partial^{j_{i,s}^1} u_i(x,t)/\partial^{j_{i,s}^1} x|_{x=1} = 0,\ u_i(x,0) = g_i(x),\ i = 1,2,\ldots,M, \end{cases} \tag{8.3}$$

where $\mathbf{L}_{i,l}^0$ and $\mathbf{L}_{i,l}^1$ represent the partial differential operators for the spatial variable x, $\mathbf{N}_{i,l}$ are nonlinear operators, and a set of appropriate values, $j_{i,s}^0$, $j_{i,s}^1$, and s makes Eq. (8.3) being a well-posed problem.

Following the discussion presented in Chap. 5, the unknown function can be expressed as

$$u_i(x,t) \approx \sum_{k=0}^{2^j} u_i(k/2^j,t)\phi_{j,k}(x), \quad i = 1,2,\ldots,M \tag{8.4}$$

and the nonlinear term in Eq. (8.3), $\mathbf{N}_{i,l}[u_1(x,t),\ldots,u_M(x,t),x,t]$, can be approximated by

$$\begin{aligned} &\mathbf{N}_{i,l}[u_1(x,t),\ldots,u_M(x,t),x,t] \\ &\quad \approx \sum_{k=0}^{2^j} \mathbf{N}_{i,l}\left[u_1\left(\frac{k}{2^j},t\right),\ldots,u_M\left(\frac{k}{2^j},t\right),\frac{k}{2^j},t\right]\phi_{j,k}(x). \end{aligned} \tag{8.5}$$

Similar to above, the function, $f_i(x,t)$, can also be expressed as

$$f_i(x,t) \approx \sum_{k=0}^{2^j} f_i(k/2^j,t)\phi_{j,k}(x). \tag{8.6}$$

Substitution of Eqs. (8.4), (8.5), and (8.6) into Eq. (8.3) yields

$$\sum_{k=0}^{2^j} \frac{du_i(k/2^j,t)}{dt}\phi_{j,k}(x)$$

$$\approx \sum_{l=1}^{M_1}\sum_{k=0}^{2^j} u_l(k/2^j,t)\mathbf{L}_{i,l}^0\phi_{j,k}(x) + \sum_{k=0}^{2^j} f_i(k/2^j,t)\phi_{j,k}(x)$$

$$+\sum_{l=1}^{M_2}\sum_{k=0}^{2^j}\mathbf{N}_{i,l}\left[u_1\left(\frac{k}{2^j},t\right),\ldots,u_M\left(\frac{k}{2^j},t\right),\frac{k}{2^j},t\right]\mathbf{L}_{i,l}^1\phi_{j,k}(x),\quad i=1,2,\ldots,M. \tag{8.7}$$

Multiplying both sides of the ith equation in Eq. (8.7) by the weight function $\phi_{j,l}(x), l = 0, 1, 2, \ldots, 2^j$, respectively, and performing integration from 0 to 1, i.e., the Galerkin method is employed, we get

$$\mathbf{A}_i d\mathbf{U}_i(t)/dt = \sum_{n=1}^{M_1}\mathbf{B}_{i,n}\mathbf{U}_n(t) + \sum_{l=n}^{M_2}\mathbf{C}_{i,n}\mathbf{D}_{i,n}(t) + \mathbf{E}_i\mathbf{F}_i(t) \tag{8.8}$$

with

$$\mathbf{A}_i = \left\{a_{lk} = \int_0^1 \phi_{j,k}(x)\phi_{j,l}(x)dx\right\},\quad \mathbf{B}_{i,n} = \left\{b_{lk}^{i,n} = \int_0^1 \mathbf{L}_{i,n}^0\phi_{j,k}(x)\phi_{j,l}(x)dx\right\}, \tag{8.9}$$

$$\mathbf{C}_{i,n} = \left\{c_{lk}^{i,n} = \int_0^1 \mathbf{L}_{i,n}^1\phi_{j,k}(x)\phi_{j,l}(x)dx\right\},\quad \mathbf{E}_i = \left\{e_{lk} = \int_0^1 \phi_{j,k}(x)\phi_{j,l}(x)dx\right\}, \tag{8.10}$$

$$\mathbf{U}_i(t) = \{u_k^i = u_i(k/2^j,t)\}^T,\quad \mathbf{F}_i(t) = \{f_k^i = f_i(k/2^j,t)\}^T, \tag{8.11}$$

$$\mathbf{D}_{i,n}(t) = \{d_k^{i,n} = \mathbf{N}_{i,n}[u_1(k/2^j,t),\ldots,u_M(k/2^j,t),k/2^j,t]\}^T \tag{8.12}$$

in which subscripts, $i = 1, 2, \ldots, M$ and $k, l = 0, 1, 2, \ldots, 2^j$, and the generalized connection coefficients, a_{lk}, $b_{lk}^{i,n}$, $c_{lk}^{i,n}$, and e_{lk}, are given in Appendix 5.1. The treatment of boundary conditions is as same as the approach in solving the boundary-value problem discussed in Chap. 7. It is evident that Eq. (8.12) is the matrix relative to the nonlinear term in Eq. (8.3). As shown in Chap. 3, it is known that such spatial calculations are closed in whole time domain.

8.1.2 Time Integration to the Induced Ordinary Differential Equations

To solve the nonlinear ordinary differential system, Eq. (8.8), many methods can be available, such as various finite difference methods, the linear multistep method, and

Runge–Kutta method. In this section, the fourth-order Runge–Kutta method with step size, Δt, is used to solve the nonlinear ordinary differential equations in Eq. (8.8). The semi-discrete system to Eq. (8.8) can be further expressed as

$$d\mathbf{U}_i(t)/dt = \mathbf{G}_i(\mathbf{U}_1, \ldots, \mathbf{U}_M, t) = \sum_{n=1}^{M_1} \mathbf{A}_i^{-1}\mathbf{B}_{i,n}\mathbf{U}_n(t) + \sum_{l=n}^{M_2} \mathbf{A}_i^{-1}\mathbf{C}_{i,n}\mathbf{D}_{i,n}(t) + \mathbf{A}_i^{-1}\mathbf{E}_i\mathbf{F}_i(t). \quad (8.13)$$

Following the Runge–Kutta formula [45], we have

$$d\mathbf{U}_i(t)/dt = \{\mathbf{U}_i[(n+1)\Delta t] - \mathbf{U}_i(n\Delta t)\}/\Delta t = (\mathbf{K}_{1,n}^i + 2\mathbf{K}_{2,n}^i + 2\mathbf{K}_{3,n}^i + \mathbf{K}_{4,n}^i)/6 \quad (8.14)$$

with

$$\begin{cases} \mathbf{K}_{1,n}^i = \mathbf{G}_i[\mathbf{U}_1(n\Delta t), \ldots, \mathbf{U}_M(n\Delta t), n\Delta t] \\ \mathbf{K}_{2,n}^i = \mathbf{G}_i[\mathbf{U}_1(n\Delta t) + \Delta t\mathbf{K}_{1,n}^1/2, \ldots, \mathbf{U}_M(n\Delta t) + \Delta t\mathbf{K}_{1,n}^M/2, n\Delta t + \Delta t/2] \\ \mathbf{K}_{3,n}^i = \mathbf{G}_i[\mathbf{U}_1(n\Delta t) + \Delta t\mathbf{K}_{2,n}^1/2, \ldots, \mathbf{U}_M(n\Delta t) + \Delta t\mathbf{K}_{2,n}^M/2, n\Delta t + \Delta t/2] \\ \mathbf{K}_{4,n}^i = \mathbf{G}_i[\mathbf{U}_1(n\Delta t) + \Delta t\mathbf{K}_{3,n}^1, \ldots, \mathbf{U}_M(n\Delta t) + \Delta t\mathbf{K}_{3,n}^M, n\Delta t + \Delta t]. \end{cases} \quad (8.15)$$

Applying Eq. (8.14) to the induced nonlinear ordinary differential equations in Eq. (8.13), a recursive integral program with the initial condition given in Eq. (8.3) can be obtained for calculating the nodal values of unknown functions, $u(k/2^j, n\Delta t)$, for $n > 0$ and $0 \le k \le 2^j$. After that, the approximate solution, $u(x, n\Delta t)$, can be reconstructed in Eq. (8.4) on the space and time domains considered.

8.1.3 Some Remarks

From the above solution program, one can find that all matrices, $\mathbf{A}_i$, $\mathbf{B}_{i,n}$, $\mathbf{C}_{i,n}$, and $\mathbf{E}_i$, generated from the spatial discretization are the $N \times N$ constant matrices which are completely independent of both the unknown vector function, $\mathbf{U}(t)$, and time, t, where $N = 2^j + 1$ is the number of nodes in space. Therefore, in subsequent time integration steps for solving ordinary differential Eq. (8.13), such matrices do not need updating.

The main difference between the wavelet-based solution method and the conventional methods mentioned above for solving the nonlinear time-dependent PDEs is that the wavelet-based solution method is closed in space for the expansion of nonlinear term(s) in Eq. (8.3), while the conventional methods for such expansion cannot be closed. In such situation, the induced ordinary differential equations in Eq. (8.8) or (8.13) is independent of the truncation errors in the space approximation. Meanwhile, the expansion coefficients of the nonlinear term in Eq. (8.12) are explicitly formulated by the node sampling of the original nonlinear function. Similar to those shown in the previous chapters, such sampling for the nonlinear function makes the corresponding calculations simple compared to the conventional methods, especially for those non-integer power nonlinear function, such as $\mathbf{N}_{i,n}(u) = \sin(u)$, can be directly conducted in the present wavelet-based solution method.

8.2 Numerical Solution to 1D Nonlinear Equations with Initial-Boundary-Value Conditions

In this section, some typical nonlinear initial-boundary-value problems are numerically solved by the above wavelet-based solution method to demonstrate the validity and efficiency of the wavelet method. To effectively estimate the accuracy of the numerical solutions, here, the following popular three error norms [18] are employed:

$$L_\infty(t) = \max_{1 \le i \le N} |u^e(x_k, t) - u^a(x_k, t)|, \tag{8.16a}$$

$$L_2(t) = \sqrt{\sum_{i=1}^{N} [u^e(x_k, t) - u^a(x_k, t)]^2} \tag{8.16b}$$

$$\mathrm{RMS}(t) = L_2/\sqrt{N} \tag{8.16c}$$

in which $u^e(x, t)$ and $u^a(x, t)$ are the exact and numerical solutions, respectively, and N is the total number of grid points in space.

8.2.1 *1D Nonlinear Klein–Gordon Equation with Initial-Boundary-Value Conditions*

For Eq. (8.3), we know that the specific form of the nonlinear operator is different for different one-dimensional initial-boundary-value problems. Here, the well-known Klein–Gordon equation usually has the following nonlinear function:

$$\mathbf{N}_i[u(x, t), x, t] = \alpha u + \beta u^2 + \gamma u^3 \tag{8.17}$$

in which α, β, γ are different constant coefficients except for different linear terms. Applying Eqs. (8.5)–(8.17), one gets

$$\alpha u+\beta u^2+\gamma u^3\approx\sum_{k=0}^{2^j}\alpha u\left(\frac{k}{2^j},t\right)\phi_{j,k}(x)+\beta u^2\left(\frac{k}{2^j},t\right)\phi_{j,k}(x)+\gamma u^3\left(\frac{k}{2^j},t\right)\phi_{j,k}(x). \tag{8.18}$$

Once the form of the Klein–Gordon equation is pre-given, there is no difficulty for one to take the above program for solving the space–time coupled equation.

Example 8.1 Consider the Klein–Gordon equation with quadratic nonlinearity [22, 25–28]. Its differential equations with the initial and boundary conditions are of the form.

$$\begin{cases}\dfrac{\partial^2 u}{\partial t^2}-\dfrac{\partial^2 u}{\partial x^2}+u^2+x\cos t-x^2\cos^2 t=0,\ x\in[-1,1]\\ u(x,0)=x,\ \partial u/\partial t|_{t=0}=0,\ u(-1,t)=-u(1,t)=-\cos t\end{cases} \tag{8.19}$$

which has the exact solution $u(x,t)=x\cos t$.

Table 8.1 shows the error norms of the numerical solutions obtained by the present wavelet-based method with spatial mesh size, $\Delta x=1/4$ and time step, $\Delta t=0.008$. Comparing them with those gained by other conventional methods, the errors of numerical solutions of the LBM [12], thin plate spline collocation method (TPSCM) [25], tension spline collocation method (TSCM) [26], cubic BSCM (CBSCM) [27], and cubic B-spline FEM (CBSFEM) [28] are also listed in this table.

From Table 8.1, one sees that when the spatial size used in the wavelet-based method is 1/4, larger than all used in other methods, and the time step 0.008 is larger than ones used in other methods except for CBSCM [27] with 0.0625, the accuracy of the numerical solutions obtained by the wavelet-based method has all in the order 10^{-10} except for one in 10^{-9}. However, TPSCM [25] has the accuracy of 10^{-4}–10^{-5} with $\Delta x=1/50$ and $\Delta t=0.0001$, TSCM [26] has accuracy of 10^{-9}–10^{-10} with the same sizes as TPSCM, LBM [12] has the accuracy of 10^{-4}–10^{-6} with $\Delta x=1/100$ and $\Delta t=0.001$, CBSCM [27] has the accuracy of 10^{-4}–10^{-5} with $\Delta x=1/8$ and $\Delta t=0.0625$, and CBSFEM [28] has the accuracy of 10^{-5} with $\Delta x=1/25$ and $\Delta t=0.001$. By means of these quantitative comparisons, one finds that the highest accuracy of the numerical solutions is gained by the wavelet-based method using the almost larger spatial size and time step. Hence, the present wavelet-based solution method also displays a powerful accuracy and low calculations in this example.

Table 8.1 Error norms of the numerical solutions at different times t for Example 8.1

t	1.0	3.0	5.0	7.0	10.0
Wavelet-based method, $\Delta x = 1/4$, $\Delta t = 0.008$					
L_∞	7.79E-10	4.50E-10	3.80E-10	3.48E-10	4.36E-10
L_2	1.05E-09	6.73E-10	6.04E-10	5.99E-10	7.21E-10
RMS	3.97E-10	2.55E-10	2.28E-10	2.27E-10	2.73E-10
TPSCM [25], $\Delta x = 1/50$, $\Delta t = 0.0001$					
L_∞	1.25E-05	1.56E-05	3.38E-05	3.78E-05	1.31E-05
L_2	6.54E-05	1.17E-04	2.20E-04	2.59E-04	7.99E-05
RMS	6.51E-06	1.17E-05	2.19E-05	2.58E-05	7.95E-06
TSCM [26], $\Delta x = 1/50$, $\Delta t = 0.0001$					
L_∞	1.03E-09	1.00E-09	2.56E-10	1.13E-09	9.46E-10
L_2	7.01E-09	6.59E-09	1.29E-09	7.47E-09	5.84E-09
RMS	6.97E-10	6.55E-10	1.28E-10	7.44E-10	5.81E-10
T	2.0	4.0	6.0	8.0	–
LBM [12], $\Delta x = 1/100$, $\Delta t = 0.001$					
L_∞	1.47E-05	1.40E-05	6.38E-06	1.77E-05	–
L_2	1.51E-04	1.39E-04	4.55E-05	1.34E-04	–
RMS	1.07E-05	9.81E-06	3.21E-06	9.44E-06	–
T	0.2	0.4	0.6	0.8	1.0
CBSCM [27], $\Delta x = 1/8$, $\Delta t = 0.0625$					
L_∞	4.20E-05	7.30E-05	6.60E-05	8.10E-05	7.70E-05
L_2	1.57E-04	2.35E-04	1.96E-04	3.05E-04	2.31E-04
RMS	3.60E-05	5.40E-05	4.50E-05	7.00E-05	5.30E-05
T	1.0	2.0	3.0	4.0	5.0
CBSFEM [28], $\Delta x = 1/25$, $\Delta t = 0.001$					
L_∞	4.60E-05	8.05E-05	1.28E-05	7.29E-05	5.13E-05

Example 8.2 Consider the Klein–Gordon equation [25–28] with the governing equations

$$\begin{cases} \dfrac{\partial^2 u}{\partial t^2} - \dfrac{\partial^2 u}{\partial x^2} + u^2 + 6t^3 x - 6tx^3 - t^6 x^6 = 0,\ x \in [0, 1] \\ u(x, 0) = \partial u/\partial t|_{t=0} = 0,\ u(0, t) = 0,\ u(1, t) = t^3. \end{cases} \tag{8.20}$$

To this example, the exact solution of Eq. (8.20) is $u(x, t) = x^3 t^3$.

Similar to Example 8.1, the error norms of the numerical solutions obtained by the wavelet-based method and other numerical methods [25–28] are listed in Table 8.2. After the $\Delta x = 1/8$ and $\Delta t = 0.005$ are used in the wavelet-based solution method, the accuracy of its approximations is within the order of 10^{-8}–10^{-9}. Comparing it

Table 8.2 Error norms of the numerical solutions at different times t for Example 8.2

t	1.0	2.0	3.0	4.0	5.0
Wavelet-based method, $\Delta x = 1/8$, $\Delta t = 0.005$					
L_∞	3.73E-09	1.18E-08	2.59E-08	4.55E-08	7.20E-08
L_2	3.79E-09	1.21E-08	2.62E-08	4.56E-08	7.21E-08
RMS	1.43E-09	4.59E-09	9.92E-09	1.72E-08	2.72E-08
TPSCM [25], $\Delta x = 1/50$, $\Delta t = 0.0001$					
L_∞	1.10E-05	1.65E-04	5.97E-04	1.83E-03	3.69E-03
L_2	5.50E-05	1.15E-03	3.26E-03	9.82E-03	1.91E-02
RMS	5.47E-06	1.15E-04	3.24E-04	9.77E-04	1.90E-03
TSCM [26], $\Delta x = 1/50$, $\Delta t = 0.0001$					
L_∞	1.13E-08	1.65E-08	1.72E-08	9.06E-09	6.70E-09
L_2	5.76E-08	8.40E-08	8.61E-08	4.42E-08	2.96E-08
RMS	8.06E-09	1.17E-08	1.20E-08	6.19E-09	4.15E-09
CBSFEM [28], $\Delta x = 1/250$, $\Delta t = 0.0001$					
L_∞	2.78E-04	2.98E-05	1.47E-04	1.17E-04	4.31E-05
T	0.2	0.4	0.6	0.8	1
CBSCM [27], $\Delta x = 1/16$, $\Delta t = 0.0625$					
L_∞	5.70E-05	5.60E-05	5.90E-05	4.50E-05	4.60E-05
L_2	1.70E-04	1.66E-04	1.48E-04	1.35E-04	1.44E-04
RMS	3.90E-05	3.80E-05	3.40E-05	3.10E-05	3.30E-05

with other methods, Table 8.2 tells us that only TSCM [26] gives the same accuracy order as the wavelet-based method when $\Delta x = 1/50$ and $\Delta t = 0.0001$ are chosen in TSCM. It is obvious that both spatial size and time step used in the wavelet-based method are larger than them employed in TSCM. Comparing to other conventional methods, the wavelet-based method has at least three to five orders of accuracy higher than them even when spatial size and time step in the wavelet-based method are larger than those used in them.

Example 8.3 Consider the Klein–Gordon equation with cubic nonlinearity [25, 29, 30] with the governing equation with Dirichlet boundary conditions

$$\begin{cases} \frac{\partial^2 u}{\partial t^2} + \alpha \frac{\partial^2 u}{\partial x^2} + \beta u + \gamma u^3 = 0, x \in [0, 1] \\ u(x, 0) = B \tan(Kx), \partial u / \partial t|_{t=0} = BcK \sec^2(Kx) \end{cases}. \tag{8.21}$$

For such case, the exact solution is of the form $u(x, t) = B \tan[K(x + ct)]$. Here, $B = \sqrt{\beta/\gamma}$, $K = \sqrt{-\beta/2(\alpha + c^2)}$, $\alpha = -5/2$, $\beta = 1$, and $\gamma = 3/2$.

Table 8.3 presents the comparison of error norms between the wavelet-based solution method and those obtained by using TPSCM [25], BSCM [29], and the Legendre wavelet spectral collocation method (LWSCM) [30]. From Table 8.3, the

Table 8.3 Error norms of the numerical solutions at different times t for Example 8.3

	$c = 0.05$				$c = 0.5$			
t	1.0	2.0	3.0	4.0	1.0	2.0	3.0	4.0
Wavelet-based method, $\Delta x = 1/64$, $\Delta t = 0.005$								
L_∞	8.87E-10	9.61E-10	1.05E-09	1.15E-09	1.37E-08	4.07E-08	2.26E-07	1.05E-05
L_2	1.42E-09	1.85E-09	2.05E-09	1.81E-09	1.44E-08	4.18E-08	2.31E-07	1.09E-05
RMS	1.79E-10	2.33E-10	2.58E-10	2.27E-10	1.81E-09	5.26E-09	2.92E-08	1.38E-06
TPSCM [16], $\Delta x = 1/100$, $\Delta t = 0.001$								
L_∞	3.65E-07	3.90E-07	4.21E-07	4.59E-07	6.00E-06	2.20E-05	9.09E-05	8.29E-04
L_2	1.79E-06	1.54E-06	1.73E-06	2.01E-06	4.08E-05	1.58E-04	6.48E-04	5.36E-03
RMS	1.78E-07	1.53E-07	1.72E-07	2.00E-07	4.06E-06	1.57E-05	6.45E-05	5.33E-04
BSCM [20], $\Delta x = 1/100$, $\Delta t = 0.001$								
L_∞	1.20E-08	2.47E-08	2.90E-08	1.99E-08	2.69E-08	8.75E-08	3.09E-07	1.94E-06
L_2	7.56E-08	1.80E-07	2.08E-07	1.41E-07	1.90E-07	6.38E-07	2.23E-06	1.34E-05
RMS	7.56E-09	1.80E-08	2.08E-08	1.41E-08	1.90E-08	6.38E-08	2.23E-07	1.34E-06
Present, $\Delta x = 1/16$, $\Delta t = 0.01$								
L_∞	1.04E-07	1.18E-07	1.31E-07	1.45E-07	4.60E-07	2.15E-06	1.96E-05	1.18E-03
L_2	1.39E-07	1.52E-07	1.50E-07	1.70E-07	5.07E-07	2.28E-06	2.05E-05	1.22E-03
RMS	3.58E-08	3.93E-08	3.87E-08	4.39E-08	1.31E-07	5.89E-07	5.28E-06	3.14E-04
LWSCM [21], $\Delta x = 1/24$, $\Delta t = 0.0001$								
L_∞	1.08E-07	1.99E-07	2.37E-07	1.92E-07	6.12E-06	2.22E-05	9.13E-05	7.79E-04
L_2	1.67E-07	3.57E-07	4.20E-07	3.24E-07	1.07E-05	4.05E-05	1.62E-04	1.35E-03
RMS	5.55E-08	1.19E-07	1.40E-07	1.08E-07	3.56E-06	1.35E-05	5.42E-05	4.48E-04

Fig. 8.1 Scaling dependency of error norm L_∞ of the wavelet solution at $t = 1$ with $\Delta t = 0.001$ on Δx

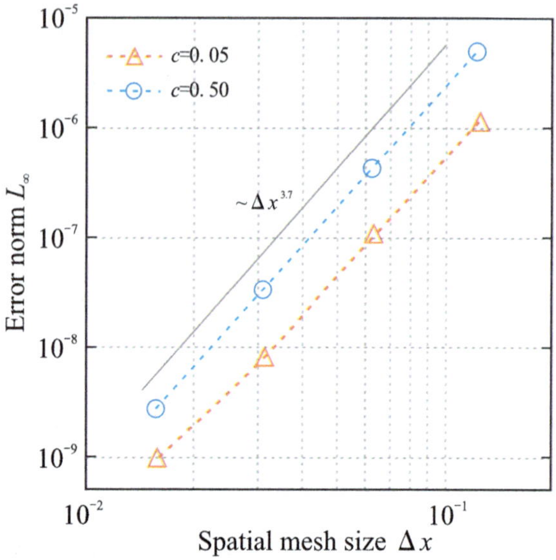

high accuracy and low calculation of the wavelet-based method are also revealed. Figure 8.1 shows the scaling dependency of the error norm, L_∞, of the wavelet-based solution at instant, $t = 1$, on spatial mesh size, Δx, when time step, $\Delta t = 0.001$ is specified.

Example 8.4 Single soliton solution in the Klein–Gordon equation [22, 29, 31, 32]. Such governing equations are of the form

$$\begin{cases} \frac{\partial^2 u}{\partial t^2} - a^2 \frac{\partial^2 u}{\partial x^2} + au - bu^3 = 0, x \in [-10, 10] \\ u(x, 0) = A\mathrm{sech}(\kappa x) \\ \frac{\partial u}{\partial t}|_{t=0} = A\kappa \, \mathrm{cosch}(\kappa x) \tanh(\kappa x) \end{cases}. \tag{8.22}$$

Under such Dirichlet boundary conditions, the exact solution is extracted by the single soliton solution, $u(x, t) = A\mathrm{sech}[\kappa(x - ct)]$. Here, parameters $A = \sqrt{2a/b}$, $\kappa = \sqrt{a/(a^2 - c^2)}$, $a = 0.3$, $b = 1$, and $c = 0.25$. The error norm L_∞ of such wavelet-based solution for different time is listed in Table 8.4 when $\Delta x = 5/128$ and $\Delta t = 0.025$, and its soliton wave form is shown in Fig. 8.2.

As a comparison, Table 8.4 also lists the error norm, L_∞, of the numerical solutions achieved by using LBM [22], TSCM [26], the second-order central difference formula (SOCDF) [31], and the fourth-order compact difference method (FOCDM) [32], and Fig. 8.3 displays the changing curves of the error norms of both wavelet-based method and FOCDM [32] with spatial mesh sizes, Δx when $\Delta t = 0.025$. From them, one sees that the wavelet-based method has more advantages over the conventional methods.

Table 8.4 Error norm L_∞ of the numerical solutions at different times t for Example 8.4

t	Wavelet based	LBM [22]	TSCM [26]	SOCDF [31]	FOCDM [32]
	$\Delta x = 5/128$	$\Delta x = 1/100$	$\Delta x = 1/200$	$\Delta x = 1/200$	$\Delta x = 1/40$
	$\Delta t = 0.025$	$\Delta t = 0.001$	$\Delta t = 0.0001$	$\Delta t = 0.0001$	$\Delta t = 0.025$
2.0	7.63E-08	2.16E-04	–	–	–
4.0	8.88E-08	4.49E-04	–	–	–
6.0	9.27E-08	1.12E-03	3.08E-06	2.10E-04	3.11E-05
8.0	2.64E-07	2.98E-03	–	–	–
9.0	4.54E-07	–	6.61E-06	7.45E-04	1.68E-04
12.0	2.31E-06	–	1.14E-05	3.17E-03	8.37E-04

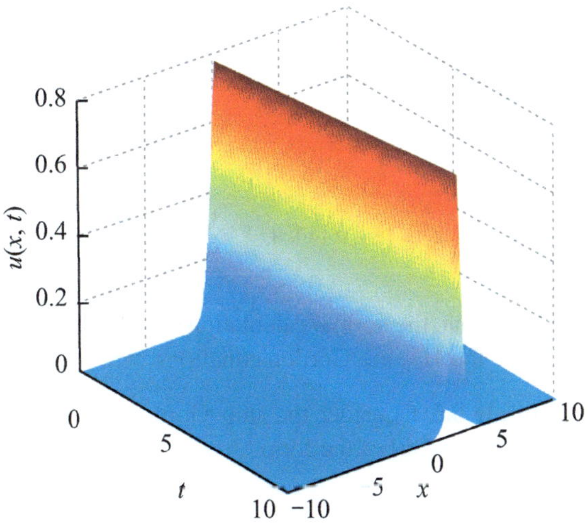

Fig. 8.2 Wavelet-based solution of the soliton wave in the Klein–Gordon equation using spatial mesh size, $\Delta x = 5/128$, and time step, $\Delta t = 0.025$, for the Example 8.4

8.2.2 *Numerical Solution to 1D sine–Gordon Equations*

For Eq. (8.3), we know that the specific form of the nonlinear operator is different for different one-dimensional initial boundary-value problems. Here, the sine-Gordon equation is used to quantitatively test the wavelet-based method. In the sine-Gordon equation, the nonlinear term is expressed by the general form

$$\mathbf{N}_i[u(x,t), x, t] = \alpha \sin(u) \tag{8.23}$$

in which α is a constant coefficient. Applying the expansion of Eqs. (8.5)–(8.23), one gets

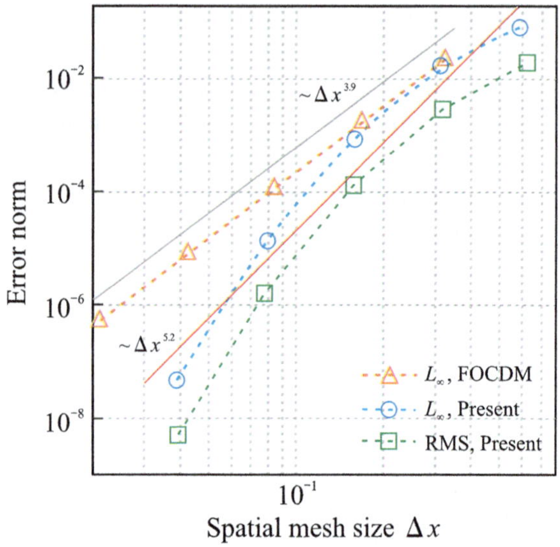

Fig. 8.3 Error norms of the numerical solutions at $t = 1$ with $\Delta t = 0.025$ varying with Δx for the soliton wave in the Klein–Gordon equation of Example 8.4. Here, "present" stands for the wavelet-based method

$$\alpha \sin(u) \approx \sum_{k=0}^{2^j} \alpha \sin\left(u\left(\frac{k}{2^j}, t\right)\right)\phi_{j,k}(x). \tag{8.24}$$

After that, the normal calculation scheme in the previous section can be employed to solve the 1D sine-Gordon equations with nonlinear term of Eq. (8.23).

Example 8.5 Consider the sine–Gordon equation with Dirichlet boundary conditions [30, 33–36] of the form

$$\begin{cases} \frac{\partial^2 u}{\partial t^2} - \frac{\partial^2 u}{\partial x^2} + \sin(u) = 0, x \in [-20, 20] \\ u(x, 0) = 0, \partial u/\partial t|_{t=0} = 4\text{sech}(x) \end{cases}. \tag{8.25}$$

For such case, the exact solution is of the form $u(x, t) = 4\arctan[t\text{sech}(x)]$.

Table 8.5 and Fig. 8.4 display the comparison of error norms of the wavelet-based solution and those obtained by using LWSCM [30], TPSCM [35], and the sixth-order compact finite difference scheme (CFD6) [36], respectively.

Table 8.5 Error norms of numerical solutions at different times t for the equation in Example 8.5

T	Wavelet based, $\Delta x = 5/64$, $\Delta t = 0.001$			TPSCM [22], $\Delta x = 1/25$, $\Delta t = 0.0001$			CFD6 [36]
	L_∞	L_2	RMS	L_∞	L_2	RMS	L_∞
0.25	8.10E-11	1.97E-10	8.73E-12	5.90E-06	3.91E-05	3.89E-06	6.91E-11
0.50	1.59E-10	5.19E-10	2.30E-11	2.01E-05	1.30E-04	1.30E-05	1.01E-09
0.75	1.81E-10	5.85E-10	2.59E-11	3.63E-05	2.36E-04	2.35E-05	8.65E-09
1.00	2.40E-10	8.59E-09	3.80E-11	5.07E-05	3.27E-04	3.25E-05	6.41E-08

CFD6 with mesh size $\Delta x = 1/25$ and $\Delta t = 0.0001$

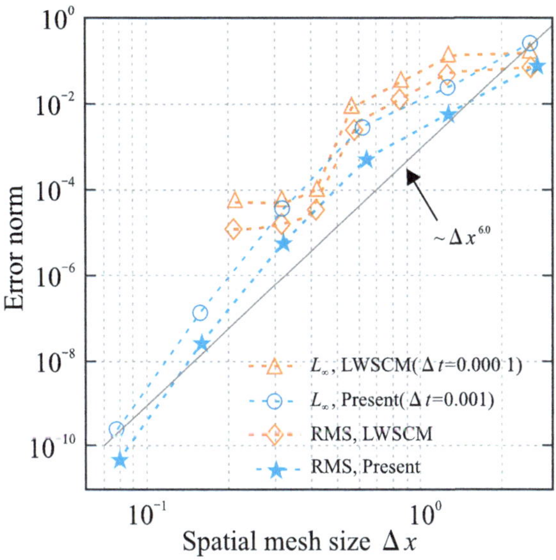

Fig. 8.4 Error norms of the numerical solutions at $t = 1$ with $\Delta t = 0.025$ varying with Δx for the soliton wave in the Klein–Gordon equation of Example 8.5. Here, "present" stands for the wavelet-based method

Example 8.6 The kink solitary wave in the sine-Gordon equation [15, 18, 20, 22]. The governing equations for such kink solitary wave are of the form

$$\begin{cases} \dfrac{\partial^2 u}{\partial t^2} - \dfrac{\partial^2 u}{\partial x^2} + \sin(u) = 0, \; x \in [-20, 20] \\ u(x, 0) = 4\arctan(e^{x/\sqrt{1-c^2}}) \\ \dfrac{\partial u}{\partial t}|_{t=0} = -\dfrac{2c\text{sech}(-x/\sqrt{1-c^2})}{\sqrt{1-c^2}} \end{cases} . \tag{8.26}$$

When the above Dirichlet boundary conditions are subjected, the exact solution of the kink solitary wave solution can be gained by $u(x, t) = 4\arctan[e^{(x-ct)/\sqrt{1-c^2}}]$, where $c = 0.5$.

The error norms of the numerical solutions obtained using various numerical methods are listed in Table 8.6, including the wavelet-based method, the Chebyshev tau meshless method (CTMM) [18], the nonlinear integral equation formulation (NIEF) [21], FDM [21], the fourth-order compact finite difference scheme (FOCFD) [33], and TPSCM [35]. Figure 8.5 illustrates the error norms of the wavelet-based solution at time, $t = 1$, with time step, $\Delta t = 0.001$, varying with the spatial mesh size, Δx.

Table 8.6 Error norms of the numerical solutions at different times t for Example 8.6

t	1.00	2.00	5.00	10.00
Present, $\Delta x = 5/32$, $\Delta t = 0.025$				
L_∞	3.75E-07	2.04E-07	2.30E-07	2.62E-07
L_2	7.27E-07	4.49E-07	5.62E-07	5.92E-07
RMS	4.55E-08	2.81E-08	3.52E-08	3.71E-08
FOCFD [33], $\Delta x = 1/15$, $\Delta t = 0.025$				
L_∞	1.42E-07	2.08E-07	2.66E-07	4.50E-07
L_2	4.52E-07	6.91E-07	1.14E-06	2.00E-06
RMS	1.84E-08	2.82E-08	4.66E-08	8.14E-08
T	1.00	2.00	3.00	4.00
NIEF [21], $\Delta x = 1/20$, $\Delta t = 0.005$				
L_∞	6.39E-05	5.61E-05	7.38E-05	1.00E-04
L_2	3.40E-04	4.16E-04	5.84E-04	9.13E-04
RMS	1.70E-05	2.08E-05	2.92E-05	4.56E-05
FDM [21], $\Delta x = 1/20$, $\Delta t = 0.005$				
L_∞	2.98E-04	6.04E-04	7.61E-04	8.12E-04
L_2	1.53E-03	3.06E-03	4.50E-03	6.64E-03
RMS	7.63E-05	1.53E-04	2.25E-04	3.32E-04
T	0.25	0.50	0.75	1.00
CTMM [18], $\Delta x = 6/34$, $\Delta t = 0.1$				
L_∞	2.46E-07	3.54E-07	6.36E-07	8.08E-07
L_2	6.87E-07	1.19E-06	1.84E-06	2.39E-06
RMS	1.21E-07	2.10E-07	3.24E-07	4.23E-07
TPSCM [35], $\Delta x = 1/25$, $\Delta t = 0.0001$				
L_∞	4.96E-06	8.43E-06	1.66E-05	2.52E-05
L_2	1.76E-05	4.31E-05	8.25E-05	1.27E-04
RMS	1.43E-05	3.51E-06	6.71E-06	1.03E-05

Example 8.7 Consider that the sine-Gordon equation [18, 37]

$$\begin{cases} \dfrac{\partial^2 u}{\partial t^2} - \dfrac{\partial^2 u}{\partial x^2} + \sin(u) = 0, x \in [-10, 10] \\ u(x, 0) = 0, \partial u/\partial t|_{t=0} = 4K\operatorname{sech}(Kx) \end{cases} \tag{8.27}$$

which is subjected to the Dirichlet boundary conditions extracted from the exact solution, $u(x, t) = 4\arctan[c^{-1}\sin(cKt)\operatorname{sech}(Kx)]$, where $K = 1/\sqrt{1+c^2}$ and $c = 0.5$. This problem is referred to as a breather, and it describes the pulse-type structure of a solution [18, 37].

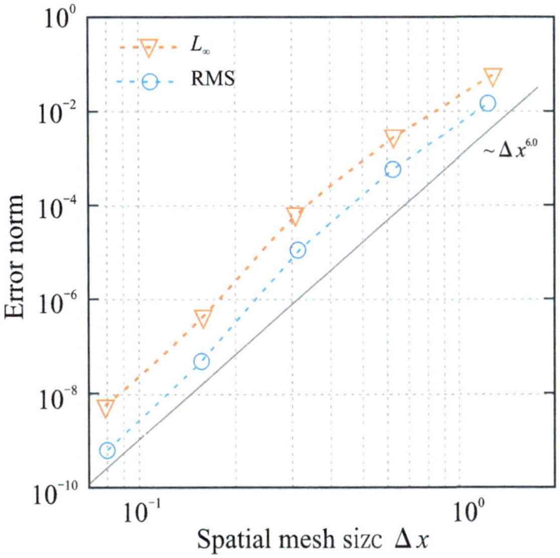

Fig. 8.5 Error norms of the wavelet solution at $T = 1$ with $\Delta t = 0.001$ varying with Δx for Example 8.6

Figure 8.6 presents the space–time graph of the wavelet-based solution when spatial mesh size $\Delta x = 5/64$ and time step $\Delta t = 0.001$. Table 8.7 shows the comparison between error norms obtained using the wavelet-based method and the central difference formula (CDF) [37]. From it, one finds that the wavelet-based solution has accuracy of five to six orders higher than the CDF [37].

Figure 8.7 plots the curves of error norm of the wavelet-based solution at time, $t = 1$, with time step, $\Delta t = 0.001$, varying with spatial mesh size, Δx. When Δx decreases, the accuracy of the wavelet-based solution to this example has a scaling order of 7 decreasing.

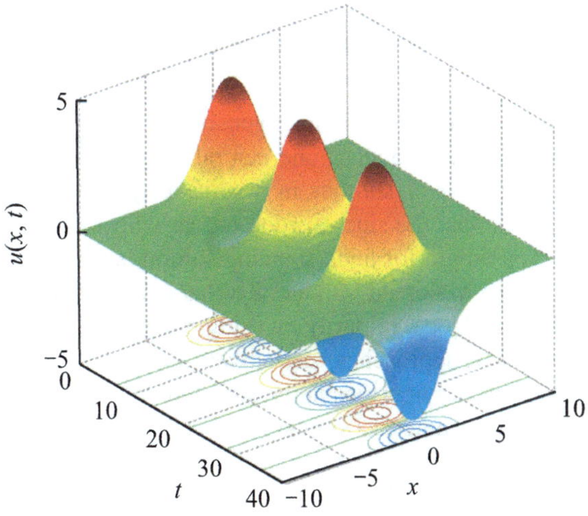

Fig. 8.6 Wavelet solution at spatial mesh size $\Delta x = 5/64$ and time step $\Delta t = 0.001$ for Example 8.7

Table 8.7 Error norms of the numerical solutions at different times t for Example 8.7

t	Present, $\Delta x = 5/64$, $\Delta t = 0.001$			CDF [37]
	L_∞	L_2	RMS	L_∞
1	4.78E-10	7.27E-10	4.55E-11	
10	1.08E-09	1.63E-09	1.02E-10	1.73E-03
20	5.18E-10	2.27E-09	1.42E-10	3.58E-04
30	7.78E-10	3.41E-09	2.14E-10	1.63E-03
40	1.12E-09	6.05E-09	3.79E-10	1.04E-03

CFD with mesh size $\Delta x = 1/100$ and $\Delta t = 0.001$

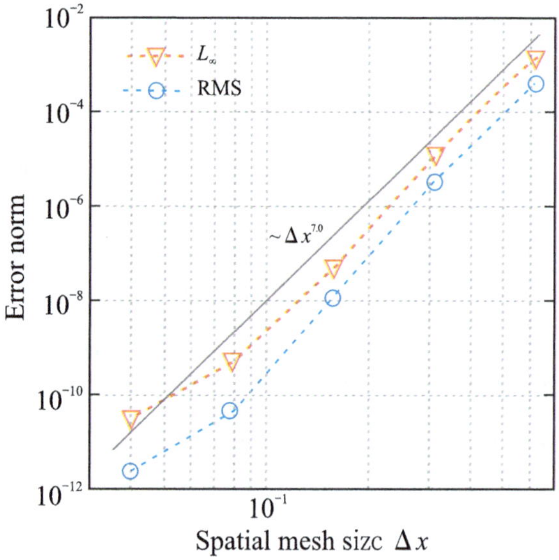

Fig. 8.7 Scaling dependence of error norms of the wavelet-based solutions at $t = 1$ with $\Delta t = 0.001$ on Δx for Example 8.7

8.2.3 Interaction Between Solitary Wave and Inclusion

Wave propagation in inhomogeneous or disordered media elicits widespread interest [46–49]. In particular, one central issue is how disorder can change the steady-state motion of solitons in nonlinear systems [46–49]. Given its broad practical applications, the sine–Gordon model with local impurity has been widely studied as a paradigm for understanding soliton behavior under perturbations [46–49]. For example, Malomed [46] and Kivshar et al. [47, 48] studied the interactions between a kink and local impurity by using the inverse scattering perturbation and collective coordinate approaches, respectively. Zhang [49] investigated the motion of a breather under weak impurities in the sine–Gordon model via collective coordinate analysis and direct numerical simulation. In this section, the interactions of a breather with a strong impurity, which can be described by the perturbed sine-Gordon equation, are considered as follows:

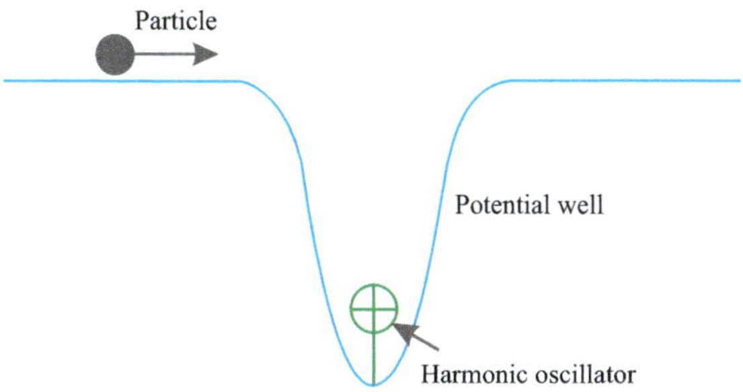

Fig. 8.8 Schematic of a particle moving in a potential well and coupled with a harmonic oscillator

$$\partial^2 u/\partial t^2 - \partial^2 u/\partial x^2 + [1 + U(x)]\sin(u) = 0, \tag{8.28}$$

where $U(x) = \varepsilon\delta(x)$ represents a local impurity. By ignoring radiation, this perturbed system can be regarded approximately as a particle placed in an attractive potential and coupled with a harmonic oscillator with undamped natural frequency $\Omega = \sqrt{1 - \varepsilon^2/4}$ [47, 48], as shown in Fig. 8.8. This analog can help understand the nonlinear responses of a system, Eq. (8.28). In particular, when parameter $\varepsilon = 0$, the unperturbed sine-Gordon equation, Eq. (8.28), has a breather solution [49]:

$$u(x,t) = 4\arctan\left\{\tan(\mu)\frac{\sin[\cos(\mu)(t - v_0 x)/\sqrt{1 - v_0^2} + \theta_0]}{\cosh[\sin(\mu)(x - v_0 t - x_0)/\sqrt{1 - v_0^2}]}\right\}, \tag{8.29}$$

where x_0, $\Omega_0 = \cos(\mu)$, v_0, and θ_0 are the initial position, frequency, velocity, and phase of the breather, respectively.

In order to dealing with $\delta(x)$ in $U(x)$ such that the numerical calculations can be performed, the local impurity can be approximately employed by the following continuous function [49]:

$$U(x) = \alpha/\cosh^2(x/\beta). \tag{8.30}$$

When $\alpha = -12.5$ and $\beta = 0.04$ are selected, Eq. (8.30) is extremely close to a δ impurity with strength $\varepsilon = 2\alpha\beta = -1$.

After the above equations are taken into account, the numeric solutions for this example can be gained by the normal wavelet-based method in Sect. 8.1. That is, the perturbed sine–Gordon system, Eq. (8.28), in the spatial interval $[-80, 80]$ with the boundary condition $u(-80) = u(80) = 0$ is solved using the wavelet-based method with spatial mesh size $\Delta x = 0.02$ and time step $\Delta t = 0.01$, and the initial condition is extracted from the breather solution, Eq. (8.29), with the initial position $x_0 = -20$ [50].

First, we give the numerical solutions of the motion evolution of the breather with a low frequency $\Omega_0 = 0.4$ and velocity $v_0 = 0.3$ under strong local impurity of Eq. (8.30), shown in Fig. 8.9. It can be found that the obtained results in Fig. 8.9 support that such a breather can either pass, be reflected, be trapped, or break into a kink–antikink pair, depending on its initial phase θ_0. Such similar nonlinear phenomena were also gained in the sine–Gordon system with weak local impurity of $|\varepsilon| = 0.5$, see [49]. To compare the interactions between a breather and impurity in the passing, reflection, and trapping cases, Fig. 8.10 presents the amplitude–frequency characteristic curves of the Fourier transforms of the evolution $u(0, t)$ (displacement of the harmonic oscillator) when $\Omega_0 = \sqrt{1 - \varepsilon^2/4} \approx 0.866$. As shown in Figs. 8.9 and 8.10, the impurity mode (vibration of the harmonic oscillator) can always be excited when the breather is close to the impurity. As shown in Fig. 8.10, such additional one, two, and more than three intense interactions (collisions) are

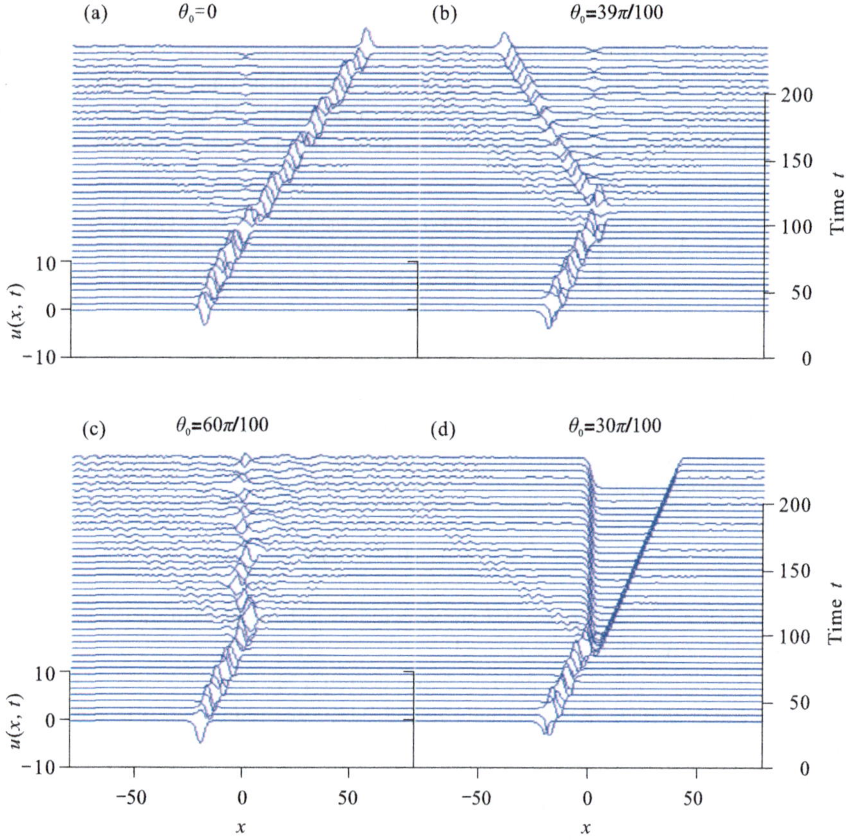

Fig. 8.9 Wavelet-based solution of evolution, $u(x, t)$, for a breather with initial frequency $\Omega_0 = 0.4$, velocity $v_0 = 0.3$, and different initial phases θ_0 under an attractive impurity. **a** passing the impurity for $\theta_0 = 0$, **b** reflected for $\theta_0 = 39\pi/100$, **c** trapped for $\theta_0 = 60\pi/100$, and **d** breaks into a kink–antikink pair for $\theta_0 = 30\pi/100$

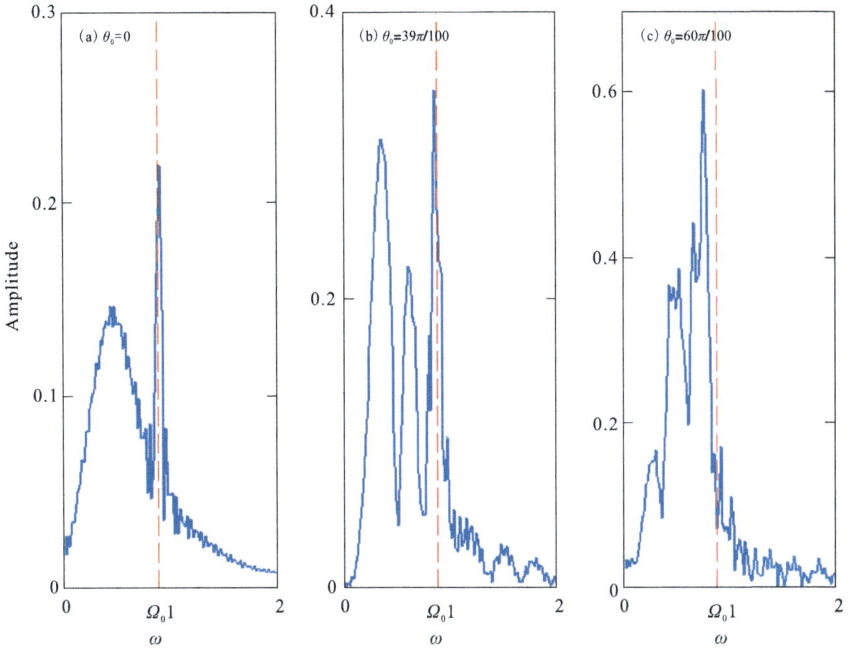

Fig. 8.10 Amplitude–frequency curves of evolution of impurity mode $u(0, t)$ corresponding to $\Omega_0 \approx 0.866$. **a** passing the impurity for $\theta_0 = 0$, **b** reflected for $\theta_0 = 39\pi/100$, and (c) trapped for $\theta_0 = 60\pi/100$

clearly observed in the passing, reflection, and trapping cases, respectively, except for vibration mode excited by the natural frequency of the harmonic oscillator.

In particular, the present observation for the reflection case in Fig. 8.10b is in good agreement with the previous hypothesis [47, 48]. That is, the particle is momentarily trapped by the attractive potential after the first collision because an excessive amount of the kinetic energy of the particle has been transferred to the oscillator during this collision for some periods of the initial phase θ_0 and will come back to interact with the oscillator again due to the constraint of the potential. If the phase at the first moment before the second collision is correct, then the particle can restore a sufficient amount of kinetic energy from the oscillator through the second collision and then escape to infinity in the opposite direction [47, 48]. Analogously, for the passing case that corresponds to several special initial phases θ_0, although part of the kinetic energy of the particle has been transferred to the oscillator during the first collision, the rest is sufficient for the particle to pass the potential and escape to infinity. Thus, only one additional response frequency can be observed, except for the undamped natural frequency Ω_0 in Fig. 8.10a. In addition, Fig. 8.10c implies that nearly all of the kinetic energy of the particle has been transferred to the oscillator after multiple collisions, and the particle is eventually trapped by the attractive potential. Moreover, Fig. 8.10c shows that the frequency of the main vibration of the oscillator is slightly smaller than

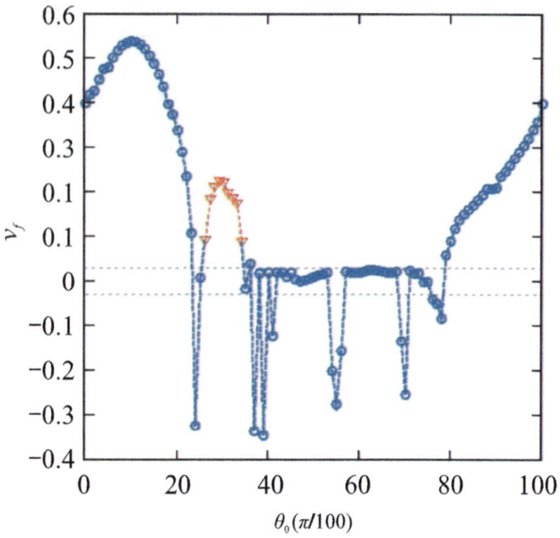

Fig. 8.11 Relation of the final velocity v_f of a breather varying with initial phase θ_0 when initial frequency $\Omega_0 = 0.4$ and velocity $v_0 = 0.3$. The triangles indicate the breaking of the breather into a kink–antikink pair, and their values are the final velocity of the kink

the undamped natural frequency Ω_0 because a high amount of radiation is generated due to multiple collisions, as shown in Fig. 8.9c. Considering that radiation dissipates the kinetic energy of the oscillator, the effect of radiation is equivalent to the effect of damping in a harmonic oscillator. Therefore, the frequency of the main vibration shown in Fig. 8.10c is actually the natural frequency of the oscillator with damping, which is slightly smaller than the undamped natural frequency Ω_0.

For the case of breaking the breather into a kink–antikink pair, the interactions of the breather with impurity cannot be investigated by analyzing the vibration of the harmonic oscillator in the analog system because its motion has been extinguished too quickly to perform effective Fourier analysis. Zhang [49] provided a simple explanation for this nonlinear phenomenon from the perspective of energy. According to results of the wavelet-based solution to this case, here, Fig. 8.11 provides the detailed relation between the final velocity v_f of the breather and its initial phase θ_0. From Fig. 8.11, the positive, negative, and approximate zero values of the final velocity v_f are clearly obtained to the cases of passing, reflection, and trapping, respectively, except for the triangle, which reveals that the breather is breaking into a kink–antikink pair and its value is the final velocity of the kink. Moreover, Fig. 8.11 also shows that the type of scattering outcome (i.e., reflection or trapping) of the breather is highly sensitive to the initial phase θ_0 when it is within the range $[35\pi/100, 41\pi/100]$. These obtained results support that the wavelet-based method can capture all nonlinear phenomena, even those that are extremely sensitive to parameters.

As we have known, the resonant interaction is a crucial issue in nonlinear dynamical systems. Further, we select the initial frequency to be the same as the undamped natural frequency, i.e., $\Omega_0 = \Omega = \sqrt{1 - \varepsilon^2/4} \approx 0.866$ to conduct this research in Eq. (8.28). The obtained results display that no matter what the initial phase is chosen, the breather will be reflected if its initial velocity is smaller than the critical velocity $v_c \approx 0.40$, but it will pass the impurity if its initial velocity exceeds the critical velocity. However, when the initial velocity is close to the critical velocity v_c, passing and reflection coexist, as shown in Fig. 8.12. The obvious difference of the wavelet-based results and Zhang [49], where the interaction between the high-frequency breather and the impurity wherein the low-velocity breather will always be trapped by the attractive impurity, is displayed by that the breather here is never trapped by the attractive impurity and will be reflected even for a minimal initial velocity $v_0 = 0.01$ probably due to severe resonance which results in the particle always escaping from the potential trap with the assistance of the harmonic oscillator.

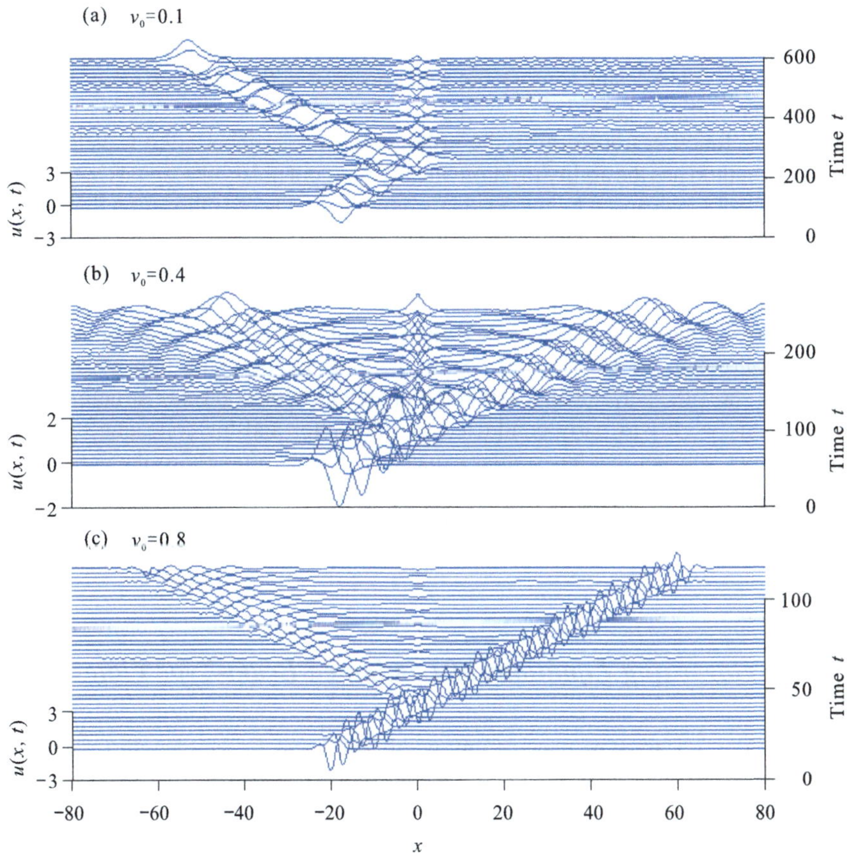

Fig. 8.12 Resonant interaction of a breather with initial frequency $\Omega_0 = \Omega = 0.866$ and initial phase $\theta_0 = 0$, with an attractive impurity for initial velocity. **a** $v_0 = 0.1$, **b** $v_0 = 0.4$, and **c** $v_0 = 0.8$

8.3 Dynamic Control of Piezoelectric Thin Beam-Type Plates with Large Deflection

The aerospace engineering has to deal with many thin plate and shell structures that have large deformation, low damping ratios, and stiffness. For these structures, any small external disturbances may cause unwanted vibrations. To suppress such undesirable vibration, the accuracy of their orientation, alignment, or pointing is thus crucially dependent on the design of relevant active vibration controllers. One promising technique of active control of structural vibration is to use piezoelectric material as sensors and actuators that are bonded to surface of the main structures. In this section, we will display that the wavelet-based method can not only be used to solve the nonlinear differential equation in the dynamic control system, but also be used to design sensors and actuators used in the feedback control system with fascinating properties. For example, owing to the low-pass filtering property of scaling function transform in the orthogonal wavelet theory, the wavelet-based control method has the ability to automatically filter out noise-like signals in the feedback control loop caused by the residual coupling effects of spillover on the sensors and actuators. Hence, the unavoidable phenomenon that such spillover coupling may excite an instability of such structure vibration with high frequency in a conventional control method can be automatically disappeared by the wavelet-based method.

As an example of the wavelet-based method applied to this kind of dynamic control in the following, we use a smart beam-type plate bonded with thin piezoelectric layers to show the wavelet approximation in the vibration control of structures. Due to that this work was early conducted by the author and his colleague using the Daubechies wavelet that has similar expansion scheme property of the generalized Coiflets [56]; here, the introduction still uses the Daubechies wavelet employed although the generalized Coiflets has the accuracy over the former, see in Chaps. 2 and 4.

8.3.1 Governing Equations of Laminated Beam-Type Plates

We consider a beam-type plate of unit width, thickness h_b, and length L as shown in Fig. 8.13. Assume that this plate is laminated with piezoelectric layers, such as PVDF, of thickness h_p on the top and bottom surfaces of the plate. Further, the thickness of piezoelectric layers is much thinner than that of the elastic layer, i.e., $h_p \ll h_b$. Denote the in-plane displacement u and the deflection w in the mid-plane of plate. After the effects of in-plane displacement u, inertial force of rotation are neglected, from the laminated plate theory, we can formulate the dynamic equations of plates with geometric nonlinearity as follows [52, 54]:

$$D\frac{\partial^4 w}{\partial x^4} - N_x\frac{\partial^2 w}{\partial x^2} + \rho\frac{\partial^2 w}{\partial t^2} = f(x,t), \quad 0 < x < L \tag{8.31}$$

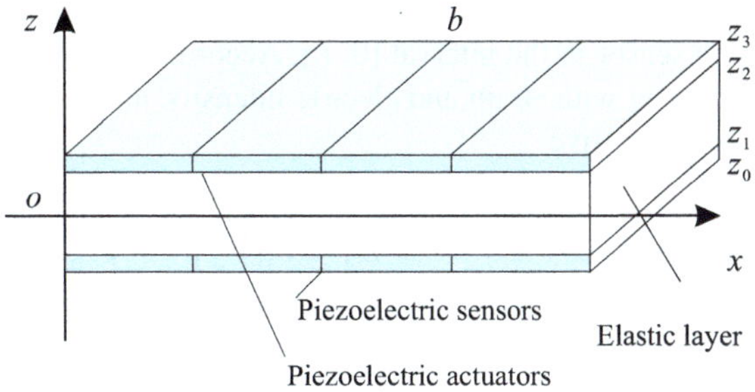

Fig. 8.13 Schematic drawing of a beam-type plate with piezoelectric sensors and actuators

in which $w = w(x, t)$ is the deflection function of plate at instant t, $f(x, t)$ stands for the control force, and the internal membrane force N_x is given by a nonlinear relation

$$N_x = \frac{1}{L}\int_0^L \left\{\frac{Y}{2}\left(\frac{\partial w}{\partial x}\right)^2 - e_{31}V(x,t)\right\}\mathrm{d}x. \tag{8.32}$$

Here, e_{31} is the piezoelectric constant. The equivalent elastic constant Y, flexible rigidity D, density of mass ρ, and thickness h are, respectively, formulated by

$$Y = \sum_{i=1}^{3} \frac{Y_i}{1-\mu_i^2}(z_i - z_{i-1}) \tag{8.33}$$

$$D = \sum_{i=1}^{3} \frac{Y_i}{3(1-\mu_i^2)}\left(z_i^3 - z_{i-1}^3\right) \tag{8.34}$$

$$\rho h = \sum_{i-1}^{3} \rho_i (z_i - z_{i-1}) \tag{8.35}$$

in which $\rho_1 = \rho_3$, $Y_1 = Y_3$, and $\mu_1 = \mu_3$ are the density of mass, Young's modulus, and Poisson's ratio of the piezoelectric layers, respectively; ρ_2, Y_2, and μ_2 are the density of mass, Young's modulus, and Poisson's ratio of the plate, respectively; z_0, z_1, z_2, and z_3 represent the coordinates of the surfaces of both plate and piezoelectric layers in z-direction as shown in Fig. 8.13; and $V = V(x, t)$ is the control voltage across piezoelectric actuators, which has the relation with control force $f(x, t)$ [52]

$$f(x,t) = -e_{31} r_{\mathrm{a}} \frac{\partial^2 V(x,t)}{\partial x^2} \tag{8.36}$$

in which r_a is the distance of piezoelectric actuator measured from the mid-plane.

Sensing equations: We denote Ω_k as the region occupied by the kth piezoelectric sensor in the interval [0, L]. According to the constitutive relation of dielectric material with strain and electric intensity, and the geometrically nonlinear relation, we can derive

$$D_3 = e_{31}\left[\frac{1}{2}\left(\frac{\partial w}{\partial x}\right)^2 - z\frac{\partial^2 w}{\partial x^2}\right]. \tag{8.37}$$

Further, we obtain the sensing design equation associated with the kth sensor

$$q_k(t) = e_{31}\int\limits_{\Omega_k}\left[\frac{1}{2}\left(\frac{\partial w}{\partial x}\right)^2 - r_s\frac{\partial^2 w}{\partial x^2}\right]\mathrm{d}x \tag{8.38}$$

in which $q_k(t)$ is the electric charge measured at the kth sensor at instant t, r_s stands for the average coordinate of the sensor measured from the mid-plane of plate. Differentiating Eq. (8.38) with respect to time variable t, we get the measured electric current $I_k(t)$ in the form

$$I_k(t) = e_{31}\int\limits_{\Omega_k}\left[\frac{\partial w}{\partial x}\frac{\partial^2 w}{\partial x\partial t} - r_\mathrm{s}\frac{\partial^3 w}{\partial x^2\hat{\partial} t}\right]\mathrm{d}x. \tag{8.39}$$

Obviously, we can see from Eqs. (8.38) and (8.39) that the sensing electric charges and currents, and the internal force N_x in Eq. (8.32) are nonlinear in terms of deflection $w(x, t)$. Once the deflection and velocity are identified in terms of the measured charges and current, a control law should be employed to suppress the vibration of the plate. The performance of the vibration control of the plate will highly depend on the effect of the control system designed. In the following, we will introduce the wavelet-based approach to realize the effective control of the structural vibration.

For simplicity, we first introduce the following dimensionless quantities:

$$\overline{w} = w/r_\mathrm{s}, \quad \overline{x} = x/L, \quad \overline{t} = t/\sqrt{\rho L^4/D}, \; \overline{V} = e_{31}r_\mathrm{a}L^2V/(r_\mathrm{s}D) \tag{8.40a}$$

$$\overline{q} = qL/\left(e_{31}r_\mathrm{s}^2\right), \quad \overline{I} = IL\sqrt{\rho L^4/D}/\left(e_{31}r_\mathrm{s}^2\right), \quad \alpha = r_\mathrm{s}^2Y/(2D) \tag{8.40b}$$

$$\overline{f} = L^4 f/(r_\mathrm{s}D), \quad \overline{N}_x = L^2N_x/D. \tag{8.40c}$$

Then, Eqs. (8.31), (8.32), (8.36), (8.38), and (8.39) can be transformed into dimensionless forms

$$\frac{\partial^4\overline{w}}{\partial\overline{x}^4} - \overline{N}_x\frac{\partial^2\overline{w}}{\partial\overline{x}^2} + \frac{\partial^2\overline{w}}{\partial\overline{t}^2} = f(\overline{x},\overline{t}), \quad 0 < \overline{x} < 1, \tag{8.41}$$

$$\overline{N}_x = \int_0^1 \left[\alpha \left(\frac{\partial \overline{w}}{\partial \overline{x}} \right)^2 + \overline{V} \right] \mathrm{d}\overline{x}, \tag{8.42}$$

$$\overline{f}(\overline{x}, \overline{t}) = -\frac{\partial^2 \overline{V}}{\partial \overline{x}^2}, \tag{8.43}$$

$$\overline{q}_k(t) = \int_{\overline{\Omega}_k} \left[\frac{1}{2} \left(\frac{\partial \overline{w}}{\partial \overline{x}} \right)^2 - \frac{\partial^2 \overline{w}}{\partial \overline{x}^2} \right] \mathrm{d}\overline{x}, \tag{8.44}$$

$$\overline{I}_k(t) = \int_{\overline{\Omega}_k} \left[\frac{\partial \overline{w}}{\partial \overline{x}} \frac{\partial^2 \overline{w}}{\partial \overline{x} \partial \overline{t}} - \frac{\partial^3 \overline{w}}{\partial \overline{x}^2 \partial \overline{t}} \right] \mathrm{d}\overline{x} \tag{8.45}$$

in which $\overline{\Omega}_k$ is the dimensionless form of Ω_k. From here on, we will only use the dimensionless variables and parameters unless otherwise indicated. Under such a situation, we may drop the bar over each dimensionless quantity for simplicity. For example, we denote $\overline{w}$ by w, etc.

8.3.2 Sensors and Actuators Designed by the Wavelet-Based Method

Take the patches of piezoelectric sensors and actuators to be overlap with the sub-regions $\Omega_k = [x_{k-1}, x_k](x_k = k2^{-n},\ \ k = 0, 1, 2, \ldots, 2^n)$. Then we have $[0, 1] = \bigcup_{k=1}^{2^n} \Omega_k$ and $\Omega_k \bigcap \Omega_l = \varnothing$ when $k \neq l$. After that, we introduce.

$$H(x, t) = \int_0^x \left[\frac{1}{2} \left(\frac{\partial w}{\partial x} \right)^2 - \frac{\partial^2 w}{\partial x^2} \right] \mathrm{d}x, \tag{8.46}$$

$$\dot{H}(x, t) = \frac{\partial H}{\partial t} = \int_0^x \left[\frac{\partial w}{\partial x} \frac{\partial^2 w}{\partial x \partial t} - \frac{\partial^3 w}{\partial x^2 \partial t} \right]. \tag{8.47}$$

Then, according to Eqs. (8.44)–(8.47), we obtain

$$H(x_k, t) = \sum_{j=1}^{k} q_j(t), \tag{8.48}$$

$$\dot{H}(x_k, t) = \sum_{j=1}^{k} I_j(t). \tag{8.49}$$

Differentiating Eqs. (8.46) and (8.47) with respect to variable x, one can obtain the following nonlinear differential equations on unknowns $w(x, t)$:

$$\frac{1}{2}\left(\frac{\partial w}{\partial x}\right)^2 - \frac{\partial^2 w}{\partial x^2} = \frac{\partial H(x,t)}{\partial x}, \tag{8.50a}$$

$$\frac{\partial w}{\partial x}\frac{\partial \dot{w}}{\partial x} - \frac{\partial^2 \dot{w}}{\partial x^2} = \frac{\partial \dot{H}(x,t)}{\partial x}. \tag{8.50b}$$

Identification of Nonlinear Deflection

The first step to us is to explicitly express the function $H(x, t)$ in terms of measured electric charges of Eq. (8.48). In order to do it, here, we apply the scaling function transform of the Daubechies wavelet on the basis of the generalized Gaussian integral to approximate a function. After the scaling function $\phi(x)$ of the Daubechies wavelet with support region $[0, 2r - 1]$ is constructed (see Chap. 2), we can write the n-level base scaling functions of the form [57]

$$\phi_{n,k}(x) = 2^{-n/2}\phi_N\left(2^n x - k\right). \tag{8.51}$$

Then, we can approximate a function $f(x)$ with the scaling function transform (see Sect. 4.1)

$$f(x) \approx 2^{-n/2} \sum_{k=1-2N}^{2^n} f\left(\frac{M_1 + k}{2^n}\right)\phi_{n,k}(x) \tag{8.52}$$

in which M_1 is the first-order moment of the Daubechies scaling function. We denote

$$\widehat{H}(x,t) = H\left(x - \frac{M_1 - [M_1]}{2^n}, t\right), \tag{8.53}$$

where $[M_1]$ represents the integer part of M_1. Application of Eq. (8.52) to $\widehat{H}(\cdot, t)$ yields

$$\widehat{H}(x,t) \approx 2^{-n/2} \sum_{k=1-2N}^{2^n} \widehat{H}\left(\frac{x_1^* + k}{2^n}, t\right)\phi_{n,k}(x), \tag{8.54}$$

or equivalently,

$$H\left(x-\frac{M_1-[M_1]}{2^n},t\right)\approx 2^{-n/2}\sum_{k=1-2N}^{2^n} H\left(\frac{[M_1]+k}{2^n},t\right)\phi_{n,k}(x). \tag{8.55}$$

Then translating the coordinate x and using Eq. (8.51), we obtain

$$H(x,t)\approx\sum_{k=1-2N}^{2^n} H\left(\frac{[M_1]+k}{2^n},t\right)\times\phi_N\left(2^n x+M_1-[M_1]-k\right). \tag{8.56}$$

According to Eq. (8.48), we have

$$H\left(\frac{[M_1]+k}{2^n},t\right)=\sum_{j=1}^{[M_1]+k} q_j(t). \tag{8.57}$$

Thus, Eq. (8.56) gives an explicit expression of $H(x, t)$ in terms of the measured electric charges $q_j(t)$. Following similar steps, we have $\dot{H}(x,t)$ in terms of the measured electric currents $I_j(t)$

$$\dot{H}(x,t)\approx\sum_{k=1-2N}^{2^n}\dot{H}\left(\frac{[M_1]+k}{2^n},t\right)\times\phi_N\left(2^n x+M_1-[M_1]-k\right) \tag{8.58}$$

in which

$$\dot{H}\left(\frac{[M_1]+k}{2^n},t\right)=\sum_{j=1}^{[M_1]+k} I_j(t). \tag{8.59}$$

Next, we determine the deflection, $w(x, t)$, from the sensing Eq. (8.50a). Applying Eq. (8.52) to the function, $w(x, t)$, and its differentiation with respect to x, we have

$$\begin{aligned} w(x,t)&\approx 2^{-n/2}\sum_{k=1-2N}^{2^n} w\left(\frac{[M_1]+k}{2^n},t\right)\phi_{n,k}(x)\\ &=\sum_{k=1-2N}^{2^n} w_k(t)\phi_N\left(2^n x-k\right), \end{aligned} \tag{8.60}$$

$$\frac{\partial w(x,t)}{\partial x}\approx 2^n\sum_{k=1-2N}^{2^n} w_k(t)\frac{\mathrm{d}\phi_N(2^n x-k)}{\mathrm{d}x}, \tag{8.61}$$

$$\left\{\frac{\partial w(x,t)}{\partial x}\right\}^2\approx\sum_{k=1-2N}^{2^n}\left\{\frac{\partial w(x,t)}{\partial x}\right\}^2\Bigg|_{x=\frac{x_i^2+k}{2^n}}\times\phi_N\left(2^n x-k\right). \tag{8.62}$$

In these equations, $w_k(t) = w\left(\frac{[M_1]+k}{2^n}, t\right)$ are the unknowns to be determined, and $\phi'_N(\hat{x}) = d\phi(\hat{x})/d\hat{x}$. We substitute Eqs. (8.61) and (8.62) into Eq. (8.50a), and apply the weighted residual method by considering the scaling base functions of the Daubechies wavelet with $r = 1$, i.e., the so-called Haar scaling function [54] with $\phi_N^H(x) = 1$, for $x \in [0,1]$, and $\phi_N^H(x) = 0$ for $x \notin [0,1]$, as the weighted functions. Eventually, we obtain a system of nonlinear algebraic equations on the unknowns $w_k(t)$

$$\begin{aligned} &\sum_{k=1-2N}^{2^n} \left\{ [2^{n-1} \sum_{i=1-2N}^{2^n} \sum_{j=1-2N}^{2^n} w_i(t) w_j(t) \phi'_N(M_1 + i - k) \phi'_N(M_1 + j - k) \right. \\ &\left. \left[\phi_N^{\int}(r + 1 - k) - \phi^{\int}(r - k)\right] - w_k(t) 2^n [\phi'_N(r + 1 - k) - \phi'_N(r - k)] \right] \Big\} \\ &= H\left(\tfrac{r+1}{2^n}, t\right) - H\left(\tfrac{r}{2^n}, t\right), \quad r = 1 - 2N + 1, 1 - 2N + 2, \ldots, 2^n - 1 \end{aligned} \tag{8.63}$$

in which $\phi_N^{\int}(x) = \int_{-\infty}^{x} \phi_N(x) dx$. The solvability of Eq. (8.63) is possible only when some boundary conditions on the deflection, $w(x, t)$, are supplemented. This can be done by considering the following boundary conditions:

$$x = 0 : w(x, t) = 0, \quad \frac{\partial w(x, t)}{\partial x} = 0, \tag{8.64}$$

or

$$\sum_{k=1-2N}^{2^n} w_k(t) \phi_N(-k) = 0 \tag{8.65}$$

$$\sum_{k=1-2N}^{2^n} w_k(t) \phi'_N(-k) = 0 \tag{8.66}$$

for a cantilevered beam-type plate and

$$x = 0 \text{ and } 1 : w(x, t) = 0, \tag{8.67}$$

or

$$\sum_{k=1-2N}^{2^n} w_k(t) \phi_N(-k) = 0 \tag{8.68}$$

$$\sum_{k=1-2N}^{2^n} w_k(t) \phi_N(2^n - k) = 0 \tag{8.69}$$

for simple supports. Solving the algebraic equations of (8.63) associated with either (8.65)–(8.66) or (8.68)–(8.69), we can get the unknowns $w_k(t)$ at instant t for the plates. Further, the deflection $w(x,t)$ is given by Eq. (8.60).

For the dynamic problems, the identification approach of velocity signal, $\dot{w}(x,t) = \partial w(x,t)/\partial t$ is needed. For such case, from the differential equation (8.50b) with unknown function $\dot{w}(x,t)$, according to the theory of differential equation, we can have its general solution as

$$\dot{w}(x,t) = \int_0^x e^{w(x',t)} \times \left[C_0 - \int_0^{x'} \dot{H}_{,x''}(x'',t) e^{-w(x'',t)} dx'' \right] dx', \tag{8.70}$$

where $\dot{H}_{,x'} = \partial \dot{H}/\partial x'$ and C_0 is an integral constant which is determined by the boundary condition of the plates. For example, we have

$$\frac{\partial \dot{w}(0,t)}{\partial x} = 0: \quad C_0 = 0. \tag{8.71}$$

Then we obtain

$$\dot{w}(x,t) = -\int_0^x \left[e^{w(x',t)} \int_0^{x'} \dot{H}_{,x''}(x'',t) \right] dx' \tag{8.72}$$

for the cantilevered beam-type plate. If $\dot{w}(0,t) = \dot{w}(1,t) = 0$, we have

$$C_0 = \left[\int_0^1 e^{w(x',t)} dx' \right]^{-1} \int_0^1 e^{w(x',t)} \int_0^{x'} \dot{H}_{,x''}(x'',t) e^{-w(x'',t)} dx'' dx'. \tag{8.73}$$

Further, we can obtain

$$\dot{w}(x,t) = \int_0^x e^{w(x',t)} \left[C_0 - \int_0^{x'} \dot{H}_{,x''}(x'',t) e^{-w(x'',t)} dx'' \right] dx' \tag{8.74}$$

for the simply supported beam-type plate.

It should be noted that the point $([M_1]+k)/2^n$ is of the dyadic points. Hence, we can determine $H(([M_1]+k)/2^n, t)$ and $\dot{H}(([M_1]+k)/2^n, t)$ at instant t using Eqs. (8.48) and (8.49) in terms of the measurable electric charge $q_k(t)$ and electric current $I_k(t)$ in the sensing elements as long as we set $q_k = I_k = 0$ for $k < 0$, or $k > 2^n$. Through this program, Eqs. (8.71) and (8.74) can be used to determine the deflection, $w(x,t)$, and velocity, $\dot{w}(x,t)$, in terms of the measurable charge,

$q_k(t)$, and the measurable current, $I_k(t)$, in the sensing elements. We will call such obtained quantities of $w(x, t)$ and $\dot{w}(x, t)$ as the identified/measured deflection and the identified/measured velocity, respectively. In order to distinct them from those corresponding quantities of real deformation of the plates, from here on, we denote them by $w^*(x, t)$ and $\dot{w}^*(x, t)$, respectively.

Design of Controller

In order to suppress the dynamic disturbance of the plate, we need to design a controller which feeds back the identified signals to the piezoelectric actuators. Here, we employ a control law of negative feedback of the measured signals of deflection and velocity to determine the distribution of applied voltage across piezoelectric actuators [54]

$$f(x, t) = -G_1 w^*(x, t) - G_2 \dot{w}^*(x, t) \tag{8.75}$$

or

$$\frac{\partial^2 V(x, t)}{\partial x^2} = G_1 w^*(x, t) + G_2 \dot{w}^*(x, t) \tag{8.76}$$

with a set of boundary conditions

$$V(0, t) = V(1, t) = 0. \tag{8.77}$$

In Eq. (8.76), G_1 and G_2 are the gain factors. Applying the expression of Eq. (8.52) to the functions $V(x, t)$, $w^*(x, t)$ and $\dot{w}^*(x, t)$, we have

$$V(x, t) \approx 2^{-n/2} \sum_{k=k_1}^{k_2} V_k(t) \phi_{n,k}(x), \tag{8.78}$$

$$w^*(x, t) \approx 2^{-n/2} \sum_{k=k_1}^{k_2} w_k^*(t) \phi_{n,k}(x), \tag{8.79}$$

$$\dot{w}^*(x, t) \approx 2^{-n/2} \sum_{k=k_1}^{k_2} \dot{w}_k^*(t) \phi_{n,k}(x), \tag{8.80}$$

where $k_1 = 1 - 2N$ and $k_2 = 2^n$, $V_k(t) = V\left(\frac{M_1+k}{2^n}, t\right)$, $w_k^*(t) = w^*\left(\frac{M_1+k}{2^n}, t\right)$, and $\dot{w}_k^*(t) = \dot{w}^*\left(\frac{M_1+k}{2^n}, t\right)$. Substituting Eqs. (8.78)–(8.80) into Eq. (8.76), and taking the base scaling functions of the Haar wavelet (see [58]) as the weighted functions of the weighted residual method to the resulting equation, we get

$$\sum_{k=k_1}^{k_2} 2^n[\phi'_N(j-k) - \phi'_N(j-k-1)]V_k(t) = \sum_{k=k_1}^{k_2} \frac{1}{2^n}[G_1 w_k^*(t) + G_2 \dot{w}_k^*(t)] \times [\phi_N^{\int}(j-k) - \phi^{\int}(j-k-1)], \tag{8.81}$$

where $j = k_1 + 1, k_2 + 2, \ldots, k_2 - 1$, and the superscript prime "′" represents the differentiation with respect to variable x. Substitution of Eq. (8.78) into Eq. (8.77) leads to

$$\sum_{k=k_1}^{k_2} V_k(t)\phi_N(-k) = 0, \tag{8.82}$$

$$\sum_{k=k_1}^{k_2} V_k(t)\phi_N(2^n - k) = 0. \tag{8.83}$$

After the system of algebraic equations of Eqs. (8.81)–(8.83) with unknowns $V_k(t)$ are solved, we get a distribution of applied voltage $V(x, t)$ at instant t by Eq. (8.78). Since the applied voltage across a piezoelectric layer is a constant without varying with x in practice, we take the control voltage applied on the kth piezoelectric actuator, $\widehat{V}_k(t)$, to be the average voltage within its corresponding interval: $[x_{k-1}, x_k]$, for $k - 1, 2, \ldots, 2^n$, i.e.,

$$\widehat{V}_k(t) = \int_{(k-1)/2^n}^{k/2^n} V(x,t)dt = \sum_{j=k_1}^{k_2} V_j(t)[\phi_N^{\int}(k-j) - \phi^{\int}(k-j-1)].$$

Thus, we give a design of controller of beam-type plates with piezoelectric sensors and actuators using the wavelet theory.

According to the wavelet theory [57, 58], we know that the scaling function transforms for approximating a function like a low-pass filter. In this case, the high-frequency parts of signal and control voltage are filtered out in the close control system when Eqs. (8.78)–(8.80) are employed. Thus, this control program with wavelet approach does not exhibit the phenomenon of control instability or spilling over, that is, some high-order vibration is excited out when a vibration modal with some low frequency is suppressed in a control system.

8.3.3 Simulation Results of Feedback Control with Piezoelectric Film Sensors and Actuators

In order to show the efficiency and ability of the above proposed approach, we consider numerical simulations of the vibration control of a cantilevered and simply supported beam-type plates. In these simulations, we take PVDF layers as the piezoelectric sensors and actuators, and a thin stainless-steel plate as the main structure. Their material and geometric parameters are listed in Table 8.8.

Identification of Deflection State

When a plate is deformed with a deflection $w(x,t)$ and a velocity $\dot{w}(x,t)$, the electric charges $q_k(t)$ and $I_k(t)$ are measured from the kth sensing element (see Eqs. 8.38 and 8.39). Then, we get $H(x_k,t)$ and $\dot{H}(x_k,t)$ from Eqs. (8.46) and (8.47), respectively. By means of the identification program introduced in Sect. 8.3.2, we obtain the identified deflection $w^*(x,t)$ and the identified velocity $\dot{w}^*(x,t)$. We denote

$$\hat{q}_k^*(t) = -\int_{x_{k-1}}^{x_k} \frac{\partial^2 w}{\partial x^2} dx \tag{8.84}$$

as the sensing equation of the plates with small deflection by piezoelectric sensors [53]. It is found that when the deflection is small, i.e., $|w| \ll 1$, we have $q_k(t) = \hat{q}_k^*(t)$. In this case, the identified deflection $w(x,t)$ from the sensing equation (8.84) approximately approaches to that from the sensing equation (8.38) or (8.44) where the effect of nonlinear deformation is considered. When the deflection is large enough, however, some notable difference between $q_k(t)$ and $\hat{q}_k^*(t)$ can be found, which leads to some differences of identified deflection $w^*(x,t)$. As we know, the sensing equation (8.38) describes practical problem of piezoelectric sensor attached to a plate with nonlinear geometrical deformation, which should be more accurate than Eq. (8.84) does. At the following numerical analysis, the parameters of scaling functions of the Daubechies wavelet are taken by $N = 5$, $n = 3$ of the supported domain and the resolution level, respectively.

From the numerical results, we find that when the deflection is small enough, there is almost no difference between the identified deflections from the linear and the nonlinear programs as shown in Fig. 8.14. As the deflection increases large enough,

Table 8.8 Parameters of materials and geometry in the case study

Material	Young's modulus (GPa)	Density of mass (Kg m^{-3})	Poison's ratio	Thickness (mm)	Length (mm)	e_{31} (N(V m)$^{-1}$)
Stainless steel	210	8000	0.3	1	600	
PVDF layer	2	1780	0.3	0.2	600	0.06

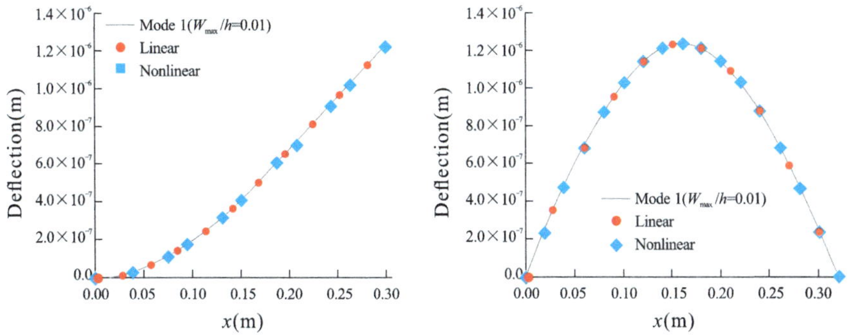

Fig. 8.14 Identified deflections of beam-type plates with deflection of modal function of order 1 by means of the piezoelectric sensors using wavelet-based approach when deflection is small (Daubechies wavelet: $N = 5$; $n = 3$): Cantilever (Left); Simply supported (Right)

e.g., $w_{\max}/h \approx 1$, there is notable difference between them (see in Fig. 8.15). Here, each original deflection is imputed by a modal function of free vibration of the plates. From the results of identified deflection with high-order mode function, it is also found that the difference varies with the order of vibration mode. For the same magnitude of maximum deflection, the higher the order, the larger the difference.

Simulation of Dynamic Control

In order to numerically simulate the responses of the dynamic control proposed in above wavelet program, we should supplement those calculations of equivalent transverse force from the control voltage and its resulting nonlinear deflection. Here, the model of equivalent bending deformation of a plate attached with piezoelectric actuators on its surfaces is of the form (see [51, 55] for example)

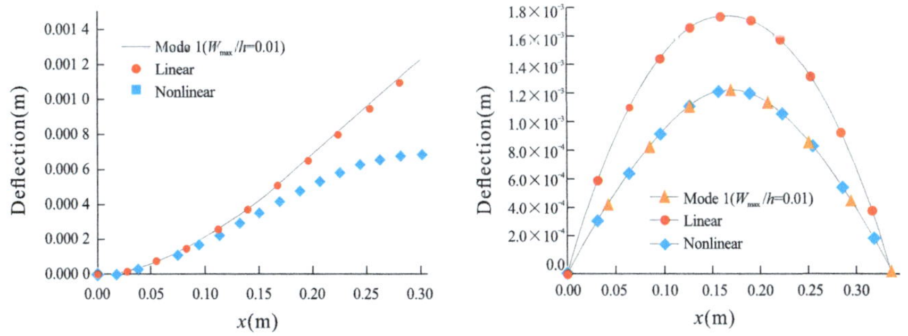

Fig. 8.15 Identified deflections of beam-type plates with large deflection of order 1 by means of the piezoelectric sensors using wavelet-based approach (Daubechies wavelet: $N = 5$; $n = 3$): Cantilever (Left); Simply supported (Right)

$$f(x,t) = -e_{31} r_a \frac{\partial^2 V(x,t)}{\partial x^2}$$

in which r_a is the distance of piezoelectric actuator measured from the mid-plane. This equation will be used to calculate the equivalent force exerted by the piezoelectric actuators.

After that, the mode expansion method is chosen to discretize the spatial part of the initial-boundary-value problem in Eq. (8.41), while the Wilson-θ method is applied to solve the resulting nonlinear initial-value problem, and Δt is chosen as the increment step of time variable. Eventually, we can get the evolution of deformation state of $w(x,t)$ and $\dot{w}(x,t)$ of the controlled plates step by step.

Usually, the natural frequencies of the control system are related to the gain G_1, while the gain G_2 influences the damping of the control system. In the following simulation examples, we fix $G_1 = 0$ and allow G_2 to vary. When initial disturbance is introduced to excite the controlled plate, numerical simulation procedures are performed following the control program mentioned above. Figure 8.16 exhibits the responses of the maximum deflection at the free end for the cantilevered plate and that at the middle point for the simply supported plate. We can see from Fig. 8.16 that amplitudes of the vibration decrease very fast, implying the great efficiency of the proposed control approach. We can also find from the simulation results that there exists notable difference between the linear and nonlinear control programs as the deflection of plates is large enough. At the same time, the control voltages in the nonlinear control program of the plates with large deflection are smaller than those applied to the corresponding actuators in the linear program of plates with small deflection as shown in Fig. 8.17.

Based on wavelet approximation of functions, we have developed the wavelet-based active strategy for the vibration control of laminated beam-type piezoelectric plates. The low-pass filtering property of scaling function transform makes the proposed control approach possess the ability of automatically filtering out noise-like

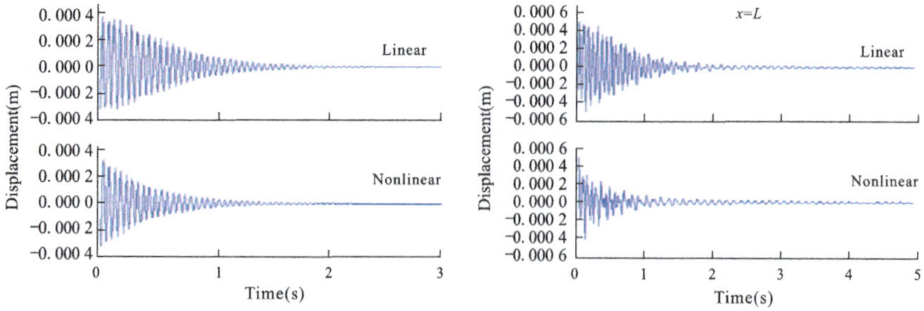

Fig. 8.16 Comparison of responses of the maximum deflection of the beam-type plates with disturbance of free vibration of mode one by means of the linear and nonlinear programs of the control system (Daubechies wavelet: $N = 5$; $n = 3$): Cantilever and initial disturbance of model 4 ($G_1 = 0$; $G_2 = 0{:}4$, $w_{\max}/h = 0{:}5$) (Left); Simply supported and initial disturbance of mode 2 ($G_1 = 0$; $G_2 = 0{:}2$,$w_{\max}/h = 0{:}3$) (Right)

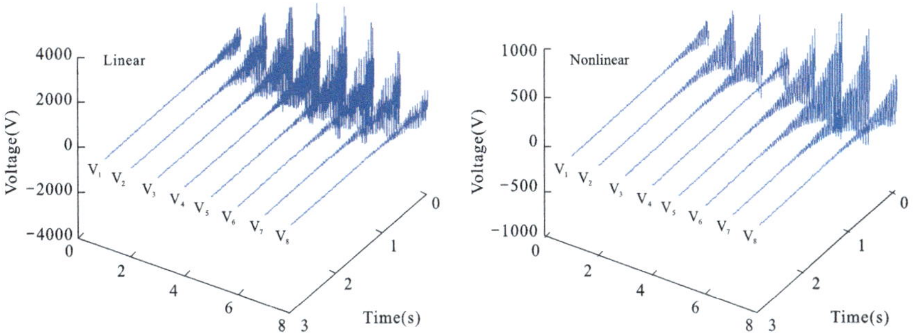

Fig. 8.17 Comparison of responses of the control voltages applied on the piezoelectric actuators attached to the simply supported beam-type plate from disturbance of free vibration mode of order 2 by means of the linear and nonlinear programs of the control system ($G_1 = 0$; $G_2 = 0.2$, $w_{\max}/h = 0.3$, Daubechies wavelet: $N = 5$; $n = 3$)

signals in the feedback control loop. Therefore, there is not any control instability occurring in the proposed dynamic control system during the control process as shown in the response of displacement and control voltages in Figs. 8.16 and 8.17.

It should be pointed out that after the simulation results of the above wavelet dynamic control program for a cantilevered beam-type plate with small deflection were published [53], it attracted the attention of the control scholars who designed an experimental prototype of such dynamic control system to show the feasibility of the dynamic control [55]. In their experiment, the suppression duration is smaller than one of the simulations of wavelet method after the identification of deflection used by the wavelet method is replaced by a technique relative to the light-ranging [55], from where the wavelet dynamic control model was experimentally demonstrated to be efficient and feasible.

8.4 Multidimensional Nonlinear Schrödinger Equations

8.4.1 Schrödinger's Governing Equation

The time-dependent Schrödinger equation with nonlinear potential functions is one of the most important mathematical models in modern science due to its wide applications in many different fields, such as nonlinear optics, plasma physics, and semiconductor industry [58–62]. In this section, we use the generalized Coiflets-based method to quantitatively solve the 3D nonlinear Schrödinger equation. Such three-dimensional generalized nonlinear Schrödinger equation (GNLSE) [63–94] can be written in the form

$$\mathrm{i}\frac{\partial\psi(\mathbf{x},t)}{\partial t} + \alpha(t)\nabla^2\psi + v(\mathbf{x},t)\psi + f(|\psi|^2)\psi + \mathrm{i}g(|\psi|^2)$$

$$+h_1(\mathbf{x},t)+\mathrm{i}h_2(\mathbf{x},t)=0,\ \mathbf{x}\in[0,\ 1]^3,\ t\geq 0 \tag{8.85}$$

with the initial condition

$$\psi(\mathbf{x},0)=R_1(\mathbf{x})+\mathrm{i}R_2(\mathbf{x}),\ \mathbf{x}\in[0,\ 1]^3 \tag{8.86}$$

and the Dirichlet boundary condition

$$\begin{cases}\psi(0,y,z,t)=r_{01}(y,z,t)+\mathrm{i}r_{02}(y,z,t),\ \psi(1,y,z,t)=s_{01}(y,z,t)+\mathrm{i}s_{02}(y,z,t)\\ \psi(x,0,z,t)=r_{11}(x,z,t)+\mathrm{i}r_{12}(x,z,t),\ \psi(x,1,z,t)=s_{11}(x,z,t)+\mathrm{i}s_{12}(x,z,t)\\ \psi(x,y,0,t)=r_{21}(x,y,t)+\mathrm{i}r_{22}(x,y,t),\ \psi(x,y,1,t)=s_{21}(x,y,t)+\mathrm{i}s_{22}(x,y,t)\end{cases}. \tag{8.87}$$

In Eqs. (8.85)–(8.87), $i=\sqrt{-1}$ is the complex symbol; $\psi(\mathbf{x},t)$ with vector $\mathbf{x}=\{x,y,z\}$ is a complex-valued unknown function; $\nabla^2=\partial^2/\partial x^2+\partial^2/\partial y^2+\partial^2/\partial z^2$ is the Laplace operator; and $\alpha(t)$, R_j, r_{ij}, and s_{ij} are known real functions for $i=0,1,2$ and $j=1,2$. In GNLSE (8.85), the external potential $v(\mathbf{x},t)$ is a given real function, whose specific form is dependent on different problems. For example, it is usually chosen as either a harmonic confining potential.

$$v(\mathbf{x},t)=-|\mathbf{x}|^2/2,$$

or an optical lattice potential

$$v(\mathbf{x},t)=A_1\cos(L_1x)+A_2\cos(L_2y)+A_3\cos(L_3z)$$

with constants A_n, L_n and $n=1,2,3$, for studying the Bose–Einstein condensation [62, 83]; real-valued smooth functions, $f(\rho)$ and $g(\rho)$, are defined through the probability density $\rho=|\psi|^2$ and determined by specific applications.

It can be found that in this study, the most popular nonlinear term $f(\rho)=\lambda\rho$ with constant λ [58–67] has been extended to several general forms, which can be chosen as arbitrary as needed, such as $f(\rho)=c_0\rho^s$, $f(\rho)=c_1(1-e^{-\rho})$, $f(\rho)=\rho/(1+\rho)$, and $f(\rho)=\ln(1+\rho)$, where c_0, s, and c_1 are real constants [83, 93]. Moreover, the general damping term $g(\rho)$ in the GNLSE (8.85) is usually ignored in many studies [65–73, 75–82, 84–91]. Because the damping effect may play an important role and should not be ignored in some physical process [64, 74, 83, 92, 93], such as the inelastic collisions in Bose–Einstein condensation [64, 83], the damping effect should be taken into account in these situations.

8.4.2 Solution of the Generalized Nonlinear Schrödinger Equation

At the beginning of the solution to the three-dimensional generalized nonlinear Schrödinger Eq. (8.85), we introduce $\psi(\mathbf{x}, t) = u_1(\mathbf{x}, t) + \mathrm{i}u_2(\mathbf{x}, t)$ where $u_1(\mathbf{x}, t)$ and $u_2(\mathbf{x}, t)$ are real-valued functions. Then, the governing equation can be reduced into the following two coupled time-dependent nonlinear partial differential equations

$$\begin{aligned}\frac{\partial u_n(\mathbf{x}, t)}{\partial t} &= (-1)^n \alpha(t)\nabla^2 u_m + (-1)^n v(\mathbf{x}, t)u_m + (-1)^n f(u_n^2 + u_m^2)u_m \\ &\quad + (-1)^n h_m(\mathbf{x}, t) - g(u_n^2 + u_m^2)u_n \end{aligned} \tag{8.88}$$

in which $n = 1, 2$ and $m = 3 - n$.

Following the generalized Coiflets-based approximations of Eqs. (3.101) and (3.113), the unknown functions $u_n(\mathbf{x}, t)$ and nonlinear terms $f(u_n^2 + u_m^2)u_m(\mathbf{x}, t)$, $g(u_n^2 + u_m^2)u_n(\mathbf{x}, t)$, $v(\mathbf{x}, t)u_m(\mathbf{x}, t)$, and $h_m(\mathbf{x}, t)$ can be expressed as follows:

$$u_n(\mathbf{x}, t) \approx \sum_{k_0=0}^{2^{j_x}} \sum_{l_0=0}^{2^{j_y}} \sum_{m_0=0}^{2^{j_z}} u_n(k_0/2^{j_x}, l_0/2^{j_y}, m_0/2^{j_z}, t)\phi_{j_x,k_0}(x)\phi_{j_y,l_0}(y)\phi_{j_z,m_0}(z) \tag{8.89}$$

$$\begin{aligned} v(\mathbf{x}, t)u_m(\mathbf{x}, t) \approx \sum_{k_0=0}^{2^{j_x}} \sum_{l_0=0}^{2^{j_y}} \sum_{m_0=0}^{2^{j_z}} & v(k_0/2^{j_x}, l_0/2^{j_y}, m_0/2^{j_z}, t) \\ & u_m(k_0/2^{j_x}, l_0/2^{j_y}, m_0/2^{j_z}, t)\phi_{j_x,k_0}(x)\phi_{j_y,l_0}(y)\phi_{j_z,m_0}(z) \end{aligned} \tag{8.90}$$

$$\begin{aligned} f(u_n^2 + u_m^2)u_m(\mathbf{x}, t) \approx \sum_{k_0=0}^{2^{j_x}} \sum_{l_0=0}^{2^{j_y}} \sum_{m_0=0}^{2^{j_z}} & f[u_n^2(k_0/2^{j_x}, l_0/2^{j_y}, m_0/2^{j_z}, t) \\ & + u_m^2(k_0/2^{j_x}, l_0/2^{j_y}, m_0/2^{j_z}, t)] \\ & \times u_m(k_0/2^{j_x}, l_0/2^{j_y}, m_0/2^{j_z}, t)\phi_{j_x,k_0}(x)\phi_{j_y,l_0}(y)\phi_{j_z,m_0}(z) \end{aligned} \tag{8.91}$$

$$\begin{aligned} g(u_n^2 + u_m^2)u_n(\mathbf{x}, t) \approx \sum_{k_0=0}^{2^{j_x}} \sum_{l_0=0}^{2^{j_y}} \sum_{m_0=0}^{2^{j_z}} & g[u_n^2(k_0/2^{j_x}, l_0/2^{j_y}, m_0/2^{j_z}, t) \\ & + u_m^2(k_0/2^{j_x}, l_0/2^{j_y}, m_0/2^{j_z}, t)] \\ & \times u_n(k_0/2^{j_x}, l_0/2^{j_y}, m_0/2^{j_z}, t)\phi_{j_x,k_0}(x)\phi_{j_y,l_0}(y)\phi_{j_z,m_0}(z) \end{aligned} \tag{8.92}$$

$$h_m(\mathbf{x}, t) \approx \sum_{k_0=0}^{2^{j_x}} \sum_{l_0=0}^{2^{j_y}} \sum_{m_0=0}^{2^{j_z}} h_m(k_0/2^{j_x}, l_0/2^{j_y}, m_0/2^{j_z}, t)\phi_{j_x,k_0}(x)\phi_{j_y,l_0}(y)\phi_{j_z,m_0}(z). \tag{8.93}$$

Substituting Eqs. (8.89)–(8.93) into Eq. (8.88), and then multiplying both sides of the resulting equation by $\phi_{j_x,k_1}(x)\phi_{j_y,l_1}(y)\phi_{j_z,m_1}(z)$, $k_1 = 1, 2, \cdots 2^{j_x} - 1$, $l_1 = 1, 2, \cdots 2^{j_y} - 1$, $m_1 = 1, 2, \cdots 2^{j_z} - 1$, respectively, and performing the integration over the domain $[0, 1]^3$, one can obtain a system of ordinary differential equations. Considering boundary conditions of Eq. (8.87), these ordinary differential equations can be written as follows:

$$\frac{d\mathbf{U}(t)}{dt} = \frac{d}{dt}\begin{bmatrix} \mathbf{U}_1(t) \\ \mathbf{U}_2(t) \end{bmatrix} \approx \mathbf{G}[\mathbf{U}(t), t] = \begin{bmatrix} \mathbf{G}_1[\mathbf{U}_1(t), \mathbf{U}_2(t), t] \\ \mathbf{G}_2[\mathbf{U}_1(t), \mathbf{U}_2(t), t] \end{bmatrix} \tag{8.94}$$

and

$$\begin{aligned} \mathbf{G}_n = {} & (-1)^n \alpha(t)\mathbf{A}^{-1}\mathbf{B}\mathbf{U}_m + (-1)^n \mathbf{V}^* \mathbf{U}_m + (-1)^n f(\mathbf{U}_n{}^* \mathbf{U}_n + \mathbf{U}_m{}^* \mathbf{U}_m)^* \mathbf{U}_m \\ & + (-1)^n \mathbf{H}_m - g(\mathbf{U}_n{}^* \mathbf{U}_n + \mathbf{U}_m{}^* \mathbf{U}_m)^* \mathbf{U}_n + \mathbf{A}^{-1}\mathbf{C}_n(t) \end{aligned} \tag{8.95}$$

in which vectors

$$\mathbf{U}_n(t) = \{u_{n,p}(t) = u_n(k_0/2^{j_x}, l_0/2^{j_y}, m_0/2^{j_z}, t)\}^{\mathrm{T}},$$

$$\mathbf{V}(t) = \{v_p(t) = v(k_0/2^{j_x}, l_0/2^{j_y}, m_0/2^{j_z}, t)\}^{\mathrm{T}},$$

$$\mathbf{H}_m(t) = \{h_{m,p}(t) = h_m(k_0/2^{j_x}, l_0/2^{j_y}, m_0/2^{j_z}, t)\}^{\mathrm{T}},$$

and matrices

$$\mathbf{A} = \{a_{op} = \Gamma_{k_1,k_0}^{j,0}\Gamma_{l_1,l_0}^{j,0}\Gamma_{m_1 m_0}^{j,0}\},$$

$$\mathbf{B} = \{b_{op} = \Gamma_{k_1,k_0}^{j,2}\Gamma_{l_1,l_0}^{j,0}\Gamma_{m_1 m_0}^{j,0} + \Gamma_{k_1,k_0}^{j,0}\Gamma_{l_1,l_0}^{j,2}\Gamma_{m_1 m_0}^{j,0} + \Gamma_{k_1,k_0}^{j,0}\Gamma_{l_1,l_0}^{j,0}\Gamma_{m_1 m_0}^{j,2}\},$$

as well as subscripts

$$k_0, k_1 = 1, 2, \cdots 2^{j_x} - 1, l_0, l_1 = 1, 2, \cdots 2^{j_y} - 1, m_0, m_1 = 1, 2, \cdots 2^{j_z} - 1,$$

$o = (2^{j_z} - 1)[(2^{j_y} - 1)(k_1 - 1) + l_1 - 1] + m_1$, $p = (2^{j_z} - 1)[(2^{j_y} - 1)(k_0 - 1) + l_0 - 1] + m_0$, $n = 1, 2$, $m = 3-n$.

In Eqs. (8.94) and (8.95), vector operational rules $\mathbf{P} * \mathbf{Q} = \{p_k q_k\}^{\mathrm{T}}$ and $f(\mathbf{P}) = \{f(p_k)\}^{\mathrm{T}}$ will always hold for vectors $\mathbf{P} = \{p_k\}^{\mathrm{T}}$ and $\mathbf{Q} = \{q_k\}^{\mathrm{T}}$, and the generalized connection coefficient $\Gamma_{k_1,k}^{j,\alpha} = \int_0^1 d^\alpha \phi_{j,k}(x)/dx^\alpha \phi_{j,k_1}(x)dx$ can be exactly and readily obtained without numerical integral but based on the database independent of the specific problems [95]. In addition, the vector $\mathbf{C}_n(t)$ determined by boundary conditions can be expressed as

$$\mathbf{C}_n(t) = \left\{ c_{n,o} = \sum_{l_0=0}^{2^{j_y}} \sum_{m_0=0}^{2^{j_z}} [\sigma_{n,o,1}(l_0, m_0, t) + \omega_{n,o,1}(l_0, m_0, t)] \right.$$
$$+ \sum_{k_0=1}^{2^{j_x}-1} \sum_{m_0=1}^{2^{j_z}-1} [\sigma_{n,o,2}(k_0, m_0, t) + \omega_{n,o,2}(k_0, m_0, t)]$$
$$\left. + \sum_{k_0=1}^{2^{j_x}-1} \sum_{l_0=1}^{2^{j_y}-1} [\sigma_{n,o,3}(k_0, l_0, t) + \omega_{n,o,3}(k_0, l_0, t)] \right\}^{\mathrm{T}} \tag{8.96}$$

$$\begin{aligned}\sigma_{n,o,1}(l_0, m_0, t) = {} & (-1)^n \alpha(t) r_{0n}(l_0/2^{j_y}, m_0/2^{j_z}, t)(\Gamma_{k_1,0}^{j,2} \Gamma_{l_1,l_0}^{j,0} \Gamma_{m_1,m_0}^{j,0} \\ & + \Gamma_{k_1,0}^{j,0} \Gamma_{l_1,l_0}^{j,2} \Gamma_{m_1,m_0}^{j,0} + \Gamma_{k_1,0}^{j,0} \Gamma_{l_1,l_0}^{j,0} \Gamma_{m_1,m_0}^{j,2}) \\ & + \Gamma_{k_1,0}^{j,0} \Gamma_{l_1,l_0}^{j,0} \Gamma_{m_1,m_0}^{j,0} \{(-1)^n \nu(0, l_0/2^{j_y}, m_0/2^{j_z}, t) r_{0m}(l_0/2^{j_y}, m_0/2^{j_z}, t) \\ & - dr_{0n}(l_0/2^{j_y}, m_0/2^{j_z}, t)/dt + (-1)^n f[r_{0n}^2(l_0/2^{j_y}, m_0/2^{j_z}, t) \\ & + r_{0m}^2(l_0/2^{j_y}, m_0/2^{j_z}, t)] r_{0m}(l_0/2^{j_y}, m_0/2^{j_z}, t) - g[r_{0n}^2(l_0/2^{j_y}, m_0/2^{j_z}, t) \\ & + r_{0m}^2(l_0/2^{j_y}, m_0/2^{j_z}, t)] r_{0n}(l_0/2^{j_y}, m_0/2^{j_z}, t) + (-1)^n h_m(0, l_0/2^{j_y}, m_0/2^{j_z}, t)\}\end{aligned} \tag{8.97}$$

$$\begin{aligned}\omega_{n,o,1}(l_0, m_0, t) = {} & (-1)^n \alpha(t) s_{0n}(l_0/2^{j_y}, m_0/2^{j_z}, t)(\Gamma_{k_1,2^{j_x}}^{j,2} \Gamma_{l_1,l_0}^{j,0} \Gamma_{m_1,m_0}^{j,0} \\ & + \Gamma_{k_1,2^{j_x}}^{j,0} \Gamma_{l_1,l_0}^{j,2} \Gamma_{m_1,m_0}^{j,0} + \Gamma_{k_1,2^{j_x}}^{j,0} \Gamma_{l_1,l_0}^{j,0} \Gamma_{m_1,m_0}^{j,2}) + \Gamma_{k_1,2^{j_x}}^{j,0} \Gamma_{l_1,l_0}^{j,0} \Gamma_{m_1,m_0}^{j,0} \\ & \{(-1)^n \nu(1, l_0/2^{j_y}, m_0/2^{j_z}, t) s_{0m}(l_0/2^{j_y}, m_0/2^{j_z}, t) \\ & - ds_{0n}(l_0/2^{j_y}, m_0/2^{j_z}, t)/dt + (-1)^n f[s_{0n}^2(l_0/2^{j_y}, m_0/2^{j_z}, t) \\ & + s_{0m}^2(l_0/2^{j_y}, m_0/2^{j_z}, t)] s_{0m}(l_0/2^{j_y}, m_0/2^{j_z}, t) \\ & - g[s_{0n}^2(l_0/2^{j_y}, m_0/2^{j_z}, t) + s_{0n}^m(l_0/2^{j_y}, m_0/2^{j_z}, t)]\end{aligned} \tag{8.98}$$

$$s_{0n}(l_0/2^{j_y}, m_0/2^{j_z}, t) + (-1)^n h_m(1, l_0/2^{j_y}, m_0/2^{j_z}, t)\},$$

$$\begin{aligned}\sigma_{n,o,2}(k_0, m_0, t) = {} & (-1)^n \alpha(t) r_{1n}(k_0/2^{j_x}, m_0/2^{j_z}, t)(\Gamma_{k_1,k_0}^{j,2} \Gamma_{l_1,0}^{j,0} \Gamma_{m_1,m_0}^{j,0} \\ & + \Gamma_{k_1,k_0}^{j,0} \Gamma_{l_1,0}^{j,2} \Gamma_{m_1,m_0}^{j,0} + \Gamma_{k_1,k_0}^{j,0} \Gamma_{l_1,0}^{j,0} \Gamma_{m_1,m_0}^{j,2}) \\ & + \Gamma_{k_1,k_0}^{j,0} \Gamma_{l_1,0}^{j,0} \Gamma_{m_1,m_0}^{j,0} \{(-1)^n \nu(k_0/2^{j_x}, 0, m_0/2^{j_z}, t) r_{1m}(k_0/2^{j_x}, m_0/2^{j_z}, t) \\ & - dr_{1n}(k_0/2^{j_x}, m_0/2^{j_z}, t)/dt + (-1)^n f[r_{1n}^2(k_0/2^{j_x}, m_0/2^{j_z}, t) \\ & + r_{1m}^2(k_0/2^{j_x}, m_0/2^{j_z}, t)] r_{1m}(k_0/2^{j_x}, m_0/2^{j_z}, t) - g[r_{1n}^2(k_0/2^{j_x}, m_0/2^{j_z}, t) \\ & + r_{1m}^2(k_0/2^{j_x}, m_0/2^{j_z}, t)] r_{1n}(k_0/2^{j_x}, m_0/2^{j_z}, t)\end{aligned}$$

$$+ (-1)^n h_m(k_0/2^{jx}, 0, m_0/2^{jz}, t)\}, \tag{8.99}$$

$$\begin{aligned}
\omega_{n,o,2}(k_0, m_0, t) = & (-1)^n \alpha(t) s_{1n}(k_0/2^{jx}, m_0/2^{jz}, t)(\Gamma^{j,2}_{k_1,k_0}\Gamma^{j,0}_{l_1,2^{jy}}\Gamma^{j,0}_{m_1,m_0} \\
& + \Gamma^{j,0}_{k_1,k_0}\Gamma^{j,2}_{l_1,2^{jy}}\Gamma^{j,0}_{m_1,m_0} + \Gamma^{j,0}_{k_1,k_0}\Gamma^{j,0}_{l_1,2^{jy}}\Gamma^{j,2}_{m_1,m_0}) + \Gamma^{j,0}_{k_1,k_0}\Gamma^{j,0}_{l_1,2^{jy}}\Gamma^{j,0}_{m_1,m_0} \\
& \{(-1)^n v(k_0/2^{jx}, 1, m_0/2^{jz}, t) s_{1m}(k_0/2^{jx}, m_0/2^{jz}, t) \\
& - ds_{1n}(k_0/2^{jx}, m_0/2^{jz}, t)/dt + (-1)^n f[s^2_{1n}(k_0/2^{jx}, m_0/2^{jz}, t) \\
& + s^2_{1m}(k_0/2^{jx}, m_0/2^{jz}, t)] s_{1m}(k_0/2^{jx}, m_0/2^{jz}, t) - g[s^2_{1n}(k_0/2^{jx}, m_0/2^{jz}, t) \\
& + s^2_{1m}(k_0/2^{jx}, m_0/2^{jz}, t)] s_{1n}(k_0/2^{jx}, m_0/2^{jz}, t) \\
& + (-1)^n h_m(k_0/2^{jx}, 1, m_0/2^{jz}, t)\},
\end{aligned} \tag{8.100}$$

$$\begin{aligned}
\sigma_{n,o,3}(k_0, l_0, t) = & (-1)^n \alpha(t) r_{2n}(k_0/2^{jx}, l_0/2^{jy}, t)(\Gamma^{j,2}_{k_1,k_0}\Gamma^{j,0}_{l_1,l_0}\Gamma^{j,0}_{m_1,0} \\
& + \Gamma^{j,0}_{k_1,k_0}\Gamma^{j,2}_{l_1,l_0}\Gamma^{j,0}_{m_1,0} + \Gamma^{j,0}_{k_1,k_0}\Gamma^{j,0}_{l_1,l_0}\Gamma^{j,2}_{m_1,0}) + \Gamma^{j,0}_{k_1,k_0}\Gamma^{j,0}_{l_1,l_0}\Gamma^{j,0}_{m_1,0} \\
& \{(-1)^n v(k_0/2^{jx}, l_0/2^{jy}, 0, t) r_{2m}(k_0/2^{jx}, l_0/2^{jy}, t) \\
& - dr_{2n}(k_0/2^{jx}, l_0/2^{jy}, t)/dt + (-1)^n f[r^2_{2n}(k_0/2^{jx}, l_0/2^{jy}, t) \\
& + r^2_{2m}(k_0/2^{jx}, l_0/2^{jy}, t)] r_{2m}(k_0/2^{jx}, l_0/2^{jy}, t) - g[r^2_{2n}(k_0/2^{jx}, l_0/2^{jy}, t) \\
& + r^2_{2m}(k_0/2^{jx}, l_0/2^{jy}, t)] r_{2n}(k_0/2^{jx}, l_0/2^{jy}, t) \\
& + (-1)^n h_m(k_0/2^{jx}, l_0/2^{jy}, 0, t)\},
\end{aligned} \tag{8.101}$$

$$\begin{aligned}
\omega_{n,o,3}(k_0, l_0, t) = & (-1)^n \alpha(t) s_{2n}(k_0/2^{jx}, l_0/2^{jy}, t)(\Gamma^{j,2}_{k_1,k_0}\Gamma^{j,0}_{l_1,l_0}\Gamma^{j,0}_{m_1,2^{jz}} \\
& + \Gamma^{j,0}_{k_1,k_0}\Gamma^{j,2}_{l_1,l_0}\Gamma^{j,0}_{m_1,2^{jz}} + \Gamma^{j,0}_{k_1,k_0}\Gamma^{j,0}_{l_1,l_0}\Gamma^{j,2}_{m_1,2^{jz}}) + \Gamma^{j,0}_{k_1,k_0}\Gamma^{j,0}_{l_1,l_0}\Gamma^{j,0}_{m_1,2^{jz}} \\
& \{(-1)^n v(k_0/2^{jx}, l_0/2^{jy}, 1, t) s_{2m}(k_0/2^{jx}, l_0/2^{jy}, t) \\
& - ds_{2n}(k_0/2^{jx}, l_0/2^{jy}, t)/dt + (-1)^n f[s^2_{2n}(k_0/2^{jx}, l_0/2^{jy}, t) \\
& + s^2_{2m}(k_0/2^{jx}, l_0/2^{jy}, t)] s_{2m}(k_0/2^{jx}, l_0/2^{jy}, t) - g[s^2_{2n}(k_0/2^{jx}, l_0/2^{jy}, t) \\
& + s^2_{2m}(k_0/2^{jx}, l_0/2^{jy}, t)] s_{2n}(k_0/2^{jx}, l_0/2^{jy}, t) \\
& + (-1)^n h_m(k_0/2^{jx}, l_0/2^{jy}, 1, t)\}.
\end{aligned} \tag{8.102}$$

Note that in the semi-discretization system Eq. (8.94) of the three-dimensional generalized nonlinear Schrödinger Eq. (8.85), both **A** and **B** are constant matrices, which are completely independent of the unknown vector $\mathbf{U}(t)$ and time t. Thus, in subsequent time integration for solving nonlinear ordinary differential Eq. (8.94), no matrix generated from spatial discretization needs updating. In addition, considering the orthogonality of the generalized Coiflets-type scaling function $\varphi(x)$ and the modified wavelet scaling function $\phi(x)$ in Sect. 5.1, it is easy to obtain that the matrix **A** is almost a diagonal matrix [95]. Therefore, calculating the inverse of matrix **A** used in Eq. (8.97) will only cost very few computing resources.

Finally, the approximate solution of Eq. (8.85) is obtained by employing a time integration scheme to solve the nonlinear ordinary differential Eq. (8.94). In this study, the classical fourth-order explicit Runge–Kutta method is adopted.

8.4.3 Numerical Examples

In this subsection, we use several numerical examples to demonstrate the accuracy and applicability of the wavelet-based method for solving two- and three-dimensional generalized nonlinear Schrödinger Eq. (8.85). To effectively evaluate the performance of numerical solutions, we introduce the error norms $E_{\max}$ and E_2, and the corresponding convergence rates in space $R_{\max}$ and R_2, which are, respectively, defined by [66, 75, 77, 80]

$$E_{\max}(N_s) = \max_k\{|\psi_{\text{num}}(\mathbf{x}_k, t) - \psi_{\text{exact}}(\mathbf{x}_k, t)|\} \tag{8.103}$$

$$E_2(N_s) = \sqrt{\sum_{k=1}^{N_s} |\psi_{\text{num}}(\mathbf{x}_k, t) - \psi_{\text{exact}}(\mathbf{x}_k, t)|^2 / N_s} \tag{8.104}$$

$$R_{\max(2)} = \frac{\ln[E_{\max(2)}(N_{s2})] - \ln[E_{\max(2)}(N_{s1})]}{\ln(N_{s2}/N_{s1})} \tag{8.105}$$

in which ψ_{num} and ψ_{exact} are the numerical and exact solutions, respectively, and $N_s = 2^j$, $N_{s1} = 2^{j_1}$ and $N_{s2} = 2^{j_2}$ are the numbers of grid points in one spatial dimension.

Example 8.8 Consider the two-dimensional linear Schrödinger equation with trapping potential [63, 67, 75, 81]

$$\mathrm{i}\frac{\partial \psi}{\partial t} + \nabla^2\psi + [3 - 2\tanh^2(x) - 2\tanh^2(y)]\psi = 0, \ \ \mathbf{x} \in [0,\ 1]^2, \ \ t \geq 0 \tag{8.106}$$

with the initial and boundary conditions extracted from its exact solution $\psi(x, y, t) = i\exp(it)/[\cosh(x)\cosh(y)]$.

In order to give a quantitative comparison of accuracy between the wavelet-based method and other methods employed in solving this example, here, the same time step $\Delta t = 10^{-4}$ is taken. Figure 8.18 shows the scaling dependence of error norm E_2 of the wavelet-based solution at time $t = 1$ on numbers of grid points of N_s, containing the results of the wavelet-based method, and those obtained by using the Peaceman–Rachford alternating direction implicit method (PR-ADI) [63, 75], semi-discrete high-order compact scheme (SD-HOC) [67, 75], high-order compact alternating direction implicit method (HOC-ADI) [75], and the linearized compact alternating direction implicit method (L-HOC-ADI) [81]. According to Fig. 8.18, one sees that the scaling dependence on the accuracy of the wavelet-based method applied in this example is about N_s^{-8}, which is much lower than all other methods, where the wavelet-based method has the highest accuracy with the lowest calculation when the error is lower than 10^{-7}.

Example 8.9 We consider the two-dimensional cubic nonlinear Schrödinger equation [91]

$$\mathrm{i}\frac{\partial \psi}{\partial t} + \frac{1}{2}\nabla^2 \psi + \psi - |\psi|^2 \psi = 0, \ \mathbf{x} \in [0,\ 2\pi]^2,\ t \geq 0 \tag{8.107}$$

subjected to the initial and boundary conditions extracted from the exact solution $\psi(x, y, t) = \exp[\mathrm{i}(x + y - t)]$.

In Table 8.9, we list the error norms E_2 of numerical solutions at time $t = 1$ under $\Delta t = 1/25{,}000$ achieved, respectively, from the wavelet-based method, and the higher order compact alternating direction implicit method (HOC-ADI) [91]. From this table, we find that the wavelet-based method has accuracy higher than the HOC-ADI.

Example 8.10 Consider the two-dimensional Gross–Pitaevskii equation [85]

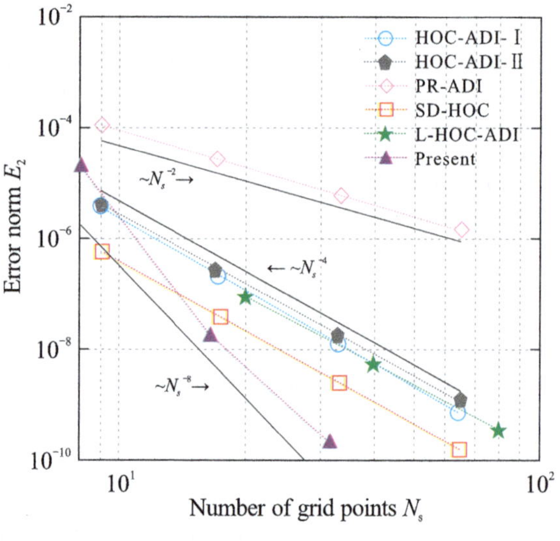

Fig. 8.18 Scaling dependence of error norm E_2 of numerical solutions at $t = 1$ on grid number N_s for solving Example 8.8 when $\Delta t = 0.0001$ is employed. "Present" stands for the wavelet-based method

Table 8.9 Error norm E_2 of numerical solutions at $t = 1$ with $\Delta t = 1/25{,}000$ for Example 8.9

N_s	Wavelet based		HOC-ADI [91]	
	E_2	R_2	E_2	R_2
8	7.10E-03	–	1.13E-03	–
16	1.95E-05	8.51	7.38E-05	3.94
32	2.02E-07	6.59	4.74E-06	3.96
64	3.04E-09	6.06	3.00E-07	3.98

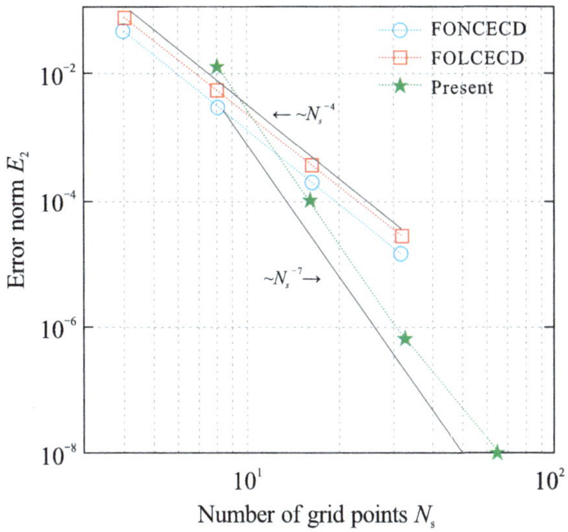

Fig. 8.19 Scaling dependence of error norm E_2 of numerical solutions at $t = 1$ on grid number N_s for solving Example 8.10 when $\Delta t = 0.001$ is employed. "Present" stands for the wavelet-based method

$$\mathrm{i}\frac{\partial \psi}{\partial t} + \nabla^2\psi + \beta|\psi|^2\psi = 0,\ \ \mathbf{x} \in [0,\ 2\pi]^2,\ t \geq 0 \tag{8.108}$$

with $\beta = 2$, and the initial and Dirichlet boundary conditions extracted from the exact solution $\psi(x, y, t) = \exp\{\mathrm{i}[x + y - (2 - \beta)t]\}$.

Figure 8.19 shows the error norm $E_{\max}$ comparison of numerical solutions at time $t = 1$ under $\Delta t = 0.001$ obtained by using the wavelet-based method, the fourth-order nonlinear compact and energy conservative difference scheme (FONCECD) [85], and the fourth-order linearized compact and energy conservative difference scheme (FOLCECD) [85]. It is also found from Fig. 8.19 that the wavelet-based method has the highest accuracy.

Example 8.11 We consider the two-dimensional nonlinear Schrödinger equation [66, 80, 81, 91]

$$\begin{cases} \mathrm{i}\dfrac{\partial \psi}{\partial t} + \dfrac{1}{2}\nabla^2\psi - [1 - \sin^2(x)\sin^2(y)]\psi(x, 0, t)\psi - |\psi|^2\psi = 0, & \mathbf{x} \in [0,\ 2\pi]^2,\ t \geq 0 \\ \psi(0, y, t) = \psi(\pi, y, t) = \psi(x, 0, t) = \psi(x, \pi, t), \ \ \psi(x, y, 0) = \sin(x)\sin(y) & \end{cases} \tag{8.109}$$

which has the exact solution $\psi(x, y, t) = \sin(x)\sin(y)\exp(-2\mathrm{i}t)$.

Figure 8.20 plots the error norms of the present wavelet solutions with $\Delta t = 1/5000$ as a function of the number of grid points in space N_s, as well as those obtained by using the split-step finite difference method (SSFD) [66], HOC-ADI [80, 81], linearized alternating direction implicit scheme (L-ADI) [81], L-HOC-ADI [81], and CCD-PRADI [91].

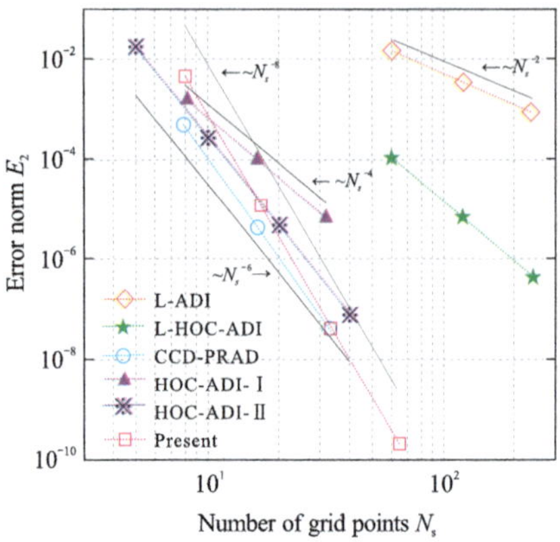

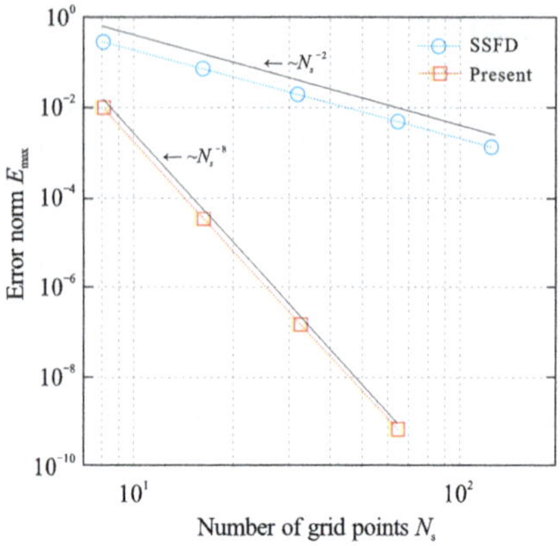

Fig. 8.20 Scaling dependence of error norm E_2 of numerical solutions at $t = 1$ on grid number N_s for solving Example 8.11 when $\Delta t = 0.0002$ is employed. "Present" stands for the wavelet-based method

Example 8.12 Consider the two-dimensional nonlinear Schrödinger equation with damping

$$\mathrm{i}\frac{\partial \psi}{\partial t} + \frac{1}{2}\nabla^2\psi + (1 + e^{-2t})\psi - |\psi|^2\psi + \mathrm{i}\psi = 0, \ \mathbf{x} \in [0,\ 2\pi]^2, \ t \geq 0 \quad (8.110)$$

subjected to the initial and boundary conditions extracted from the exact solution $\psi(x, y, t) = \exp[\mathrm{i}(x + y) - t]$.

The error norms of the present wavelet solutions at time $t = 1$ and 3 under different space meshes are listed in Table 8.10.

Example 8.13 We consider the three-dimensional nonlinear Schrödinger equation [66].

$$\begin{cases} \mathrm{i}\dfrac{\partial \psi}{\partial t} + \dfrac{1}{2}\nabla^2\psi - [1 - \sin^2(x)\sin^2(y)\sin^2(z)]\psi - |\psi|^2\psi = 0, \ \mathbf{x} \in [0,\ 2\pi]^3, \ t \geq 0 \\ \psi(0, y, z, t) = \psi(2\pi, y, z, t) = \psi(x, 0, z, t) = \psi(x, 2\pi, z, t) = 0 \\ \psi(x, y, z, t) = \psi(x, y, 2\pi, t) = 0, \ \psi(x, y, z, 0) = \sin(x)\sin(y)\sin(z) \end{cases} \quad (8.111)$$

Table 8.10 Error norms of wavelet solutions at $t = 1$ and 3 with $\Delta t = 0.001$ for Example 8.12

N_s	$t = 1$				$t = 3$			
	E_{max}	R_{max}	E_2	R_2	E_{max}	R_{max}	E_2	R_2
8	5.82e-03	–	2.94e-03	–	8.63e-04	–	4.35e-04	–
16	2.59e-05	7.81	9.01e-06	8.35	3.84e-06	7.81	1.39e-06	8.29
32	2.37e-07	6.78	8.42e-08	6.74	7.54e-08	5.67	1.56e-08	6.47
64	3.82e-09	5.95	1.04e-09	6.34	6.05e-10	6.96	1.56e-10	6.65

Table 8.11 Error norm $E_{\max}$ and convergence rate $R_{\max}$ of numerical solutions at $t = 2$ for Example 8.13

N_s	Present ($\Delta t = 0.0002$)		SSFD (Δt=0.0001) [66]	
	$E_{\max}$	$R_{\max}$	$E_{\max}$	$R_{\max}$
8	1.28E-02	–	8.40E-01	–
16	4.56E-05	8.13	2.13E-01	1.98
32	1.44E-07	8.31	5.36E-02	1.99
64	–	–	1.34E-02	2.00
128	–	–	3.35E-03	2.00

whose exact solution is $\psi(x, y, z, t) = \sin(x)\sin(y)\sin(z)\exp(-\mathrm{i}5t/2)$.

In Table 8.11, we make a comparison of the error norm $E_{\max}$ between the wavelet-based solutions (marked by "Present") and those achieved from SSFD [66].

It can be seen from Tables 8.9, 8.10 and 8.11 and Figs. 8.18, 8.19 and 8.20 that the proposed space–time fully decoupled wavelet-based Galerkin method has a good accuracy, efficiency and stability. Moreover, these results also demonstrate that its order of convergence in space can go up to 6. In addition, we can see from the comparisons shown in Figs. 8.18, 8.19 and 8.20 and Tables 8.8, 8.9 and 8.10 that the present wavelet-based solutions under coarser mesh have a much better numerical accuracy than those obtained by many other existing numerical methods, including PR-ADI [63, 75], SD-HOC [67, 75], HOC-ADI [75, 80, 91], L-HOC-ADI [81], FONCECD [85], FOLCECD [85], SSFD [66], and L-ADI [81].

References

1. Hu HY (2006) Applied nonlinear dynamics. Aviation Industry Press, Beijing (in Chinese)
2. Baumann G (2005) Mathematica for theoretical physics: classical mechanics and nonlinear dynamics. Springer Science and Business Media Inc., New York
3. Cole JD (1968) Perturbation method in applied mathematics. Blaisdell Publishing Company, Waltham
4. Mei C (1972) Finite element displacement method for larger amplitude free oscillations of beams and plates. Comput Struct 3:163–174
5. Feng Y, Bert CW (1992) Application of the quadrature method to flexural vibration analysis of a geometrically nonlinear beam. Nonlinear Dyn 3:13–18
6. Bhashyam GR, Prathap G (1980) Galerkin finite element method for non-linear beam vibrations. J Sound Vib 72:191–203
7. Singh G, Sharma AK, Rao GV (1990) Large-amplitude free vibrations of beams-a discussion on various formulations and assumptions J. Sound Vib 142:77–85
8. Guo Q (2004) Further Development of spline-based differential quadrature method and nonlinear vibration analysis of beam (in Chinese). Tsinghua University, Beijing
9. Chen SH (2007) Theorem analysis method of strongly nonlinear vibration system. Science Press, Beijing (in Chinese)
10. Wazwaz AM (2005) The tanh and the sine-cosine methods for compact and noncompact solutions of the nonlinear Klein-Gordon equation. Appl Math Comput 167:1179–1195

11. Wazwaz AM (2005) The tanh and the sine-cosine methods for compact and noncompact solutions of the nonlinear Klein-Gordon equation. Appl Math Comput 167:1196–1210
12. Kaya D, El-Sayed SM (2004) A numerical solution of the Klein-Gordon equation and convergence of the decomposition method. Appl Math Comput 156:341–353
13. Basak KC, Ray PC, Bera RK (2009) Solution of non-linear Klein-Gordon equation with a quadratic non-linear term by Adomian decomposition method. Commun Nonlinear Sci Numer Simul 14:718–723
14. Chowdhury MSH, Hashim I (2009) Application of homotopy-perturbation method to Klein-Gordon and sine-Gordon equations. Chaos Solitons Fractals 39:1928–1935
15. Kanth A, Aruna K (2009) Differential transform method for solving the linear and nonlinear Klein-Gordon equation. Comput Phys Commun 180:708–711
16. Shakeri F, Dehghan M (2008) Numerical solution of the Klein-Gordon equation via He's variational iteration method. Nonlinear Dyn 51:89–97
17. Yin F, Song J, Lu F (2014) A coupled method of Laplace trans-form and Legendre wavelets for nonlinear Klein-Gordon equations. Math Methods Appl Sci 37:781–792
18. Shao W, Wu X (2014) The numerical solution of the nonlinear Klein-Gordon and sine-Gordon equations using the Chebyshev tau meshless method. Comput Phys Commun 185:1399–1409
19. Pekmen B, Tezer-Sezgin M (2012) Differential quadrature solution of nonlinear Klein-Gordon and sine-Gordon equations. Comput Phys Commun 183:1702–1713
20. Bulbul B, Sezer M (2013) A new approach to numerical solution of nonlinear Klein-Gordon equation. Math Prob Eng 2013:869749.
21. Jang TS (2014) An integral equation formalism for solving the nonlinear Klein-Gordon equation. Appl Math Comput 243:322–338
22. Lai H, Ma C (2011) Lattice Boltzmann model for generalized nonlinear wave equations. Phys Rev E 84:046708
23. Kuo PY, Luis V (1984) Numerical solution of a nonlinear wave equation in polar coordinates. Appl Math Comput 14:313–329
24. Argyris J, Haase M (1987) An engineer's guide to soliton phenomena: application of the finite element method. Comput Methods Appl Mech Eng 61:71–122
25. Dehghan M, Shokri A (2009) Numerical solution of the nonlinear Klein-Gordon equation using radial basis functions. J Comput Appl Math 230:400–410
26. Rashidinia J, Mohammadi R (2010) Tension spline approach for the numerical solution of nonlinear Klein-Gordon equation. Comput Phys Commun 181:78–91
27. Lakestani M, Dehghan M (2010) Collocation and finite difference-collocation methods for the solution of nonlinear Klein-Gordon equation. Comput Phys Commun 181:1392–1401
28. Khuri SA, Sayfy A (2010) A spline collocation approach for the numerical solution of ageneralized nonlinear Klein-Gordon equation. Appl Math Comput 216:1047–1056
29. Rashidinia J, Ghasemi M, Jalilian R (2010) Numerical solution of the nonlinear Klein-Gordon equation. J Comput Appl Math 233:1866–1878
30. Yin F, Tian T, Song J, Zhu M (2015) Spectral methods using Legendre wavelets for nonlinear Klein\Sine-Gordon equations. J Comput Appl Math 275:321–334
31. Bratsos AG (2009) On the numerical solution of the Klein-Gordon equation. Numer Methods Part Diff Eq 25:939–951
32. Dehghan M, Mohebbi A, Asgari Z (2009) Fourth-order compact solution of the nonlinear Klein-Gordon equation. Numer Alg 52:523–540
33. Mohebbi A, Dehghan M (2010) High-order solution of one-dimensional sine-Gordon equation using compact finite difference and DIRKN methods. Math Comput Model 51:537–549
34. Wei GW (2000) Discrete singular convolution for the sine-Gordon equation. Physica D 137:247–259
35. Dehghan M, Shokri A (2008) A numerical method for one-dimensional nonlinear sine-gordon equation using collocation and radial basis functions. Numer Methods Part Diff Eq 24:687–698
36. Sari M, Gurarslan G (2011) A sixth-order compact finite difference method for the one-dimensional sine-Gordon equation. Int J Numer Methods Biomed Eng. 27:1126–1138

37. Bratsos AG (2008) A numerical method for the one-dimensional sine-Gordon equation. Numer Methods Part Diff Eq 24:833–844
38. Dehghan M, Ghesmati A (2010) Application of the dual reciprocity boundary integral equation technique to solve the nonlinear Klein-Gordon equation. Comput Phys Commun 181:1410–1418
39. Cui M (2009) Fourth-order compact scheme for the one-dimensional sine-Gordon equation. Numer Methods Part Diff Eq 25:685–711
40. Chen JB (2006) Symplectic and multisymplectic Fourier pseudo spectral discretizations for the Klein-Gordon equation. Lett Math Phys 75:293–305
41. Wang Q, Cheng D (2005) Numerical solution of damped non-linear Klein-Gordon equations using variational method and finite element approach. Appl Math Comput 162:381–401
42. Guo PF, Liew KM, Zhu P (2015) Numerical solution of nonlinear Klein-Gordon equation using the element-free kp-Ritz method. Appl Math Model 39:2917–2928
43. Pontaza JP, Reddy JN (2004) Space-time coupled spectral/hp least-squares finite element formulation for the incompressible Navier-Stokes equations. J Comput Phys 197:418–459
44. Dorao CA, Jakobsen HA (2007) A parallel time–space least-squares spectral element solver for incompressible flow problems. Appl Math Comput 185:45–58
45. Fletcher CAJ (1984) Computational Galerkin methods. Springer, Berlin
46. Malomed BA (1985) Inelastic interactions of solitons in nearly integrable systems II. Physica D 15:385–401
47. Kivshar YS, Fei Z, Vazquez L (1991) Resonant soliton-impurity interactions. Phys Rev Lett 67:1177–1180
48. Fei Z, Kivshar YS, Vazquez L (1992) Resonant kink-impurity interactions in the sine-Gordon model. Phys Rev A 45:6019–6030
49. Zhang F (1998) Breather scattering by impurities in the sine-Gordon model. Phys Rev E 58:2558–2563
50. Liu XJ, Wang JZ, Zhou YH (2017) A space-time fully decoupled wavelet Galerkin method for solving a class of nonlinear wave problems. Nonlinear Dyn 90:599–616
51. Tzou HS, Anderson GL (eds) (1992) Intelligent structural systems. Kluwer Academic Publishers, Boston
52. Lee CK (1992) Piezoelectric laminates: theory and experiments for distributed sensors and actuators. In: Tzou HS, Anderson GL (eds) Intelligent structural systems. Academic Publishers and Kluwer, Boston
53. Zhou YH, Wang JZ (1998) A dynamic control model of piezoelectric cantilevered beam-plate based on wavelet theory. Acta Mechanica Sinica 30(6):719–727 (in Chinese)
54. Zhou YH, Wang JZ, Zheng XJ, Jiang Q (2000) Vibration control of variable thickness plates with piezoelectric sensors and actuators by wavelet theory. J Sound Vib 237(3):395–410
55. Guo G, Chai TY (2002) Predictive control strategy for a cantilever. Proc CSEE (12):81–85 (in Chinese)
56. Tzou HS, Zhou YH (1995) Dynamics and control of piezoelectric circular plates with geometrical nonlinearity. J Sound Vib 188(2):189–207
57. William JR, Amaratunga K (1994) Introduction to wavelet in engineering. Int J Numer Meth Eng 37:2365–2388
58. Motard RL, Joseph B (1994) Wavelet applications in chemical engineering. Kluwer Academic Publishers, Boston
59. Kivshar YS, Malomed BA (1989) Dynamics of solitons in nearly integrable systems. Rev Mod Phys 61(4):763–915
60. Dalfovo F, Giorgini S, Pitaevskii LP et al (1999) Theory of Bose-Einstein condensation in trapped gases. Rev Mod Phys 71(3):463
61. Agrawal GP (2007) Nonlinear fiber optics, 4th edn. Academic Press, San Diego
62. Solli DR, Ropers C, Koonath P et al (2007) Optical rogue waves. Nature 450(7172):1054–1057
63. Ye Y (2009) Self-similar solutions for nonlinear Schrödinger equations. Math Prob Eng
64. Subaşi M (2002) On the finite differences schemes for the numerical solution of two dimensional Schrödinger equation. Numer Methods Part Diff Eq: Int J 18(6):752–758

65. Bao W, Jaksch D (2003) An explicit unconditionally stable numerical method for solving damped nonlinear Schrödinger equations with a focusing nonlinearity. SIAM J Numer Anal 41(4):1406–1426
66. Bao W, Jaksch D, Markowich PA (2003) Numerical solution of the Gross-Pitaevskii equation for Bose-Einstein condensation. J Comput Phys 187(1):318–342
67. Wang H (2005) Numerical studies on the split-step finite difference method for nonlinear Schrödinger equations. Appl Math Comput 170(1):17–35
68. Kalita JC, Chhabra P, Kumar S (2006) A semi-discrete higher order compact scheme for the unsteady two-dimensional Schrödinger equation. J Comput Appl Math 197(1):141–149
69. Dehghan M, Shokri A (2007) A numerical method for two-dimensional Schrödinger equation using collocation and radial basis functions. Comput Math Appl 54(1):136–146
70. Sweilam NH (2007) Variational iteration method for solving cubic nonlinear Schrödinger equation. J Comput Appl Math 207(1):155–163
71. Dehghan M, Mirzaei D (2008) Numerical solution to the unsteady two-dimensional Schrödinger equation using meshless local boundary integral equation method. Int J Numer Meth Eng 76(4):501–520
72. Alomari AK, Noorani MSM, Nazar R (2009) Explicit series solutions of some linear and nonlinear Schrodinger equations via the homotopy analysis method. Commun Nonlinear Sci Numer Simul 14(4):1196–1207
73. Liao H, Sun Z, Shi H (2010) Error estimate of fourth-order compact scheme for linear Schrödinger equations. SIAM J Numer Anal 47(6):4381–4401
74. Mertens FG, Quintero NR, Bishop AR (2010) Nonlinear Schrödinger equation with spatiotemporal perturbations. Phys Rev E 81(1):016608
75. Tian ZF, Yu PX (2010) High-order compact ADI (HOC-ADI) method for solving unsteady 2D Schrödinger equation. Comput Phys Commun 181(5):861–868
76. Gao Z, Xie S (2011) Fourth-order alternating direction implicit compact finite difference schemes for two-dimensional Schrödinger equations. Appl Numer Math 61(4):593–614
77. He D, Pan K (2017) An unconditionally stable linearized CCD–ADI method for generalized nonlinear Schrödinger equations with variable coefficients in two and three dimensions. Comput Math Appl 73(11):2360–2374
78. Wang S, Zhang L (2011) Split-step orthogonal spline collocation methods for nonlinear Schrödinger equations in one, two, and three dimensions. Appl Math Comput 218(5):1903–1916
79. Zhu H, Chen Y, Song S et al (2011) Symplectic and multi-symplectic wavelet collocation methods for two-dimensional Schrödinger equations. Appl Numer Math 61(3):308–321
80. Kong L, Duan Y, Wang L, Yin X, Ma Y (2012) Spectral-like resolution compact ADI finite difference method for the multi-dimensional Schrödinger equations. Math Comput Model 55(5–6):1798–1812
81. Xu Y, Zhang L (2012) Alternating direction implicit method for solving two-dimensional cubic nonlinear Schrödinger equation. Comput Phys Commun 183(5):1082–1093
82. Abbasbandy S, Ghehsareh HR, Hashim I (2013) A meshfree method for the solution of two-dimensional cubic nonlinear Schrödinger equation. Eng Anal Boundary Elem 37(6):885–898
83. Antoine X, Bao W, Besse C (2013) Computational methods for the dynamics of the nonlinear Schrödinger/Gross-Pitaevskii equations. Comput Phys Commun 184(12):2621–2633
84. Dehghan M, Emami-Naeini F (2013) The Sinc-collocation and Sinc-Galerkin methods for solving the two-dimensional Schrödinger equation with nonhomogeneous boundary conditions. Appl Math Model 37(22):9379–9397
85. Wang T, Guo B, Xu Q (2013) Fourth-order compact and energy conservative difference schemes for the nonlinear Schrödinger equation in two dimensions. J Comput Phys 243:382–399
86. Taleei A, Dehghan M (2014) Time-splitting pseudo-spectral domain decomposition method for the soliton solutions of the one-and multi-dimensional nonlinear Schrödinger equations. Comput Phys Commun 185(6):1515–1528
87. Tay WC, Tan EL (2014) Pentadiagonal alternating-direction-implicit finite-difference time-domain method for two-dimensional Schrödinger equation. Comput Phys Commun 185(7):1886–1892

88. Zhang LW, Deng YJ, Liew KM et al (2014) The improved complex variable element-free Galerkin method for two-dimensional Schrödinger equation. Comput Math Appl 68(10):1093–1106
89. Zhang LW, Liew KM (2014) An element-free based solution for nonlinear Schrödinger equations using the ICVMLS-Ritz method. Appl Math Comput 249:333–345
90. Bhrawy AH, Abdelkawy MA (2015) A fully spectral collocation approximation for multi-dimensional fractional Schrödinger equations. J Comput Phys 294:462–483
91. Li LZ, Sun HW, Tam SC (2015) A spatial sixth-order alternating direction implicit method for two-dimensional cubic nonlinear Schrödinger equations. Comput Phys Commun 187:38–48
92. Mocz P, Succi S (2015) Numerical solution of the nonlinear Schrödinger equation using smoothed-particle hydrodynamics. Phys Rev E 91(5):053304
93. Quintero NR, Mertens FG, Bishop AR (2015) Soliton stability criterion for generalized nonlinear Schrödinger equations. Phys Rev E 91(1):012905
94. Sheu TWH, Lin L (2015) Dispersion relation equation preserving FDTD method for nonlinear cubic Schrödinger equation. J Comput Phys 299:1–21
95. Liu XJ, Wang JZ, Zhou YH (2016) A space–time fully decoupled wavelet Galerkin method for solving two-dimensional Burgers' equations. Comput Math Appl 72(12):2908–2919

Chapter 9
Applications to Nonlinear Solid Mechanics

Solid mechanics is essential and important to a variety of structures, which are widely used in almost all engineering areas, e.g., bridge structures, architecture structures, ship structures, mechanical structures, aerospace structures, rail structures, etc. The nonlinear solid problems can mainly be classified into two categories. One is generated by the constitutive relationships in which the stress state is formulated in terms of the strain state, or vice versa, and it is called as the material nonlinearity, e.g., nonlinear elasticity and plasticity. Another one is the geometrical nonlinearity, i.e., at least one strain is nonlinearly formulated by displacements in the solid. In a solid structure, such nonlinear characteristics are mainly dependent on external loads or environment factors. When the externals are small, both the deformation and stress in the structure are small too so as both the constitutive relationships and the geometrical relations are simultaneously linear. In such case, the solid mechanics is referred to the linear and elastic category. Once one of the above two nonlinear categories or both is entered by the externals, the solid problem becomes nonlinear. As we have known, the investigation of the nonlinear solid mechanics has initiated one of the nonlinear sciences, which has lasted more than 100 years. Till today, this nonlinear research is still a significant area in scientific committee of solid mechanics.

In this chapter, some typical solid problems either with geometrical nonlinearity or with material nonlinearity are numerically solved by the established wavelet-based method to show the accuracy, the convergence rate, as well as the advantages compared with other existing methods.

Y.-H. Zhou, *Wavelet Numerical Method and Its Applications in Nonlinear Problems*, Engineering Applications of Computational Methods 6,
https://doi.org/10.1007/978-981-33-6643-5_9

9.1 Large Deflection and Post-buckling Tracking of Beams

9.1.1 *Strong Nonlinear Solution of Deflection in Post-buckling Path*

Various flexible structures have been widely used in engineering fields, such as microelectromechanical systems, aerospace, intelligent structures, and bionics [1]. To efficiently perform specific functions, components with flexible structures in such systems usually need to be considerably deformed [1]. The thin beam is one of such basic components, whose bending problem with large deflection has elicited extensive attention [1–7]. For example, Bisshop and Drucker [2] used an elliptic integral to obtain the exact solution of the large deflection bending of a cantilever beam subjected to a concentrated load. For large deflection bending of flexible structures under complex loading conditions, a few numerical methods have been proposed, such as the FEM [3], the shooting method [4], and the Adomian decomposition method [5]. However, in spite of the great success, these methods face challenges in dealing with the problem of excessive deflection bending of beams, as the error and calculation cost will increase sharply when the deflection becomes very large [3–10].

Consider a cantilever beam with length L subjected to the concentrated force P and Q, and bending moment M_0 at the free end, as shown in Fig. 9.1. Based on the relation between bending moment and curvature, the governing equation is [1–4]

$$EI\frac{d\theta}{ds} = Q(a - x) + P(b - y) + M_0, \tag{9.1}$$

where EI is the bending stiffness of the beam, s is the arc length of the point (x, y) to the fixed end, and θ is the angle between the beam axis and the x-axis along the beam with no deflection.

Taking derivative on both sides of Eq. (9.1) with respect to s and considering the geometric relations, $dx/ds = \cos\theta$ and $dy/ds = \sin\theta$, we have

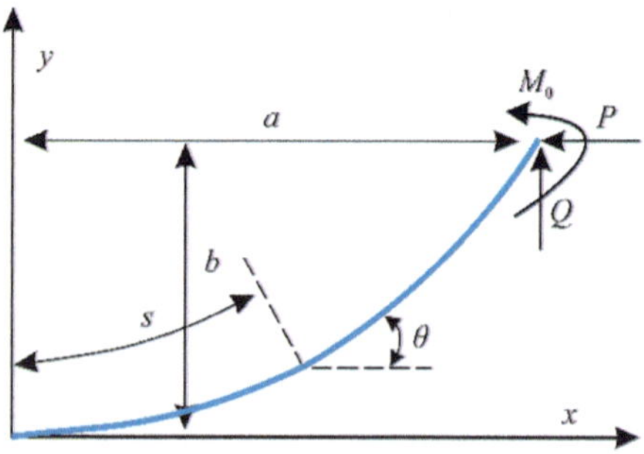

Fig. 9.1 Diagram of a flexible cantilever beam and coordinate system under concentrated loads at the free end

$$\frac{d^2\theta}{ds^2} = -\frac{Q}{EI}\cos\theta - \frac{P}{EI}\sin\theta \tag{9.2}$$

with the boundary conditions

$$\theta(0) = 0, \quad EI\frac{d\theta}{ds}|_{s=L} = M_0. \tag{9.3}$$

By utilizing the dimensionless variable $\overline{s} = s/L$ and ignoring the bar over the variable $\overline{s}$ in the dimensionless equations, the dimensionless governing equation for this problem can be rewritten as

$$\begin{cases} d^2\theta/ds^2 = -(\alpha\cos\theta + \beta\sin\theta) = -\delta\sin(\theta+\eta) \\ \theta(0) = 0,\ d\theta/ds|_{s=1} = \gamma \end{cases}, \tag{9.4}$$

where the dimensionless quantities $\alpha = QL^2/EI$, $\beta = PL^2/EI$, $\gamma = M_0L/EI$, and $\delta = \sqrt{\alpha^2+\beta^2}$, and the angle $\eta = \arcsin(\alpha/\delta)$.

From Eq. (9.4), one sees that the governing equation is a nonlinear differential equation with the second-type nonhomogeneous boundary condition. To directly solve this problem by using the wavelet method introduced in Sect. 3.5, we define

$$\omega(s) = \theta(s) - \gamma s. \tag{9.5}$$

Substitution of Eq. (9.5) into Eq. (9.4) leads to an equivalent problem with homogeneous boundary conditions, i.e.,

$$\begin{cases} \frac{d^2\omega}{ds^2} = -\delta\sin(\omega + \gamma s + \eta), \\ \omega(0) = 0,\ \frac{d\omega}{ds}|_{s=1} = 0. \end{cases} \tag{9.6}$$

Following the standard solution scheme described in Sect. 3.5 or Chap. 5, we can obtain

$$\omega(s) \approx \sum_{k=1}^{2^j} \omega_{j,k}\phi_{j,k}(s), \tag{9.7}$$

$$\sin(\omega + \gamma s + \eta) \approx \sum_{k=1}^{2^j} \sin(\omega_{j,k} + \gamma k/2^j + \eta)\phi_{j,k}(s) + \sin\eta\phi_{j,0}(s), \tag{9.8}$$

where $\omega_{j,k} = \omega(k/2^j)$. Substituting Eqs. (9.7) and (9.8) into Eq. (9.6), one gets

$$\sum_{k=1}^{2^j} \omega_{j,k} d^2\phi_{j,k}(s)/ds^2 + \delta \sum_{k=1}^{2^j} \sin(\omega_{j,k} + \gamma k/2^j + \eta)\phi_{j,k}(s)$$
$$+ \delta \sin \eta \phi_{j,0}(s) \approx 0. \tag{9.9}$$

Multiplying both sides of Eq. (9.9) by the weight function $\phi_{j,l}(s)$, $l = 1, 2, \ldots, 2^j$, and then integrating them from 0 to 1, one can obtain the algebraic equations of the matrix form

$$\mathbf{AW} + \mathbf{BS} + \mathbf{E} = \mathbf{0}, \tag{9.10}$$

where the matrices and vectors

$$\mathbf{A} = \{a_{lk} = \Gamma_{k,l}^{2,0}\}, \mathbf{B} = \{b_{lk} = \overline{\Gamma}_{k,l}^{0,0}\},$$
$$\mathbf{W} = \{\omega_k = \omega_{j,k}\}^{\mathrm{T}}, \mathbf{E} = \{e_l = \delta \sin \eta \, \overline{\Gamma}_{0,l}^{0,0}\}^{\mathrm{T}}.$$
$$\mathbf{S} = \{s_k = \delta \sin(\omega_{j,k} + \gamma k/2^j + \eta)\}^{\mathrm{T}}$$

Here, the connection coefficients $\Gamma_{k,l}^{2,0} = \int_0^1 d^2\phi_{j,k}(s)/ds^2 \phi_{j,l}(s)dx$ and $\overline{\Gamma}_{k,l}^{0,0} = \int_0^1 \phi_{j,k}(s)\phi_{j,l}(s)dx$ can be accurately obtained following the method described in Appendix 5.1, and $k, l = 1, 2, \ldots, 2^j$. According to Sect. 5.1, we can directly discretize the boundary conditions and replace the corresponding equations in (9.10).

Applying the Newton iteration method to solve the nonlinear algebraic equations in Eq. (9.10), we can obtain $\omega_{j,k}$, $k = 1, 2, \ldots, 2^j$, and then $\omega(s)$ can be reconstructed by Eq. (9.7). Eventually, the solution $\theta(s)$ for Eq. (9.4) can be obtained through Eq. (9.5).

For the spatial configuration in coordinates (x, y), in accordance with the geometric relations, $dx/ds = \cos\theta$ and $dy/ds = \sin\theta$, we can obtain

$$x(s) = \int_0^s \cos\theta(t)dt, \ y(s) = \int_0^s \sin\theta(t)dt. \tag{9.11}$$

Using the wavelet numerical integration method introduced in Appendix 9.1, we can derive

$$x(s) \approx \sum_{k=0}^{2^j} \cos\left[\theta\left(\frac{k}{2^j}\right)\right]\widetilde{\theta}_{j,k}(s), \ y(s) \approx \sum_{k=0}^{2^j} \sin\left[\theta\left(\frac{k}{2^j}\right)\right]\widetilde{\theta}_{j,k}(s), \tag{9.12}$$

where $\widetilde{\theta}_{j,k}(s)$ can be obtained by using the method introduced in Appendix 9.1.

In the following, we give some numerical tests for the deflection of cantilevered beam under different loads, where the resolution level $j = 4$ is chosen.

Case 1: Bending deflection of the cantilevered beam under a transverse concentrated load, i.e., $\alpha \neq 0$ and $\beta = \gamma = 0$.

For this case, when the external application α is small, the deflection w and deflection angle θ are small too. In such situation, the governing equation can be degenerated into linear one, where the output w or θ varies with α linearly. As the external load α increases, the relation of deflection w or θ varying with α becomes nonlinear. Figure 9.2 (left) displays the characteristic curve of the nonlinear relation of real deflection w/h at the free end varying with α, where a comparison of it with the linear solution is also given. From this figure, one sees that when $\alpha < 1$ or $w/h < 0.3$, the nonlinear solution is well coincident with its linear solution. After that, the nonlinear solution is smaller than its linear one. The external load is larger, the difference between them is larger too. When α approaches to and over 6, the increase in deflection takes place slowly. At the same time, with increase in α, the calculation time has some change. Figure 9.2 (right) shows such change of calculation time varying with α, from where one sees that the consumed computation time of the wavelet-based method does not dramatically increase with the enhancement of nonlinearity, namely, with the increase in load parameter α, calculation time to each load is stably located in the region 0.04–0.06 s. Thus, this wavelet-based method can provide us a high computational efficiency and stable calculations, particularly for strong nonlinear systems.

Case 2: Strong nonlinear bending deflection of cantilevered beam under joint loads.

In this case, at least two of external loads α, β, and γ are not equal to zero. Table 9.1 lists the numerical solution results of the cantilevered beam under four loads. In this table, the numerical solutions of the wavelet-based method and the shooting method [4] based on the fifth-order Runge–Kutta method, and the Adomian decomposition [4] method retained to the eighth-order approximation as well as the exact solution

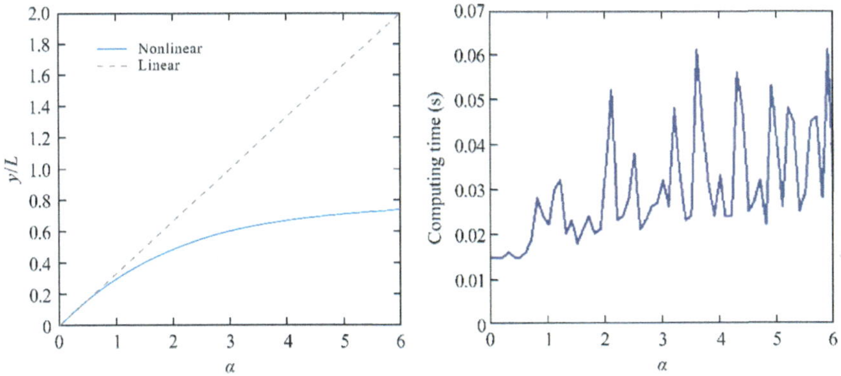

Fig. 9.2 Characteristic curve of deflection at free end varying with transverse concentrated load (Left), and variation of calculation time (Right) of the wavelet-based solution with the external load when the resolution level $j = 4$

Table 9.1 Numerical solutions of deflection y/L at the free end of cantilevered beam obtained by using three methods under four different loads and their relative error E to exact solution

Parameters			Exact	Wavelet based		Shooting [4]		Adomian [4]	
α	β	γ	y/L	y/L	$E(\%)$	y/L	$E(\%)$	y/L	$E(\%)$
1.0	1.0	0.0	0.42921	0.42922	0.002	0.42953	0.074	0.42764	− 0.365
1.0	1.0	0.2	0.51390	0.51391	0.001	0.51429	0.075	0.51204	− 0.361
1.0	1.0	− 0.6	0.04565	0.04565	0.000	0.04560	− 0.109	0.04586	0.460
0.2	0.1	− 0.6	− 0.24187	− 0.24188	0.004	− 0.24212	0.103	− 0.24063	− 0.512

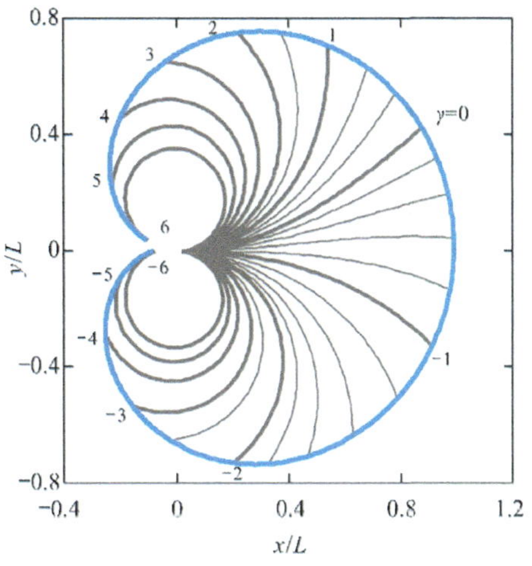

Fig. 9.3 Strong nonlinear deflection of cantilevered beam under joint loads: Spatial configuration of deflection and end trajectory of the flexible bean varying with load γ when the loads $\alpha = \beta = 1$

results are displayed, containing their relative errors of the numerical results. From Table 9.1, one sees that all solution results from the wavelet-based method have the relative error with 0.004%, while relative errors of the shooting method and the Adomian method are within 0.103% and 0.512%, respectively. Thus, the wavelet-based method has accuracy of two orders higher than them.

When $\alpha = \beta = 1$, furthermore, Fig. 9.3 shows the configuration of deflection and free-end trajectory of the flexible beam when γ has taken different values and the wavelet-based method is employed. From this figure, we find that the wavelet-based method can effectively solve the problems of strong geometric nonlinear mechanics caused by exceptionally large deformation.

Case 3: Post-buckling tracking of the cantilevered beam.

Usually, when $\alpha = \gamma = 0$ and $\beta > 0$ (i.e., compressed load), Eq. (9.4) behaves for the beam-buckling and post-buckling problems. In such case, $\theta(x) \equiv 0$ or $w(x) = 0$ is always a solution of the cantilevered beam governed by Eq. (9.4), which is called as a trivial solution. When the application β is not over the critical value, such trivial solution is stable. When β approaches the critical value, the solution of the problem may appear bifurcated. After that, for this beam, two stable solutions will occur. Such solutions when the external application is over the critical value are referred to the research area of post-buckling. If one desires to get such a post-buckling solution, the nonlinear governing equation established under the deflection configuration has to be employed, as displayed in Eq. (9.2) or (9.4). Recently, such original buckling and post-buckling concepts have been extended to containing those original deflection of a structure under some loads rather than a zero trivial solution. When some of the external loads increase such that the original deflection becomes instable, the generalized buckling and post-buckling occur. It should be noted that whether or not such post-buckling solution can be tracked is usually and strongly dependent on the characteristic of the post-buckling solution, associated with a numerical solution

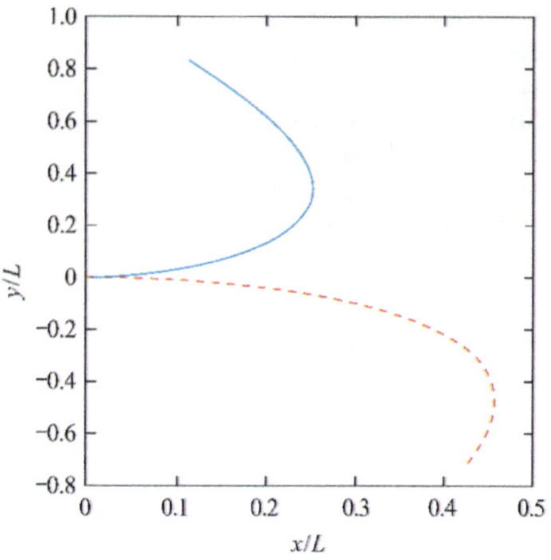

Fig. 9.4 Two solutions of post-buckling configuration of deflected cantilevered beam under loads $\alpha = \beta = 5, \gamma = 0$

method employed. As shown in Ref. [4] for the case of both $\alpha \neq 0$ and $\beta \neq 0$ subjected to the cantilevered beam, i.e., the beam is deflected before bucking, the deflection solution will become instable when the external load approaches to a certain critical value, and after the external load is over the critical value, there are two stable deformation states in the loaded cantilevered beam. Here, we give one case that the cantilevered beam under one set of specified loads has two stable solutions on the basis of the wavelet-based method. Figure 9.4 shows the two solutions when $\alpha = \beta = 5, \gamma = 0$, which correspond to the two equilibrium configurations of the deflected beam.

From above results shown in this subsection, we can conclude that the wavelet-based method can effectively and uniformly deal with the large deflection bending problems of flexible beams, containing linear and nonlinear deflection, buckling and post-buckling problems, where accuracy and efficiency are significantly guaranteed no matter how the deformation increases. In addition, the results also show that the wavelet-based method is suitable for searching multiple equilibrium configurations existing in a strong nonlinear mechanics problem.

9.1.2 Solution of Large Deflection of Flexible Beam with Immovably Simple Supports

Here, we consider a simply supported elastic beam with length, l, subjected to a transverse distributed load, $p(x)$. By means of the relations between exact curvature and displacement, and taking into account the influence of the mid-surface tensile force, the differential equation that describes the large deformation bending problem of the beam can be written as [8, 9]

$$\begin{cases} EI[\frac{w_{,xx}}{(1+w_{,x}^2)^{3/2}}]_{,xx} - [\frac{EA}{l}\int_0^l (\sqrt{1+w_{,x}^2} - 1)dx]w_{,xx} = p(x), x \in (0,l) \\ w(0) = w(l) = 0, w_{,xx}(0) = w_{,xx}(l) = 0 \end{cases}, \tag{9.13}$$

where $w(x)$ is the deflection of the beam, E is the elastic modulus of the material, A and I are the cross-sectional area and the moment of inertia, respectively, and the operator $(\bullet)_{,x}$ stands for the derivative of x. In Eq. (9.13), the exact expression of the curvature can be expressed as

$$\frac{1}{\rho} = \frac{w_{,xx}}{(1+w_{,x}^2)^{3/2}}, \tag{9.14}$$

where ρ is the radius of curvature. In many studies, researchers usually use the approximate expression of the curvature [5–9] of the form $1/\rho \approx w_{,xx}$ to simplify the calculations. Once this approximation is used, Eq. (9.13) can be simplified as follows:

$$\begin{cases} EIw_{,xxxx} - \left[\frac{EA}{l}\int_0^l \left(\sqrt{1+w_{,x}^2} - 1\right)dx\right]w_{,xx} = p(x), x \in (0,l), \\ w(0) = w(l) = 0, w_{,xx}(0) = w_{,xx}(l) = 0. \end{cases} \tag{9.15}$$

In the following, the differences between Eqs. (9.13) and (9.15) will be discussed when the wavelet-based solution method is employed. If one takes a further approximation of $\sqrt{1+w_{,x}^2} \approx 1$, it is obvious that the differential equation in Eq. (9.15) is reduced into the well-known Euler–Bernoulli equation of beam bending with small deflection.

For Eq. (9.13), we introduce the dimensionless quantities of $\overline{x} = x/l$, $\overline{w} = w/l$, $\overline{p} = pl^3/EI$, and $H = Al^2/I$ such that Eq. (9.13) can be rewritten by the dimensionless form

$$\begin{cases} [\frac{w_{,xx}}{(1+w_{,x}^2)^{3/2}}]_{,xx} - H[\int_0^1 (\sqrt{1+w_{,x}^2} - 1)dx]w_{,xx} = p(x), x \in (0,1) \\ w(0) = w(1) = 0, w_{,xx}(0) = w_{,xx}(1) = 0 \end{cases}, \tag{9.16}$$

where the bar over each quantity is discarded for simplifying the expression without affecting the understanding. For example, $\overline{w}$ is denoted as w.

Before we use the wavelet-based method to solve such typically nonlinear differential equation shown in Eq. (9.16), let us see how a conventional solution method is used to deal with the nonlinear terms in the equation. Usually, such nonlinear terms cannot be directly solved by the conventional methods when one of the nonlinear terms contains a non-integer power nonlinearity. In the conventional solution methods, a common approach is to use the Taylor expansion to convert such non-integer power nonlinear term into a series of the integer powers. For example [8, 9],

$$\sqrt{1+w_{,x}^2} = 1 + \frac{1}{2}w_{,x}^2 - \frac{1}{8}w_{,x}^4 + \cdots \tag{9.17}$$

Then, the first one or two terms will be intercepted in their calculations. It is obvious that such treatment is necessary in the conventional methods which makes some additional errors a priori. For some situation, these additional errors can even be unacceptable. However, in the wavelet-based method this kind of error does not appear, as pointed out in Sect. 3.5.

Now, let us introduce the solution steps of the wavelet-based method to Eq. (9.16). Here, we denote

$$f(x) = w_{,x}, \tag{9.18}$$

$$g(x) = w_{,xx}. \tag{9.19}$$

Then, Eq. (9.16) can be rewritten as

$$\begin{cases} [\frac{g(x)}{(1+f^2(x))^{3/2}}]_{,xx} - H[\int\limits_0^1 (\sqrt{1+f^2(x)}-1)dx]w_{,xx} = p(x), x \in (0,1), \\ w(0) = w(1) = 0, w_{,xx}(0) = w_{,xx}(1) = 0. \end{cases} \tag{9.20}$$

Taking the generalized Coiflet scaling base functions, we can write the approximate of unknown deflection, $w(x)$, of the form

$$w(x) \approx \sum_{k=1}^{2^j-1} w_{j,k}\phi_{j,k}(x), \tag{9.21}$$

where $w_{j,k} = w(k/2^j)$. Similar to the above done to others in Eq. (9.20), we have

$$f(x) \approx \sum_{k=0}^{2^j} f_{j,k}\phi_{j,k}(x), \tag{9.22}$$

$$g(x) \approx \sum_{k=0}^{2^j} g_{j,k}\phi_{j,k}(x), \tag{9.23}$$

$$\frac{g(x)}{[1+f^2(x)]^{3/2}} \approx \sum_{k=0}^{2^j} \frac{g_{j,k}}{(1+f_{j,k}^2)^{3/2}}\phi_{j,k}(x), \tag{9.24}$$

$$\sqrt{1+f^2(x)}-1 \approx \sum_{k=0}^{2^j}(\sqrt{1+f_{j,k}^2}-1)\phi_{j,k}(x), \tag{9.25}$$

$$p(x) \approx \sum_{k=0}^{2^j} p_{j,k}\phi_{j,k}(x). \tag{9.26}$$

Substitution of Eqs. (9.21), (9.22), and (9.23) into Eqs. (9.18) and (9.19) leads to

$$\sum_{k=0}^{2^j} f_{j,k}\phi_{j,k}(x) \approx \sum_{k=1}^{2^j-1} w_{j,k}d\phi_{j,k}(x)/dx, \tag{9.27}$$

$$\sum_{k=0}^{2^j} g_{j,k}\phi_{j,k}(x) \approx \sum_{k=1}^{2^j-1} w_{j,k}d^2\phi_{j,k}(x)/dx^2. \tag{9.28}$$

To both sides of Eqs. (9.27) and (9.28), multiplying the weight function $\phi_{j,l}(x)$, $l = 0,\ 1, \ldots, 2^j$, and integrating them from 0 to 1, we obtain the following algebraic equations of the matrix form:

$$\mathbf{F} \approx \mathbf{A}^{-1}\mathbf{B}\mathbf{W}, \tag{9.29}$$

$$\mathbf{G} \approx \mathbf{A}^{-1}\mathbf{C}\mathbf{W}, \tag{9.30}$$

where matrices

$$\mathbf{A} = \{a_{lk} = \int_0^1 \phi_{j,k}(x)\phi_{j,l}(x)dx\},$$

$$\mathbf{B} = \{b_{lm} = \int_0^1 d\phi_{j,m}(x)/dx\phi_{j,l}(x)dx\},$$

$$\mathbf{C} = \{c_{lm} = \int_0^1 d^2\phi_{j,m}(x)/dx^2\phi_{j,l}(x)dx\},$$

vectors $\mathbf{F} = \{f_k = f(k/2^j)\}$, $\mathbf{G} = \{g_k = g(k/2^j)\}$, and $\mathbf{W} = \{w_m = w(m/2^j)\}$, and subscripts $k, l = 0, 1, \ldots, 2^j$ and $m = 1, 2, \ldots, 2^j - 1$. Here, the generalized connection coefficients a_{lk}, b_{lm}, and c_{lm} can be obtained by using the method described in Appendix 5.1.

Integrating Eq. (9.25) from 0 to 1, one gets

$$\int_0^1 (\sqrt{1+f^2(x)}-1)dx \approx \sum_{k=0}^{2^j} (\sqrt{1+f_{j,k}^2}-1)\widetilde{\theta}_{j,k}(1). \tag{9.31}$$

The definition and calculation method of the first integral $\widetilde{\theta}_{j,k}(1)$ of the scaling function, $\varphi_{j,k}(x)$, can be found in Appendix 9.1.

After that, substituting Eqs. (9.21), (9.24), (9.26), and (9.31) into Eq. (9.20), we gain

$$\sum_{k=0}^{2^j} \frac{g_{j,k}}{(1+f_{j,k}^2)^{3/2}} \frac{d^2\phi_{j,k}(x)}{dx^2} - H\sum_{n=0}^{2^j} (\sqrt{1+f_{j,n}^2}-1)\widetilde{\theta}_{j,n}(1),$$
$$\times \sum_{k=1}^{2^j-1} w_{j,k} \frac{d^2\phi_{j,k}(x)}{dx^2} \approx \sum_{k=0}^{2^j} p_{j,k}\phi_{j,k}(x). \tag{9.32}$$

Inserting Eqs. (9.29) and (9.30) into Eq. (9.32), we further have

$$\sum_{k=0}^{2^j} \frac{[\mathbf{A}^{-1}\mathbf{CW}]_k}{(1+[\mathbf{A}^{-1}\mathbf{BW}]_k^2)^{3/2}} \frac{d^2\phi_{j,k}(x)}{dx^2} - H\sum_{n=0}^{2^j} (\sqrt{1+[\mathbf{A}^{-1}\mathbf{BW}]_n^2}-1)\widetilde{\theta}_{j,n}(1)$$
$$\times \sum_{k=1}^{2^j-1} w_{j,k} \frac{d^2\phi_{j,k}(x)}{dx^2} \approx \sum_{k=0}^{2^j} p_{j,k}\phi_{j,k}(x), \tag{9.33}$$

where $[\mathbf{F}]_k$ represents f_k, the kth element of vector $\mathbf{F}$. Multiplying both sides of Eq. (9.33) by the weight function $\phi_{j,l}(x)$ and then performing integration from 0 to 1, we obtain the final algebraic equations of the form

$$\mathbf{D}\overline{\mathbf{W}}(\mathbf{W}) - H\Lambda(\mathbf{W})\mathbf{EW} \approx \mathbf{QP}, \tag{9.34}$$

where matrices

$$\mathbf{D} = \{d_{lk} = \int_0^1 d^2\phi_{j,k}(x)/dx^2\phi_{j,l}(x)dx\},$$

$$\mathbf{E} = \{e_{lm} = \int_0^1 d^2\phi_{j,m}(x)/dx^2\phi_{j,l}(x)dx\},$$

$$\mathbf{Q} = \{q_{lk} = \int_0^1 \phi_{j,k}(x)\phi_{j,l}(x)dx\},$$

vectors $\mathbf{P} = \{p_k = p(k/2^j)\}$, $\mathbf{W} = \{w_m = w(m/2^j)\}$, and

$$\overline{\mathbf{W}}(\mathbf{W}) = \{\overline{w}_k = \frac{[\mathbf{A}^{-1}\mathbf{CW}]_k}{(1 + [\mathbf{A}^{-1}\mathbf{BW}]_k^2)^{3/2}}\}. \tag{9.35}$$

The coefficient

$$\Lambda(\mathbf{W}) = \sum_{n=0}^{2^j} (\sqrt{1 + [\mathbf{A}^{-1}\mathbf{BW}]_n^2} - 1)\widetilde{\theta}_{j,n}(1). \tag{9.36}$$

According to Sect. 5.1, we can directly discretize the boundary conditions and replace the corresponding equations in Eq. (9.34).

Equation (9.34) is clearly a $2^j - 1$-dimensional nonlinear algebraic system of unknown quantities $w(m/2^j)$, and $m = 1, 2, \ldots, 2^j - 1$. Thus, the unknown nodal values, $w(m/2^j), m = 1, 2, \ldots, 2^j - 1$ can be obtained by using the classical Newton iteration method; further, the approximate solution for the problem, Eq. (9.13), can be reconstructed from Eq. (9.21).

In order to test the validity and accuracy of the above wavelet-based solution method, after the dimensionless quantity $H = 100$ in Eq. (9.16) is taken into account, and when the dimensionless load is taken by the form

$$\begin{aligned} p(x) = & \frac{r\pi^4 \sin(\pi x)}{[1 + r^2\pi^2 \cos^2(\pi x)]^{3/2}} - \frac{15r^5\pi^8 \cos^2(\pi x)\sin^3(\pi x)}{[1 + r^2\pi^2 \cos^2(\pi x)]^{7/2}} \\ & + \frac{3r^3\pi^6[\sin^3(\pi x) - 3\cos^2(\pi x)\sin(\pi x)]}{[1 + r^2\pi^2 \cos^2(\pi x)]^{5/2}} \\ & + 100r\pi^2 \sin(\pi x) \int_0^1 (\sqrt{1 + r^2\pi^2 \cos^2(\pi x)} - 1)dx. \end{aligned} \tag{9.37}$$

Equation (9.16) has the exact solution $w(x) = r\sin(\pi x)$. It is obvious that the dimensionless quantity, r, represents the maximum dimensionless deflection of the beam, i.e., $r = w_{\max}/l$. In the subsequent studies, we used this parameter to characterize the strength of the beam deformation of the problem with nonlinearity.

When $r = 0.5$, first of all, we give a comparison of the wavelet-based solutions to the flexible beam with immovably simple supports governed by the fine Eq. (9.13) and by an approximation Eq. (9.15) as shown in Fig. 9.5 where the resolution levels $j = 4$ and 5. From Fig. 9.5, one sees that all numerical solutions of this case are in good agreement with its exact solution, while the results of fine model of Eq. (9.13) are more closed to the exact solution. To quantitatively display the difference between the numerical solutions and the exact solution, we use their relative errors distributed in the region of the beam as shown in Fig. 9.6. From Fig. 9.6, one sees that the relative errors from Eq. (9.13) is uniformly distributed in [0,1], and the relative error does not exceed 0.42% and 0.02% when the resolution level is $j = 4$ and 5, respectively.

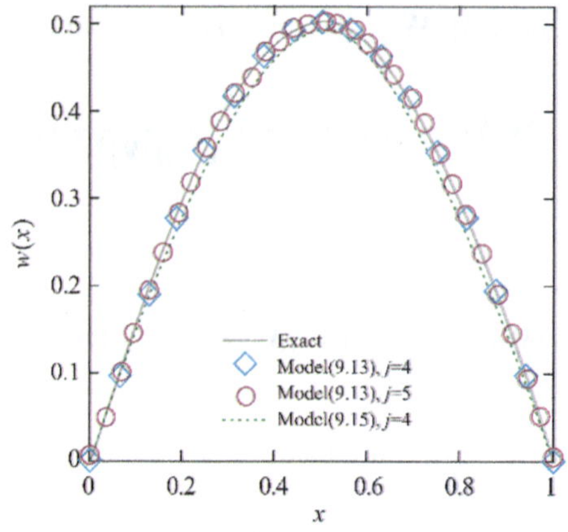

Fig. 9.5 Comparison of the deformation configurations of flexible beam with immovably simple supports governed by different models when the wavelet-based solution method is used and $r = 0.5$. Here, Eq. (9.13) presents exact model, and Eq. (9.15) has an approximation to Eq. (9.13)

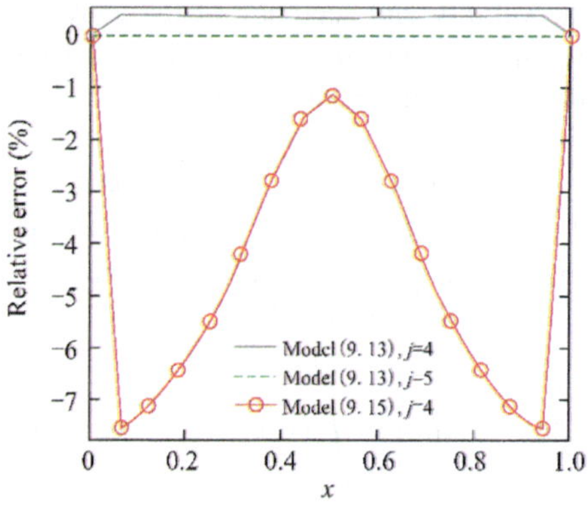

Fig. 9.6 Distribution of relative errors of deflection solutions for the flexible beam with immovably simple supports governed by the fine Eq. (9.13) and an approximation Eq. (9.15) solved by the wavelet-based method

To the approximate Eq. (9.15), however, there are some jumps in the distribution of its relative error near to the ends, where the maximum absolute value is about 7.53% when the resolution level is $j = 4$. Compressing to 0.42% for the case of the fine Eq. (9.13) employed with the same resolution level, the value 7.53% of relative error from Eq. (9.15) is about 18 times of the relative error from Eq. (9.13).

When the parameter r increases, it is evident that both the application load and the deflection increase too. Next, we display the numerical results of the wavelet-based method for solving the flexible beam with immovably simple supports. Figure 9.7 shows those curves of the maximum relative errors in the numerical solutions varying with the parameter r from 0.01 to 0.5. As a comparison, such kind of error variation

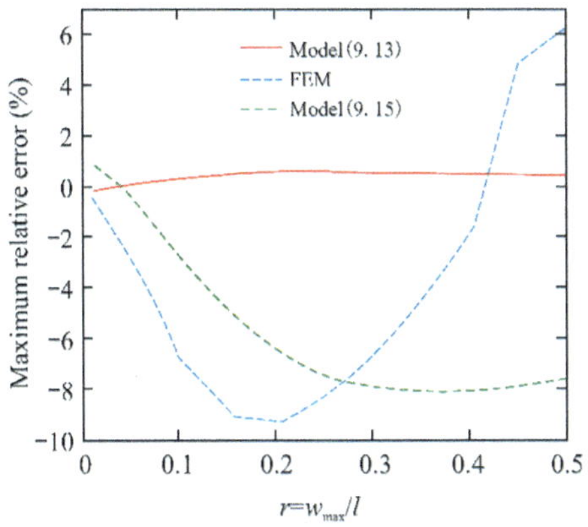

Fig. 9.7 Variation curves of the maximum error of numerical solutions varying with the maximum dimensionless deflection r of the flexible beam with immovably simple supports. Here, Model (9.13) and Model (9.15) correspond to the solutions of the wavelet-based method when the resolution level $j = 4$, while FEM is calculated by 15 nodes

from the FEM with 15 nodal points is also given in this figure. According to Fig. 9.7, we find that the fine Eq. (9.13) has a litter increasing of the relative error with r, while the approximate Eq. (9.15) has a sustaining increase of the error in absolute value with r increasing. This reveals that the approximation in Eq. (9.15) may lead to an increase of the difference between Eqs. (9.15) and (9.13) when the deformation or nonlinear effect increases. When $r < 0.1$, the maximum error from Eq. (9.15) is within 2%. Hence, Eq. (9.15) generates obvious difference deviated from Eq. (9.13) when $r > 0.1$.

From Fig. 9.7, in addition, one can see that the maximum relative error from FEM has sustaining decrease with negative value up to −9.2% when r increases from $r = 0.01$ to about 0.2, and then increase to about 6.4% when r increases to 0.5. Compared to the wavelet-based method with $j = 4$ (corresponding 15 nodes) that has a steady accuracy within 0.5%, the FEM used in this problem emerges as unsteady in accuracy, and the relative error of 9.2% is about 18 times of 0.5% in the wavelet-based method. From the discussion in Sect. 3.5, the wavelet-based method is closed and will not suffer from the problem of cumulative error caused by a numerical tracking technology. Therefore, its accuracy only depends on the resolution level or the number of nodes used in the calculation and is relatively independent of the nonlinear strength of a system. By contrast, what the traditional methods such as FEM have relative low accuracy is reasonable due to no closure characteristic contained in them.

To compare the computational efficiency of the wavelet-based method with that of the FEM, further, Table 9.2 lists such comparison of their computational time required to solve this bending problem of the beams when 15 nodals are used. Here, the wavelet-based method and the classical Newton iteration method are applied to solve the nonlinear algebraic equation, Eq. (9.34), where all the initial iterative values

Table 9.2 Computation time of the wavelet method and FEM with same nodes

$r = w_{\max}/l$	Wavelet based	FEM	$r = w_{\max}/l$	Wavelet based	FEM
0.01	0.72	0.86	0.30	1.89	2.47
0.05	0.98	0.92	0.35	1.91	2.81
0.10	1.22	1.19	0.40	2.14	3.55
0.15	1.49	1.30	0.45	2.14	11.11
0.20	1.66	1.36	0.50	2.14	–
0.25	1.67	1.42			

are selected by zero. For the FEM, the mature commercial software of ANSYS is adopted on the basis of the incremental principle in the solution of large deformation problems. As we have known, a proper setting of the loading step is necessary in the solution methods when the load increases, which is directly related to the time consumed in the calculation and influences the convergence of the solution. However, there is no firm theoretical basis on such setting of the loading step. In this situation, the selection of such a loading step will heavily rely on experiences of the software users. Here, the same load step is employed in those two methods. In comparison of the two methods, when a default value in the calculations of a method appeared, the calculations of the method are stopped. In the numerical tests, it is found that the FEM software failure occurs at $r = 0.5$. However, the wavelet-based method can be still performed when $r \geq 0.5$ without increasing the resolution level and decreasing the load step. Here, the accuracy of all numerical solutions is set in the same order.

From Table 9.2, one finds that when r increases from 0.01 to 0.45, the computation time of the wavelet-based method increases from 0.72 to 2.14 s, while the computation time of FEM increases from 0.86 to 11.11 s. When $r \geq 0.4$, the computation time of the wavelet-based method is steady at 2.14 s but FEM consumes time from 3.55 to 11.11 s which is about 5 times of 2.14 s. In addition, as the deformation or nonlinear effect increases in this region, the computation time in the wavelet-based method has about 2 times increase, while FEM has about 12 times increase. These facts tell us that wavelet-based method is more efficient and stable than FEM in solving the nonlinear problem, particularly strong nonlinearity. It should be noted that all wavelet-based solutions in this problem are conducted by selecting zero solution in the initial step of solving nonlinear algebraic solutions at each load application. If the initial input for latter load is chosen as the solution of the former step load, liking the incremental method used in FEM, the computation time in the wavelet-based solution should be largely decreased. In such case, the wavelet-based method will consume computation much lower than those in Table 9.2.

9.1.3 *Solution Results to Elastic Line Equation of Flexible Rod with Material Nonlinearity*

The material nonlinearity is one of the important nonlinear effects in mechanics. In this subsection, we consider a nonlinear elastic rod made by the material obeying the Ludwick constitutive law. This nonlinear elastic constitutive law is one of the generalizations of Hooke's law that can describe the nonlinear elastic behavior of materials, such as polymeric and metallic fibers.

The stress–strain relationship of the Ludwick-type constitutive law can be expressed through the following nonlinear function [11]:

$$\sigma = \mathrm{sign}(\varepsilon)E|\varepsilon|^{\tilde{n}}, \tag{9.38}$$

where σ and ε are the stress and strain, respectively, $\tilde{n}$ represents a material parameter that indicates the degree of material nonlinearity, and E is Young's modulus. When $\tilde{n} = 1$, Eq. (9.38) can be reduced to the Hookean elastic law. But, when $\tilde{n} < 1$, and the strain value reaches to zero, the stress gradient $d\sigma/d\varepsilon$ will approach infinity. To eliminate this deficit, the modified variation of this model is improved as [12]

$$\sigma = sign(\varepsilon)E[(|\varepsilon| + \varepsilon_0)^{\tilde{n}} - \varepsilon_0^{\tilde{n}}],\ E_0 = \left.\frac{d\sigma}{d\varepsilon}\right|_{\varepsilon=0} = \tilde{n}\varepsilon_0^{\tilde{n}-1}E, \tag{9.39}$$

where ε_0 is an additional material parameter and E_0 is the stress gradient at $\varepsilon = 0$. Although the Ludwick constitutive law is unrealistic, this law is still important in theoretical analysis because of its simplicity.

Consider a cantilevered rod loaded at its free end with a combined loading of an axial load, Q, and a vertical load, P. The axial displacement and deflection of the free end are denoted as δ_h and δ_v, respectively. The equilibrium equation can be written as

$$\frac{dM(s)}{ds} = -P\cos\theta - Q\sin\theta, \tag{9.40}$$

where $M(s)$ is the bending moment and θ is the bending angle. The relationship between bending moment, M, and curvature, κ, can be derived based on the constitutive law given in Eq. (9.39) and the Euler–Bernoulli beam theory. Differentiating this relation with respect to s, we get

$$\frac{dM}{ds} = \frac{Eb}{\kappa^3}[-A\varepsilon_0^{\tilde{n}} + (\kappa h/2 + \varepsilon_0)^{\tilde{n}}(A - B\kappa h + C\kappa^2 h^2)]\frac{d\kappa}{ds}, \tag{9.41}$$

where $A = 4\varepsilon_0^2/[(\tilde{n}+1)(\tilde{n}+2)]$, $B = 2\tilde{n}\varepsilon_0/[(\tilde{n}+1)(\tilde{n}+2)]$, and $C = \tilde{n}/[2(\tilde{n}+2)]$ are constants in terms of the material parameters, and $\kappa = d\theta/ds$ is the curvature.

For the beam with a rectangular section, several dimensionless quantities are introduced as follows:

$$\bar{s} = \frac{s}{L}, \ \beta = \frac{h}{L}, \ \lambda_{\tilde{n}} = \frac{PL^2}{EI_0}, \ \alpha_{\tilde{n}} = \frac{QL^2}{EI_0}, \tag{9.42}$$

where $I_0 = bh^3/12$, b and h are the width and height of the section, and L is the rod length.

By substituting Eq. (9.41) into the equilibrium equation and utilizing the induced equation with dimensionless variables in Eq. (9.42), we get the governing equation of the rod of the form

$$\begin{aligned} &\{-A\varepsilon_0^{\tilde{n}} + (\frac{d\theta}{d\bar{s}}\frac{\beta}{2} + \varepsilon_0)^{\tilde{n}}[A - B\beta\frac{d\theta}{d\bar{s}} + C(\beta\frac{d\theta}{d\bar{s}})^2]\}\frac{d^2\theta}{d\bar{s}^2} \\ &= -(\lambda_{\tilde{n}}\cos\theta + \alpha_{\tilde{n}}\sin\theta)(\frac{d\theta}{d\bar{s}})^3\frac{\beta^3}{12} \end{aligned} \tag{9.43}$$

in which the bars over the variables are discarded. To further simplify the governing equation, a constant $r = \varepsilon_0/\beta$ is introduced. Thus, Eq. (9.43) can be rewritten as

$$\frac{d^2\theta}{ds^2} = -\frac{(d\theta/ds)^3(\lambda_1\cos\theta + \alpha_1\sin\theta)}{[d\theta/(2ds) + r]^{\tilde{n}}[a_{\tilde{n}} - b_{\tilde{n}}d\theta/ds + c_{\tilde{n}}(d\theta/ds)^2] - a_{\tilde{n}}r^{\tilde{n}}} \tag{9.44}$$

where $\lambda_1 = \lambda_{\tilde{n}}\beta^{1-\tilde{n}}/12$, $\alpha_1 = \alpha_{\tilde{n}}\beta^{1-\tilde{n}}/12$, $a_{\tilde{n}} = A/\beta^2$, and $b_{\tilde{n}} = B/\beta$, $c_{\tilde{n}} = C$.

For the unmodified cases of the Ludwick constitutive law with $\varepsilon_0 = 0$, the governing equation can be reduced to

$$\tilde{n}\frac{d^2\theta}{ds^2} = -(\lambda\cos\theta + \alpha\sin\theta)(\frac{d\theta}{ds})^{1-\tilde{n}}, \tag{9.45}$$

where $\lambda = PL^{1+\tilde{n}}/(EI_{\tilde{n}})$, $\alpha = QL^{1+\tilde{n}}/(EI_{\tilde{n}})$, and $I_{\tilde{n}} = bh(h/2)^{\tilde{n}+1}/(\tilde{n}+2)$.

Equations (9.44) and (9.45) can be unified by the form

$$\begin{aligned} &\frac{d^2\theta}{ds^2} = -(\lambda\cos\theta + \alpha\sin\theta)f(\kappa), \\ &\theta(0) = 0, \ d\theta/ds|_{s=1} = 0. \end{aligned} \tag{9.46}$$

Following the wavelet-based method, the unknown functions in Eq. (9.46) can be approximated as follows:

$$\theta(s) \approx \sum_{k=1}^{2^j-1} \theta(\frac{k}{2^j})\phi_{j,k}(s), \tag{9.47}$$

$$\kappa(s) \approx \sum_{k=0}^{2^j-1} \kappa_{j,k}\phi_{j,k}(s), \tag{9.48}$$

$$[\lambda \cos\theta(s) + \alpha \sin\theta(s)]f[\kappa(s)] \approx \sum_{k=0}^{2^j} [\lambda \cos\theta(\frac{k}{2^j}) + \alpha \sin\theta(\frac{k}{2^j})]f[\kappa(\frac{k}{2^j})]\phi_{j,k}(s). \tag{9.49}$$

After the expressions of Eqs. (9.47)–(9.49) are substituted into Eq. (9.46), then applying the wavelet Galerkin method to the induced equation, one gets the following nonlinear algebraic equations of the matrix form:

$$\mathbf{F}(\boldsymbol{\theta}, \alpha) = \mathbf{A}\boldsymbol{\theta} + \mathbf{B}diag(\lambda \cos\boldsymbol{\theta} + \alpha \sin\boldsymbol{\theta}) f(\mathbf{B}^{-1}\mathbf{B}_1\boldsymbol{\theta}) = 0, \tag{9.50}$$

where the matrices

$$\mathbf{A} = \{a_{l,k} = \int_0^1 \phi''_{j,k}(s)\phi_{j,l}(s)ds\},$$

$$\mathbf{B} = \{b_{l,k} = \int_0^1 \phi_{j,k}(s)\phi_{j,l}(s)ds\},$$

$$\mathbf{B}_1 = \{b_{1,l,k} = \int_0^1 \phi'_{j,k}(s)\Phi\phi_{j,l}(s)ds\}$$

and vectors are $\boldsymbol{\theta} = \{\theta_{j,k} = \theta(k/2^j)\}$ and $f(\boldsymbol{\kappa}) = \{f(\kappa_{j,k}) = f[\kappa(k/2^j)]\}$, and diag($\mathbf{U}$) represents the diagonal matrix with diagonal elements as elements of vector $\mathbf{U}$.

Then, the numerical solution of the bending and post-buckling problems of the rods, except for the buckling solution under the limit or bifurcation loads, can be directly obtained by solving Eq. (9.50). After an additional technique is used to search for the critical value of the bifurcation load under Eq. (9.50), the post-buckling solution can be gained from Eq. (9.50).

As we know, search for critical value of the limit and bifurcation loads is important in the buckling researches of structures. In general, there exist two popular approaches to search for the critical value. Here, the extended system approach is adopted.

In general, the Jacobian matrix, $D_\theta\mathbf{F}(\boldsymbol{\theta}, \alpha)$, of the nonlinear algebraic equations, $\mathbf{F}(\boldsymbol{\theta}, \alpha)$, in Eq. (9.50) becomes singular when load reaches to critical value of the limit or the bifurcation loads. The extended system approach provides an augmentative set of algebraic equations, which transform the singular solution for $\mathbf{F}(\boldsymbol{\theta}, \alpha)$ at the critical load into the regular solution for the new one. Such extended set of algebraic equations for the limit load can be expressed by [13]

$$\begin{cases} \mathbf{F}(\boldsymbol{\theta}, \alpha) = 0, \\ D_\theta \mathbf{F}(\boldsymbol{\theta}, \alpha)\varphi = 0, \\ \mathbf{l}^T \varphi - 1 = 0. \end{cases} \tag{9.51}$$

Following this scheme, the further extended set of equations for the bifurcation load is of the form [14]

$$\begin{cases} \mathbf{F}(\boldsymbol{\theta}, \alpha) + \mu\varphi = 0, \\ D_\theta \mathbf{F}(\boldsymbol{\theta}, \alpha)\varphi = 0, \\ D_\theta \mathbf{F}(\boldsymbol{\theta}, \alpha)\mathbf{v} + D_\alpha \mathbf{F}(\boldsymbol{\theta}, \alpha) = 0, \\ \varphi^T \varphi - 1 = 0, \\ \varphi^T \mathbf{v} = 0, \end{cases} \tag{9.52}$$

where $\boldsymbol{\theta}$ is the bending angle, $\boldsymbol{\varphi}$ and $\boldsymbol{v}$ are eigenvectors, α is the load, $\boldsymbol{l}$ is an arbitrary vector, and μ is an unknown parameter. $\{\boldsymbol{\theta}, \boldsymbol{\varphi}, \alpha\}$ and $\{\boldsymbol{\theta}, \boldsymbol{\varphi}, \boldsymbol{v}, \alpha, \mu\}$ are unknown variables in Eqs. (9.51) and (9.52), respectively. All these nonlinear algebraic systems can be solved by using the Newton–Raphson algorithm.

In accordance with small deformation theory, the governing equation for materials made of slightly buckled rod that follows the constitutive law in Eq. (9.39) becomes [15]

$$E_0 I_0 y'' + Qy = 0. \tag{9.53}$$

To such case, the rod has the non-trivial solution for linear eigenvalue problems when the buckling load is approached by the form

$$Q_{cr} = \frac{\pi^2 E_0 I_0}{4l^2}. \tag{9.54}$$

Then, the normalized buckling load can be rewritten as [15]

$$\alpha_n = \frac{\tilde{n}\varepsilon_0^{\tilde{n}-1}\pi^2}{4}. \tag{9.55}$$

To the cases of the unmodified Ludwick constitutive law, we have known that the elastic modulus E_0 at $\varepsilon = 0$ is an infinite number. Therefore, the buckling load of such rod is inexistent. However, rods may buckle at the bifurcation buckling load when the nonlinear constitutive laws in Eqs. (9.38) and (9.39) are considered in the numerical tracking. In the literature, such theoretical analysis of searching the limit force is extremely complicated and rare.

Case 1: Bending deflection of the rods with nonlinear materials when $\alpha = 0$.

In this case, the rods are subjected by a transverse concentrated load identified by λ. At the following numerical solutions to the rods with nonlinear materials, the index parameters employed in the nonlinear materials are listed in Table 9.4.

For such bending problem of nonlinear rod, the analytical solution of Eq. (9.45) is gained, see in Appendix 9.2. For the thin rod or beam, we take $\beta = 0.02$ in the numerical solutions. When $\tilde{n} = 1/2.16$, the end deflections, δ_v, varying with λ before buckling, in the numerical solutions of the governing Eq. (9.45) by means of the wavelet-based method with the resolution level $j = 6$ and other conventional methods are listed in Table 9.4 as a comparison. From this table, one sees that the accuracy of the wavelet-based method is the highest among the numerical solutions compared with the analytical solution.

In order to find the relative error of curvature, κ, in the calculations conducted by the wavelet-based method, the distributions of the relative error on the deflection angle $\theta_L = 36°$ at the free end varying with the arc length coordinate are plotted in Fig. 9.8 for the resolution levels, $j = 4$, 5, and 6. From this figure, one sees that the relative error is sharply decreased when the resolution level is equal to 6, which leads to the solution of the wavelet-based method for Eq. (9.45) having accuracy higher than the Euler method and FEM, as revealed in Table 9.3.

The applied load λ changes and the rod with different nonlinear parameter $\tilde{n}$ listed in Table 9.3, Fig. 9.9a and b, respectively, display the end deflection-load curves and the relative error of the end deflection angle, θ, of the wavelet-based solutions with $j = 6$.

From Fig. 9.9 (Left), one sees that there is a transition value of λ in the characteristic curves, which is about 1.5. That is, when $0 < \lambda < 1.5$, the end deflection increases with increasing $\tilde{n}$, while when $\lambda > 1.5$, however, the end deflection decreases with $\tilde{n}$. Figure 9.9 (Right) tells us that when $\tilde{n} < 1$, the smaller the $\tilde{n}$, the bigger the relative error. As the load increases, such error decreases monotonously.

Table 9.3 Material constant $\tilde{n}$ for different materials

Material type	Aluminum alloy	Polypropylene plastic	Acetal plastic	Annealed copper	Glass fiber
$\tilde{n}$	1/4.79	1/4.00	1/3.00	1/2.16	1/2.00

Table 9.4 Deflection δ_v obtained using Eq. (9.45) under various loads with $\tilde{n} = 1/2.16$

$\lambda^{1/\tilde{n}}$	Analytical method	Wavelet method	Euler method [15]	Runge–Kutta method [11]	FEM [15]
0.25	0.05974	0.05974	0.05989	0.05973	0.08041
1.0	0.22084	0.22085	0.22125	0.22085	0.22675
4.0	0.53434	0.53434	0.53440	0.53433	0.53199
7.0	0.64418	0.64419	0.64391	0.64414	0.63883
10.0	0.70065	0.70065	0.70028	0.70065	0.69334

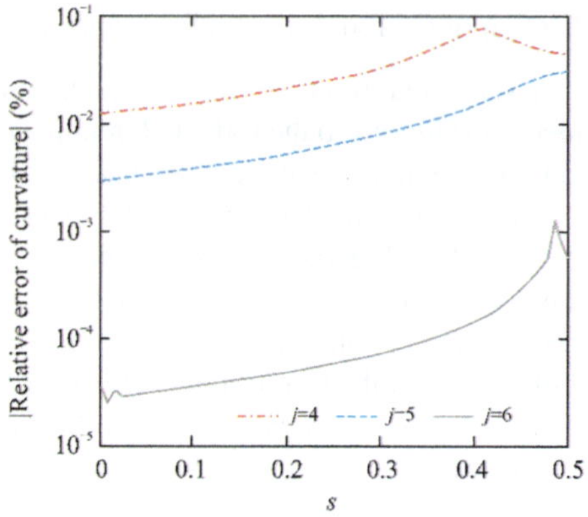

Fig. 9.8 Distribution of the relative error of the wavelet-based solutions with different resolution levels for curvature, κ. Here, s stands for the arc length coordinate along the rod

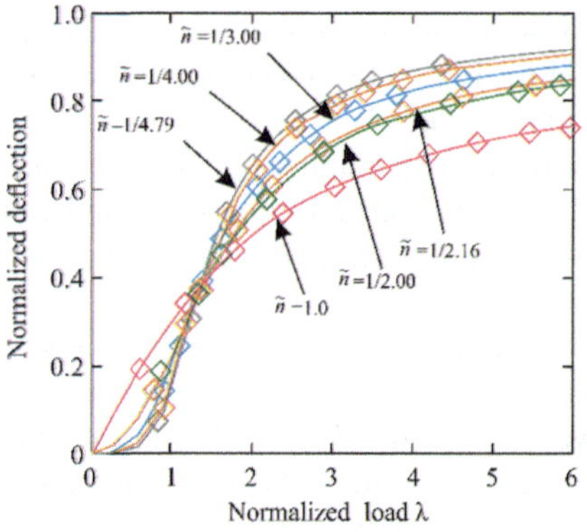

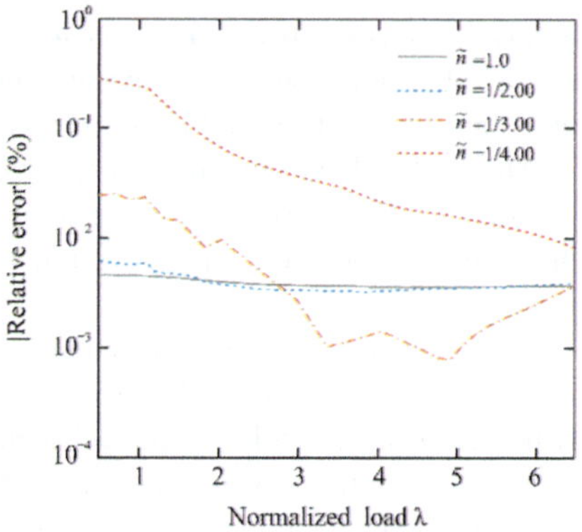

Fig. 9.9 Characteristic curves of end deflection, δ_v, varying with the normalized load, λ, for rods with different material constants, $\tilde{n}$, where the solid lines and the circles represent the wavelet-based solutions and analytical solutions, respectively (Left); Relative error of the end deflection angle varies with the load on the wavelet-based method (Right). Here, $j = 6$

Now, let us see the solutions based on Eq. (9.44) where the modified Ludwick constitutive law is considered. When $j = 6$, the wavelet-based solutions to such governing equation are obtained. The obtained characteristic curves of $\delta_v - \lambda$ are drawn in Fig. 9.10 (Left) for different E/E_0 when $\tilde{n} = 1/2.16$ and Fig. 9.10 (Right) for different $\tilde{n}$. Due to the reason that the rods used by the modified Ludwick-type material are softer than those used by the original Ludwick-type material with the same $\tilde{n}$, it is feasible that the deflection of the rod for the former is larger than the later when the same load is applied.

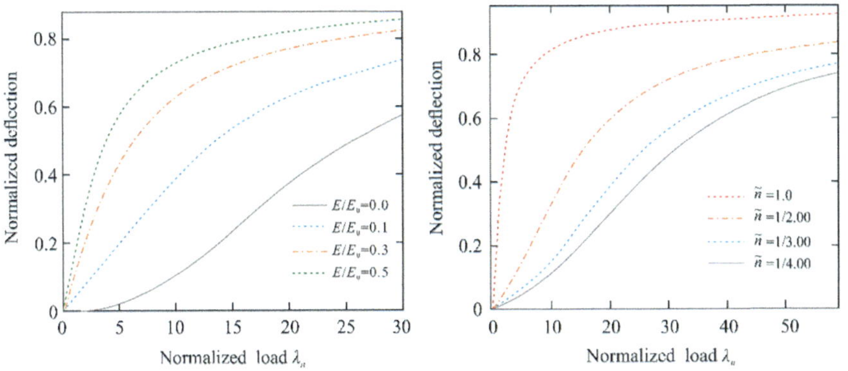

Fig. 9.10 Characteristic curves of $\delta_y - \lambda$ of Eq. (9.44) solved by the wavelet-based method with $j = 6$: $\tilde{n} = 1/2.16$ and different $\varepsilon_0 = (\tilde{n}E/E_0)^{1/(1-\tilde{n})}$ (Left); $\varepsilon_0 = 0.001$ and different $\tilde{n}$ (Right)

From both Fig. 9.10 (Left) and Fig. 9.10 (Right), one sees that there is no transition value of λ in all characteristic curves. When E/E_0 and $\tilde{n}$ increase, the end deflection increases too. And all end deflections increase with the load λ monotonously.

Case 2: Bucking and post-buckling solution of the rod under pressure load.

To the buckling and post-buckling problems of the rods, the load parameters should be taken by $\lambda = 0$ and $\alpha \neq 0$ in Eq. (9.45) and $\lambda_1 = 0$ $\alpha_1 \neq 0$ in Eq. (9.44). For this case, an analytical solution is provided only in the elliptic integral for rods made of a linearly elastic material.

The equilibrium path of the linearly elastic rod occurring buckling is plotted in Fig. 9.11 by using the wavelet-based method with a resolution level of $j = 5$, which displays that the numerical path agrees well with its analytical one.

As a comparison, Fig. 9.12 illustrates the equilibrium path of the rods using the Ludwick-type material with different $\tilde{n}$. Here, the wavelet-based solution is

Fig. 9.11 Comparison of the wavelet-based solution (solid line) with $j = 5$ and analytical solution (circles) on equilibrium path of the rod with $\tilde{n}$=1 before bucking and post-buckling

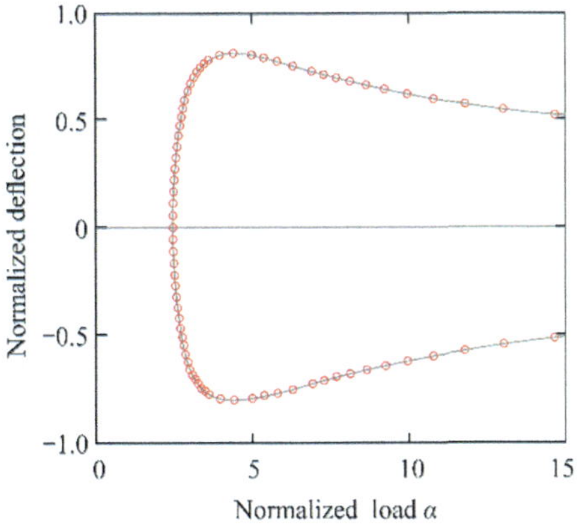

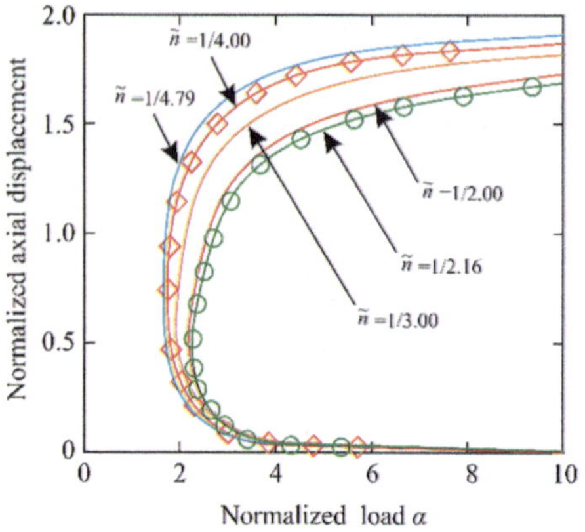

Fig. 9.12 Comparison of the wavelet-based solution (solid line) with $j = 5$ and the shooting method with Runge–Kutta [16] (hollows) on equilibrium path of the rod using the Ludwick-type material with different $\tilde{n}$=1 before bucking and post-buckling

conducted by the resolution level $j = 5$, and the results from the shooting method with Runge–Kutta [16] are also given. From Fig. 9.12, we find that when $\tilde{n} = 0.2$ and 0.25, the wavelet-based solutions are consistent with those obtained by using the shooting method. In addition, after searching for the post-buckling solution within a large load range, no new bifurcation load is found. At the same time, the limit loads of the buckling in Eq. (9.45) are changed with parameters $\tilde{n}$. By taking different resolution levels, such limit loads are obtained by using the wavelet-based Galerkin method with resolution level Eq. (5.2), respectively, see in Table 9.5, where the limit values feasibly obtained by the shooting technique embedded with the fourth-order Runge–Kutta method are also listed.

From Table 9.5, one sees that the tracked limit loads by the wavelet-based method have high accuracy when the resolution level is equal to 5. When the material constant parameters $\tilde{n} = 0.9$ and $\varepsilon_0 = (\tilde{n}/2)^{1/(1-\tilde{n})}$, the characteristic deformation quantities varying with the pressure load parameter α_{n} subjected to the rods governed by Eq. (9.44) during the post-buckling are listed in Table 9.6, where the results of both

Table 9.5 Normalized limit load for Eq. (9.54) with different $\tilde{n}$ values

n	Wavelet-based method			Shooting method [17]
	$j = 3$	$j = 4$	$j = 5$	
1/4.79	1.6429	1.6440	1.6436	–
1/4.00	1.7435	1.7428	1.7425	1.7395
1/3.00	1.9320	1.9308	1.9306	–
1/2.16	2.1951	2.1938	2.1937	–
1/2.00	2.2631	2.2619	2.2617	2.2615
1/1.5	2.5191	2.5180	2.5179	2.5180

Table 9.6 Deformation of rods made of the modified Ludwick-type material with $\tilde{n} = 0.9$ and $\varepsilon_0 = (\tilde{n}/2)^{1/(1-\tilde{n})}$

δ_v	Shooting method [2]			Wavelet method		
	δ_h	θ_L (rad)	α_n	δ_h	θ_L (rad)	α_n
0.1116978	0.0077064	0.1745329	4.42473	0.0077020	0.1741410	4.45954
0.3275932	0.0688380	0.5235988	4.21915	0.0688201	0.5229870	4.21157
0.5186768	0.1867682	0.8726646	4.30857	0.1866340	0.8706506	4.34044
0.6701200	0.3527672	1.2217305	4.61609	0.3524014	1.2187002	4.64783
0.7272017	0.4501450	1.3962634	4.86167	0.4495529	1.3925805	4.89313

the wavelet-based method with $j = 7$ and the shooting method with the fourth-order Runge–Kutta method are also presented as a comparison. Figure 9.13 displays their compassion of the equilibrium path, which also tells us that in the post-buckling of the rods made of the modified Ludwick-type material, bifurcation and limit load of buckling exist.

In accordance with small deformation theory, an analytical formula for the bifurcation load is given by Eq. (9.55) [15]. From this analytical formula, the normalized limit or bifurcation load for the rods with $\varepsilon_0 = 0.01$ and different $\tilde{n}$ values is shown in Table 9.7, where the relative error between the bifurcation loads from the wavelet-based method and from the analytical formula is all within 1% for different materials. This accuracy tells us that the analytical formula is suitable to the case when the rod deformation is small. In addition, the equilibrium paths of these rods from the analytical formula are plotted in Fig. 9.14.

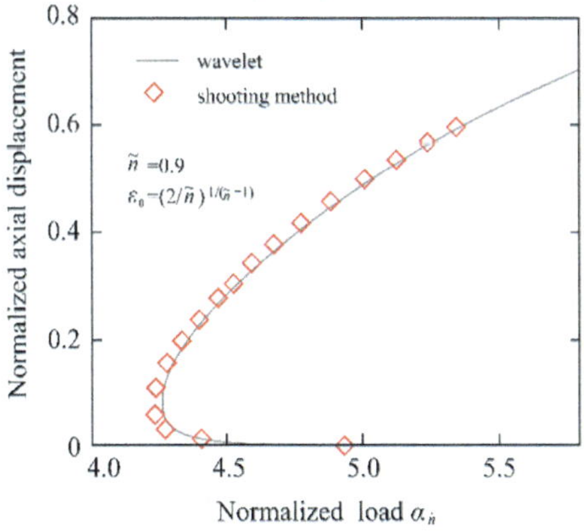

Fig. 9.13 Equilibrium path of the rods made of the modified Ludwick-type material, where the hollow marks represent the numerical solution obtained by using the shooting method with Runge–Kutta [15], and the solid line stands for the wavelet-based solution with $j = 7$

Table 9.7 Normalized bifurcation and limit loads for Eq. (9.54) with $\varepsilon_0 = 0.01$

$\tilde{n}$	Bifurcation load			Limit load
	Analytical	Wavelet	Relative error (%)	Wavelet
1/2.00	12.33701	12.37623	0.31797	11.49021
1/2.16	13.54753	13.61538	0.50085	12.48637
1/3.00	17.71952	17.81428	0.53482	15.71896
1/4.00	19.50652	19.61795	0.57123	16.84478
1/4.79	19.69553	19.70166	0.03112	16.77847

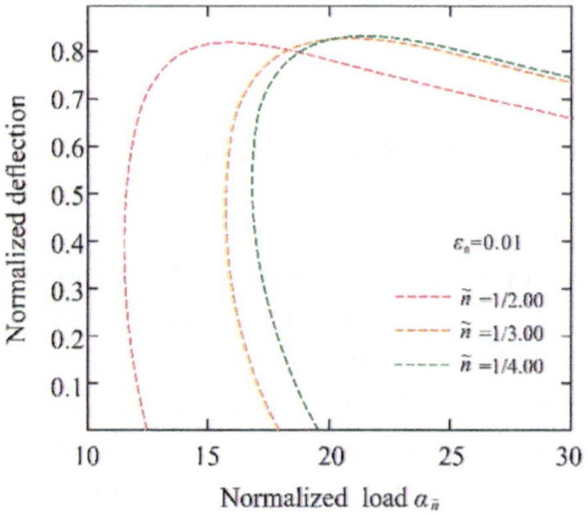

Fig. 9.14 Equilibrium path of the rods made of the modified Ludwick-type material with different $\tilde{n}$ values when the analytical formula of the solution is employed

9.2 Solution of Axisymmetric Deflection of Von Kármán Circular Plate with Strong Nonlinearity

9.2.1 Essential Equations with Two Coupled Unknown Functions

For the bending of a circular plate under an axisymmetric transverse load, the nonlinear von Kármán plate equations can be written as follows:

$$D\frac{1}{r}\frac{d}{dr}r\frac{d}{dr}\frac{1}{r}\frac{d}{dr}r\frac{dw(r)}{dr} = \frac{1}{r}\frac{d}{dr}\left[rN_r(r)\frac{dw(r)}{dr}\right] + q(r), \tag{9.56}$$

$$r\frac{d}{dr}\frac{1}{r}\frac{d}{dr}\left[r^2N_r(r)\right] + \frac{Eh}{2}\left(\frac{dw(r)}{dr}\right)^2 = 0, \tag{9.57}$$

where $D = Eh^3/12(1-\nu^2)$ is the flexural rigidity of the plate; r is the radial coordinate; h is the thickness of the plate; $w(r)$ is the deflection of the plate; $N_r(r)$ is the radial stress per unit length; $q(r)$ is the distribution function of the load exerted on the plate; and E and v are Young's modulus and Poisson's ratio, respectively.

The boundary conditions of the plate can be expressed in the general form

$$r = a: \; w = 0, \; N_r = -k_1 u, \; M_r = -k_2 \frac{dw}{dr}; \tag{9.58}$$

$$r = 0: \; \frac{dw}{dr} = 0, \; N_r = \text{finite}, \tag{9.59}$$

where $M_r(r) = -D\left[w(r)_{,rr} + vw(r)_{,r}/r\right]$ is the radial bending moment of the plate, $u(r) = \left[rN_r(r)_{,r} + (1-v)N_r(r)\right]/Eh$ is the radial displacement of the plate, and k_1 and k_2 are the constants of springs used as constraints at the edge of the plate, and they are dependent on the form the edge support. When the plate is simply supported at the edge, for example, $k_1 = k_2 = 0$.

With the aid of the following dimensionless parameters:

$$y = \left(\frac{r}{a}\right)^2, \; W = \sqrt{3(1-v)^2}\frac{w}{h}, \; S(y) = 3\left(1-v^2\right)\frac{a^2 y N_r}{Eh^3},$$

$$\Phi(y) = y\frac{dW}{dy}, \; pf(y) = \frac{3\left(1-v^2\right)\sqrt{3\left(1-v^2\right)}a^2}{4\pi Eh^4 y}\int_0^r q(r)2\pi r dr, \tag{9.60}$$

where dimensionless p and $f(y)$ are the intensity and distribution function of the load, respectively, Eqs. (9.56)–(9.59) can be reduced to the dimensionless form

$$y^2 d^2\Phi(y)/dy^2 = \Phi(y)S(y) + pf(y), \tag{9.61}$$

$$y^2 d^2 S(y)/dy^2 = -\Phi^2(y)/2 \tag{9.62}$$

and

$$y = 0: \quad \Phi(y) = 0, \; S(y) = 0, \tag{9.63}$$

$$y = 1: \quad \Phi(y) = \frac{\lambda}{\lambda - 1}\frac{d\Phi}{dy}, \; S(y) = \frac{\mu}{\mu - 1}\frac{dS}{dy}, \tag{9.64}$$

where

$$\lambda = \frac{2}{(1+v) + \frac{k_2 r}{D}}, \; \mu = \frac{2}{(1-v) + \frac{Eh}{k_1}}. \tag{9.65}$$

9.2.2 Wavelet-Based Solution Arithmetic

Here, the uniformly distributed load, $q(r) = q =$ constant, is considered for an example. In such case, Eqs. (9.61) and (9.62) of the von Kármán equations become [18–20]

$$y^2 d^2\Phi(y)/dy^2 = \Phi(x)S(y) + y^2 p, \tag{9.66}$$

$$y^2 d^2 S(y)/dy^2 = -\Phi^2(y)/2\,, 0 < y < 1. \tag{9.67}$$

Here, $p = [3(1-\nu^2)]^{3/2} a^4 q/Eh^4$. The boundary conditions associated with Eqs. (9.66) and (9.67) are

$$\Phi(y) = 0,\ S(y) = 0, \quad \text{at} \quad y = 0; \tag{9.68}$$

$$(\lambda - 1)\Phi(y) = \lambda d\Phi/dy \text{ and } (\mu - 1)S(y) = \mu dS(y)/dy, \text{ at } y = 1, \tag{9.69}$$

where λ and μ are parameters related to the boundary condition at $y = 1$. In particular, for the four types of common boundary conditions, we have [18]

(a) Unmovable clamped supports: $\lambda = 0$ and $\mu = 2/(1 - \nu)$.
(b) Movable clamped supports: $\lambda = 0$ and $\mu = 0$.
(c) Simple supports: $\lambda = 2/(1 + \nu)$ and $\mu = 0$.
(d) Simple hinged supports: $\lambda = 2/(1 + \nu)$ and $\lambda = 2/(1 + \nu)$.

Equations (9.66) and (9.67) associated with the boundary conditions, Eqs. (9.68) and (9.69), form the complete mathematical model for the large deflection bending problem of circular thin plates. By solving these equations, the unknown functions $\Phi(y)$ and $S(y)$ can be obtained. Then, other mechanical quantities of the plate in dimensionless form can be calculated through the following relations [18]:

(A) Deflection $W(y) = -\int_y^1 \Phi(\xi)/\xi d\xi$.
(B) Radial membrane force $S_{N_r}(y) = 3(1 - \nu^2)a^2 N_r(y)/Eh^3 = S(y)/y$.
(C) Circumferential membrane force $S_{N_\theta}(y) = 2dS(y)/dy - S(y)/y$.
(D) Radial bending force of the convex $S_{W_r}(y) = (1 - \nu)\Phi(y)/y - 2d\Phi(y)/dy$.
(E) Circumferential bending force of the convex $S_{W_\theta} = (\nu - 1)\Phi(y)/y - 2\nu d\Phi(y)/dy$.

In order to apply the wavelet-based method to solve these two coupled unknowns, here, we introduce two composite functions of the form

$$\widehat{f}(y) = f(y, d^2\Phi/dy^2) = y^2 d^2\Phi/dy^2, \tag{9.70}$$

$$\hat{g}(y) = g(y, d^2S/dy^2) = y^2 d^2 S(y)/dy^2. \tag{9.71}$$

Then, the von Kármán equations can be rewritten as

$$\hat{f}(y) = \Phi(y)S(y) + py^2, \tag{9.72}$$

$$\hat{g}(y) = -\Phi^2(y)/2. \tag{9.73}$$

As an example, the case with movable clamped supports, which requires $\Phi(0) = S(0) = 0$ and $\Phi(1) = S(1) = 0$, is considered. Thus, the unknown functions $\Phi(y)$ and $S(y)$ can be approximated as

$$\Phi(y) \approx \sum_{k=1}^{2^n-1} \Phi_{n,k}\phi_{n,k}(y), \tag{9.74}$$

$$S(y) \approx \sum_{k=1}^{2^n-1} S_{n,k}\phi_{n,k}(y), \tag{9.75}$$

where $f_{n,k} = f(k/2^n)$. From Eqs. (9.74) and (9.75), we can derive

$$d^2\Phi(y)/dy^2 \approx \sum_{k=1}^{2^n-1} \Phi_{n,k}d^2\phi_{n,k}(y)/dy^2, \tag{9.76}$$

$$d^2S(y)/dy^2 \approx \sum_{k=1}^{2^n-1} S_{n,k}d^2\phi_{n,k}(y)/dy^2. \tag{9.77}$$

Denote them by the matrix forms

$$d^2\Phi(y)/dy^2 \approx \mathbf{A}(y)\mathbf{\Psi}, \tag{9.78}$$

$$d^2S(y)/dy^2 \approx \mathbf{A}(y)\mathbf{S}, \tag{9.79}$$

where vectors $A(y) = a_k = d^2\phi_{n,k}(y)/dy^2$, $\mathbf{\Psi} = \{\Phi_{n,k}\}^{\mathrm{T}}$, $\mathbf{S} = \{S_{n,k}\}^{\mathrm{T}}$, and $k = 1, 2, \ldots, 2^n - 1$. The coefficients a_k can be obtained exactly by using the procedure suggested in the Appendix 5.1. and the composite functions $\hat{f}(y)$ and $\hat{g}(y)$ can be expressed as

$$\hat{f}(y) \approx \sum_{i=0}^{2^n} \hat{f}_{n,i}\phi_{n,i}(y) = \sum_{i=0}^{2^n} f\left(y_{n,i}^2, \mathbf{A}_{n,i}\mathbf{\Psi}\right)\phi_{n,i}(y), \tag{9.80}$$

$$\hat{g}(y) \approx \sum_{i=0}^{2^n} \hat{g}_{n,i}\phi_{n,i}(y) = \sum_{i=0}^{2^n} g\left(y_{n,i}^2, \mathbf{A}_{n,i}\mathbf{S}\right)\phi_{n,j}(y). \tag{9.81}$$

Similarly, we have

$$\Phi(y)S(y) \approx \sum_{0}^{2^n} \Phi_{n,i} S_{n,i} \phi_{n,i}(y), \tag{9.82}$$

$$\Phi^2(y) \approx \sum_{i=0}^{2^n} \Phi_{n,i}^2 \phi_{n,i}(y), \tag{9.83}$$

$$y^2 \approx \sum_{i=0}^{2^n} y_{n,i}^2 \phi_{n,i}(y). \tag{9.84}$$

Substituting Eqs. (9.82)–(9.84) into Eqs. (9.66) and (9.67), we get

$$\sum_{i=0}^{2^n} f\left(y_{n,i}^2, \mathbf{A}_{n,i}\boldsymbol{\Psi}\right)\phi_{n,i}(y) \approx \sum_{i=0}^{2} \Phi_{n,i} S_{n,i} \phi_{n,i}(y) + p \sum_{i=0}^{2} y_{n,i}^2 \phi_{n,i}(y), \tag{9.85}$$

$$\sum_{i=0}^{2^x} g\left(y_{n,i}^2, \mathbf{A}_{n,i}\mathbf{S}\right)\phi_{n,i}(y) \approx -\frac{1}{2} \sum_{i=0}^{2^n} \Phi_{n,j}^2 \phi_{n,i}(y). \tag{9.86}$$

Multiplying both sides of Eqs. (9.85) and (9.86) by $\phi_{n,l}(y)$, $l = 1, 2, \ldots, 2^{n-1}$ and then integrating them in the interval [0, 1], one obtains two sets of nonlinear algebraic equations in the matrix form

$$\mathbf{CF}(\boldsymbol{\Psi}) - \mathbf{HR} \approx \mathbf{Q}, \tag{9.87}$$

$$\mathbf{CG}(\mathbf{S}) + \mathbf{HB} \approx 0, \tag{9.88}$$

where matrices

$$\mathbf{C} = \left\{ c_{li} = \int_0^1 \phi_{n,i}(x)\phi_{n,l}(x)dx \right\},$$

$$\mathbf{H} = \left\{ h_{li} = \int_0^1 \phi_{n,l}(x)d^2\phi_{n,i}(x)/dx^2 dx \right\}$$

and vectors $\mathbf{F}(\boldsymbol{\Psi}) = \{f(y_{n,i}^2, \mathbf{A}_{n,i}\boldsymbol{\Psi})\}^{\mathbf{T}}$, $\mathbf{G}(\mathbf{S}) = \{g(y_{n,j}^2, \mathbf{A}_{n,i}\mathbf{S})\}^{\mathbf{T}}$, $\mathbf{R} = \{r_i = \varphi_{n,i} S_{n,i}\}^{\mathbf{T}}$, $\mathbf{B} = \{b_i = \varphi_{n,j}^2/2\}^{\mathbf{T}}$, and $Q = \{q_1 = p\sum_{i=0}^{2^n} y_{n,i}^2 c_{li}\}$; with $i =$

$0, 1, 2, \ldots 2^n$ and $l = 1, 2, \ldots, 2^n - 1$. Notably, the generalized connection coefficients c_{li} and h_{li} can be obtained exactly by using the procedure suggested in Appendix 5.1.

By solving above two sets of nonlinear algebraic equations, Eqs. (9.87) and (9.88), the nodal values of unknown functions $\Phi(k/2^n)$ and $S(k/2^n)$, $k = 1, 2, \ldots, 2^n - 1$ can be obtained, which can be used to reconstruct $\Phi(y)$ and $S(y)$ by means of Eqs. (9.74) and (9.75). After that, the mechanical quantities of the thin plate can be obtained on the basis of the relations shown in the previous subsection.

9.2.3 Numerical Results and Discussions

In this subsection, the large bending behavior of a circular plate under four types of boundary conditions is systemically studied. Poisson's ratio $v = 0.3$, and the classical Newton iteration method is used to numerically solve the nonlinear algebraic equations, Eqs. (9.77) and (9.78), where the initial values $\Phi(k/2^n) = S(k/2^n) = 0$, $k = 1, 2, \ldots, 2^n - 1$ are adopted.

For the clamped support, Fig. 9.15 shows the solution results of deflection configuration of the wavelet-based method for the von Karman circular plate under various dimensionless loads, and the relative error of the center deflection under resolution level $n = 4, 5, 6$, respectively, comparing with the numerical solution of this problem by means of the interpolation iterative method [18].

From Fig. 9.15 (Left), one sees that when the dimensionless load $Q = qa^4/Eh^4$ increases from 10 to 1000, the central deflection w/h of the circular plate increases

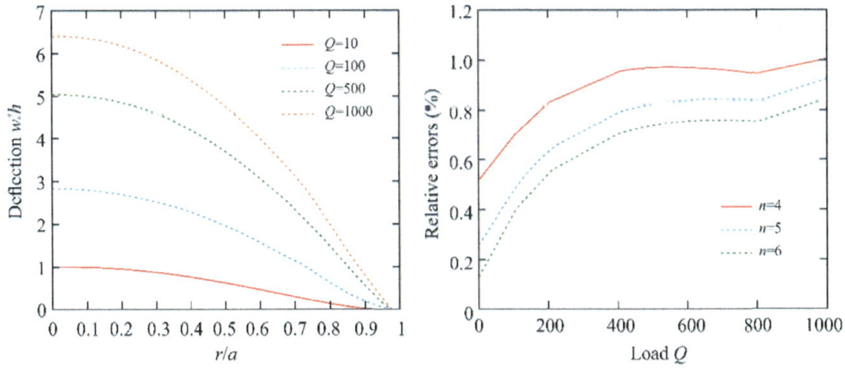

Fig. 9.15 Wavelet solutions for the bending deflection of the von Kármán circular plate with clamped support under different uniform dimensionless loads, $Q = qa^4/Eh^4$: Deflection configurations of w/h varying with r/a when the resolution level, $n = 6$ (Left); Relative errors of deflection at $r = 0$ (center of plate) at resolution levels, $n = 4$, 5, and 6 corresponding to the solution of the interpolation iterative method [18] (Right)

from 1 to 6.3. According to the theory of nonlinear plates, we know that such deflection belongs to the scope of strong nonlinearity. To the central or maximum deflection, its relative error curves shown in Fig. 9.15 (Right) tell us that when the resolution level increases from 4 to 6, the relative errors decrease from the region of about [0.5, 1%]–[0.1, 0.82%]. As the applied load increases from 0 to 1000, the increase of the relative error is not over 1%, which reveals that the wavelet-based solution is steady no matter how the nonlinearity is stronger. For other supports of the nonlinear circular plate, the same properties are ensured in the wavelet-based method.

For different supports of the nonlinear circular plate at the edge, Fig. 9.16 illustrates wavelet-based solutions with $j = 6$ for the bending angle, $\theta = dW/dy$, of the deformed plate under two dimensionless loads: one is corresponding to the case of the plate with small deformation and another is to the strong nonlinearity. All results shown in Fig. 9.16 are in good agreement with the "exact" solutions [18]. It should be noted that the deflection angle $\theta = dW/dy$ displayed here is not its practical value expression due to $y = (r/a)^2$. To the practical deflection angle in dimensionless loads, it is calculated by $\widetilde{\theta} = 2(r/a)dW/dy$. After $\theta = dW/dy$ is obtained, there is no difficulty for one to gain the practical deflection angle in dimensionless case. According to Fig. 9.16, we find that the maximum deflection angle for the clamped circular plates transfers to the edge as the load increases, while it is always located at the edge for the simple supports, and the deflection angles for movable supports are greater than those of the corresponding clamped supports except for the case of clamped supports when the deformation is small, i.e., $Q = 1$.

To quantitatively identify the applicable scopes of linear theory of plates with small deflection and the member theory, we take the circular plate with clamped edge as an example. The comparison of them and the wavelet-based solution for the von Kármán equations is shown in Fig. 9.17, where the central deflection and relative error are given. As shown in this figure, when the dimensionless deflection, $w(0)/h < 0.2$, the relative error between the linear solution and the nonlinear solution from the von Kármán equations can be less than 3%, and the same relative error scope

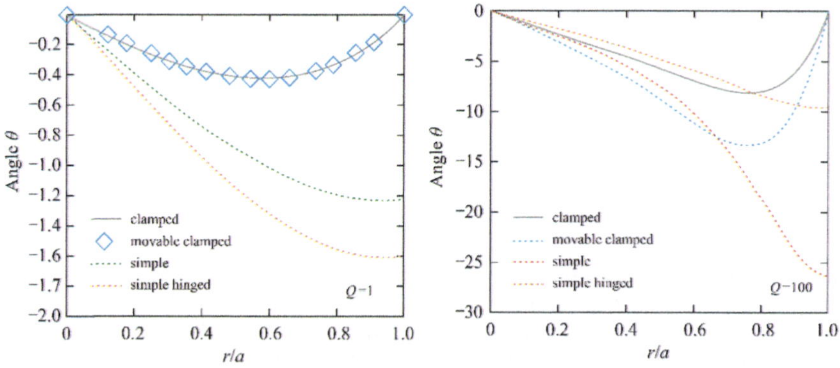

Fig. 9.16 Deflection angle curves of $\theta = dW/dy$ varying with the normalized radial coordinate of the von Kármán circular plates with four common supports: $Q = 1$ (Left); $Q = 100$ (Right)

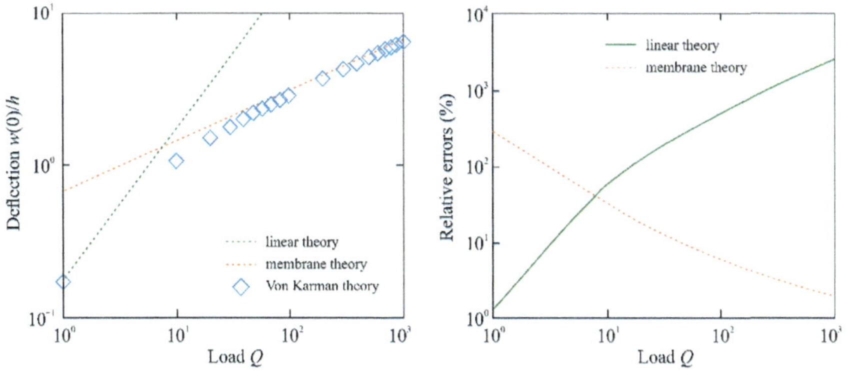

Fig. 9.17 Comparison of the central deflections obtained by using linear, membrane, and von Kármán plate theories for the bending of a circular plate with a clamped boundary varying with applied load: central deflection (Left); relative errors of the solutions from the linear theory and membrane theory corresponding to the von Kármán plate theory (Right)

between the membrane solution and the von Kármán solution requires $w(0)/h > 5$. Since the geometrically nonlinear plate theory was proposed by von Kármán [20], it has attracted much more attention of the researches in the nonlinear scientific committee of solid mechanics [21–36]. As early as 1950s, the researchers solving the von Kármán equations know that such nonlinear equations can be effectively solved by the regular perturbation methods from the linear solutions for the weak nonlinearity and by the irregular perturbation methods from the membrane solutions for the strong nonlinearity [21–24]. However, such two kinds of solution methods cannot be intersected. In other words, once the applied load is located at the scope over the feasible scopes of these two methods, e.g., $Q = 100$, the solutions from the two perturbation methods all lose efficiency. This problem which appeared in solving the von Kármán plate equations was called as the "transition problem" [25]. In 1990s, the author and his colleague using the interpolation iteration method with variable load step and interpolation factor solved this problem, that is, the solution from the von Kármán plate equations can be got from the linear solution of the plate with small deformation to the membrane solution of the plate with very large deflection except for difference at the edge when the plate is clamped. Here, the wavelet-based method successfully provides a unified solution approach to solve the nonlinear equations from linear through weak nonlinear to strong nonlinear problems contained in the von Kármán plate equations.

Based on the above wavelet-based solutions, Fig. 9.18a–f shows other concentrated internal force quantities in the circular plate with different supports varying with the applied load, containing the radial membrane forces and the circumferential membrane forces, etc. at the center and edge, respectively.

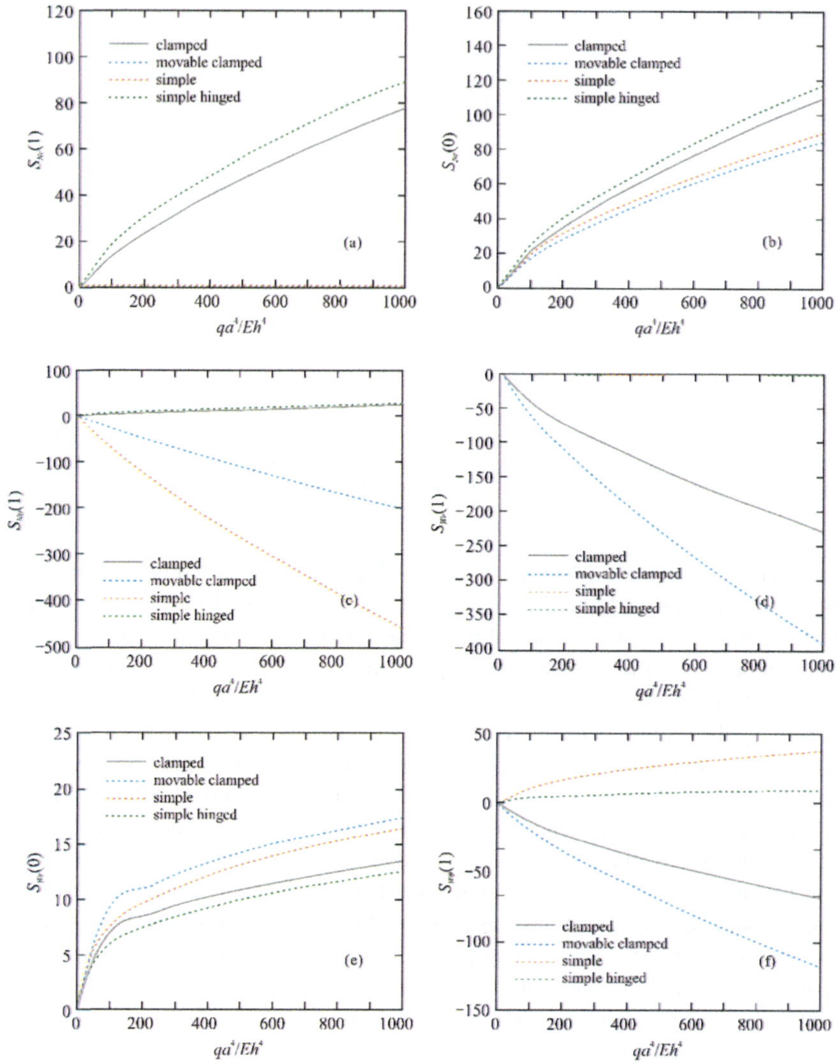

Fig. 9.18 Characteristic curves of internal forces varying the applied load of the von Kármán plates with different supports: **a** Radial membrane force at the edge; **b** Radial membrane force at the center; **c** Circumferential membrane force at the edge; **d** Radial bending force at the edge; **e** Radial bending force at the center; **f** Circumferential bending force at the edge

9.3 Wavelet-Based Solution to Von Kármán Equations of Rectangular Thin Plates

9.3.1 Essential Equations

Such von Kármán theory for the rectangular thin plates with large deflection is a typical set of 2D nonlinear problems with two coupled nonlinear partial differential equations of two unknown functions. Considering the rectangular plates occupied in the space region of [0, a][0, b], we can write the von Kármán equations [37] as follows:

$$\begin{aligned} & D\nabla^4 w = q(x, y) + h\mathrm{H}(w, f) \quad x \in [0, a], \\ & \nabla^4 f = -\tfrac{E}{2}\mathrm{H}(w, w) \qquad\qquad y \in [0, b], \end{aligned} \tag{9.89}$$

where $w(x, y)$ is the deflection function of the plate, $f(x, y)$ stands for the stress function, h is the thickness of the plate, E is Young's modulus, $D = Eh^3/12(1 - \nu^2)$ is the bending stiffness, ν is Poisson's ratio, ∇^4 is the biharmonic operator, and the operator $\mathrm{H}(w, f)$ is defined as

$$\mathrm{H}(w, f) = \frac{\partial^2 f}{\partial x^2}\frac{\partial^2 w}{\partial y^2} + \frac{\partial^2 f}{\partial y^2}\frac{\partial^2 w}{\partial x^2} - 2\frac{\partial^2 f}{\partial x \partial y}\frac{\partial^2 w}{\partial x \partial y}. \tag{9.90}$$

In this study, the plate is assumed to be in-plane movable clamped or in-plane movable simply supported, and thus, the corresponding boundary conditions can be given by

$$w = \frac{\partial w}{\partial x} = 0, \; N_x = N_{xy} = 0 \,\text{at}\, x = 0, a\,; \tag{9.91a}$$

$$w = \frac{\partial w}{\partial y} = 0, \; N_y = N_{xy} = 0 \,\text{at}\, y = 0, b\,; \tag{9.91b}$$

$$w = \frac{\partial^2 w}{\partial x^2} = 0, \; N_x = v = 0 \,\text{at}\, x = 0, a\,; \tag{9.92a}$$

$$w = \frac{\partial^2 w}{\partial y^2} = 0, \; N_y = u = 0 \,\text{at}\; y = 0, b\,; \tag{9.92b}$$

where N_x, N_y, and N_{xy} are in-plane membrane forces; and u and v are displacements along the x- and y-axes. Following a simple derivation, Eqs. (9.91a), (9.91b) and (9.92a), (9.92b) can be equivalently replaced by [40]

$$w = \frac{\partial w}{\partial x} = 0,\ f = \frac{\partial f}{\partial x} = 0, \text{ at } x = 0,a; w = \frac{\partial w}{\partial y} = 0,\ f = \frac{\partial f}{\partial y} = 0, \text{ at } y = 0,b; \tag{9.93}$$

$$w = \frac{\partial^2 w}{\partial x^2} = 0,\ f = \frac{\partial^2 f}{\partial x^2} = 0, \text{ at } x = 0,a; w = \frac{\partial^2 w}{\partial y^2} = 0,\ f = \frac{\partial^2 f}{\partial y^2} = 0, \text{ at } y = 0,b. \tag{9.94}$$

By introducing the following dimensionless quantities:

$$\overline{w} = \frac{w}{h},\ \overline{f} = \frac{f}{Eh^2},\ \overline{x} = \frac{x}{a},\ \overline{y} = \frac{y}{b},\ \beta = \frac{a}{b} \text{ and } \overline{q}(\overline{x}, \overline{y}) = \frac{q(x, y)a^4}{Eh^4}, \tag{9.95}$$

Equations (9.89)–(9.94) can be rewritten as the dimensionless form

$$\begin{aligned} \overline{\nabla}^4 \overline{w} &= \overline{q}(\overline{x}, \overline{y})\lambda + \beta^2 \lambda L(\overline{w}, \overline{f}) \quad && \overline{x} \in [0, 1], \\ \overline{\nabla}^4 \overline{f} &= -\tfrac{\beta^2}{2} L(\overline{w}, \overline{w}) && \overline{y} \in [0, 1] \end{aligned} \tag{9.96}$$

and

$$\begin{aligned} \overline{w}|_{\overline{x}=0,1} &= \left.\frac{\partial^2 \overline{w}}{\partial \overline{x}^2}\right|_{\overline{x}=0,1} = 0,\quad \overline{f}|_{\overline{x}=0,1} = \left.\frac{\partial^2 \overline{f}}{\partial \overline{x}^2}\right|_{\overline{x}=0,1} = 0, \\ \overline{w}|_{\overline{y}=0,1} &= \left.\frac{\partial^2 \overline{w}}{\partial \overline{y}^2}\right|_{\overline{y}=0,1} = 0,\quad \overline{f}|_{\overline{y}=0,1} = \left.\frac{\partial^2 \overline{f}}{\partial \overline{y}^2}\right|_{\overline{y}=0,1} = 0, \end{aligned} \tag{9.97}$$

where $\lambda = 12(1 - \nu^2)$ and operator $\overline{\nabla}^4 = \partial^4/\partial\overline{x}^4 + 2\beta^2\partial^4/\partial\overline{x}^2\partial\overline{y}^2 + \beta^4\partial^4/\partial\overline{y}^4$. For simplicity, the bar over each dimensionless quantity is dropped, except for the operator $\overline{\nabla}^4$.

We further define

$$w_1(x, y) = w_{,xx}(x, y),\ w_2(x, y) = w_{,xy}(x, y),\ w_3(x, y) = w_{,yy}(x, y), \tag{9.98a}$$

$$f_1(x, y) = f_{,xx}(x, y),\ f_2(x, y) = f_{,xy}(x, y),\ f_3(x, y) = f_{,yy}(x, y). \tag{9.98b}$$

Here, $w_{,xy}(x, y) = \partial^2 w(x, y)/\partial x \partial y$ and similar symbols of partial derivatives to others in Eqs. (9.98a) and (9.98b). Further, Eq. (9.89) can be rewritten as

$$\begin{aligned} \overline{\nabla}^4 w &= \beta^2\lambda(f_1 w_3 + f_3 w_1 - 2 f_2 w_2) + \lambda q(x, y), \\ \overline{\nabla}^4 f &= \beta^2(w_2^2 - w_1 w_3). \end{aligned} \tag{9.99}$$

9.3.2 Arithmetic of Wavelet-Based Solution Method

When the wavelet Galerkin method is used to solve nonlinear differential equations in Eq. (9.99), we firstly have the following series expansions:

$$w(x,y) \approx \sum_{i=1}^{2^n-1}\sum_{j=1}^{2^n-1} w(\frac{i}{2^n},\frac{j}{2^n})\phi_{n,i}(x)\phi_{n,j}(y), \tag{9.100a}$$

$$f(x,y) \approx \sum_{i=1}^{2^n-1}\sum_{j=1}^{2^n-1} f(\frac{i}{2^n},\frac{j}{2^n})\phi_{n,i}(x)\phi_{n,j}(y), \tag{9.100b}$$

$$w_r(x,y) \approx \sum_{i=0}^{2^n}\sum_{j=0}^{2^n} w_r(\frac{i}{2^n},\frac{j}{2^n})\phi_{n,i}(x)\phi_{n,j}(y), \tag{9.100c}$$

$$f_s(x,y) \approx \sum_{i=0}^{2^n}\sum_{j=0}^{2^n} f_s(\frac{i}{2^n},\frac{j}{2^n})\phi_{n,i}(x)\phi_{n,j}(y), \tag{9.100d}$$

$$w_r f_s \approx \sum_{i=0}^{2^n}\sum_{j=0}^{2^n} w_r(\frac{i}{2^n},\frac{j}{2^n}) f_s(\frac{i}{2^n},\frac{j}{2^n})\phi_{n,i}(x)\phi_{n,j}(y),$$

$$w_r w_s \approx \sum_{i=0}^{2^n}\sum_{j=0}^{2^n} w_r(\frac{i}{2^n},\frac{j}{2^n}) w_s(\frac{i}{2^n},\frac{j}{2^n})\phi_{n,i}(x)\phi_{n,j}(y),$$

$$q(x,y) \approx \sum_{i=0}^{2^n}\sum_{j=0}^{2^n} q_{ij}\phi_{n,i}(x)\phi_{n,j}(y). \tag{9.100e}$$

Here, $r, s = 1, 2, 3$. Substituting the expansions of Eqs. (9.100a), (9.100b), (9.100c), (9.100d) and (9.100e) into Eqs. (9.98a), (9.98b) and applying the Galerkin method to the induced equations, we get the algebraic equations of the form

$$\mathrm{A}_1\mathbf{W}_r = \mathbf{B}_r\mathbf{W},\ \mathrm{A}_1\mathbf{F}_r = \mathbf{B}_r\mathbf{F}, \tag{9.101}$$

where matrices and vectors

$$\begin{gathered}
\mathbf{A}_1 = \{a_{1,kl,ij} = \Omega_{kl,ij}\}, \mathbf{B}_1 = \{b_{1,kl,ij} = \Upsilon_{kl,ij}^{20,00}\}, \\
\mathbf{B}_2 = \{b_{2,kl,ij} = \Upsilon_{kl,ij}^{11,00}\}, \mathbf{B}_3 = \{b_{3,kl,ij} = \Upsilon_{kl,ij}^{02,00}\}, \\
\mathbf{W}_r = \{w_{r,ij} = w_r(i/2^n, j/2^n)\}, \mathbf{W} = \{w_{ij} = w(i/2^n, j/2^n)\}, \\
\mathbf{F} = \{f_{ij} = f(i/2^n, j/2^n)\}, \mathbf{F}_r = \{f_{r,ij} = f_r(i/2^m, j/2^m)\}
\end{gathered}$$

and $r = 1, 2, 3$, in which

$$\Omega_{kl,ij} = \int_0^1 \phi_{n,i}(x)\phi_{n,k}(x)dx \int_0^1 \phi_{n,j}(y)\phi_{n,l}(y)dy,$$

$$\Upsilon_{kl,ij}^{de,00} = \int_0^1 \phi_{n,i}^{(d)}(x)\phi_{n,k}(x)dx \int_0^1 \phi_{n,j}^{(e)}(y)\phi_{n,l}(y)dy.$$

Then, substitution of the wavelet series expansions of Eqs. (9.100a), (9.100b), (9.100c), (9.100d) and (9.100e) into Eq. (9.99) yields

$$\begin{aligned}&\sum_{i=1}^{2^n-1}\sum_{j=1}^{2^n-1} w_{ij}\overline{\nabla}^4[\phi_{n,i}(x)\phi_{n,j}(y)]\\&=\lambda\sum_{i=1}^{2^n}\sum_{j=1}^{2^n}[\beta^2(f_{1,ij}w_{3,ij}-2f_{2,ij}w_{2,ij}+f_{3,ij}w_{1,ij})+q_{ij}]\phi_{n,i}(x)\phi_{n,j}(y)\end{aligned} \tag{9.102a}$$

$$\begin{aligned}&\sum_{i=1}^{2^n-1}\sum_{j=1}^{2^n-1} f_{ij}\overline{\nabla}^4[\phi_{n,i}(x)\phi_{n,j}(y)]\\&=\beta^2\sum_{i=0}^{2^n}\sum_{j=0}^{2^n}(w_{2,ij}^2-w_{1,ij}w_{3,ij})\phi_{n,i}(x)\phi_{n,j}(y).\end{aligned} \tag{9.102b}$$

Multiplying both sides of Eqs. (9.102a) and (9.102b) by the scaling bases functions $\phi_{n,k}(x)\phi_{n,l}(y)$, $k, l = 1, 2, \ldots, 2^n - 1$, and then performing integration over $[0, 1]^2$, we can get

$$\begin{aligned}\mathbf{AW}=&\lambda\mathbf{A}_2[\beta^2\mathrm{diag}(\mathbf{A}_1^{-1}\mathbf{B}_1\mathbf{F})(\mathbf{A}_1^{-1}\mathbf{B}_3\mathbf{W})-2\beta^2\mathrm{diag}(\mathbf{A}_1^{-1}\mathbf{B}_2\mathbf{F})(\mathbf{A}_1^{-1}\mathbf{B}_2\mathbf{W})\\&+\beta^2\mathrm{diag}(\mathbf{A}_1^{-1}\mathbf{B}_3\mathbf{F})(\mathbf{A}_1^{-1}\mathbf{B}_1\mathbf{W})+\mathbf{Q}],\end{aligned} \tag{9.103a}$$

$$\mathbf{AF}=\beta^2\mathbf{A}_2[\mathrm{diag}(\mathbf{A}_1^{-1}\mathbf{B}_2\mathbf{W})(\mathbf{A}_1^{-1}\mathbf{B}_2\mathbf{W})-\mathrm{diag}(\mathbf{A}_1^{-1}\mathbf{B}_1\mathbf{W})(\mathbf{A}_1^{-1}\mathbf{B}_3\mathbf{W})], \tag{9.103b}$$

where

$$\begin{aligned}\mathbf{A}&=\{a_{kl,ij}=\Upsilon_{kl,ij}^{40,00}+2\beta^2\Upsilon_{kl,ij}^{22,00}+\beta^4\Upsilon_{kl,ij}^{04,00}\},\\\mathbf{A}_2&=\{a_{2,kl,ij}=\Upsilon_{ij,kl}^{00,00}\},\\\mathbf{Q}&=\{q_{ij}=q(i/2^n, j/2^n)\}\end{aligned}$$

and diag(**U**) represents the diagonal matrix with the diagonal elements to be the elements of vector **U**. The generalized connect coefficients, $\Upsilon_{kl,ij}^{de,00}$, and $\Omega_{kl,ij}$ can be obtained exactly by using the procedure suggested in Appendix 5.1. According to Sect. 5.1, we can directly discretize the boundary conditions to replace the corresponding equations in equation group (9.103a) and (9.103b).

Equations (9.103a) and (9.103b) is a set of nonlinear algebraic equations with $2 \times (2^n - 1)^2$ unknown variables. By adopting the classical Newton iteration method to solve the nonlinear algebraic equations, Eqs. (9.103a and 9.103b), the nodal values of the unknown functions $w(i/2^n, j/2^n)$ and $f(i/2^n, j/2^n)$, $i, j = 1, 2, \ldots, 2^n - 1$ can be obtained. After that, the solutions of $w(x, y)$ and $f(x, y)$ can be reconstructed by their expansions.

9.3.3 Examples of Numerical Solution

To evaluate the accuracy and convergence of the proposed wavelet Galerkin method, a similar equation with the same boundary conditions as the von Kármán equations in Eqs. (9.97) and (9.99) is first considered as follows:

$$\begin{aligned} \overline{\nabla}^4 w &= \beta^2 \lambda (f_1 w_3 + f_3 w_1 - 2 f_2 w_2) + \lambda q(x, y), \\ \overline{\nabla}^4 f &= \beta^2 (w_2^2 - w_1 w_3) + h(x, y) \end{aligned} \tag{9.104}$$

where

$$\begin{aligned} q(x, y) &= \frac{w_0 \pi^4}{\lambda}(1 + 2\beta^2 + \beta^4)\sin(\pi x)\sin(\pi y) + w_0 f_0 \beta^2 \pi^4 [\cos(2\pi x) + \cos(2\pi y)], \\ h(x, y) &= f_0 \pi^4 (1 + 2\beta^2 + \beta^4)\sin(\pi x)\sin(\pi y) - \frac{w_0^2 \beta^2 \pi^4}{2}[[\cos(2\pi x) + \cos(2\pi y)]]. \end{aligned} \tag{9.105}$$

For this case, the analytic solutions can be given by

$$w(x, y) = w_0 \sin \pi x \sin \pi y, \ f(x, y) = f_0 \sin \pi x \sin \pi y. \tag{9.106}$$

To Eqs. (9.104) and (9.105), the numerical solutions are obtained by above program of the wavelet-based method with resolution levels $n = 4$, 5, and 6, respectively. Figure 9.19 illustrates the relative error distribution along x-direction of the numerical solution for $w(x, 0.5)$. From Fig. 9.19 (Left), one finds that each relative error distribution is located in a very small region, and the relative error decreases with one order when the resolution level increases each one. When the resolution level $n = 6$, the wavelet-based solution has an accuracy within 0.004%. As shown in Fig. 9.19 (Right), such decreasing error has the scaling dependence of ~$2^{-3.2n}$ on the number of grid points $(2^n - 1)^2$ to the absolute error of the wavelet solution $w(0.5, 0.5)$. When the plate aspect ratio $\beta(=a/b)$ changes from 1 to 10, further, Fig. 9.20 plots

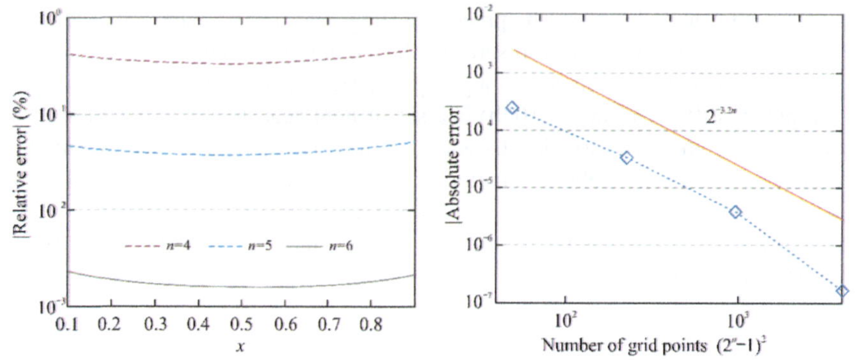

Fig. 9.19 Error distribution and its scaling dependence of the wavelet-based solutions for the nonlinear Eq. (9.104) with exact solution shown in Eq. (9.106) when $w_0 = f_0 = 0.01$, $\beta = 1$ and the resolution level $n = 4, 5$, and 6 (corresponding to the number of grid points $(2^n - 1)^2$), respectively: Relative error distribution of the solution $w(x, 0.5)$ (Left); scaling dependence of absolute error of solution $w(0.5, 0.5)$ on the grid points (Right)

Fig. 9.20 Absolute values of relative error of the wavelet-based solution of Eqs. (9.104) and (9.105) for $w(0.5, 0.5)$ as a function of the plate aspect ratio β when the resolution levels, $n = 4$, 5, and 6, respectively. Here $w_0 = f_0 = 0.01$

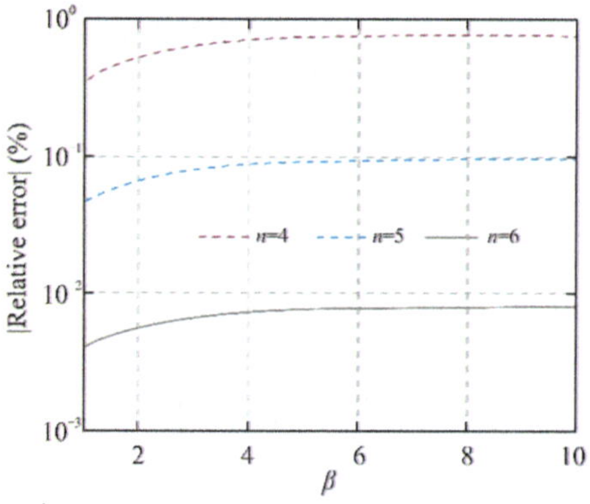

the relative error of the wavelet-based solution $w(0.5, 0.5)$ of the nonlinear equations with resolution levels 4, 5, and 6, respectively. As shown in Fig. 9.20, we find that to each resolution level, the relative error steadily changes with a small increase as the aspect ratio increases. As we know, when $\beta = 1$, the region of $[0, a][0, b]$ is a square in the (x, y)-plane; when $\beta = 10$, the region of $[0, a][0, b]$ becomes a band linking a corridor. To all these different rectangular regions, the wavelet-based method has all satisfied solutions to the nonlinear and coupled differential equations. When $n = 6$, for example, such relative errors all are within 0.01%.

Now, let us give the numerical results of the wavelet-based method for solving the von Kármán plate equations in the rectangular form subjected to a uniform transverse load. For the resaving space, here, we only give them in the case of square plate as an example. The obtained characteristic deflection-load curves obtained by the wavelet-based solution with the resolution levels 3, 4, and 5, respectively, are drawn in Fig. 9.21a for the simply supports and Fig. 9.21c for the clamped supports at the edges when $\beta = 1$, $\lambda = 12 \times (1 - 0.316^2)$. The corresponding consumption time of the calculations in their numerical solutions varies with the grid number or resolution level is drawn in Fig. 9.21b and d, respectively. As a comparison, those conventional methods, such as FEM with shell elements [41] of S4R, S4R5, S8R, and S8R5, respectively, are employed to numerically solve the same rectangular plate of the von Kármán plate equations, which can be found by the commercial software packages of the Abaqus. Here, the Abaqus is used to perform FEM analysis on this problem and the same computer is employed. The FEM results are also plotted in Fig. 9.21a–d.

From Fig. 9.21a and d, one sees that the FEM solution results have obvious difference at the stage of the normalized load being greater than 100 when FEM is taken by different grids. Here, there is almost no such notable difference in the wavelet-based solutions with the resolution levels 3, 4, 5, respectively. When the elements of FEM are taken as 32*32, its solution approaches to the wavelet-based solution. From Fig. 9.21b and d, however, the calculation time consumption for FEM

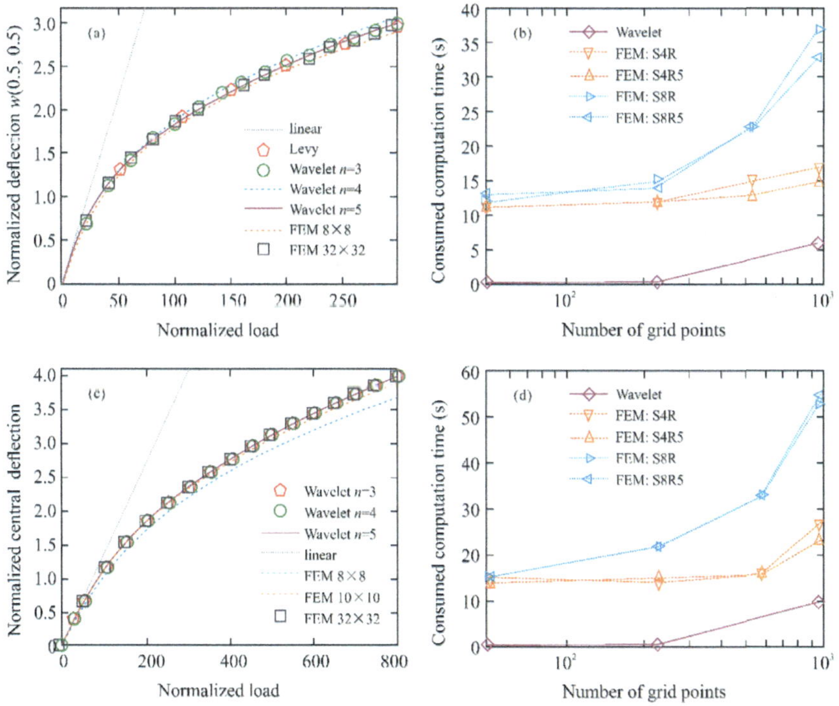

Fig. 9.21 Deflection–load characteristic curves and calculation time varying with grids of the numerical methods employed in solving the von Kármán equation of square plates under uniform transverse load: **a** and **b** for simple supports; **c** and **d** for clamped supports. All results are taken in a same PC

Table 9.8 Central deflection of a square plate under uniformly distributed loads

q	FEM (32 × 32)	Wavelet (15 × 15)	RPIM [39] (11 × 11)	RBF [38] (9 × 9 + 36)
10	0.41289	0.4062	0.42892	0.43636
20	0.72269	0.7204	0.74655	0.75331
30	0.95377	0.9592	0.98039	1.00341
40	1.13770	1.1506	1.15196	–
50	1.29092	1.3111	1.32353	–

is about 10 s; to about 35 s for the simply supports; and to about 55 s for the clamped supports when the grids increase from 8*8 to 32*32. However, the calculation time consumption for wavelet-based method has only a litter increase of about 5 s for the simple supports and of about 10 s for the clamped supports for the same increasing number of grids. These results tell us that the wavelet-based solutions have accuracy and efficiency higher than FEM.

Table 9.8 presents a quantitative comparison of the central deflections of the simply supported plate subjected to different transverse uniform loads among the numerical solutions obtained by the wavelet-based method, the FEM, and other two different meshless methods [38, 39] with different numbers of grid points. When the grids of 15 × 15 are employed in the wavelet-based method, the obtained results are all close to the results of FEM with 32 × 32 grids. However, the results from the meshless methods [38, 39] all have notable deviation from them.

As the FEM uses some shell elements to solve extra-large deflection problems of von Kármán plate equations, such numerical solution method has some discrepancy caused by the shear locking numerical problems of FEM [39] when the ratio of length to thickness increases up to 10^6 and the ratio is less than 10. However, the wavelet-based method has no such discrepancy. When the central deflection of the simply supported square plate approaches to about 1.8 times of the thickness of the plate, Fig. 9.22 illustrates the central deflection of the simply supported plate varying with the length-to-thickness ratio. As shown in Fig. 9.22, the wavelet-based solution

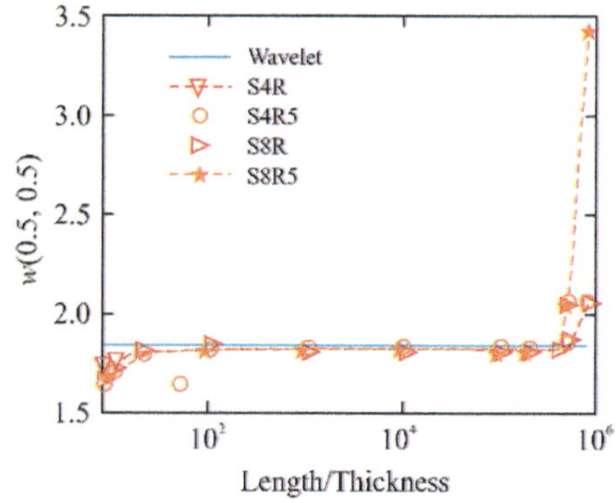

Fig. 9.22 Comparison between the normalized central deflections obtained by using the FEM with mesh 8 × 8 and the wavelet-based method with $n = 3$ to the simply supported square plate varying with the length–thickness ratio

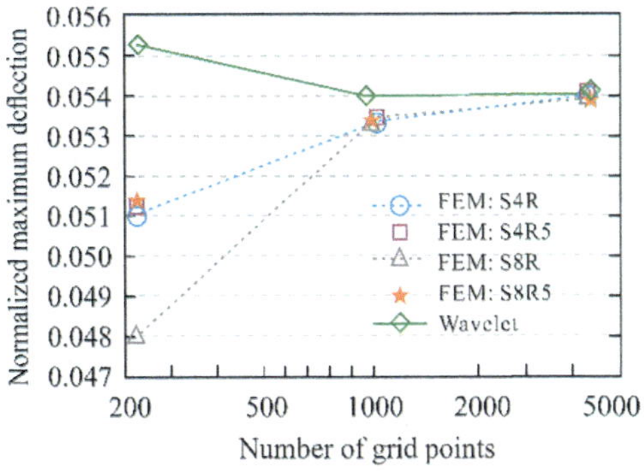

Fig. 9.23 Normalized maximum deflection of a simply supported square plate under the normalized load, $q(x, y) = 500\sin(4\pi x)\sin(4\pi y)$, varying with the number of grid points

has a constant output of the normalized deflection even when the resolution level is taken as 3, while the FEM has a notable or large discrepancy.

Finally, we give the example used in literature for the simply supported square plate governed by the von Kármán equations, where the external load is containing high spatial frequency, e.g., $q(x, y) = 500\sin(4\pi x)\sin(4\pi y)$. Figure 9.23 plots the normalized central deflection of the numerical solutions by the wavelet-based method and the FEM using different grids. From this figure, it is notably found that if one expects to obtain a steady solution for this problem with high spatial frequency, both the wavelet-based method and the FEM should take much more grids in the calculations, which is coincident with the quantitative analysis of the spectrum characteristic for the wavelet-based method in Chap. 2.

9.4 Solution to Other Nonlinear Problems of Beam and Plate Structures

9.4.1 Deflection of Rectangular Thin Plate with Third-Order Power Nonlinearity

This nonlinear problem is similar to a plate over a foundation with nonlinear elastic stiffness. Such nonlinear governing equation for a simply supported rectangular plate can be written as [42, 43]

$$\nabla^4 w + k_1 w + k_2 w^3 = q(x, y) \tag{9.107a}$$

associated with the boundary conditions

$$w(0, y) = 0, w(1, y) = 0,$$
$$w(x, 0) = 0, w(x, 1) = 0,$$

$$w_{,xx}(0,y)=0,\ w_{,xx}(1,y)=0,$$
$$w_{,yy}(x,0)=0,\ w_{,yy}(x,1)=0. \tag{9.107b}$$

Applying Eqs. (5.2), (5.5), and (5.6) to functions in Eqs. (9.107a) and (9.107b), we get

$$w(x,y)\approx\sum_{i=1}^{2^n-1}\sum_{j=1}^{2^n-1}w_{ij}\phi_{n,i}(x)\phi_{n,j}(y), \tag{9.108}$$

$$w^3(x,y)\approx\sum_{i=1}^{2^n-1}\sum_{j=1}^{2^n-1}w_{ij}^3\phi_{n,i}(x)\phi_{n,j}(y), \tag{9.109}$$

where $w_{ij}=w\left(\frac{i}{2^n},\frac{j}{2^n}\right)$. Substitution of above two functions into Eqs. (9.107a) and (9.107b) yields

$$\sum_{i=1}^{2^n-1}\sum_{j=1}^{2^n-1}w_{ij}\nabla^4[\phi_{n,i}(x)\phi_{n,j}(y)]+\sum_{i=1}^{2^n-1}\sum_{j=1}^{2^n-1}(k_1w_{ij}+k_2w_{ij}^3)\phi_{n,i}(x)\phi_{n,j}(y)\approx q(x,y). \tag{9.110}$$

Multiplying both sides of Eq. (9.110) by the weight function $\phi_{n,i'}(x)\phi_{n,j'}(y)$ taking the integration in the region, we get the following algebraic equations in the matrix form:

$$\mathbf{AW}+k_2\mathbf{GW}_3=\mathbf{Q}, \tag{9.111}$$

where $\mathbf{W}=\{w_k\}$, $\mathbf{W}_3=\{w_k^3\}$, $w_k=w_{i,j}$, $k=(i-1)2^n+j$, coefficient matrices $\mathbf{A}$ and $\mathbf{G}$, and loading vector $\mathbf{Q}$ are

$$\begin{aligned}\mathbf{A}&=\{\int_0^1\int_0^1\phi_{n,i'}(x)\phi_{n,j'}(y)\{k_1\phi_{n,i}(x)\phi_{n,j}(y)+\nabla^4[\phi_{n,i}(x)\phi_{n,j}(y)]dxdy\}\}\\ \mathbf{G}&=\{\int_0^1\int_0^1\phi_{n,i'}(x)\phi_{n,j'}(y)\phi_{n,i}(x)\phi_{n,j}(y)dxdy\},\\ \mathbf{Q}&=\{\int_0^1\int_0^1\phi_{n,i'}(x)\phi_{n,j'}(y)q(x,y)dxdy\},\\ &i',j',ij=1,2,\ldots\ldots,2^n-1.\end{aligned} \tag{9.112}$$

When $q(x,y)=(25\pi^4+k_1)\sin(2\pi x)\sin(\pi y)+k_2[\sin(2\pi x)\sin(\pi y)]^3$, one gets exact solution $w(x,y)=\sin(2\pi x)\sin(\pi y)$. In the following numerical example to this case, we take $k_1=0.1$ and $k_2=0.2$. After w_{ij} are obtained by solving Eq. (9.111) by using the Newton–Raphson method, then the approximate solution can be obtained

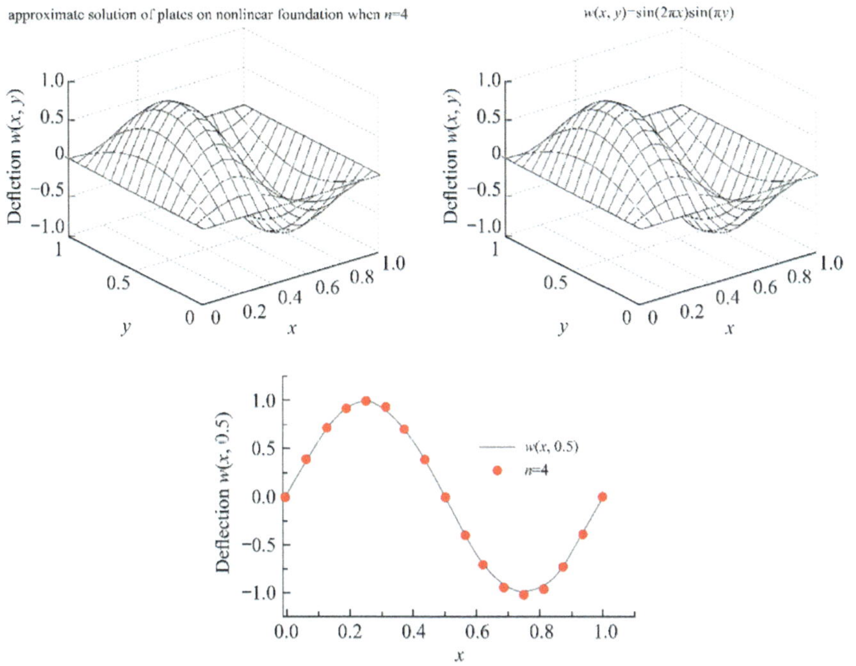

Fig. 9.24 Comparison between the approximate solution of the modified Coiflets-based Galerkin method with resolution level $n = 4$ and the exact solution for the bending plate under a nonlinear foundation: deflection configuration of the approximate solution (Upper left); exact solution (Upper right); direct comparison between the exact and approximate solutions in the midline along x-direction of the plate (Lower)

by the reconstruction of Eq. (9.108). Figure 9.24 shows the comparison of the exact and numerical solutions obtained by using the Coiflets with $\tilde{N}=17$ ($M_1 = 7$)and resolution level $n = 4$. As shown in the figure, the numerical solution is almost identical to the exact solution when $n = 4$.

9.4.2 Nonlinear Free Vibration of Beam with Immovably Simple Supports

The nonlinear vibration of a beam is a classical nonlinear dynamic problem. Researchers have proposed numerous methods to quantitatively study this kind of problems, such as FEM [44], the differential quadrature method [6], and the traditional Galerkin method [7], etc. However, these techniques still exhibit certain limitations, i.e., the amplitude should not be too large. In general, most of these methods can only provide the result when the amplitude is less than or equal to twice the radius of gyration [6–10, 44]. For example, the amplitude should be less than one-third of the height for the rectangular and circular sections. Here, we use the closed wavelet-based method introduced earlier to quantitatively study the nonlinear vibration of simply supported beams with immovable supports. When amplitude is extra large, the relationship between the characteristic frequency and the amplitude of the nonlinear free vibration of beams can be provided.

Consider an immovable simply supported elastic beam with length l, subjected to a transverse distribution load $p(x, t)$. Following the large deflection bending problem studied in Sect. 9.1.2, the fine differential equation that describes the nonlinear vibration problem can be expressed as [8–10]

$$\begin{cases} \rho A w_{,tt} + \mu w_{,t} + EI[\frac{w_{,xx}}{(1+w_{,x}^2)^{3/2}}]_{,xx} - [\frac{EA}{l}\int_0^l(\sqrt{1+w_{,x}^2}-1)dx]w_{,xx} = p, x \in (0,l) \\ w(0,t) = w(l,t) = 0, w_{,xx}(0,t) = w_{,xx}(l,t) = 0, w(x,0) = \chi(x), w_{,t}(x,0) = \chi_{,t}(x) \end{cases}. \tag{9.113}$$

Here, ρ is the material density per length, μ represents the damping coefficient, and $w(x, t)$ stands for the deflection function. The other parameters are defined as the same as Eq. (9.13). Recall that $\overline{x} = x/l$, $\overline{w} = w/l$, and $\overline{p} = p/\rho Al$, and the bar over each quantity is disregarded without affecting understanding. Further, the dimensionless equation except for time variable can be written as

$$\begin{cases} w_{,tt} + Jw_{,t} + M[\frac{w_{,xx}}{(1+w_{,x}^2)^{3/2}}]_{,xx} - L[\int_0^1(\sqrt{1+w_{,x}^2}-1)dx]w_{,xx} = p, x \in (0,1) \\ w(0,t) = w(1,t) = 0, w_{,xx}(0,t) = w_{,xx}(1,t) = 0, w(x,0) = \chi(x), w_{,t}(x,0) = \chi_{,t}(x) \end{cases}, \tag{9.114}$$

where parameters $J = \mu/\rho A$, $M = EI/\rho Al^4$, and $L = E/\rho l^2$.

Following the generalized solution scheme of the wavelet-based method, we get

$$\begin{aligned} &\sum_{k=1}^{2^j-1} \ddot{w}_{j,k}\phi_{j,k}(x) + J\sum_{k=1}^{2^j-1} \dot{w}_{j,k}\phi_{j,k}(x) + M\sum_{k=0}^{2^j} \frac{[\mathbf{A}^{-1}\mathbf{CW}]_k}{(1+[\mathbf{A}^{-1}\mathbf{BW}]_k^2)^{3/2}} \frac{d^2\phi_{j,k}(x)}{dx^2} \\ &\quad - L\sum_{n=0}^{2^j}(\sqrt{1+[\mathbf{A}^{-1}\mathbf{BW}]_n^2}-1)\widetilde{\theta}_{j,n}(1)\sum_{k=1}^{2^j-1} w_{j,k}\frac{d^2\phi_{j,k}(x)}{d^2x} \approx \sum_{k=0}^{2^j} p_{j,k}\phi_{j,k}(x), \end{aligned} \tag{9.115}$$

where $(\dot{\bullet})$ represents the first derivative with respect to t and the vector $\mathbf{W} = \{w_k = w_{j,k} = (k/2^j, t)\}^T$. The definitions of the other quantities are the same as those in Eq. (9.33).

Multiplying both sides of Eq. (9.115) by the weight function $\phi_{j,l}(x)$, $l = 1, 2, \ldots, 2^j - 1$ and then integrating it with respect to x from 0 to 1, we obtain the following nonlinear ordinary differential equations on the variable time t in the matrix form

$$\mathbf{O}\ddot{\mathbf{W}} + J\mathbf{O}\dot{\mathbf{W}} + M\mathbf{D}\overline{\mathbf{W}}(\mathbf{W}) - L\Lambda(\mathbf{W})\mathbf{E}\mathbf{W} \approx \mathbf{Q}\mathbf{P}, \tag{9.116}$$

where matrix $\mathbf{O} = \{o_{lk} = \int_0^1 \phi_{j,k}(x)\phi_{j,l}(x)dx\}$, $k, l = 1, 2, \ldots, 2^j - 1$ can be accurately derived using the method described in Appendix 5.1. In addition, the other quantities in Eq. (9.116) are defined in the same manner as those in Eq. (9.34).

By multiplying both sides of Eq. (9.116) by the inverse matrix $\mathbf{O}^{-1}$ and denoting that $\mathbf{V} = \dot{\mathbf{W}}$, Eq. (9.116) can be rewritten as

$$\begin{cases} \dot{\mathbf{V}} = \mathbf{Z}(t, \mathbf{W}, \mathbf{V}) = \mathbf{O}^{-1}\mathbf{Q}\mathbf{P} + L\Lambda(\mathbf{W})\mathbf{O}^{-1}\mathbf{E}\mathbf{W} - J\mathbf{V} - M\mathbf{O}^{-1}\mathbf{D}\overline{\mathbf{W}}(\mathbf{W}) \\ \dot{\mathbf{W}} = \mathbf{V}. \end{cases} \tag{9.117}$$

Here, the fourth-order Runge–Kutta method is used to solve the system of nonlinear ordinary differential equations in Eq. (9.117). Following the standard Runge–Kutta procedure, an iterative format is obtained as follows:

$$\begin{cases} \mathbf{W}_{n+1} = \mathbf{W}_n + \frac{h}{6}(\mathbf{K}_{w1} + 2\mathbf{K}_{w2} + 2\mathbf{K}_{w3} + \mathbf{K}_{w4}) \\ \mathbf{V}_{n+1} = \mathbf{V}_{n+1} + \frac{h}{6}(\mathbf{K}_{v1} + 2\mathbf{K}_{v2} + 2\mathbf{K}_{v3} + \mathbf{K}_{v4}) \end{cases}. \tag{9.118}$$

Here, the vectors or columns in Eq. (9.118) are of the form

$$\begin{cases} \mathbf{K}_{w1} = \mathbf{V}_n,\ \mathbf{K}_{w2} = \mathbf{V}_n + \dfrac{\Delta t}{2}\mathbf{K}_{w1}, \\ \mathbf{K}_{w3} = \mathbf{V}_n + \dfrac{\Delta t}{2}\mathbf{K}_{w2},\ \mathbf{K}_{w4} = \mathbf{V}_n + \Delta t\mathbf{K}_{w3}, \\ \mathbf{K}_{v1} = \mathbf{Z}(t_n, \mathbf{W}_n, \mathbf{V}_n), \\ \mathbf{K}_{v2} = \mathbf{Z}(t_n + \dfrac{\Delta t}{2}, \mathbf{W}_n + \dfrac{\Delta t}{2}\mathbf{K}_{w1}, \mathbf{V}_n + \dfrac{\Delta t}{2}\mathbf{K}_{v1}), \\ \mathbf{K}_{v3} = \mathbf{Z}(t_n + \dfrac{\Delta t}{2}, \mathbf{W}_n + \dfrac{\Delta t}{2}\mathbf{K}_{w2}, \mathbf{V}_n + \dfrac{\Delta t}{2}\mathbf{K}_{v2}), \\ \mathbf{K}_{v4} = \mathbf{Z}(t_n + \Delta t, \mathbf{W}_n + \Delta t\mathbf{K}_{w3}, \mathbf{V}_n + \Delta t\mathbf{K}_{v3}), \end{cases} \tag{9.119}$$

where Δt is the step size of the time integration, $t_n = n\Delta t$, and vector $\mathbf{W}_n = \mathbf{W}(n\Delta t) = \{w(k/2)^j, n\Delta t\}^T$, $k = 1, 2, \ldots, 2^j - 1$. Therefore, given the time step Δt and combined with the initial conditions, vector $\mathbf{W}_n$ can be obtained, and the approximate solution $w(x, n\Delta t)$ of the problem, Eq. (9.113), can be further reconstructed by Eq. (9.21).

To test the validity and accuracy of the above wavelet-based method, we consider $J = 1$, $M = 1 \times 10^{-5}$, $L = 1 \times 10^{-4}$, the initial condition $\chi(x) = 0.5\sin(\pi x)$, and load

$$p(x, t) = e^{-t}\pi\sin(2\pi t)\sin(\pi x) - 2e^{-t}\pi^2\cos(2\pi t)\sin(\pi x)$$

$$+\frac{e^{-t}\pi^4\cos(2\pi t)\sin(\pi x)}{2\times10^5[1+0.25e^{-2t}\pi^2\cos^2(2\pi t)\cos^2(\pi x)]^{3/2}}$$
$$+\frac{3e^{-3t}\pi^6\cos^3(2\pi t)\sin(\pi x)[\sin^2(\pi x)-3\cos^2(\pi x)]}{8\times10^5[1+0.25e^{-2t}\pi^2\cos^2(2\pi t)\cos^2(\pi x)]^{5/2}}$$
$$-\frac{3e^{-5t}\pi^8\cos^5(2\pi t)\sin^3(\pi x)\cos^2(\pi x)}{6.4\times10^5[1+0.25e^{-2t}\pi^2\cos^2(2\pi t)\cos^2(\pi x)]^{7/2}}$$
$$+\frac{e^{-t}\pi^2\cos(2\pi t)\sin(\pi x)}{20000}\int_0^1[\sqrt{1+0.25e^{-2t}\pi^2\cos^2(2\pi t)\cos^2(\pi x)}-1]dx. \tag{9.120}$$

Then, the exact solution of this problem is gained by $w(x,t) = 0.5e^{-t}\cos(2\pi t)\sin(\pi x)$.

Figures 9.25, 9.26, 9.27 and 9.28 show the comparison between the exact solution and the solution obtained by the wavelet closed method. From them, we find that the approximate solution obtained using the proposed wavelet method quantitatively coincides with the exact solution well, and the absolute value of the maximum relative error is not over 0.01% except a few short moments when the deflection is extremely close to zero.

When we note that the normalized deflection w is relative to the length of the beam, we know that the deflection greater than 0.1 is so large corresponding to the ratio of thickness to length being less than 1/8. In such situation, one can see from the above-obtained results that the wavelet-based solution scheme for analyzing the nonlinear vibration of beams still exhibits high accuracy even when the amplitude is so large. When the maximum amplitude reaches half of the beam length, for example, the relative error can still be within 0.01%.

Next, we use the wavelet method to study the nonlinear free vibration of beams without exact solution. Here, the parameters in Eq. (9.114) are chosen as $J = 0,\ M =$

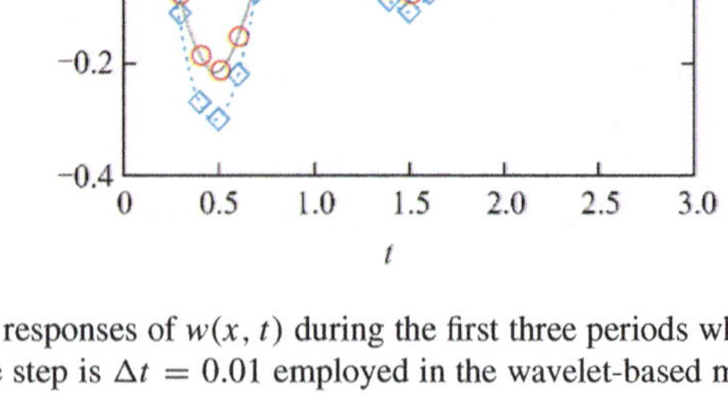

Fig. 9.25 Deflection responses of $w(x,t)$ during the first three periods when the resolution level is $j = 4$ and the time step is $\Delta t = 0.01$ employed in the wavelet-based method for the damping free vibration of nonlinear beam with unmovably simple supports

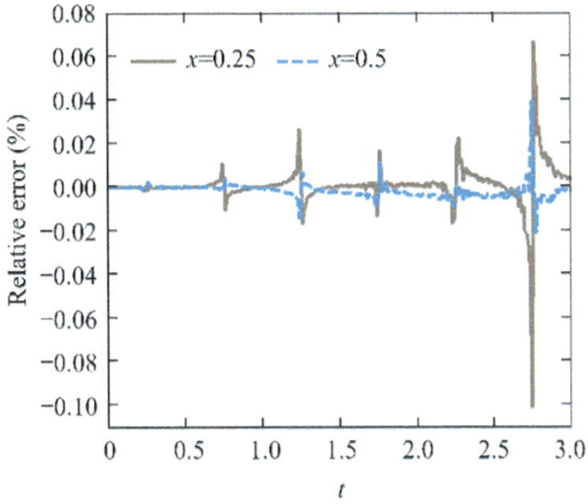

Fig. 9.26 Relative error of the deflection responses during the first three periods of the wavelet-based solution with the resolution level is $j = 4$ and the time step $\Delta t = 0.01$ corresponding to its exact solution for the damping free vibration of nonlinear beam with unmovably simple supports

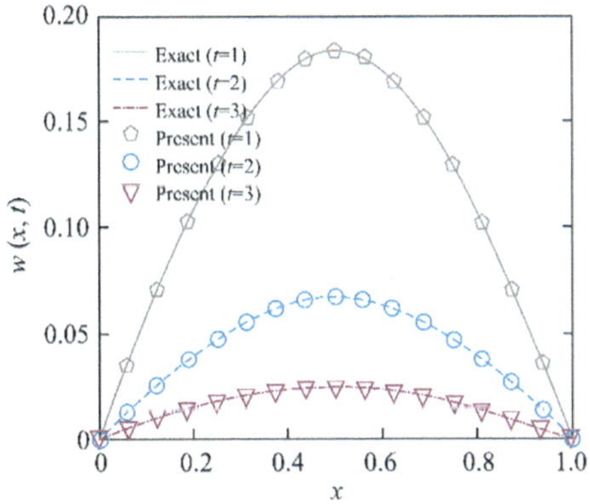

Fig. 9.27 Deflection configuration of *w(x, t)* at different time instants when the resolution level $j = 4$ and the time step $\Delta t = 0.01$ are used in the wavelet-based solution method for the damping free vibration of nonlinear beam with unmovably simple supports

6.25×10^{-4}, $L = 25/3\ \mathrm{s}^{-2}$, $p = 0$, and the initial conditions $\dot{\chi}(x) = 0$ and $\chi(x) = W_0 \sin(\pi x)$, where the parameter $W_0 = w_{max}/l$ is the dimensionless amplitude. Then, the wavelet algorithm introduced in the preceding section with resolution level $j = 6$ is employed in the numerical simulation of the nonlinear free vibration of the beam with Eq. (9.114) under different values of W_0.

Figures 9.29, 9.30, 9.31, 9.32 and 9.33 show the dynamic responses of the normalized deflection at the midpoint of the beam, i.e., $w(0.5, t)$, and their Fourier spectrums for the initial amplitude $W_0 = 0.1, 0.2, 0.3, 0.4, 0.5$, respectively. From these figures, one sees that to each initial amplitude, the amplitude in its dynamic response remains unchanged, and except for the main frequency, there is or are some

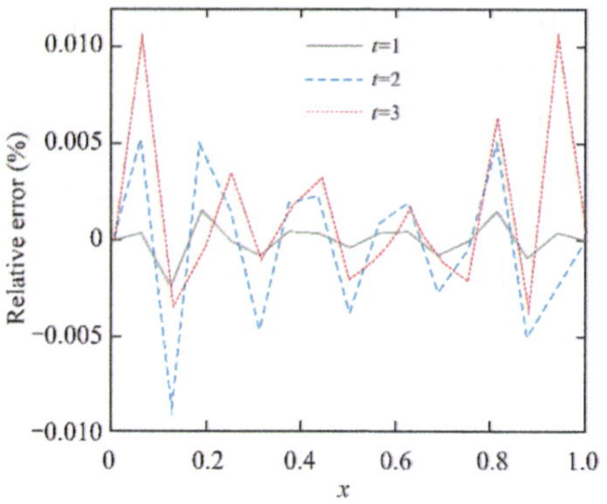

Fig. 9.28 Spatial distributions of the relative error of deflection at different instant t when the resolution level $j = 4$ and the time step $\Delta t = 0.01$ are used in the wavelet-based method for the damping free vibration of nonlinear beam with unmovably simple supports

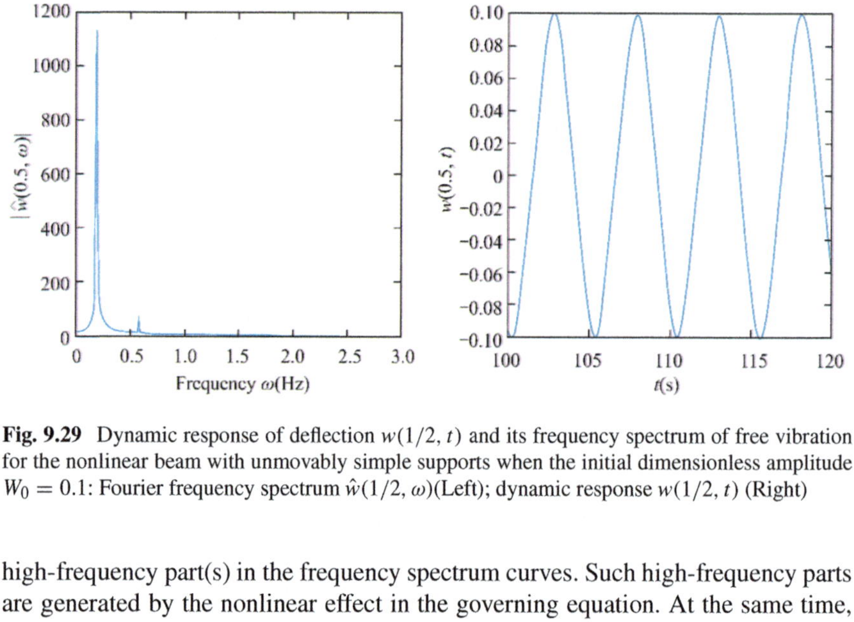

Fig. 9.29 Dynamic response of deflection $w(1/2, t)$ and its frequency spectrum of free vibration for the nonlinear beam with unmovably simple supports when the initial dimensionless amplitude $W_0 = 0.1$: Fourier frequency spectrum $\hat{w}(1/2, \omega)$(Left); dynamic response $w(1/2, t)$ (Right)

high-frequency part(s) in the frequency spectrum curves. Such high-frequency parts are generated by the nonlinear effect in the governing equation. At the same time, the main frequency in the system increases with the initial amplitude. For example, when $W_0 = 0.1$, the main frequency is about 0.2 Hz, while it increases up to 0.867 Hz when $W_0 = 0.5$. These results quantitatively reveal the property that a nonlinear free vibration system has no natural frequency and there are high-frequency responses in the dynamic system even when the initial excitation is a single first-order mode corresponding to its linear dynamic system.

Corresponding to the multiple frequency dynamic response of free vibration of the nonlinear beam with simple supports under different initial amplitudes, we know that the actual mode of the free vibration of the beam is no longer a single mode of

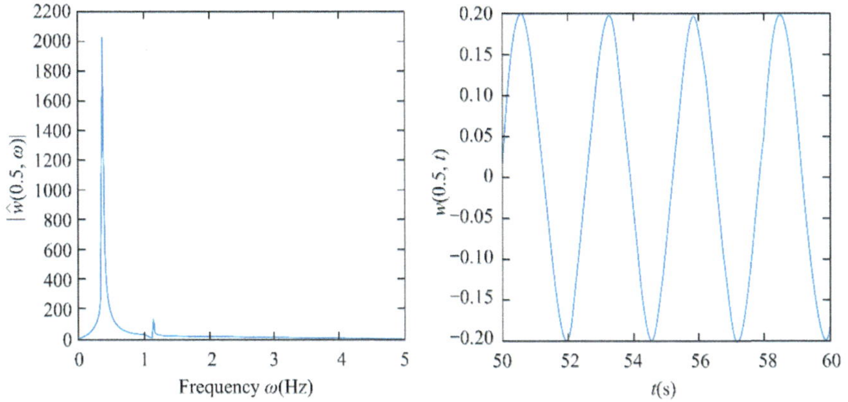

Fig. 9.30 Dynamic response of deflection $w(1/2, t)$ and its frequency spectrum curve of free vibration for the nonlinear beam with unmovably simple supports when the initial dimensionless amplitude $W_0 = 0.2$: Fourier frequency spectrum $\hat{w}(1/2, \omega)$(Left); dynamic response $w(1/2, t)$ (Right)

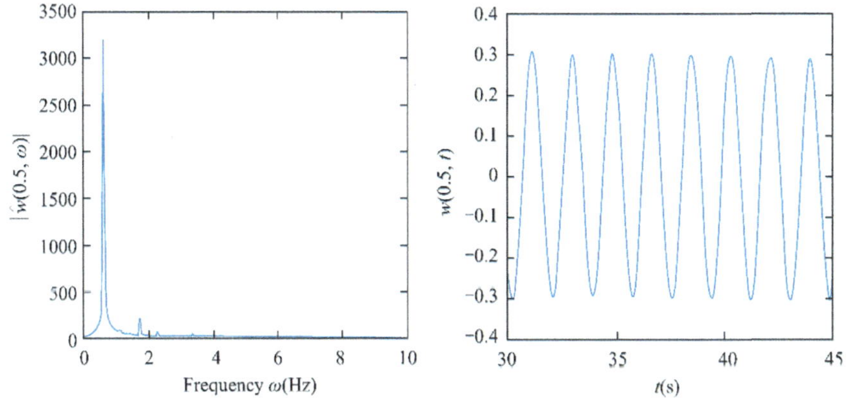

Fig. 9.31 Dynamic response of deflection $w(1/2, t)$ and its frequency spectrum curve of free vibration for the nonlinear beam with unmovably simple supports when the initial dimensionless amplitude $W_0 = 0.3$: Fourier frequency spectrum $\hat{w}(1/2, \omega)$(Left); dynamic response $w(1/2, t)$ (Right)

the first order. When $W_0 = 0.5$, for example, Fig. 9.34 plots the deflection configurations of the beam at different instants, which explicitly displays that the deflection configuration at each instant is not a standard sinusoid form. In addition, this result indicates that the assumption of a single space mode response shown in Appendix 9.3 is not suitable to this study.

Based on the above results shown in Figs. 9.29, 9.30, 9.31, 9.32, 9.33 and 9.34, we can obtain the relation between the main frequency and the initial amplitude, see in Fig. 9.35, where the results from the perturbation method in Appendix

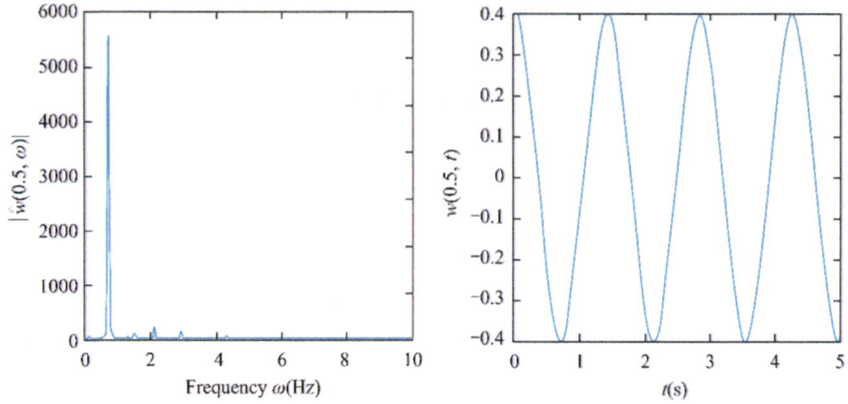

Fig. 9.32 Dynamic response of deflection $w(1/2, t)$ and its frequency spectrum curve of free vibration for the nonlinear beam with unmovably simple supports when the initial dimensionless amplitude $W_0 = 0.4$: Fourier frequency spectrum $\hat{w}(1/2, \omega)$(Left); dynamic response $w(1/2, t)$ (Right)

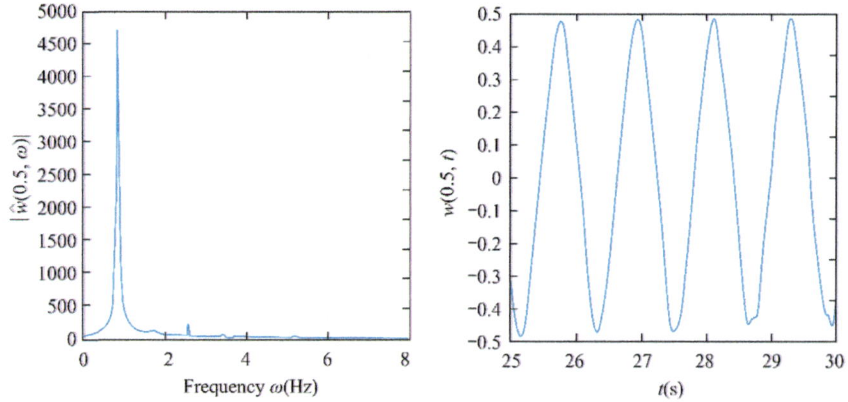

Fig. 9.33 Dynamic response of deflection $w(1/2, t)$ and its frequency spectrum curve of free vibration for the nonlinear beam with unmovably simple supports when the initial dimensionless amplitude $W_0 = 0.5$: Fourier frequency spectrum $\hat{w}(1/2, \omega)$(Left); dynamic response $w(1/2, t)$ (Right)

9.3 are also plotted and the natural frequency of the corresponding linear system $\omega_0 = \sqrt{\pi^4 EI/\rho A l^4}$ is used. According to Fig. 9.35, one sees that when initial amplitude is relatively small, the perturbation results can consistent well with ones of the wavelet-based solutions. When the initial dimensionless amplitude $W_0 \leq 0.1$, for example, the maximum relative error of the perturbation results corresponding to one of the wavelet-based method can approach to about 4%. Such notable difference can be found in the inset figure of Fig. 9.35. With the subsequent increase of the

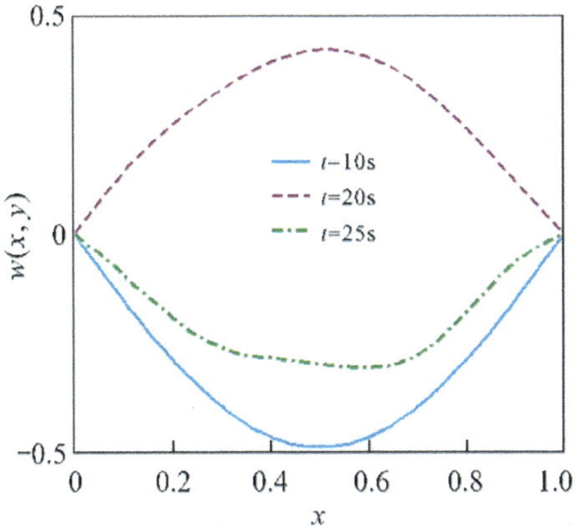

Fig. 9.34 Deflection configurations of free vibration of the nonlinear beams with unmovably simple supports at different instants when the initial dimensionless amplitude $W_0 = 0.5$

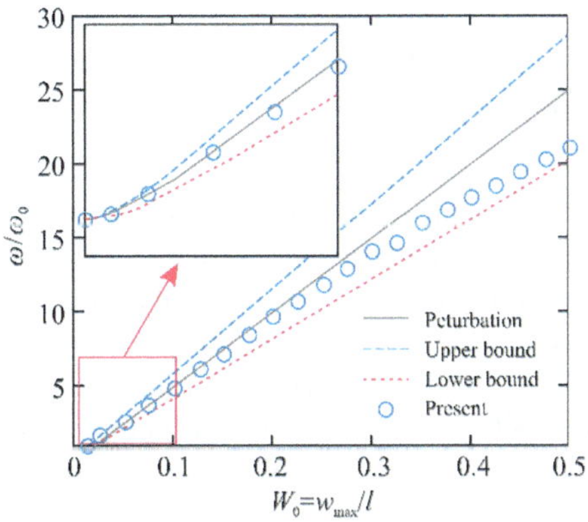

Fig. 9.35 Dependence of the main frequency on the initial amplitude for the free vibration of nonlinear beam with unmovably simple supports

initial amplitude, such difference increases larger. As we have known, such perturbation is usually suitable for a weak nonlinear problem, where a small perturbation parameter related to the amplitude in this case is employed. As the initial amplitude increases, the nonlinear free vibration becomes a strong nonlinear problem such that the perturbation method loses efficiency in accuracy. However, the wavelet-based method can provide high accuracy of solving a nonlinear problem, either weakly or strongly, with a unified calculation program, as displayed by all the previous examples, quantitatively.

In addition, Fig. 9.35 also reveals that the curvature of such frequency–amplitude curve is greater than zero for the amplitude $W_0 < 0.25$, which is qualitatively as

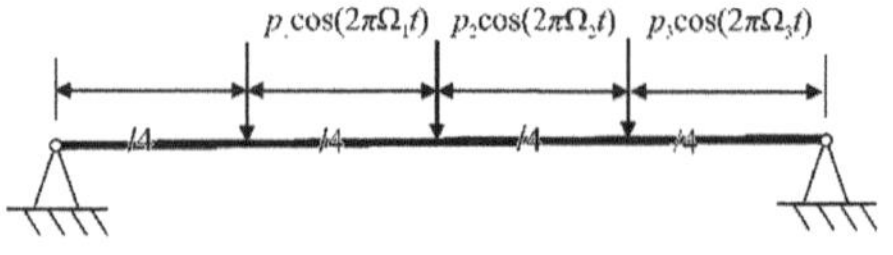

Fig. 9.36 Schematic of the forced vibration of nonlinear beam with unmovably simple supports

the same as the existing observation [9]. However, when $W_0 > 0.25$, the curvature of the frequency–amplitude curve obtained by the wavelet-based method becomes a negative value. Such a phenomenon has not been mentioned in existing study, because almost all existing works investigate the free vibration of this kind of beams only for $W_0 \leq 0.2$ [8, 9]. The main reason why no investigation is conducted in the conventional methods for $W_0 > 0.25$ is that the solution results with a satisfaction accuracy cannot be possibly provided by their methods, where their suitable scopes are overturned, as shown in the appendix and the earlier discussion.

9.4.3 Nonlinear Forced Vibration of Beam with Immovably Simple Supports

Here, the dynamic system of the forced vibration beam under applied loads is shown in Fig. 9.36, where the beam is supported by the unmovably simple supports at the two ends and is subjected to three harmonic concentrated excitation forces. The geometric parameters of the beam are length $l = 1m$, cross-sectional area $A = 0.1\text{m}^2$, and moment of inertia $I = 8.1 \times 10^{-4}\text{m}^4$. The material parameters are elastic modulus $E = 2 \times 10^5\text{Pa}$ and density $\rho = 2000\text{kg/m}^3$. The damping is disregarded. The load parameters are $p_1 = p_3 = p_2/2 = 72.9\text{N}$ with excitation frequencies of the form $\Omega_1 = \Omega_3 = \Omega_2/2 = 0.36\text{Hz}$.

After the wavelet-based method is used to solve this forced vibration of nonlinear beam with the unmovably simple supports, some characteristic results are plotted in Figs. 9.37, 9.38, 9.39 and 9.40. In Fig. 9.37, the dynamic response of $w(1/2, t)$ is displayed, while its deflection configurations at two instants are plotted in Fig. 9.38 after the response reaches to the stable state. From Fig. 9.37, we find that the maximum amplitude is about 0.2 m, and the response has multiple frequencies. Figure 9.39 gives the frequency spectrum of the dynamic response, while its phase diagram is shown in Fig. 9.40.

From the application excitation of periodic loads, we know that only one composition of two frequencies is contained. Figure 9.39, however, clearly tells us that there are at least five frequency motions contained in the dynamic response. That is, the first five periodic frequencies are $\omega_1 \approx 0.36\text{Hz}$, $\omega_2 \approx 0.72\text{Hz}$, $\omega_3 \approx 1.44\text{Hz}$, $\omega_4 \approx 1.80\text{Hz}$, and $\omega_5 \approx 2.16\text{Hz}$. It is evident that such response frequencies have the relations of the form $\omega_1 = \Omega_1 = \Omega_3$, $\omega_2 = \Omega_2 = 2\Omega_1 = 2\Omega_3 = \Omega_1 + \Omega_3$, $\omega_3 = \Omega_1 + \Omega_2 + \Omega_3$, $\omega_4 = 2\Omega_1 + \Omega_2 + \Omega_3$, and $\omega_5 = 2\Omega_1 + \Omega_2 + 2\Omega_3$, which

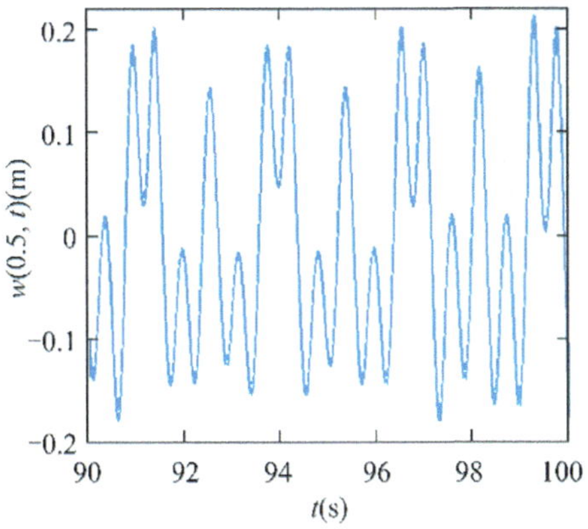

Fig. 9.37 Dynamic response of deflection $w(1/2, t)$ of the forced vibration of the nonlinear beam at the midpoint of the beam with unmovably simple supports subjected to periodic loads

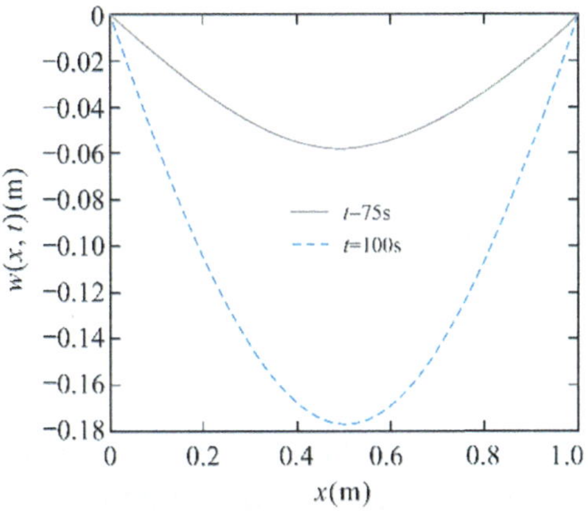

Fig. 9.38 Deflection configurations the forced vibration of the nonlinear beam at the midpoint of the beam with unmovably simple supports subjected to periodic loads

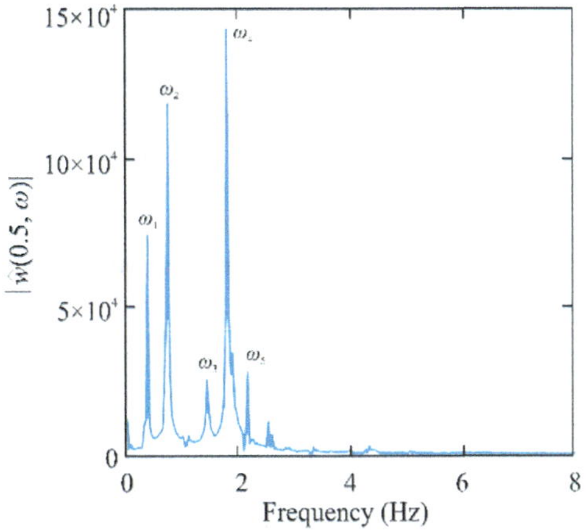

Fig. 9.39 Fourier frequency spectrum $\hat{w}(1/2, \omega)$ of dynamic response of deflection $w(1/2, t)$ of the forced vibration of the nonlinear beam at the midpoint of the beam with unmovably simple supports subjected to periodic loads

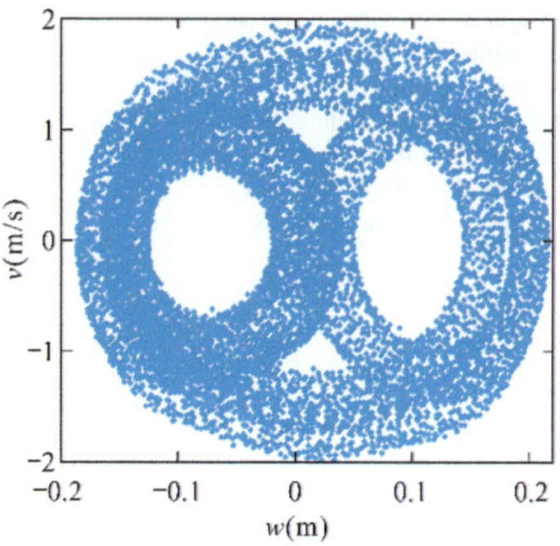

Fig. 9.40 Phase diagram of the dynamic response of deflection $w(1/2, t)$ of the forced vibration of the nonlinear beam with unmovably simple supports subjected to periodic loads

are corresponding to either the applied frequencies or their combination of harmonic responses. These quantitative observations are in good agreement with the theory of nonlinear vibration. In other words, when a nonlinear system is subjected to the periodic excitation forces with frequencies $\overline{\Omega}_1$ and $\overline{\Omega}_2$, the response of the nonlinear system includes not only the fundamental harmonic response of frequencies $\overline{\Omega}_1$ and $\overline{\Omega}_2$, but also a combined harmonic response with a frequency of $|\pm m\,\overline{\Omega}_1 \pm n\,\overline{\Omega}_2|$ (m, n are integers) [10].

Appendix 9.1 Wavelet Numerical Integration Method with Modified Scaling Function

Consider the integral

$$g(x) = \int_0^x f(y)dy, \quad x \in [0, 1]. \tag{9.121}$$

When the approximation scheme, Eq. (5.6), is substituted into the preceding formula, one gets

$$g(x) \approx \tilde{g}(x) = \sum_{k=0}^{2^j} f_{n,k}\tilde{\theta}_{n,k}(x), \quad x \in [0, 1], \tag{9.122}$$

in which

$$\tilde{\theta}_{n,k}(x) = 2^{-j}\begin{cases} \sum\limits_{l=1-\tilde{N}+M_1}^{-1} u_{0,k}(\frac{l}{2^n})\theta|_{M_1-l}^{2^n x-l+M_1} + \theta|_{M_1-k}^{2^n x-k+M_1} & k \in [0,\ r-1] \\ \theta|_{M_1-k}^{2^n x-k+M_1} & k \in [r,\ 2^n - r] \\ \sum\limits_{l=2^n+1}^{2^n-1+M_1} u_{1,2^n-k}(\frac{l}{2^n})\theta|_{M_1-l}^{2^n x-l+M_1} + \theta|_{M_1-k}^{2^n x-k+M_1} & k \in [2^n - r + 1,\ 2^n] \end{cases}, \tag{9.123}$$

where $\theta|_y^x = \theta(x) - \theta(y)$, and the integral of the scaling function is

$$\theta(x) = \int_0^x \varphi(y)dy. \tag{9.124}$$

In accordance with the normalization condition, the compact support of the scale function is $\left[0,\ \tilde{N}\right]$. From Sect. 2.5.2, we obtain that the integral in Eq. (9.124) of the scale function has properties

$$\theta(x) = \begin{cases} 0, & x \le 0 \\ 1, & x \ge \tilde{N} \end{cases}. \tag{9.125}$$

Appendix 9.2 An Analytical Solution of Elastic Line Equation of Flexible Rod

For such case, when $\alpha = 0$ in Eq. (9.45), the governing equation for the bending problems of the cantilevered rod or beam can be written as

$$n\frac{d^2\theta}{ds^2} = -\lambda\cos\theta(\frac{d\theta}{ds})^{1-n}, \tag{9.126}$$

where the curvature is $\kappa = d\theta/ds$ and θ is the bending angle. Then, the governing equation can then be rewritten into

$$n\kappa^n d\kappa = -\lambda\cos\theta d\theta. \tag{9.127}$$

By integrating the preceding equation once and considering the boundary condition at the free end, $\kappa(1) = 0$, we get the following integration:

$$\frac{n}{n+1}\kappa^{n+1} = \lambda\sin\theta_L - \lambda\sin\theta, \tag{9.128}$$

where θ_L is the bending angle at the free end.

Substitution of the expression of the curvature into Eq. (9.128) leads to

$$(\sin\theta_L - \sin\theta)^{-1/(n+1)}d\theta = \left(\frac{n+1}{n}\lambda\right)^{-1/(n+1)} ds. \tag{9.129}$$

In accordance with the non-extensibility assumption in Euler–Bernoulli beam theory, integration of Eq. (9.129) from the fixed end to the free end yields

$$\lambda = \frac{n}{n+1}\left[\int_0^{\theta_L} g(\theta)d\theta\right]^{n+1}, \tag{9.130}$$

where $g(\theta) = (\sin\theta_L - \sin\theta)^{-\frac{1}{n+1}}$. This equation provides an analytical relation between the bending angle, θ_L, and the normalized load, λ.

Further, the curvature, κ, can then be expressed as a function of the bending angle, θ_L, as follows:

$$\kappa = \frac{\int_0^{\theta_L} g(\theta)d\theta}{Lg(\theta)}. \tag{9.131}$$

Due to the geometric relationship

$$dx = \cos\theta ds = \cos\theta\frac{d\theta}{\kappa}, \tag{9.132}$$

substituting the curvature of Eq. (9.131) into Eq. (9.132) and integrating the induced equation from the fixed end, we obtain the following formula:

$$x = \frac{\int_0^{\theta} \cos\theta g(\theta)d\theta}{\int_0^{\theta_L} g(\theta)d\theta}. \tag{9.133}$$

Further, the axial displacement of the free end is formulated by

$$\delta_h = 1 - \frac{\int_0^{\theta_L} \cos\theta g(\theta)d\theta}{\int_0^{\theta_L} g(\theta)d\theta}. \tag{9.134}$$

With the similar steps, the deflection of every point of the rod can be obtained as

$$y = \frac{\int_0^{\theta} \sin\theta g(\theta)d\theta}{\int_0^{\theta_L} g(\theta)d\theta}. \tag{9.135}$$

Then, the deflection of the free end is denoted by $\delta_v = y(\theta_L)$.

Appendix 9.3 Approximate Theoretical Analyses on the Nonlinear Free Vibration of Beams

When the beam is undergoing large amplitude vibration, the following nonlinear strain–displacement relation is employed [8]:

$$\varepsilon_x = u_{,x} + 0.5w_{,x}^2 - zw_{,xx}, \tag{9.136}$$

where u, w, and z are the axial displacement, the transverse displacement or deflection, and the distance from the neutral plane along the thickness direction of the beam, respectively.

Based on Eq. (9.136), the strain energy and kinetic energy of the elastic beam are written by the form [8]

$$U = \frac{1}{2}\int_0^l [EA(u_{,x}^2 + u_{,x}w_{,x}^2 + \frac{1}{4}w_{,x}^4) + EIw_{,xx}^2]dx, \tag{9.137}$$

$$T = \int_0^l mw_{,t}^2 dx/2, \tag{9.138}$$

where E, A, I, and m are the elastic modulus, area of cross section, inertia moment of cross section, and mass per unit length of the beam, respectively. Here, the axial motion is disregarded in calculating kinetic energy.

Assume

$$u = u_0(t)\sin(2\pi x/l), \tag{9.139}$$

$$w = w_0(t)\sin(\pi x/l), \tag{9.140}$$

which satisfy the kinematic boundary conditions of a simply supported beam with immovable ends. Then, substituting them into Eqs. (9.137) and (9.138), the Lagrange dynamic equation of the free vibration of beams can be written as [8]

$$m\ddot{w}_0 + \alpha w_0 + \beta w_0^3 = 0, \tag{9.141}$$

where parameters $\alpha = \pi^4 EI/l^4$ and $\beta = \pi^4 EA/4l^4$.

The energy conservation equation [8] can be further obtained by multiplying both sides of Eq. (9.141) by $\dot{w}_0$ and integrating with respect to time such that

$$m\dot{w}_0^2 + \alpha w_0^2 + \beta w_0^4/2 = T + U = \text{constant}. \tag{9.142}$$

Assume that the nonlinear free vibration of the elastic beam satisfies the single harmonic response relationship $w_0 = W_0 \cos(\omega t)$, where W_0 and ω are dealt with by the amplitude and characteristic frequency of the nonlinear free vibration of a beam, respectively. Further, the dynamic equation (9.141) and the energy Eq. (9.127), respectively, yield [8].

$$\omega/\omega_0 = \sqrt{1 + (W_0/\xi)^2/4}, \tag{9.143}$$

$$\omega/\omega_0 = \sqrt{1 + (W_0/\xi)^2/8}, \tag{9.144}$$

where $\omega_0 = \sqrt{\alpha/m}$ is the natural frequency of the linear free vibration of the same beam, and parameter $\xi = \sqrt{I/A}$ is the radius of gyration of the beam's cross section.

It is obvious that these two relations of Eqs. (9.143) and 9.144) are conflicted, which means that on the basis of assumption of the harmonic oscillations, such induced results cannot satisfy the equation of motion and the energy balance equation simultaneously. Singh et al. [26] pointed out that Eqs. (9.143) and (9.144) are, respectively, the upper and lower bounds on the actual solution, as displayed in Fig. 9.35.

In addition, the governing equation of motion can also be solved by using the perturbation method without the assumption of harmonic response, in which the following relation is obtained [8]:

$$\omega/\omega_0 = \sqrt{1 + \frac{3}{16}(W_0/\xi)^2}. \tag{9.145}$$

As pointed out previously, such perturbation solution is suitable only to the small initial amplitude.

References

1. Howell LL (2001) Compliant mechanisms. Wiley.
2. Bisshop KE, Drucker DC (1945) Large deflection of cantilever beams. Q Appl Math 3:272–275
3. Cardona A, Geradin M (1988) A beam finite element non-linear theory with finite rotations. Int J Numer Meth Eng 26:2403–2438
4. Banerjee A, Bhattacharya B, Mallik AK (2008) Large deflection of cantilever beams with geometric non-linearity: analytical and numerical approaches. Int J Non-Linear Mech 43:366–376
5. Mei C (1973) Finite element displacement method for larger amplitude free oscillations of beams and plates. Comput Struct 3:163–174
6. Feng Y, Bert CW (1992) Application of the quadrature method to flexural vibration analysis of a geometrically nonlinear beam. Nonlinear Dyn 3:13–18
7. Bhashyam GR, Prathap G (1980) Galerkin finite element method for non-linear beam vibrations. J Sound Vib 72:191–203
8. Singh G, Sharma AK, Rao GV (1990) Large-amplitude free vibrations of beams-a discussion on various formulations and assumptions. J Sound Vib 142:77–85

9. Guo Q (2001) Further development of spline-based differential quadrature method and nonlinear vibration analysis of beams. Tsinghua University, China
10. Chen SH (2007) The definite quantitative methods for strongly non-linear vibration. Science Press, Beijing
11. Lewis G, Monasa F (1981) Large deflections of cantilever beams of nonlinear materials. Comput Struct 14:357–360
12. Jung JH, Kang TJ (2005) Large deflection analysis of fibers with nonlinear elastic properties. Text Res J 75:715–723
13. Moore G, Spence A (1980) The Calculation of turning points of nonlinear equations. SIAM J Numer Anal 17:567–576
14. Moore G (1980) The numerical treatment of non-trivial bifurcation points. Numer Funct Anal Optim 2:441–472
15. Saetiew W, Chucheepsakul S (2012) Post-buckling of simply supported column made of nonlinear elastic materials obeying the generalized Ludwick constitutive law. ZAMM J Appl Math Mech 92:479–489
16. Brojan M, Puksic A, Kosel F (2007) Buckling and post-buckling of a nonlinearly elastic column. ZAMM J Appl Math Mech 87:518–527
17. Lacarbonara W (2008) Buckling and post-buckling of non-uniform non-linearly elastic rods. Int J Mech Sci 50:1316–1325
18. Zheng XJ (1990) Large deflection theory of circular thin plate and its application. Jilin Science Technology Press, Jilin
19. Wang XM, Liu XJ, Wang J, Zhou YH (2015) A wavelet method for bending of circular plate with large deflection. Acta Mech Sin 28:83–90
20. Von Kármán T (1910) Festigkeits probleme im maschinenbau. Encykl Math Wiss 4:311–385
21. Vincent JJ (1931) The bending of a thin circular plate. Lond Edinb Dublin Philos Mag J Sci 12:185–196
22. Chien WZ (1947) Large deflection of a circular clamped plate under uniform pressure. Acta Phys Sin 11:102–107
23. Chien WZ, Yeh KY (1954) On the large deflection of circular plate. Acta Phys Sin 10:209–238
24. Chien WZ (1948) Asymptotic behavior of a thin clamped circular plate under uniform normal pressure at very large deflection, vol 5. Science Reports (A) of National Tsinghua University, pp 1–24
25. Chien WC, Lin H, Hu H, Yeh KY (1956) Large deflection problems of circular thin plates. Science Press, Beijing
26. Zheng XJ, Zhou YH (1991) On the problem of transition from von Kármán plate equations to theory of membrane. J Lanzhou Univ (Nat Sci) 27:18–25
27. Zheng XJ, Zhou YH (1989) On convergence of interpolated iterative method of geometrically nonlinear equations of circular plates. Sci China (Ser A). 32:316–327
28. Zheng XJ, Lee J (1995) On the convergence of the Chien's perturbation method for von Kármán plate equations. Int J Eng Sci 33:1085–1094
29. Patrício da Silva P, Krauth W (1997) Numerical solution of the von Kármán equations for a thin plate. Int J Mod Phys C 8: 427–434
30. Waidemam L, Venturini WS (2009) BEM formulation for von Kármán plates. Eng Anal Bound Elem 33:1223–1230
31. Altekin M, Yukseler RF (2011) Large deflection analysis of clamped circular plates. In: Proceedings of the world congress on engineering, London
32. Yeh KY, Zheng XJ, Zhou YH (1989) An analytical formula of the exact solution to von Kármán's equations of a circular plate under a concentrated load. Int J Non-Linear Mech 24:551–560
33. Zheng XJ, Zhou YH (1990) Analytical formulas of solutions of geometrically nonlinear equations of axisymmetric plates and shallow shells. Acta Mech Sin 6:69–80
34. Zhou YH, Zheng XJ (1989) On the range of applicability of von Kármán plate equations. J Appl Mech 56:724–726
35. Yeh KY, Zheng XJ, Wang XZ (1990) On some properties and calculation of the exact solution to von Kármán's equations of circular plates under a concentrated load. Int J Non-Linear Mech 25:17–26

36. Zheng XJ, Zhou YH (1987) On exact solution of Kármán's equations of rigid clamped circular plate and shallow spherical shell under a concentrated load. Appl Math Mech (Engl Ed) 8:1057–1068
37. Timoshenko SP, Woinowsky-Krieger S (1959) Theory of plates and shells. McGraw-hill, New York, pp 415–425
38. Naffa M, Al-Gahtani HJ (2007) RBF-based meshless method for large deflection of thin plates. Eng Anal Bound Elem 31:311–317
39. Zhao X, Liu GR, Dai KY, Zhong ZH, Li GY (2008) Geometric nonlinear analysis of plates and cylindrical shells via a linearly conforming radial point interpolation method. Comput Mech 42:133–144
40. Zhou YH, Zheng XJ (1992) Exact solution of large deflection problem of simple supported thin rectangular plate with variable thickness (in Chinese). J Lanzhou Univ (Nat Sci) 28:21–28
41. Zhuang Z, You XC, Liao JH (2009) The finite element analysis and applications based on ABAQUS (in Chinese). Tsinghua University Press, Beijing
42. Shen HS, Zhang JW (1997) Thermal post-buckling analysis of rectangular plates based on nonlinear elasticity. Chin J Appl Mech 14: 29–35(in Chinese)
43. Xu ZL (1992) Elastic mechanics. Higher Education Press, Beijing (in Chinese)
44. Mei C (1972) Finite element displacement method for larger amplitude free oscillations of beams and plates. Comput Struct 3:163–174

Chapter 10
Applications to Laminar Flows in Nonlinear Fluid Mechanics

The previous chapter has discussed the closed wavelet-based method to solve the nonlinear problems of solid mechanics. The proposed wavelet-based method can also be applied to problems of fluid mechanics, because it basically provides an approach for discretizing partial differential equations (PDEs) in the spatial domain. This chapter presents the detailed instruction on how to implement the proposed closed wavelet-based method for computational fluid mechanics problems. As we have been known, the finite difference method (FDM), the finite volume method (FVM), and the finite element method (FEM) have been widely applied to practical problems with different features differential equations in mathematics. Usually, the FEM is more suitable to the solid mechanics, while FDM and FVM associated with the spectrum method are chosen in the fluid mechanics in priority. This state of using solution methods implies that even these popular methods, i.e., their suitability in solving PDEs has some limitation of accuracy. In addition, an important feature of these methods is that a coupling between the approximate solution and the truncation error exists in the approximation of nonlinear terms, as discussed in Chaps. 1 and 3. This trouble results in a significant decrease of accuracy when the nonlinearity in the PDEs becomes strong, which can also cause problems related to stability and convergence. The closed wavelet-based method, which employs a closed expansion of nonlinear terms, can offer a promising alternative unified method for solving these problems. At the following sections, we introduce those typical problems of laminar fluid flows solved by the wavelet-based method, where the detailed derivations are omitted for saving space if it is not necessary.

Y.-H. Zhou, *Wavelet Numerical Method and Its Applications in Nonlinear Problems*, Engineering Applications of Computational Methods 6,
https://doi.org/10.1007/978-981-33-6643-5_10

10.1 Burgers' Equation

10.1.1 Governing Equations with Initial-Boundary-Value Conditions

The two-dimensional Burgers' equation, which incorporates both nonlinear convection and viscous diffusion, can be expressed in the conservative form

$$\frac{\partial u}{\partial t}+\frac{1}{2}\frac{\partial u^2}{\partial x}+\frac{1}{2}\frac{\partial u^2}{\partial y}=\frac{1}{\mathrm{Re}}\left(\frac{\partial^2 u}{\partial x^2}+\frac{\partial^2 u}{\partial y^2}\right), x, y \in [0, 1] \tag{10.1}$$

in which $u(x, y, t)$ is the dependent variable, and Re is the Reynolds number characterizing the size of viscosity. Without loss of generality, we consider that the periodic boundary condition and the Dirichlet boundary conditions $u(x, 0, t) = u_0(x, t)$ and $u(x, 1, t) = u_1(x, t)$ are imposed in the x- and y-directions, respectively. The initial condition is $u(x, y, 0) = \overline{u}(x, y)$.

Burgers' equation (10.1) which is a fundamental and typical partial differential equation in fluid mechanics was first introduced by Bateman [1] who also found its steady solutions. Subsequently, it has been widely used as a mathematical model for studying free turbulence, gas dynamics, traffic flow, wave propagation in acoustics, etc. [2–4]. On the other hand, Burgers' equation (10.1) is very similar to the Navier–Stokes equations and a shock wave may also arise at high Reynolds number [5–7]. Therefore, the study on the numerical solution of Burgers' equation has attracted so much attention of scientific community due to not only its wide applications in the various fields [1–8] but also its role as a numerical test for various numerical methods [8–16] , such as the finite element method [8] , the finite difference method [9] , the collocation method [10], the boundary element approach [11], the group explicit method [12], the Petrov–Galerkin method [13] , the multiquadric method [14], the differential quadrature method [15], the lattice Boltzmann method [16], etc.

10.1.2 Ordinary Differential Equations Gained by Wavelet-Based Space Expansion

Based on the wavelet approximations developed in Chap. 5, and considering the boundary conditions, the unknown function $u(x, y, t)$ and nonlinear terms $u^2(x, y, t)$ in Eq. (10.1) can be written as

$$u(x, y, t) \approx \sum_{k=0}^{2^j-1}\sum_{l=1}^{2^j-1} u_{k,l}(t)\phi_{j,k}^{P}(x)\phi_{j,l}(y)$$

$$+\sum_{k=0}^{2^j-1}[u_{0,k}(t)\phi_{j,k}^P(x)\phi_{j,0}(y)+u_{1,k}(t)\phi_{j,k}^P(x)\phi_{j,2^j}(y)] \tag{10.2}$$

$$u^2(x,y,t)\approx\sum_{k=0}^{2^j-1}\sum_{l=1}^{2^j-1}u_{k,l}^2(t)\phi_{j,k}^P(x)\phi_{j,l}(y)$$
$$+\sum_{k=0}^{2^j-1}[u_{0,k}^2(t)\phi_{j,k}^P(x)\phi_{j,0}(y)+u_{1,k}^2(t)\phi_{j,k}^P(x)\phi_{j,2^j}(y)] \tag{10.3}$$

in which $u_{k,l}(t)=u(k/2^j,l/2^j,t)$, $u_{0,k}(t)=u_0(k/2^j,t)$, and $u_{1,k}(t)=u_1(k/2^j,t)$. Substituting Eqs. (10.2) and (10.3) into Eq. (10.1), and then multiplying both sides of the resulting equation by $\phi_{j,n}^P(x)\phi_{j,m}(y)$, $n=0,2,...,2^j-1$, $m=1,2,...,2^j-1$, respectively, and performing integration over the interval $[0,1]^2$, we get the system of ordinary differential equations in the matrix form

$$\mathbf{A}\frac{d\mathbf{U}(t)}{dt}+\frac{1}{2}\mathbf{B}(\mathbf{U}*\mathbf{U})\approx\frac{1}{\mathrm{Re}}\mathbf{D}\mathbf{U}+\mathbf{F}(t) \tag{10.4}$$

in which matrixes $\mathbf{A}=\{a_{oq}=\Gamma_{n,k}^{0,0}\overline{\Gamma}_{m,l}^{0,0}\}$, $\mathbf{B}=\{b_{oq}=\Gamma_{n,k}^{0,1}\overline{\Gamma}_{m,l}^{0,0}+\Gamma_{n,k}^{0,0}\overline{\Gamma}_{m,l}^{0,1}\}$, $\mathbf{D}=\{d_{oq}=\Gamma_{n,k}^{0,2}\overline{\Gamma}_{m,l}^{0,0}+\Gamma_{n,k}^{0,0}\overline{\Gamma}_{m,l}^{0,2}\}$, vectors $\mathbf{U}(t)=\{u_q=u(k/2^j,l/2^j,t)\}^{\mathrm{T}}$, subscripts $o=(2^j-1)n+m$, $q=(2^j-1)k+l$, and $k,n=0,2,\ldots,2^j-1$, $l,m=1,2,\ldots,2^j-1$, the operational rule $\mathbf{U}*\mathbf{U}=\{u_qu_q\}^{\mathrm{T}}$, and the vector

$$\begin{aligned}\mathbf{F}(t)=\{f_o-&\frac{1}{\mathrm{Re}}\sum_{k=0}^{2^j-1}[(\Gamma_{n,k}^{0,2}\overline{\Gamma}_{m,0}^{0,0}+\Gamma_{n,k}^{0,0}\overline{\Gamma}_{m,0}^{0,2})u_{0,k}(t)+(\Gamma_{n,k}^{0,2}\overline{\Gamma}_{m,2^j}^{0,0}+\Gamma_{n,k}^{0,0}\overline{\Gamma}_{m,2^j}^{0,2})u_{1,k}(t)]\\&-\sum_{k=0}^{2^j-1}(\Gamma_{n,k}^{0,0}\overline{\Gamma}_{m,0}^{0,0}du_{0,k}(t)/dt+\Gamma_{n,k}^{0,0}\overline{\Gamma}_{m,2^j}^{0,0}du_{1,k}(t)/dt)\\&-\frac{1}{2}\sum_{k=0}^{2^j-1}[(\Gamma_{n,k}^{0,1}\overline{\Gamma}_{m,0}^{0,0}+\Gamma_{n,k}^{0,0}\overline{\Gamma}_{m,0}^{0,1})u_{0,k}^2(t)+(\Gamma_{n,k}^{0,1}\overline{\Gamma}_{m,2^j}^{0,0}+\Gamma_{n,k}^{0,0}\overline{\Gamma}_{m,2^j}^{0,1})u_{1,k}^2(t)]\}^{\mathrm{T}}.\end{aligned} \tag{10.5}$$

Here, the integrations $\Gamma_{n,k}^{0,c}=\int_0^1 d^c\phi_{j,k}^P(x)/dx^c\phi_{j,n}^P(x)dx$ and $\overline{\Gamma}_{m,l}^{0,c}=\int_0^1 d^c\phi_{j,l}(y)/dy^c\phi_{j,m}(y)dy$ are given in Chap. 5.

The nonlinear ordinary differential equations in Eq. (10.4) can be directly solved by the time integration scheme, such as the finite difference method and the Runge–Kutta method. In the following, a mixed explicit–implicit scheme [17] which is a popular method in solving problems of fluid mechanics is employed to solve the semi-discretization system.

Applying the mixed explicit–implicit scheme to Eq. (10.4), one gets

Table 10.1 Coefficients of J-order mixed explicit–implicit scheme for $J \le 4$ [17]

Coefficients	$J = 1$	$J = 2$	$J = 3$	$J = 4$
γ_0	1	3/2	11/6	25/12
α_0	1	2	3	4
α_1		−1/2	−3/2	−3
α_2			1/3	4/3
α_3				−1/4
β_0	1	2	3	4
β_1		−1	−3	−6
β_2			1	4
β_3				−1

$$(\gamma_0 \mathbf{A} - \Delta t \mathbf{D}/\mathrm{Re})\mathbf{U}^{n+1} \approx \mathbf{A}\sum_{i=0}^{J-1} \alpha_i \mathbf{U}^{n-i} + \Delta t \mathbf{F}^{n+1} - \frac{\Delta t \mathbf{B}}{2}\sum_{i=0}^{J-1} \beta_i (\mathbf{U}^{n-i} * \mathbf{U}^{n-i}) \tag{10.6}$$

in which $\mathbf{U}^n = \mathbf{U}(n\Delta t)$; Δt is time step; and coefficients α_i, β_i, and γ_0 for the J-order mixed explicit–implicit scheme are listed in Table 10.1.

Applying Eq. (10.6) recursively and considering the initial conditions, we can obtain the nodal values of unknown function $u(k/2^j, l/2^j, n\Delta t)$, from which the approximate solution $u(x, y, n\Delta t)$ can be reconstructed by Eq. (10.2).

10.1.3 Solution Results to 1D and 2D Burgers' Equations

A computation code has been developed based on the formulations proposed above. The examples given in this section are mainly designed to test and benchmark the closed wavelet-based method for solving the nonlinear convection–diffusion problems. To effectively evaluate the numerical solution, the maximum error norm $E_{\max}$ and error E_2 are considered, which are, respectively, defined as

$$E_{\max} = \max_k\{|u_{\mathrm{num}}(x_k, t) - u_{\mathrm{exact}}(x_k, t)|\}, \tag{10.7}$$

$$E_2 = \sqrt{\sum_k |u_{\mathrm{num}}(x_k, t) - u_{\mathrm{exact}}(x_k, t)|^2/N}, \tag{10.8}$$

where u_{num} and u_{exact} are the numerical and exact solutions, respectively, and N is the number of spatial partitions.

Example 10.1 Consider the 1D Burgers equation subjected to the boundary and initial conditions.

$$\text{Case I}:\begin{cases}u(0,t)=u(1,t)=0,\\ u(x,0)=\frac{2\pi}{\text{Re}}\frac{\sin(\pi x)}{100+\cos(\pi x)}\end{cases}.\text{ and Case II}:\begin{cases}u(0,t)=u(1,t)=0,\\ u(x,0)=\sin(\pi x).\end{cases}\quad(10.9)$$

The exact solutions are [18, 19]

$$\text{Case I}: u(x,t)=\frac{2\pi}{\text{Re}}\frac{\exp(-\pi^2 t/\text{Re})\sin(\pi x)}{100+\exp(-\pi^2 t/\text{Re})\cos(\pi x)},\quad(10.10)$$

$$\text{Case II}: u(x,t)=\frac{2\pi}{\text{Re}}\frac{\sum_{n=1}^{+\infty}c_n\exp(-n^2\pi^2 t/\text{Re})n\sin(n\pi x)}{c_0+\sum_{n=1}^{+\infty}c_n\exp(-n^2\pi^2 t/\text{Re})\cos(n\pi x)}\quad(10.11)$$

in which

$$c_0=\int_0^1\exp\left(-\text{Re}\frac{1-\cos(\pi x)}{2\pi}\right)dx,\ c_n=2\int_0^1\exp\left(-\text{Re}\frac{1-\cos(\pi x)}{2\pi}\right)\cos(n\pi x)dx.\quad(10.12)$$

Tables 10.2 and 10.3 show the comparison of the maximum error norm $E_{\max}$ between solutions obtained by using different numerical methods, containing the wavelet-based method. From them, it can be seen that the proposed wavelet-based method is much more accurate and efficient than many other existing methods,

Table 10.2 Errors of numerical solutions at Re = 200 and time step, $\Delta t = 0.01$, for various number of spatial partitions N at time, $t = 1.0$, for 1D Burgers' equation in Case I

Error	Wavelet based		MCBSCM [20]		DQM [15]	
	$N = 16$	$N = 32$	$N = 20$	$N = 40$	$N = 20$	$N = 40$
$E_{\max}$	2.060E-09	6.898E-11	3.062E-08	7.644E-09	1.091E-08	1.980E-09
E_2	7.250E-10	1.709E-11	2.153E-08	5.378E-09	4.465E-09	2.786E-10

Table 10.3 Maximum error norm $E_{\max}$ of the numerical solutions at Re = 10 and 100 for Case II

Re	Wavelet $N = 16$ $\Delta t = 10^{-3}$	MQ [19] $N = 10$ $\Delta t = 10^{-3}$	MQQI [21] $N = 100$ $\Delta t = 10^{-3}$	BSQI [22] $N = 100$ $\Delta t = 10^{-3}$	MQRBF [14] $N = 10$ $\Delta t = 10^{-3}$	FEM [8] $N = 80$ $\Delta t = 10^{-4}$	DQM [15] $N = 25$ $\Delta t = 10^{-4}$	LBM [16] $N = 10$ $\Delta t = 10^{-3}$
10	3.80E-5	3.20E-4	1.59E-2	3.50E-4	2.20E-3	2.69E-3	5.00E-5	1.26E-3
100	2.39E-4	7.00E-4	3.04E-2	4.50E-4	2.43E-2	6.05E-3	8.00E-5	—

including the multiquadric method [14], the weighted average differential quadrature method [15], the lattice Boltzmann method [16], the multiquadric method [19], the modified cubic B-splines collocation method [20], the multiquadric quasi-interpolation method [21], the cubic B-spline quasi-interpolation method [22], and the B-spline finite element method [8]. For example, as shown in Table 10.2 when the same time step is taken in the three methods and the number of spatial partitions employed in the wavelet-based method is less than other methods, the accuracy of the wavelet-based method has at least one order higher than others.

Figure 10.1 displays the scaling relation between the maximum errors $E_{\max}$ and the number of spatial partitions N. From it, one can find that the order of convergence of the proposed wavelet method is about 5, which obviously exceeds the order of convergence 2 for the weighted average differential quadrature method [15]. Figure 10.2 shows the present wavelet-based solution for high Reynolds number Re = 10,000. The wavelet-based solution is in good agreement with the solution given by Hon

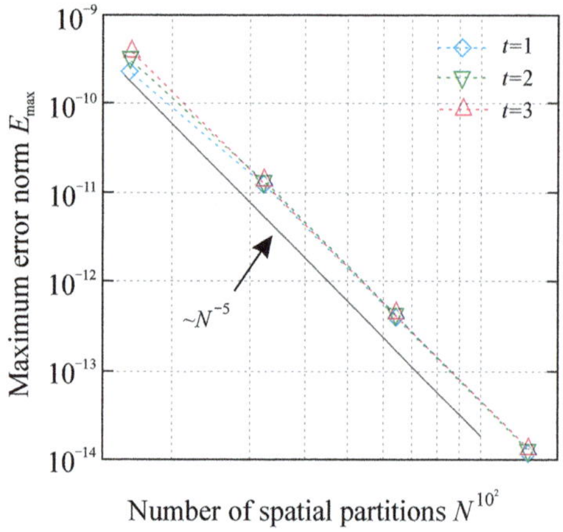

Fig. 10.1 Maximum error norm $E_{\max}$ of the wavelet-based solution for 1D Burgers' equation of case I under Reynolds number, Re = 1000, and time step, $\Delta t = 0.0001$, at different times, t, varying with the number of spatial partitions, N

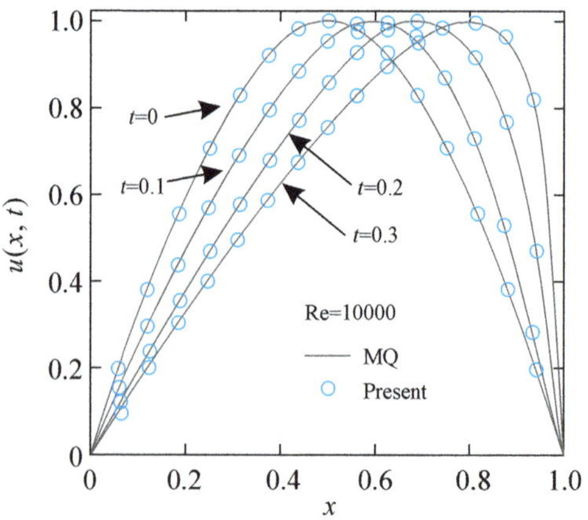

Fig. 10.2 Numerical solutions of 1D Burgers' equation under $\Delta t = 0.001$ and $N = 16$ for Reynolds number, Re = 10,000, at different times, t, for Case II. "Present" stands for the solutions from the wavelet-based method

Table 10.4 Maximum error norm $E_{\max}$ for the numerical solutions of 1D Burgers' equation with exact solution of $u(x, t) = v(x, t) = e^{-t}\sin(\pi x)$ using time step $\Delta t = 0.01$

Time t	Wavelet ($N = 7$)	Wavelet ($N = 15$)	DQM ($N = 21$) [6]
0.5	1.34072E-04	4.33522E-06	1.51688E-04
1.0	8.20725E-05	2.62682E-06	1.83970E-04
2.0	2.99970E-05	9.59266E-07	1.35250E-04
3.0	1.09437E-05	3.50049E-07	7.46014E-05

and Mao [19], which is considered as an excellent approximation of the exact solution, implying that the present wavelet method has the ability for solving Burgers' equation with high Reynolds numbers.

Example 10.2 Consider the 1D coupled viscous Burgers' equations.

$$\begin{cases} \frac{\partial u}{\partial t} = \frac{1}{\pi^2}\frac{\partial^2 u}{\partial x^2} + \frac{1}{\pi}\frac{\partial u^2}{\partial x} - \frac{1}{\pi}\frac{\partial (uv)}{\partial x} \\ \frac{\partial v}{\partial t} = \frac{1}{\pi^2}\frac{\partial^2 v}{\partial x^2} + \frac{1}{\pi}\frac{\partial v^2}{\partial x} - \frac{1}{\pi}\frac{\partial (uv)}{\partial x} \end{cases}, \quad x \in (0,\ 1), t \geq 0 \tag{10.13}$$

which is subjected to the boundary and initial conditions $u(x, 0) = v(x, 0) = \sin(\pi x)$. The exact solution to this problem is $u(x, t) = v(x, t) = e^{-t}\sin(\pi x)$ [6, 23].

The errors of approximate solutions are shown in Table 10.4 and Fig. 10.3. From them, one can see that the present wavelet-based method has a better numerical accuracy than the differential quadrature method (DQM). That is, the wavelet-based method using much smaller number of nodes N can get the solution with either same or higher accuracy, which clearly demonstrated the accuracy and efficiency of the proposed wavelet-based method.

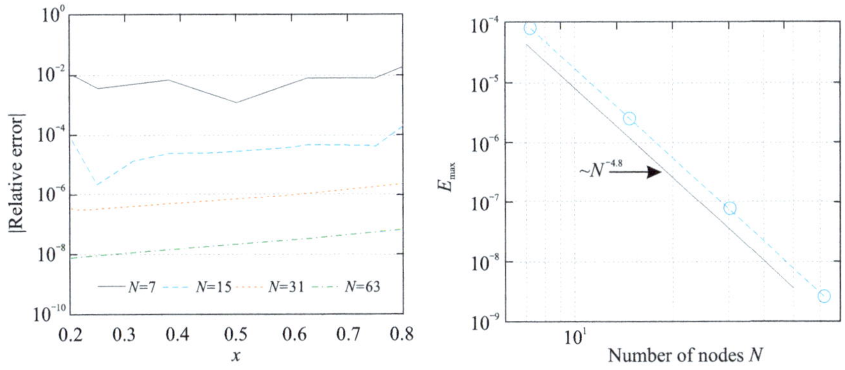

Fig. 10.3 Error of the wavelet-based numerical solutions for 1D Burgers' equation with exact solution $u(x, t) = v(x, t) = e^{-t}\sin(\pi x)$ using $\Delta t = 0.0005$ at $t = 1$: Spatial distributions of the relative errors (Left), scaling dependence of maximum error norm $E_{\max}$ on the number of nodes (Right)

Example 10.3 Consider the 2D Burgers' equation over the domain $[0, 1] \times [0, 1]$, which is subjected to initial and Dirichlet boundary conditions taken from the exact solution [24]

$$u(x, y, t) = \frac{1}{1 + e^{\mathrm{Re}(x+y-t)/2}}. \tag{10.14}$$

Tables 10.5, 10.6, 10.7 show the error norms of numerical solutions obtained respectively by using the proposed wavelet-based method, the fourth-order compact finite difference scheme (CFD4) [24], the Eulerian–Lagrangian method (ELM) [25], and the lattice Boltzmann method (LBM) [26]. It can be seen from these results that solutions obtained by using the wavelet-based method with more coarse space–time mesh have a better numerical accuracy than those achieved from LBM, CFD4 and ELM.

Table 10.5 Absolute errors of numerical solutions under $\Delta t = 0.01$ and $\mathrm{Re} = 1$ at some specific points

Wavelet based ($N = 8 \times 8$)					CFD4 ($N = 10 \times 10$) [24]				
x	y	$t = 10$	$t = 15$	$t = 20$	x	Y	$t = 10$	$t = 15$	$t = 20$
0.125	0.125	2.28E-09	7.44E-10	1.03E-09	0.1	0.1	1.22E-04	1.03E-05	8.47E-07
	0.5	1.77E-09	6.22E-10	8.49E-10		0.5	3.26E-04	2.76E-05	2.27E-06
	0.875	3.43E-09	6.25E-10	1.03E-09		0.9	1.53E-04	1.30E-05	1.07E-06
0.5	0.125	1.77E-09	6.22E-10	8.49E-10	0.5	0.1	3.26E-04	2.76E-05	2.27E-06
	0.5	4.38E-10	6.23E-10	6.42E-10		0.5	1.02E-03	8.61E-05	7.09E-06
	0.875	2.44E-09	5.34E-10	8.44E-10		0.9	6.71E-04	5.70E-05	4.69E-06
0.875	0.125	3.43E-09	6.25E-10	1.03E-09	0.9	0.1	1.58E-04	1.34E-05	1.10E-06
	0.5	2.44E-09	5.34E-10	8.44E-10		0.5	5.83E-04	4.96E-05	4.08E-06
	0.875	7.79E-09	4.51E-10	1.02E-09		0.9	1.67E-03	1.42E-04	1.17E-05

Table 10.6 Absolute errors of numerical solutions under $\Delta t = 0.001$ and $\mathrm{Re} = 20$ at some specific points

Wavelet based ($N = 16 \times 16$)					ELM ($N = 441$) [25]				
x	y	$t = 0.5$	$t = 0.75$	$t = 1.0$	x	y	$t = 0.5$	$t = 0.75$	$t = 1.0$
0.125	0.125	1.56E-05	1.21E-06	3.93E-07	0.1	0.1	6.07E-04	1.70E-04	6.08E-06
	0.5	6.68E-05	1.33E-05	3.39E-06		0.5	4.66E-04	1.05E-03	1.95E-04
	0.875	1.01E-05	1.60E-05	1.87E-04		0.9	1.59E-06	6.60E-05	9.45E-04
0.5	0.125	6.68E-05	1.33E-05	3.39E-06	0.5	0.1	4.66E-04	1.05E-03	1.95E-04
	0.5	7.69E-07	9.33E-06	1.92E-05		0.5	8.07E-06	5.18E-04	2.99E-03
	0.875	7.91E-08	1.12E-06	5.82E-06		0.9	2.00E-07	2.93E-06	1.18E-04
0.875	0.125	1.01E-05	1.60E-05	1.87E-04	0.9	0.1	1.59E-06	6.60E-05	9.45E-04
	0.5	7.91E-08	1.12E-06	5.82E-06		0.5	2.00E-07	3.93E-06	1.18E-04
	0.875	3.47E-09	4.80E-08	5.23E-07		0.9	3.92E-08	2.40E-07	1.96E-06

Table 10.7 Error norms of numerical solutions at $t = 0.25$ under $\Delta t = \Delta x/10$

Re	Wavelet based			LBM [26]		
	Grid	E_2	E_{max}	Grid	E_2	E_{max}
1	8 × 8	6.93E-08	1.20E-07	20 × 20	2.65E-05	1.48E-04
	16 × 16	1.75E-08	3.26E-08	40 × 40	3.68E-05	2.92E-04
	32 × 32	4.59E-09	8.75E-09	80 × 80	5.18E-05	5.82E-04
10	8 × 8	8.14E-05	2.30E-04	20 × 20	4.54E-04	3.08E-03
	16 × 16	7.58E-06	3.24E-05	40 × 40	1.88E-03	1.63E-02
	32 × 32	1.31E-06	2.73E-06	80 × 80	3.13E-03	3.82E-02
100	8 × 8	8.43E-03	4.00E-02	80 × 80	7.61E-03	5.89E-02
	16 × 16	3.82E-03	3.75E-02	200 × 200	4.37E-03	5.71E-02
	32 × 32	4.84E-04	4.52E-03	–	–	–

The present wavelet solutions at $t = 1$ for high Reynolds number Re = 200 and 300 are presented in Fig. 10.4, in which the spatial grid 32 × 32 and time step Δt = 0.01 are used. It can be found that the proposed wavelet method even under a coarse space–time mesh also has the capability to capture the shock wave, although there are some slight oscillations near the sharp transition. By contrast, under the same space–time mesh, very conspicuous oscillations are apparent in numerical solutions obtained by using the global meshless collocation method (CM), the Haar wavelet CM, and the Legendre wavelet CM [27]. The profiles of these solutions are highly distorted, and some even have lost the basic shape of the shock wave [27]. To investigate the spatial convergence, Fig. 10.5 plots the error norms of wavelet solutions at $t = 2$ for Re = 10 and time step $\Delta t = 10^{-4}$ as a function of the spatial mesh size $\Delta x(=\Delta y)$. It can be seen from Fig. 10.5 that the convergence rate in space

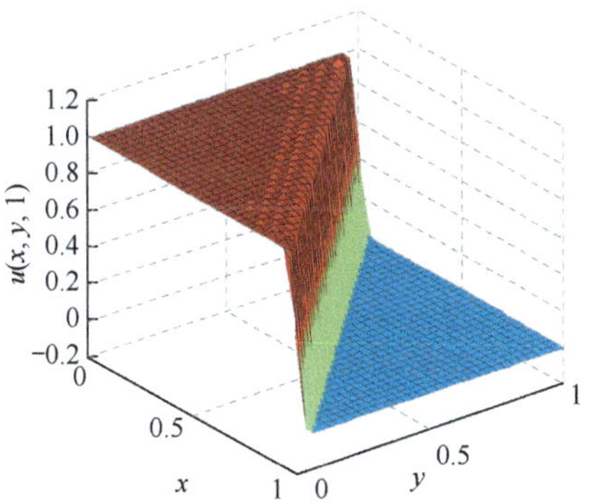

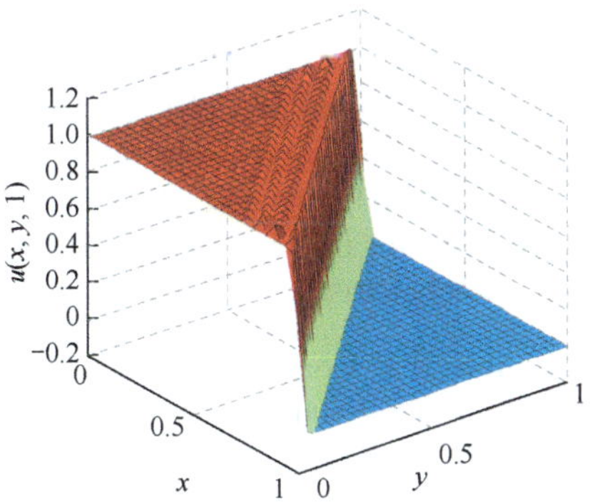

Fig. 10.4 Shock waves of the wavelet-based solutions at $t = 1$ under grid 32 × 32 and $\Delta t = 0.01$ for 2D Burgers' equation with exact solution of $u(x, y, t) = [1 + e^{\mathrm{Re}(x+y-t)/2}]^{-1}$: Re = 200 (Left), Re = 300 (Right)

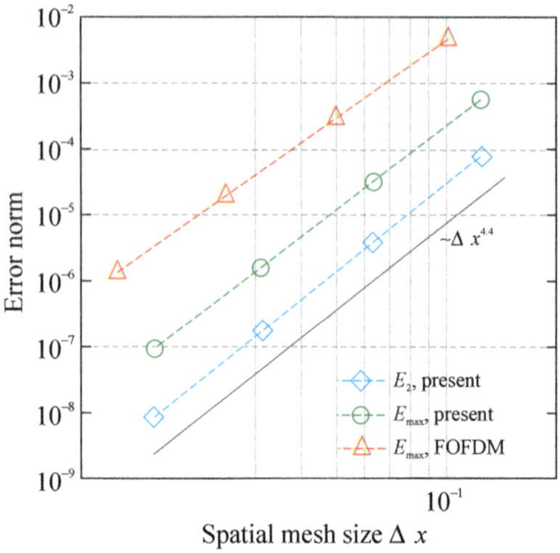

Fig. 10.5 Scaling dependence of error norms of numerical solutions at $t = 2$ under Re = 10 and $\Delta t = 10^{-4}$ on the $\Delta x(=\Delta y)$ for the 2D Burgers' equation with exact solution of $u(x, y, t) = [1 + e^{\text{Re}(x+y-t)/2}]^{-1}$. Here, "present" stands for the wavelet-based method

of the present wavelet method is about order 4, and its accuracy is better than the fourth-order finite difference method (FOFDM) [28].

Example 10.4 Consider the 2D Burgers equation over the domain $[-0.5, 0.5] \times [-0.5, 0.5]$, and initial and Dirichlet boundary conditions are extracted from the exact solution [26]

$$u(x, y, t) = \frac{1}{2} - \tanh\left[\frac{\text{Re}(x + y - t)}{2}\right]. \tag{10.15}$$

To this shock wave problem, the solution from the wavelet-based method under grid 256 × 256 and $\Delta t = 10^{-4}$ is plotted in Fig. 10.6 for shock wave forms with Re = 500 and 1000, respectively, at $t = 0.1$. From Fig. 10.6, one sees that the proposed wavelet method is capable of finding a convergent and stable solutions without any oscillations for the large values of Reynolds number where a steep oblique shock wave without arises.

Example 10.5 Consider the 2D Burgers' equation over the domain $[0, 1] \times [0, 1]$, which is subjected to the periodic boundary condition and the initial value $u(x, y, 0) = \sin(2\pi x) \cos(2\pi y)$ [29].

Figure 10.7 gives the contours of the wavelet solutions at $t = 0.1$ under grid, 32 × 32, and $\Delta t = 0.001$ for such 2D Burgers' equation with Reynolds numbers, Re = 100 and 500, respectively. From it, one can see that these numerical solutions are very smooth without any undesirable oscillations, and the solution for Re = 100 is in good agreement with the precise LBM solution given by Zhang and Yan [29]. When

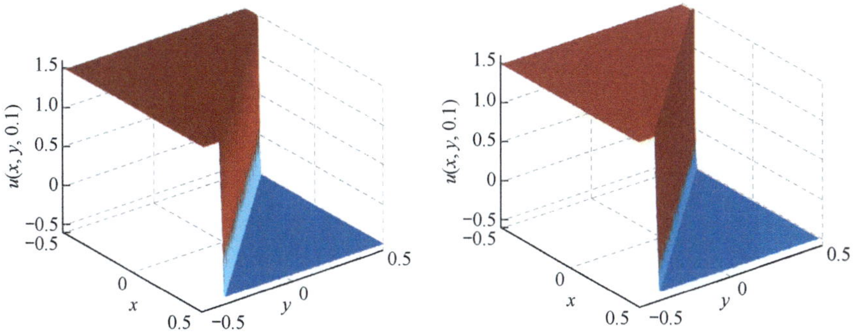

Fig. 10.6 Shock waves of the wavelet-based solutions at $t = 0.1$ under grid 256×256 and $\Delta t = 10^{-4}$ for 2D Burgers' equation with exact solution of $u(x, y, t) = 0.5 - \tan\mathrm{h}[0.5\mathrm{Re}(x + y - t)]$: Re = 500 (Left), Re = 1000 (Right)

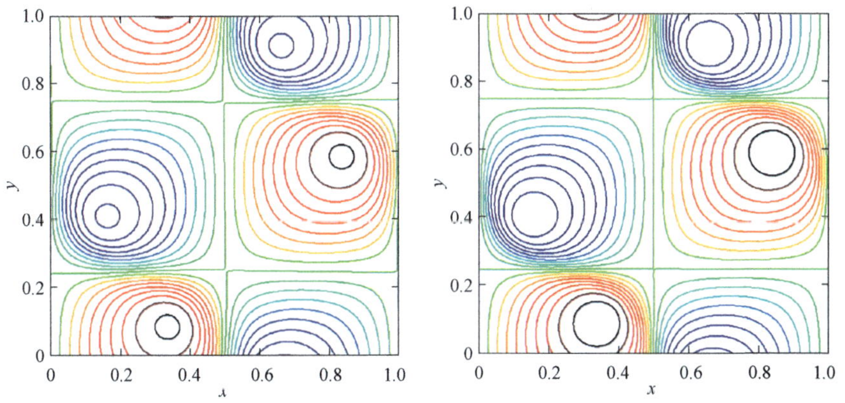

Fig. 10.7 Contours of the wavelet-based solutions at $t = 0.1$ under grid 32×32 and $\Delta t = 10^{-3}$ for 2D Burgers' equation with exact solution of $u(x, y, 0) = \sin(2\pi x)\cos(2\pi y)$: Re = 100 (Left), Re = 500 (Right)

$t = 0.125$, a comparison of the contours obtained by the wavelet-based solution and the LBM for the 2D Burgers' equation with Re = 100 is shown in Fig. 10.8. From the evaluation criterion of accuracy to such problem, we know that each central contour is more approached to a circular form, and the solution is more accurate. According to Fig. 10.8, one finds that all central contours from the wavelet-based solutions are more rounded than those from the LBM when the wavelet-based method uses the grids at least three orders lower than that in LBM.

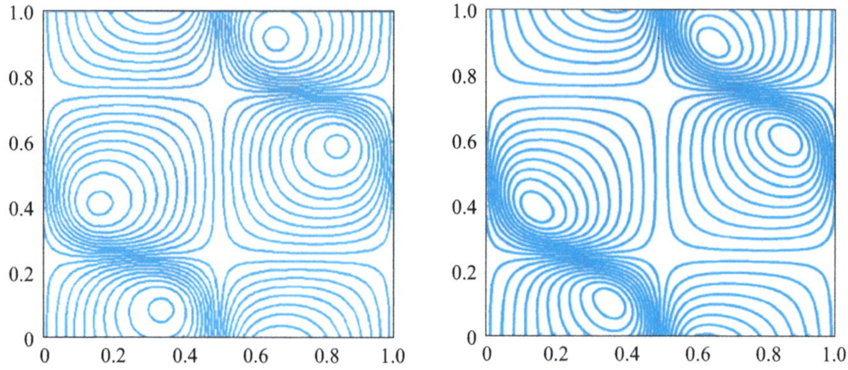

Fig. 10.8 Comparison of contours from the wavelet-based solution using 32 × 32 grid (Left) and the LBM using 2048 × 2048 grid [29] (Right) for 2D Burgers' equation with exact solution of $u(x, y, 0) = \sin(2\pi x)\cos(2\pi y)$ when Re = 100

10.2 2D Poiseuille and Couette Laminar Flow

10.2.1 Governing Equations with Initial-Boundary-Value Conditions

For such typical problems of both Poiseuille and Couette fluid flows, the diagram is shown in Fig. 10.9, where the fluid is located between two parallel infinite plates. The Poiseuille flow is generated by a body force acting in the longitudinal direction while both plates remain stationary. On the contrary, the Couette flow is generated by movement of the upper plate with a constant velocity in longitudinal direction [30]. Due to that both Poiseuille and Couette laminar flows can be solved analytically, here, we use them as examples by using the closed wavelet-based method to numerically solve them.

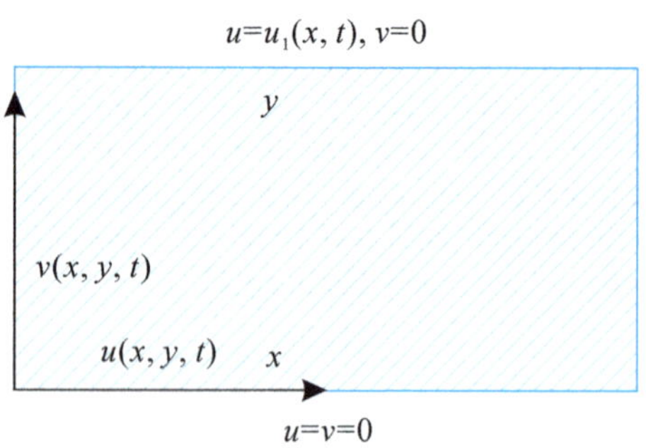

Fig. 10.9 Diagram of plane channel flow

The governing equations of Poiseuille and Couette flows can be written into [31]

$$\begin{cases} \frac{\partial u}{\partial x} + \frac{\partial v}{\partial y} = 0, \\ \frac{\partial u}{\partial t} + \frac{\partial (u^2)}{\partial x} + \frac{\partial (uv)}{\partial y} = -\frac{\partial p}{\partial x} + \frac{1}{\mathrm{Re}}\frac{\partial^2 u}{\partial x^2} + f_x,\ x \in [0, 2\pi],\ y \in [0, 2], \\ \frac{\partial v}{\partial t} + \frac{\partial (uv)}{\partial x} + \frac{\partial (v^2)}{\partial y} = -\frac{\partial p}{\partial y} + \frac{1}{\mathrm{Re}}\frac{\partial^2 v}{\partial y^2} + f_y, \end{cases} \tag{10.16}$$

which is subjected to the initial condition $u = v = 0$, the periodic boundary condition in x-direction, and the Dirichlet boundary conditions in y-direction of $u(2, 0, t) = u_1(x, t)$ and $u(x, 0, t) = v(x, 0, t) = v(x, 2, t) = 0$.

Specifically, there are $u_1(x, t) = 0,\ f_y = 0,\ f_x = 1$ for the Poiseuille flow, and $u_1(x, t) = 1,\ f_x = f_y = 0$ for the Couette flow. Both Poiseuille and Couette flows will be in the steady state after some time. Their stationary solutions are as follows [31]:

$$u = \frac{\mathrm{Re}}{2} y(2 - y),\ v = 0,\ p = 0 \text{ for Poiseuille flow,} \tag{10.17}$$

$$u = \frac{y}{2},\ v = 0,\ p = 0 \text{ for Couette flow.} \tag{10.18}$$

10.2.2 High-Order Splitting Methods for Incompressible Flows

The pressure in incompressible flows plays a very particular role, which should always be in equilibrium with a divergence-free velocity field. However, the pressure does not appear explicitly in the governing equation (10.16) imposing such a divergence condition. Therefore, a specialized procedure is needed to deal with this issue. Here, a high-order splitting method [32] is introduced to solve the problems of incompressible flows by combining the closed wavelet-based method.

Using the mixed explicit–implicit scheme to integrate the momentum equations given in Eq. (10.16) over time step Δt, we get

$$\begin{aligned} &\gamma_0 u^{n+1} - \sum_{\lambda=0}^{J-1} \alpha_\lambda u^{n-\lambda} + \Delta t \sum_{\lambda=0}^{J-1} \beta_\lambda \left(\frac{\partial u^2}{\partial x} + \frac{\partial (uv)}{\partial y} \right)^{n-\lambda} \\ &= -\Delta t \frac{\partial p^{n+1}}{\partial x} + \frac{\Delta t}{\mathrm{Re}} \frac{\partial^2 u^{n+1}}{\partial x^2} + \Delta t f_x^{n+1}, \end{aligned} \tag{10.19}$$

$$\gamma_0 v^{n+1} - \sum_{\lambda=0}^{J-1} \alpha_\lambda v^{n-\lambda} + \Delta t \sum_{\lambda=0}^{J-1} \beta_\lambda \left(\frac{\partial(uv)}{\partial x} + \frac{\partial v^2}{\partial y} \right)^{n-\lambda} = -\Delta t \frac{\partial p^{n+1}}{\partial y} + \frac{\Delta t}{\text{Re}} \frac{\partial^2 v^{n+1}}{\partial y^2} + \Delta t f_y^{n+1} \tag{10.20}$$

in which the coefficients α_i, β_i, and γ_0 for the J-order mixed explicit–implicit scheme are given in Table 10.1.

In the essential Eqs. (10.19) and (10.20) of calculation program, one can see that such two equations on unknowns u and v contain also other unknown pressure p. In the theory of fluid mechanics, the unknown pressure is explicitly contained in the mass conservation equation and the state equation when the fluid is compressible. When the fluid is taken into account as incompressible, the pressure is inexplicitly contained by the mass conservation equation or the divergence of the fluid velocity. Following it, we add such equation as follows.

Taking the derivatives of Eqs. (10.19) and (10.20) with respect to x and y, respectively, and summing the resulting equations, we have

$$\frac{\partial^2 p^{n+1}}{\partial x^2} + \frac{\partial^2 p^{n+1}}{\partial y^2} = -\sum_{\lambda=0}^{J-1} \beta_\lambda \left(\frac{\partial^2 u^2}{\partial x^2} + 2\frac{\partial^2 (uv)}{\partial x \partial y} + \frac{\partial^2 v^2}{\partial y^2} \right)^{n-\lambda} - \frac{\partial f_x^{n+1}}{\partial x} - \frac{\partial f_y^{n+1}}{\partial y} \tag{10.21}$$

in which the divergence-free condition of velocity field has been used and it constitutes a basic equation for calculating p in the spatial domain. To its boundary condition calculations, using the explicit scheme to integrate the momentum equations over each time step Δt, one can obtain

$$\begin{cases} \frac{\partial p^{n+1}}{\partial x} = -\frac{\gamma_0}{\Delta t} u^{n+1} + \sum_{\lambda=0}^{J-1} \frac{\alpha_\lambda}{\Delta t} u^{n-\lambda} - \sum_{\lambda=0}^{J-1} \beta_\lambda \left(\frac{\partial u^2}{\partial x} + \frac{\partial (uv)}{\partial y} - \frac{1}{\text{Re}} \frac{\partial^2 u}{\partial x^2} \right)^{n-\lambda} + f_x^{n+1}, \\ \frac{\partial p^{n+1}}{\partial y} = -\frac{\gamma_0}{\Delta t} v^{n+1} + \sum_{\lambda=0}^{J-1} \frac{\alpha_\lambda}{\Delta t} v^{n-\lambda} - \sum_{\lambda=0}^{J-1} \beta_\lambda \left(\frac{\partial (uv)}{\partial x} + \frac{\partial v^2}{\partial y} - \frac{1}{\text{Re}} \frac{\partial^2 v}{\partial y^2} \right)^{n-\lambda} + f_y^{n+1}, \end{cases} \tag{10.22}$$

which is a kind of Neumann boundary conditions. It is obvious that the combination of differential Eq. (10.21) and boundary conditions of Eq. (10.22) constitutes a standard Poisson problem, which can be solved by the wavelet-based method proposed in Chap. 5. Once the pressure is obtained by solving the Poisson problem, the momentum Eq. (10.16) or its discretization schemes (10.19) and (10.20) become a standard Burgers' problem. Then, the velocity field can be also obtained by using the proposed closed wavelet-based method to solve such a Burgers' problem.

10.2.3 Numerical Solution Results

Figure 10.10 shows the fields of velocity and pressure of the Poiseuille flow with Re = 1000 by using the wavelet-based method. Table 10.8 lists the maximum error norm E_{max} of velocity numerical solutions for both Poiseuille flow and Couette flow solved by the wavelet-based method relative to their exact solutions, respectively. From Table 10.8, one can see that the wavelet-based method can use a coarse grid to find a solution with high accuracy for plane channel laminar flow, e.g., when grids are taken by 16 × 16, the accuracy of the solution is 10^{-7} for the Poiseuille flow and 10^{-8} for the Couette flow, respectively.

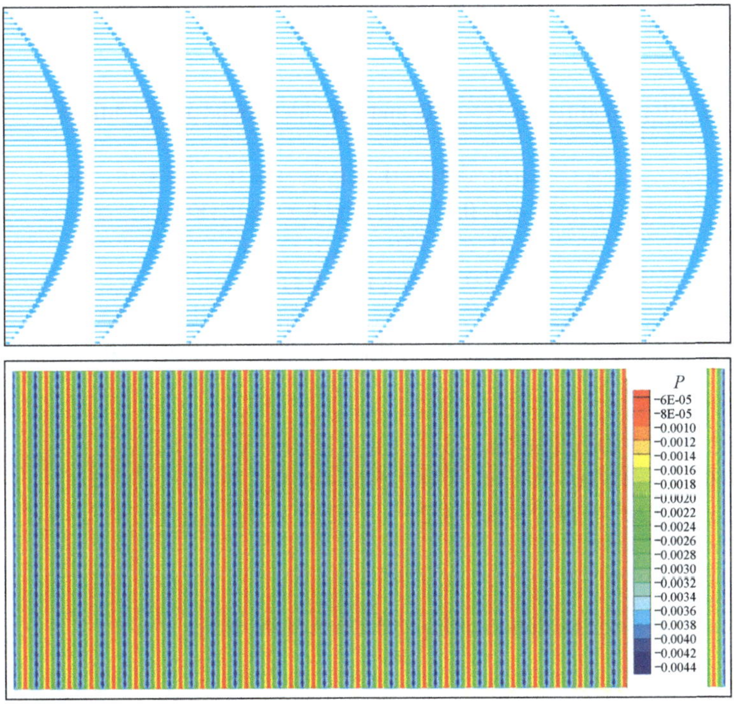

Fig. 10.10 The fields of velocity (upper) and pressure (lower) for the Poiseuille flow with Re = 1000 solved by the wavelet-based method

Table 10.8 The maximum error norm E_{max} of velocity

Grid	Poiseuille	Couette
16 × 16	1.08E-07	6.96E-08
32 × 32	9.20E-09	6.24E-09

10.3 2D Cavity Laminar Flow

10.3.1 Essential Equations in the Calculations

As shown in Fig. 10.11, a 2D cavity of size 1×1 is filled with the viscous fluid, where the top wall moves following the velocity profile $u_1(x)$, and the rest walls are stationary. To such fluid problem, the governing equations of this problem can be written as [33]

$$\begin{cases} \frac{\partial u}{\partial x} + \frac{\partial v}{\partial y} = 0, \\ \frac{\partial u}{\partial t} + \frac{\partial (u^2)}{\partial x} + \frac{\partial (uv)}{\partial y} = -\frac{\partial p}{\partial x} + \frac{1}{\text{Re}}\frac{\partial^2 u}{\partial x^2} + f_x,\ x, y \in [0, 1], \\ \frac{\partial v}{\partial t} + \frac{\partial (uv)}{\partial x} + \frac{\partial (v^2)}{\partial y} = -\frac{\partial p}{\partial y} + \frac{1}{\text{Re}}\frac{\partial^2 v}{\partial y^2} + f_y \end{cases} \tag{10.23}$$

with the initial condition $u = v = 0$, and Dirichlet boundary conditions

$$\begin{cases} u(0, y, t) = 0, u(1, y, t) = 0, \\ u(x, 0, t) = 0, u(x, 1, t) = u_1(x), \\ v(0, y, t) = 0, v(1, y, t) = 0, \\ v(x, 0, t) = 0, v(x, 1, t) = 0. \end{cases} \tag{10.24}$$

When the velocity profile $u_1(x) = 16(x^4 - 2x^3 + x^2)$ and the body force [33]

$$\begin{cases} f_x(x, y, t) = 0, \\ f_y(x, y, t) = \{96x^3[3x(2x - 5) + 10] + 1440y^2(y^2 - 1)(2x - 1), \\ +960x(x - 1)(2x - 1)(6y^2 - 1)\}/15\text{Re} + [64(y^4 - y^2)(4y^3 - 2y), \\ \times(4x^6 - 12x^5 + 14x^4 - 8x^3 + 2x^2) - 32(x^4 - 2x^3 + x^2)^2(24y^5 - 8y^3 + 4y)] \end{cases} \tag{10.25}$$

are chosen, the exact solution of Eq. (10.23) is of the form

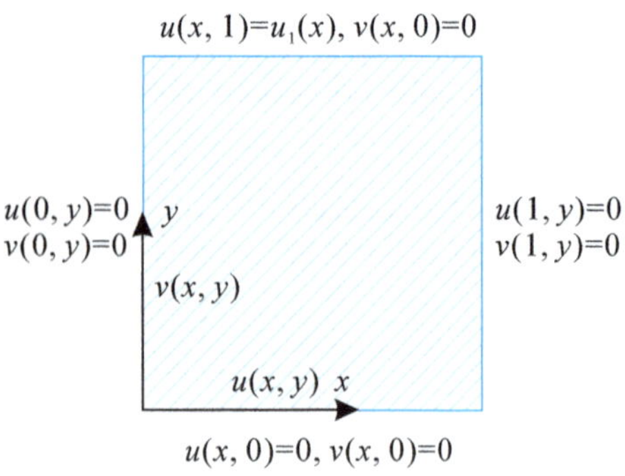

Fig. 10.11 Diagram of 2D plane cavity flow

$$\begin{cases} u(x, y, t) = 8(x^4 - 2x^3 + x^2)(4y^3 - 2y), \\ v(x, y, t) = -8(4x^3 - 6x^2 + 2x)(y^4 - y^2). \end{cases} \tag{10.26}$$

10.3.2 Numerical Solution Results

Applying the wavelet-based method to solve Eq. (10.23) with boundary conditions of Eq. (10.24), the numerical solution results are plotted in Figs. 10.12 and 10.13 for the streamline and vorticity fields when Re = 10. From them, one can see that the

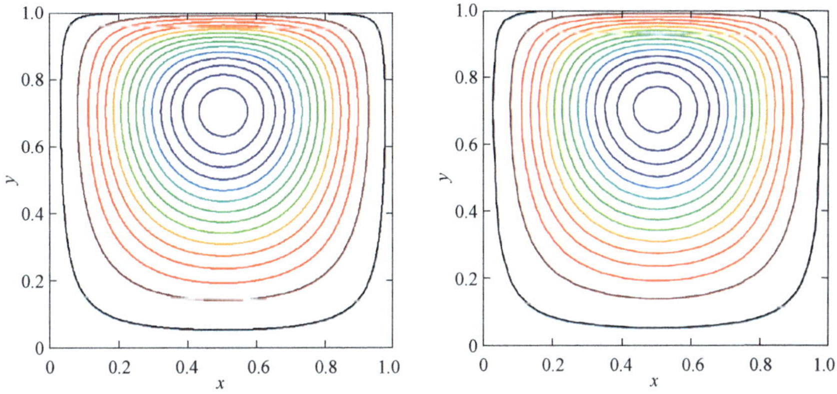

Fig. 10.12 Streamlines of 2D cavity flow for the case with exact solution in Eq. (10.26) (Left) and its wavelet-based numerical solution (Right) when Re = 10

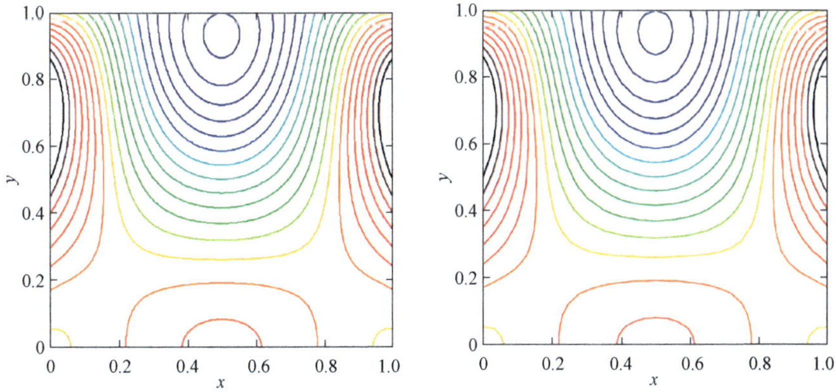

Fig. 10.13 Vorticity contours of 2D cavity flow for the case with exact solution in Eq. (10.26) (Left) and its wavelet-based numerical solution (Right) when Re = 10

Table 10.9 Accuracy comparison between the wavelet-based method and lattice Boltzmann method

Grid	Wavelet		LBM [33]		
	E_g	R	Grid	E_g	R
8 × 8	3.36E-03	–	10 × 10	6.37E-02	–
16 × 16	1.33E-04	−4.66	20 × 20	1.65E-02	−1.95
32 × 32	5.20E-06	−4.68	40 × 40	4.21E-03	−1.97
64 × 64	2.42E-07	−4.43	80 × 80	1.09E-03	−1.95

numerical solutions are in good agreement with the exact ones. When the relative global error, E_g, and the rate of convergence, R, are, respectively, employed by

$$E_g = \sqrt{\frac{\sum_k \{[u_{\text{num}}(x_k) - u_{\text{exact}}(x_k)]^2 + [v_{\text{num}}(x_k) - v_{\text{exact}}(x_k)]^2\}}{\sum\limits_k [u_{\text{exact}}(x_k)^2 + v_{\text{exact}}(x_k)^2]}} \tag{10.27}$$

and

$$R = \frac{\lg[E_g(N_2)] - \lg[E_g(N_1)]}{\lg(N_2) - \lg(N_1)}. \tag{10.28}$$

Table 10.9 presents a comparison between the proposed closed wavelet-based method and the lattice Boltzmann method, which also reveals that the wavelet-based method is much more accurate and efficient than the lattice Boltzmann method [33].

10.4 Strong Nonlinear Problem of 2D Vortex Emerging Interaction

10.4.1 Governing Equations for the Problem

In this section, the merger of two positive vortices pushed together with a weaker negative vortex is chosen to test the proposed closed wavelet-based method, because it is a typical fluid flow with high nonlinearity [34].

In this simulation, the domain is taken to be $\Omega = [0, 2\pi] \times [0, 2\pi]$, i.e., a square box of side length 2π, in which the periodic boundary conditions are imposed on all boundaries. The governing equations for this problem in velocity–vorticity formulation can be written as

$$\frac{\partial \omega}{\partial t} + \frac{\partial (u\omega)}{\partial x} + \frac{\partial (v\omega)}{\partial y} = \upsilon\left(\frac{\partial^2 \omega}{\partial x^2} + \frac{\partial^2 \omega}{\partial y^2}\right), \tag{10.29}$$

$$\frac{\partial^2 \psi}{\partial x^2} + \frac{\partial^2 \psi}{\partial y^2} = -\omega, \tag{10.30}$$

$$u = \frac{\partial \psi}{\partial y}, v = -\frac{\partial \psi}{\partial x}, \tag{10.31}$$

in which ω is the vorticity, ψ is the stream function, u and v are the velocity fields, and $\upsilon = 5 \times 10^{-5}$ is the kinematic viscosity. The initial condition is taken to be as [34]

$$\omega(x, y) = \sum_{i=1}^{3} A_i \exp\{-\pi^2[(x - x_i)^2 + (y - y_i)^2]\} \tag{10.32}$$

with $A_1 = A_2 = -2A_3 = \pi$, $x_1 = 3\pi/4$, $x_2 = x_3 = 5\pi/4$, $y_1 = y_2 = \pi$, and $y_2 = \pi(1 + 1/\sqrt{8})$.

The kinetic energy, $E(t)$, and enstrophy, $Z(t)$, of the system are, respectively, defined as

$$E(t) = \frac{1}{2} \int_{\Omega} (u^2 + v^2) d\Omega \tag{10.33}$$

and

$$Z(t) = \frac{1}{2} \int_{\Omega} \omega^2 d\Omega. \tag{10.34}$$

10.4.2 Arithmetic of Wavelet-Based Solution Method

By defining $\overline{x} = x/2\pi$ and $\overline{y} = y/2\pi$, the governing equations (10.29)–(10.31) are firstly transformed as

$$\frac{\partial \omega}{\partial t} + \frac{1}{2\pi} \frac{\partial (u\omega)}{\partial x} + \frac{1}{2\pi} \frac{\partial (v\omega)}{\partial y} = \frac{\upsilon}{4\pi^2} \left(\frac{\partial^2 \omega}{\partial x^2} + \frac{\partial^2 \omega}{\partial y^2} \right), \tag{10.35}$$

$$\frac{1}{4\pi^2} \left(\frac{\partial^2 \psi}{\partial x^2} + \frac{\partial^2 \psi}{\partial y^2} \right) = -\omega, \tag{10.36}$$

$$u = \frac{1}{2\pi} \frac{\partial \psi}{\partial y}, v = -\frac{1}{2\pi} \frac{\partial \psi}{\partial x}, \tag{10.37}$$

where the overbar has been omitted without confusion for simplicity.

From Eqs. (10.35)–(10.37), one can see that the vorticity transport equation is almost same as the 2D Burgers' equation (10.1), Eq. (10.36) is a standard Poisson problem, and Eq. (10.37) is the same as Eq. (9.18). Thus, by using the corresponding algorithms proposed in Sects. 9.1 and 10.1, one obtains

$$\gamma_0 \mathbf{A}\boldsymbol{\omega}^{n+1} - \mathbf{A}\sum_{\lambda=0}^{J-1} \alpha_\lambda \boldsymbol{\omega}^{n-\lambda} + \Delta t \sum_{\lambda=0}^{J-1} \beta_\lambda [\mathbf{B}(\mathbf{U}^{n-\lambda} * \boldsymbol{\omega}^{n-\lambda}) + \mathbf{C}(\mathbf{V}^{n-\lambda} * \boldsymbol{\omega}^{n-\lambda})]$$
$$= \Delta t \upsilon \mathbf{D}\boldsymbol{\omega}^{n+1}, \tag{10.38}$$

$$\mathbf{D}\boldsymbol{\psi}^{n+1} = -\mathbf{A}\boldsymbol{\omega}^{n+1}, \tag{10.39}$$

$$\mathbf{A}\mathbf{U}^{n+1} = \mathbf{C}\boldsymbol{\psi}^{n+1}, \mathbf{A}\mathbf{V}^{n+1} = -\mathbf{B}\boldsymbol{\psi}^{n+1} \tag{10.40}$$

in which the operation $\mathbf{U}^n * \boldsymbol{\omega}^n = \{u_p^n \omega_p^n\}$; the matrices

$$\begin{cases} \mathbf{A} = \{a_{op} = \Gamma_{n,k}^{0,0}\Gamma_{m,l}^{0,0}\}, \mathbf{B} = \{b_{op} = \Gamma_{n,k}^{0,1}\Gamma_{m,l}^{0,0}/2\pi\}, \\ \mathbf{C} = \{c_{op} = \Gamma_{n,k}^{0,0}\Gamma_{m,l}^{0,1}/2\pi\}, \mathbf{D} = \{d_{op} = (\Gamma_{n,k}^{0,2}\Gamma_{m,l}^{0,0} + \Gamma_{n,k}^{0,0}\Gamma_{m,l}^{0,2})/4\pi^2\}; \end{cases} \tag{10.41}$$

and the vectors

$$\begin{cases} \boldsymbol{\omega}^n = \{\omega_p^n = \omega(k/2^j, l/2^j, n\Delta t)\}, \boldsymbol{\psi}^n = \{\psi_p^n = \psi(k/2^j, l/2^j, n\Delta t)\}, \\ \mathbf{U}^n = \{u_p^n = u(k/2^j, l/2^j, n\Delta t)\}, \mathbf{V}^n = \{v_p^n = v(k/2^j, l/2^j, n\Delta t)\}. \end{cases} \tag{10.42}$$

In Eqs. (10.41) and (10.42), the subscripts $o = 2^j n + m$, $p = 2^j k + l$, $k, l, n, m = 0, 1, 2, \ldots, 2^{j-1}$, and the integrations $\Gamma_{n,k}^{0,c} = \int_0^1 d^c\phi_{j,k}^P(x)/dx^c \phi_{j,n}^P(x)dx$ are given in Eq. (10.4).

Then, by solving the system of algebraic equations in Eqs. (10.38)–(10.40) in turn, we obtain the vorticity and velocity field at each time step. The kinetic energy, $E(t)$, and enstrophy, $Z(t)$, of the system can be further calculated through their definitions of Eqs. (10.33) and (10.34).

10.4.3 Results of Numerical Solution

Figure 10.14 plots the vorticity of the numerical results at different instants obtained by the wavelet-based method with a grid mesh of 128 × 128. From this figure, it can be seen that the present wavelet solution is in well agreement with the reference ones [34], although there are some slightly numerical oscillations. This observation can be further verified by Fig. 10.15 which shows a direct comparison of vorticity

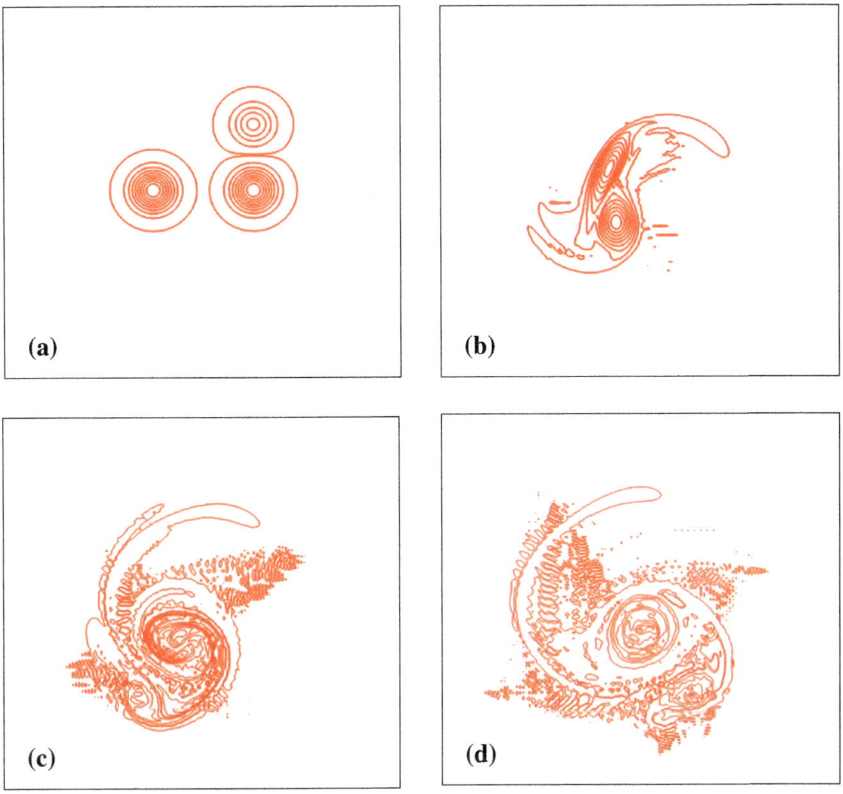

Fig. 10.14 Contours of vorticity at **a** $t = 0$, **b** $t = 10$, **c** $t = 20$, and **d** $t = 30$ of the wavelet-based solution XE by using grid of 128×128 to the 2D vortex emerging interaction

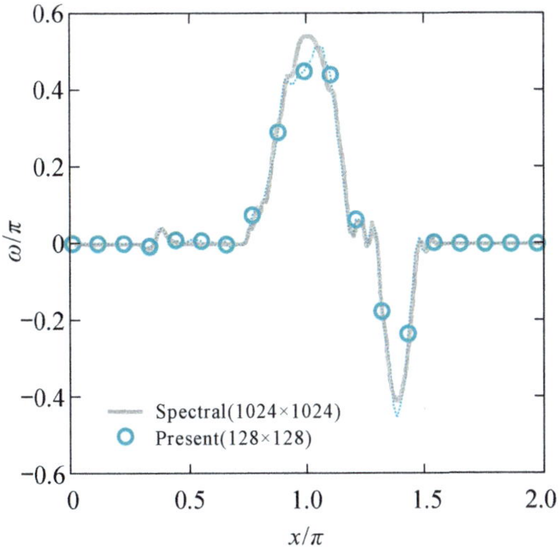

Fig. 10.15 Comparison of vorticity distribution along line $y = \pi$ when $t = 40$ for the 2D vortex emerging interaction solved by the wavelet-based method and the pseudo-spectral reference, respectively

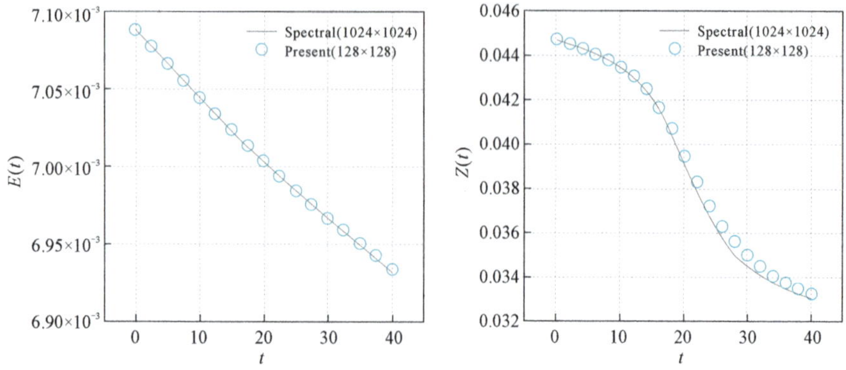

Fig. 10.16 The evolution of kinetic energy, $E(t)$ (Left), and enstrophy, $Z(t)$ (Right) for the 2D vortex emerging interaction solved by the wavelet-based method and the pseudo-spectral reference, respectively

distribution along line $y = \pi$ at $t = 40$ gained from the wavelet-based solution using grid 128×128 and one obtained by the pseudo-spectral reference method using grid 1024×1024.

Further, Fig. 10.16 displays the evolution of kinetic energy $E(t)$ and enstrophy $Z(t)$ of both the wavelet-based solution and the pseudo-spectrum reference solution as a comparison. From it, one can also see that the present wavelet-based solutions are quantitatively consistent with the pseudo-spectral reference method very well.

As pointed out in Ref. [34], since the 2D vortex merging interaction is the strongest nonlinear interaction of typical two-dimensional turbulence and if a numerical solution method can solve it with enough accuracy, such a method can also be used to solve a full 2D turbulent flow. In such point of views, hence, the closed wavelet-based method can be possibly expected to work well for simulating a full two-dimensional turbulent flow.

References

1. Bateman H (1915) Some recent researches on the motion of fluids. Mon Weather Rev 43:163–170
2. Burgers JM (1995) Mathematical examples illustrating relations occurring in the theory of turbulent fluid motion. Springer, Dordrecht, pp 281–334
3. Burgers JM (1948) A mathematical model illustrating the theory of turbulence. Adv Appl Mech 1:171–199
4. Lai LL, Cheng RJ, Li ZP, Ge HX (2013) The KdV–Burgers equation in a modified speed gradient continuum model. Chin Phys B 22:060511
5. Zhang XH, Ouyang J, Zhang L (2009) Element-free characteristic Galerkin method for Burgers' equation. Eng Anal Bound Elem 33:356–362
6. Mittal RC, Jiwari R (2012) Differential quadrature method for numerical solution of coupled viscous Burgers' equations. Int J Comput Methods Eng Sci Mech 13:88–92

7. Liu XJ, Wang JZ, Zhou YH (2016) A space-time fully decoupled wavelet Galerkin method for solving two-dimensional Burgers' equations. Comput Math Appl 72:2908–2919
8. Kutluay S, Esen A, Dag I (2004) Numerical solutions of the Burgers' equation by the least-squares quadratic B-spline finite element method. J Comput Appl Math 167:21–33
9. Hassanien IA, Salama AA, Hosham HA (2005) Fourth-order finite difference method for solving Burgers' equation. Appl Math Comput 170:781–800
10. Saka B, Dag I (2007) Quartic B-spline collocation method to the numerical solutions of the Burgers' equation. Chaos Soliton and Fractals 32:1125–1137
11. Kakuda K, Tosaka N (1990) The generalized boundary element approach to Burgers' equation. Int J Numer Methods Eng 29:245–261
12. Evans DJ, Abdullah AR (1984) The group explicit method for the solution of Burgers' equation. Computing 32:239–253
13. Gardner LRT, Gardner GA, Dogan A (1997) A Petrov-Galerkin finite element scheme for Burgers' equation. Arab J Sci Eng 22:99–109
14. Xie H, Li D (2013) A meshless method for Burgers' equation using MQ-RBF and high-order temporal approximation. Appl Math Model 37:9215–9222
15. Jiwari R, Mittal RC, Sharma KK (2013) A numerical scheme based on weighted average differential quadrature method for the numerical solution of Burgers' equation. Appl Math Comput 219:6680–6691
16. Gao Y, Le LH, Shi BC (2013) Numerical solution of Burgers' equation by lattice Boltzmann method. Appl Math Comput 219:7685–7692
17. Deville MO, Fischer PF, Mund EH (2002) High-order methods for incompressible fluid flow. Cambridge University Press, Cambridge
18. Liu XJ, Zhou YH, Zhang L, Wang JZ (2014) Wavelet solutions of Burgers' equation with high Reynolds numbers. Sci China Technol Sci 57:1285–1292
19. Hon YC, Mao XZ (1998) An efficient numerical scheme for Burgers' equation. Appl Math Comput 95:37–50
20. Mittal RC, Jain RK (2012) Numerical solutions of nonlinear Burgers' equation with modified cubic B-splines collocation method. Appl Math Comput 218:7839–7855
21. Chen R, Wu Z (2006) Applying multiquadric quasi-interpolation to solve Burgers' equation. Appl Math Comput 172:472–484
22. Zhu CG, Wang RH (2009) Numerical solution of Burgers' equation by cubic B-spline quasi-interpolation. Appl Math Comput 208:260–272
23. Liu XJ, Wang JZ, Zhou YH (2013) A wavelet method for solving coupled viscous Burgers' equations. In: AIP conference proceedings, vol 1558, pp 935–937
24. Hammad DA, El-Azab MS (2015) 2N order compact finite difference scheme with collocation method for solving the generalized Burger's–Huxley and Burgers–Fisher equations. Appl Math Comput 258:296–311
25. Young DL, Fan CM, Hu SP, Atluri SN (2008) The Eulerian-Lagrangian method of fundamental solutions for two-dimensional unsteady Burgers' equations. Eng Anal Bound Elem 32:395–412
26. Duan YL, Liu RX (2007) Lattice Boltzmann model for two-dimensional unsteady Burgers' equation. J Comput Appl Math 206:432–439
27. Islam S, Aziz I, Al-Fhaid AS, Shah A (2013) A numerical assessment of parabolic partial differential equations using Haar and Legendre wavelets. Appl Math Model 37:9455–9481
28. Liao W (2010) A fourth-order finite-difference method for solving the system of two-dimensional Burgers' equations. Int J Numer Methods Fluids 64:565–590
29. Zhang J, Yan G (2008) Lattice Boltzmann method for one and two-dimensional Burgers equation. Phys A 387:4771–4786
30. Lobovsky L, Kren J (2007) Smoothed particle hydrodynamics modelling of fluids and solids. Appl Comput Mech 1:521–530
31. Kundu PK, Cohen IM, Dowling DR (2012) Fluid mechanics, 5th edn. Elsevier, New York
32. Karniadakis GE, Israeli M, Orszag SA (1991) High-order splitting methods for the incompressible Navier-Stokes equations. J Comput Phys 97:414–443

33. Guo ZL, Zheng CG, Shi BC (2002) Non-equilibrium extrapolation method for velocity and pressure boundary conditions in the lattice Boltzmann method. Chin Phys 11:366–374
34. Schneider K, Kevlahan NKR, Farge M (1997) Comparison of an adaptive wavelet method and nonlinearly filtered pseudospectral methods for two-dimensional turbulence. Theoret Comput Fluid Dyn 9:191–206

Chapter 11
Extended Wavelet Methods to 2D Irregular Domain and Local Refinement

In the previous chapters, one can find that the definition domains for all the problems are restricted to regular ones when the wavelet-based method is employed, such as a rectangle in two-dimension and a cube in three-dimension. In fact, the difficulty of handling the irregular domain has always been one of the main obstacles in the development of various wavelet-based methods, which will restrict the applications in practice.

It is well known that the wavelet analysis is originally defined on the whole space [1]. If one directly applies wavelets series to approximate functions defined on finite domains by simply truncating the corresponding series, the consistency near the boundary will lose, resulting in a significant drop in accuracy and a very low rate of convergence near the boundary, as discussed in the Chap. 4. To resolve this problem, the technique of boundary extension and concept of interval wavelets are proposed, respectively. However, both the existing technique of boundary extension and concept of interval wavelets are developed for one-dimensional cases. And the high-dimensional wavelet bases are generated by the tensor products of one-dimensional wavelet bases which are the interval ones or associated with the boundary extension, as shown in Chap. 4. Thus, the corresponding high-dimensional wavelet approximations can only work well on regular domains.

To deal with the regular domain, the so-called wavelet finite element methods have been developed by using the wavelet approximation to replace the polynomial interpolation used in the standard finite element method [2, 3]. However, the wavelet elements are normally non-conforming elements. Moreover, in general it is very difficult to calculate the integration of the product of the Jacobian matrix, wavelet basis, and its derivative [2, 3]. As another attempt, the classical wavelet collocation method (WCM) and Wavelet-Galerkin method (WGM) have also been extended to handle the regular domain by directly truncating the wavelet series. Unsurprisingly, the accuracy of these methods near the boundary is undesired, even unacceptable, because of the loss of the consistency near the boundary [4–6]. Moreover, the wavelet basis functions are short of the delta function property, resulting in considerable trouble

Y.-H. Zhou, *Wavelet Numerical Method and Its Applications in Nonlinear Problems*, Engineering Applications of Computational Methods 6,
https://doi.org/10.1007/978-981-33-6643-5_11

in enforcing essential boundary conditions [7]. Even for the so-called interpolating wavelet approximation [8], it only can exactly pass through the function values when nodes are distributed evenly throughout the whole domain. Such a completely uniform distribution of nodes is nowhere near good enough for solving problems with complex domains or localized steep gradients. Last, evaluating integrals in complex domains is usually a complex and expensive task in the WGM [4–8], because most of the popular wavelets cannot be expressed by low-order polynomials, and even have no explicit analytical expressions [1, 2, 4]. Although the WCM can circumvent these integrals [6], but it commonly possesses relatively poor stability compared with the WGM [2, 4, 9]. In addition, applying natural boundary conditions is often a daunting challenge in the WCM, particularly for the problems with complex domains [2, 9].

In this chapter, a technique of boundary extension based on the Lagrange interpolation is proposed to restore the consistency near the boundary of the truncated wavelet approximation defined on an arbitrarily irregular two-dimensional domain. Then, by reforming the classical interpolating wavelet multiresolution analysis, a novel formulation of the wavelet multiresolution approximation is developed to represent functions. In such a wavelet approximation, arbitrarily specified nodal coefficients can be exactly the function values at the corresponding nodes. Using the proposed wavelet multiresolution interpolant to construct trial and weight functions, here, a wavelet multiresolution interpolation Galerkin method (WMIGM) is proposed for solving elasticity problems with irregular domains as examples [10, 11].

11.1 Wavelet Multiresolution Approximation with Targeted Interpolation

11.1.1 Brief Overview of the Interpolating Wavelet Approximation

As discussed in Chap. 2, the Daubechies wavelets are a family of orthogonal wavelets which have the highest number $r = \gamma/2$ of vanishing moments for given support width, $\gamma - 1$, where γ is an even number. Each Daubechies wavelet member is governed by a set of corresponding coefficients, h_k, $k = 0, 1, \ldots, \gamma - 1$, through the two-scale relations [12]

$$\phi(x) = \sqrt{2}\sum_{k=0}^{\gamma-1} h_k\phi(2x-k), \text{ and } \psi(x) = \sqrt{2}\sum_{k=0}^{\gamma-1}(-1)^k h_{N-1-k}\phi(2x-k) \tag{11.1}$$

in which $\phi(x)$ and $\psi(x)$ are again referred to as the scaling function and wavelet function, respectively. And they hold the following properties [12, 13]

$$\int \phi(x)\phi(x-k)dx = \delta_{0,k}, \int \psi(x)\psi(x-k)dx = \delta_{0,k}, \tag{11.2}$$

$$M_n^{\psi} = \int x^n \psi(x)\,dx = 0, \text{ for } n = 0, 1, \ldots, \gamma/2 - 1, \tag{11.3}$$

$$\sum_{k\in\mathbb{Z}} k^n \phi(x-k) = \sum_{m=0}^{n} (-1)^m C_n^m M_m^{\phi} x^{n-m}, \text{ for } n = 0, 1, \ldots, \gamma/2 - 1 \tag{11.4}$$

where $M_m^{\phi} = \int x^m \phi(x)dx$ is the mth moment of $\phi(x)$ with $M_0^{\phi} = 1$, C_n^m is the binomial coefficient, and $\delta_{i,j}$ is the Kronecker delta function with the property $\delta_{i,j} = 1$ for $i = j$ and $\delta_{i,j} = 0$ otherwise.

The auto-correlation functions of the scaling function $\phi(x)$ and wavelet function $\psi(x)$ are defined as [8, 14]

$$\theta(x) = \int \phi(y)\phi(y-x)\,dy, \text{ and } \vartheta(x) = \int \psi(y)\psi(y-x)dy. \tag{11.5}$$

Substituting the two-scale Eqs. (11.1) into (11.5), we get the following two-scale relations

$$\theta(x) = \sum_{k=1-\tilde{N}}^{\gamma-1} p_k \theta(2x-k), \text{ and } \vartheta(x) = \sum_{k=1-\tilde{N}}^{\gamma-1} (-1)^k p_k \theta(2x-k) \tag{11.6}$$

in which the auto-correlation coefficients

$$p_k = \begin{cases} \delta_{0,k}, & k \in Even \text{ or } |k| \geq \gamma, \\ p_{-k} = \sum_{l=0}^{\gamma-1} h_l h_{l-k}, & k \in Odd \end{cases} \tag{11.7}$$

with $Even = \{2k, k \in \mathbb{Z}\}$ and $Odd = \{2k+1, k \in \mathbb{Z}\}$. Table 11.1 shows the auto-correlation coefficients p_k, $k = 1, 3, \ldots, \gamma - 1$, for $\gamma = 2, 4, 6,$ and 8.

Table 11.1 Auto-correlation coefficients p_k for $\gamma = 2, 4, 6,$ and 8

	k	$p_k = p_{-k}$		k	$p_k = p_{-k}$
$\gamma = 2$	1	5.00000000000000E-01	$\gamma = 6$	5	1.17187500000000E-02
$\gamma = 4$	1	5.62500000000000E-01	$\gamma = 8$	1	5.98144531250000E-01
	3	−6.25000000000000E-02		3	−1.19628906250000E-01
$\gamma = 6$	1	5.85937500000000E-01		5	2.39257812500000E-02
	3	−9.76562500000000E-02		7	−2.44140625000000E-03

By direct examination of definition in Eq. (11.6) and considering the scaling function $\phi(x) \in C_0^{\gamma/2-1}[0, \gamma - 1]$ [12, 13], it is easy to check that the auto-correlation function $\theta(x) \in C_0^{\gamma/2-1}[1 - \gamma, \gamma - 1]$ and its derivative of order $\gamma/2$ is absolutely integrable [8, 14]. Here, the notation $C_0^m[a, b]$ with non-negative integer, m, represents the vector space consisting of all functions, $u(x)$, and all their derivatives $u^{(\alpha)}(x) = d^\alpha u/dx^\alpha$, $0 \le \alpha \le m$, which are continuous and compactly supported within the interval $[a, b]$, and satisfy $u^{(\alpha)}(a) = u^{(\alpha)}(b) = 0$ for $0 \le \alpha \le m$.

Moreover, following the orthogonality relation of Eq. (10.2) and definition in Eq. (10.6), one can see that the function $\theta(x)$ holds the delta function property $\theta(k) = \delta_{0,k}$ for all $k \in \mathbb{Z}$. And there are the following additional properties for the auto-correlation function $\theta(x)$.

Proposition 11.1 For the auto-correlation functions $\theta(x)$ with supported interval $[1 - \gamma, \gamma - 1]$, there are relations

$$M_n^\theta = \int x^n \theta(x) dx = \delta_{0,n}, \tag{11.8}$$

$$\sum_{k \in \mathbb{Z}} p_{2k}(2k)^n = \sum_{k \in \mathbb{Z}} p_{2k+1}(2k+1)^n = \delta_{0,n}, \tag{11.9}$$

$$\sum_{k \in \mathbb{Z}} (x - k)^n \theta^{(m)}(x - k) = (-1)^n \delta_{n,m} n! \tag{11.10}$$

for $n, m = 0, 1, \ldots, \gamma - 1$. The proof of this proposition is given in Appendix 11.1.

We note that the relations similar to Eqs. (11.8) and (11.9) are also given by Beylkin and Saito [14], but without any proof. More importantly, the work in [14] stated that the relation of Eq. (11.8) is also valid for $n = \gamma$, which is incorrect. In addition, they only pointed out that Eq. (11.9) is true when n is an even number [14].

Since a set of function bases for two-dimensional space can be directly generated by the tensor products of one-dimensional wavelet bases, a continuous two-dimensional function $f(\mathbf{x}) \equiv f(x, y)$ can be approximated by [8, 14]

$$f(\mathbf{x}) \approx S^j f(\mathbf{x}) = \sum_{k \in \mathbb{Z}} \sum_{l \in \mathbb{Z}} f(\mathbf{x}_{j,k,l}) \phi_n \theta_{j,k,l}(\mathbf{x}), \tag{11.11}$$

where $\mathbf{x}_{j,k,l} = (k/2^j, l/2^j)$ and wavelet basis $\theta_{j,k,l}(\mathbf{x}) = \theta(2^j x - k)\theta(2^j y - l)$. The methods for calculating the function value, derivatives, and integral of $\theta_{j,k,l}(\mathbf{x})$ can be found in Chap. 2. Since $\theta(x) \in C_0^{\gamma/2-1}[1-\gamma, \gamma-1]$, there is $\theta_{j,k,l}(\mathbf{x}) \in C_0^{\gamma/2-1}\Omega_{j,k,l}^{\text{inf}}$ where $\Omega_{j,k,l}^{\text{inf}} = [2^{-j}(1-\gamma+k), 2^{-j}(\gamma+k-1)] \times [2^{-j}(1-\gamma+l), 2^{-j}(\gamma+l-1)]$. Moreover, it is easy to check the relation $f(\mathbf{x}_{j,k,l}) \equiv S^j f(\mathbf{x}_{j,k,l})$ for all $k, l, j \in \mathbb{Z}$ based on the property $\theta(k) = \delta_{0,k}$ for $k \in \mathbb{Z}$.

Moreover, there is the multiresolution decomposition

$$S^{j+1} f(\mathbf{x}) = S^j f(\mathbf{x}) + R^j f(\mathbf{x}) \tag{11.12}$$

with

$$R^j f(\mathbf{x}) = \sum_{(k,l)\in\mathbb{R}} [f(\mathbf{x}_{j+1,k,l}) - S^j f(\mathbf{x}_{j+1,k,l})]\theta_{j+1,k,l}(\mathbf{x}) \tag{11.13}$$

in which $\mathbb{R}$ is the set of all integer pair (k, l) satisfying that at least one element k or l is an odd number, i.e., $\mathbb{R} = \{(k, l), k, l \in \mathbb{Z} \text{ and } k \text{ or } l \in Odd\}$. The proof of the decomposition in Eq. (11.12) is presented in Appendix 11.2.

By using the relation of Eq. (11.12) recursively, one can obtain a multiresolution approximation

$$f(\mathbf{x}) = \lim_{j\to\infty} S^j f(\mathbf{x}) = S^{j_0} f(\mathbf{x}) + \sum_{j\geq j_0} R^j f(\mathbf{x}). \tag{11.14}$$

Theorem 11.1 For all functions $f(\mathbf{x}) \in C^0(R^2)$, there is a constant $C_{0,\infty}$ satisfying.

$$\left\| f(\mathbf{x}) - S^j f(\mathbf{x}) \right\|_{\infty} \leq C_{0,\infty} 2^{-j}. \tag{11.15}$$

Moreover, if $f(\mathbf{x}) \in C^{\mu}(R^2)$ with $\mu \geq 1$, there are the relations

$$\left\| f(\mathbf{x}) - S^j f(\mathbf{x}) \right\|_{\infty} \leq C_{1,\infty} 2^{-j\lambda} \tag{11.16}$$

and

$$\left\| f(\mathbf{x}) - S^j f(\mathbf{x}) \right\|_{L^2} \leq C_{1,L} 2^{-j\lambda} \tag{11.17}$$

where $\lambda = \min\{\mu, \gamma\}$, and constants $C_{1,\infty}$ and $C_{1,L}$ are dependent on the γ th order partial derivatives $D^{\alpha}_{\lambda} f(\mathbf{x}) = \partial^{\lambda} f / \partial^{\alpha} x \partial^{\lambda-\alpha} y$ for $\alpha = 0, 1, \ldots, \lambda$, but independent of the decomposition level j. Specially, if $D^{\alpha}_{\lambda} f(\mathbf{x}) \equiv 0$ for $\alpha = 0, 1, \ldots, \lambda$, we have $C_{1,\infty} = C_{1,L} = 0$. The space $C^{\eta}(\Omega)$ denotes the vector space consisting of all functions $u(\mathbf{x}) \in L^2(\Omega)$ which are η times continuously differentiable. The proof of this theorem is presented in Appendix 11.3.

11.1.2 Wavelet Approximation of Functions Defined on Finite Domain

Since the interpolation Eq. (11.11) is originally developed for a function defined over the whole two-dimensional space [8, 14], some troubles will appear when Eq. (11.11)

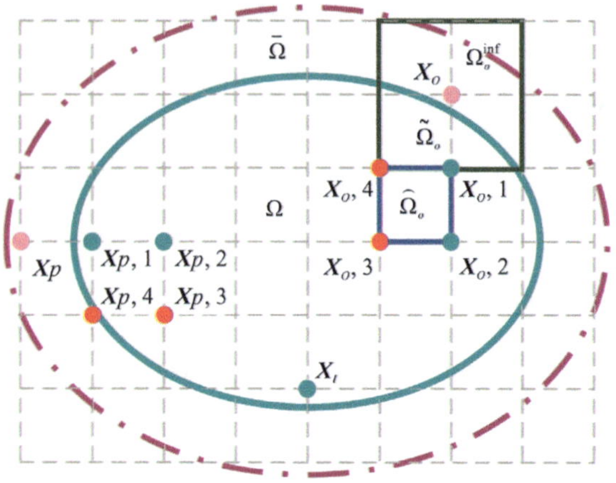

Fig. 11.1 Schematic diagram of distribution of nodes for the approximation Eq. (11.11) in domain Ω when $\gamma = \eta_o = 2$

is used to approximate a function defined in a finite domain. Specifically, if the interpolation Eq. (11.11) is applied to approximate a function $f(\mathbf{x}) \in C^0(R^2)$ in a limited domain Ω, by considering $\theta_{\rho(n),k(n),l(n)}(\mathbf{x}) \in C_0^{\gamma/2-1}\Omega_n^{\inf}$, we have

$$S^j f(\mathbf{x}) = \sum_{n=1}^{N} f(\mathbf{x}_l)\theta_{j,k(n),k(n)}(\mathbf{x}) + \sum_{n=N+1}^{N+N_e} f(\mathbf{x}_l)\theta_{j,k(n),k(n)}(\mathbf{x}) \tag{11.18}$$

where $\mathbf{x}_l$ for $l = 1, 2, \ldots, N$ are all nodes located in Ω, and $\mathbf{x}_l$ for $l = N+1, N+2, \ldots, N+N_e$ are all nodes located in the domain $\overline{\Omega}$, as shown in Fig. 11.1, where the domain $\overline{\Omega}$ is the set of all points $\mathbf{x} \notin \Omega$ verifying $\Omega_{\mathbf{x}}^{\inf} \cap \Omega \neq \emptyset$ in which $\Omega_{\mathbf{x}}^{\inf}$ is a square with central point $\mathbf{x}$ and edge length $2^{-j+1}(\gamma - 1)$. Here, the functions $k(n) = 2^{\rho(n)}x_n$ and $l(n) = 2^j y_n$ are employed, respectively, in which the function $\rho(n) \geq j_0$ is used to return the minimum integer j which can make both $2^j x_n$ and $2^j y_n$ become integers. Namely in such notations, there is always the relation $\mathbf{x}_n = (k(n)/2^{\rho(n)}, l(n)/2^{\rho(n)})$ for any nodes, $\mathbf{x}_n$.

It can be seen from Eq. (11.18) that the function values at some nodes located outside of the domain Ω are needed, when the function is approximated only in this domain. If one simply throws away the last term of Eq. (11.18), the error near the boundary of Ω will increase sharply, and the high-order accuracy shown in Eq. (11.16) cannot be expected even that the function is smooth enough in Ω. Unfortunately, these nodal values are virtually nonexistent, when approximating a continuous function defined in the domain Ω. Therefore, a suitable substitute of those nodal values should be provided. In the following, a technique based on the Lagrange interpolation for creating such a substitution is proposed.

Considering a function $f(\mathbf{x}) \in C^0(\Omega)$, as shown in Fig. 11.1, the function values $f(\mathbf{x}s_o)$ at external nodes, $\mathbf{x}_o$, for $o = N+1, N+2, \ldots, N+N_e$ are defined by the Lagrange interpolation

$$f(\mathbf{x}_o) = \sum_{n=1}^{\eta_o^2} f(\mathbf{x}_{I(o,n)})\Pi_n^o(\mathbf{x}_o) \tag{11.19}$$

in which $\Pi_n^o(\mathbf{x}) = L_n^o(x)M_n^o(y)$ with

$$L_n^o(x) = \prod_{\substack{i=1 \\ i\neq n}}^{\eta_o} \frac{x - \tilde{x}_{o,i}}{x_{I(o,n)} - \tilde{x}_{o,i}} \text{ and } M_n^o(y) = \prod_{\substack{i=1 \\ i\neq n}}^{\eta_o} \frac{y - \tilde{y}_{o,i}}{y_{I(o,n)} - \tilde{y}_{o,i}}. \tag{11.20}$$

In Eqs. (11.19) and (11.20), the nodal group $\mathbf{x}_{I(o,n)}$ for $n = 1, 2, \ldots \eta_o^2$ is selected from interior nodes, $\mathbf{x}_l \in \Omega$, such that they form a uniform rectangular grid and the average distance between $\mathbf{x}_{I(o,n)}$ and $\mathbf{x}_o$ is the smallest, and the sets $\{\tilde{x}_{o,i}, i = 1, 2, \ldots, \eta_o\}$ and $\{\tilde{y}_{o,i}, i = 1, 2, \ldots, \eta_o\}$ are all x- and y-coordinates of nodal group $\mathbf{x}_{o,n}$, respectively. The function $I(o, n)$ returns the serial number of the corresponding node.

Substitution of Eqs. (11.19) into (11.18) yields

$$\begin{aligned} S_L^{j_0} f(\mathbf{x}) &= \sum_{n=1}^{N} f(\mathbf{x}_n)\theta_{\rho(n),k(n),l(n)}(\mathbf{x}) \\ &+ \sum_{n=N+1}^{N+N_e} \sum_{m=1}^{\eta_n^2} f(\mathbf{x}_{I(n,m)})\Pi_m^n(\mathbf{x}_n)\theta_{\rho(n),k(n),l(n)}(\mathbf{x}) \end{aligned}, \text{ for } \mathbf{x} \in \Omega \tag{11.21}$$

which can be further rewritten as

$$S_L^{j_0} f(\mathbf{x}) - \sum_{n=1}^{N} f(\mathbf{x}_n)\phi_n(\mathbf{x}) \tag{11.22}$$

with the modified wavelet basis

$$\phi_n(\mathbf{x}) = \theta_{\rho(n),k(n),l(n)}(\mathbf{x}) + \sum_{m\in\Xi_m} \Pi_n^m(v_m)\theta_{\rho(m),k(m),l(m)}(\mathbf{x}) \tag{11.23}$$

in which Ξ_n is the set of serial number, m, such that $f_m(\mathbf{x})$ for $m \in \Xi_n$ are exactly all the Lagrange interpolations defined by Eq. (11.19) using the node $\mathbf{x}_n$. On the basis of the property $\theta(k) = \delta_{0,k}$ for $k \in \mathbb{Z}$, it is easy to check the relations $\phi_n(\mathbf{x}_k) = \delta_{n,k}$ for $k, n = 1, 2, \ldots, N$. Namely, there is the relation $S_L^{j_0} f(\mathbf{x}_l) \equiv f(\mathbf{x}_l)$ for all nodes $\mathbf{x}_l, l = 1, 2, \ldots, N$.

In order to guarantee the accuracy of approximation Eq. (11.22), we always use a reasonable decomposition level $\rho(n) = j_0$ for $n = 1, 2, \ldots, N + N_e$, which is small enough to make the distance between any point at the boundary and the nearest node being less than $\sqrt{2}\Delta$ with gird size $\Delta = 2^{-j_0}$. Then considering the definition of domain, $\overline{\Omega}$, there are the relations $\left|x_l - \tilde{x}_{l,i}\right| \le \alpha_1 2^{-j_0}$ and $\left|y_l - \tilde{y}_{l,i}\right| \le \alpha_2 2^{-j_0}$ with constants α_1 and α_2 for $l = 1, 2, \ldots, N_{\overline{\Omega}}$ at least for $\eta_l = 1$. Based on the above relations, the accuracy of approximation in Eq. (11.22) can be estimated by the following theorem.

Theorem 11.2 For all functions $f(\mathbf{x}) \in C^{\mu}(\Omega)$ with $\mu \ge 1$ and $\eta = \min\{\eta_l\} \ge 2$, there are constants $C_{2,\infty}$ and $C_{2,L}$, respectively, satisfying the relations.

$$\left\| f(\mathbf{x}) - S_L^j f(\mathbf{x}) \right\|_{\infty} \le C_{2,\infty} 2^{-j\lambda}, \tag{11.24}$$

$$\left\| f(\mathbf{x}) - S_L^j f(\mathbf{x}) \right\|_{L^2(\Omega)} \le C_{2,L} 2^{-j\lambda}, \tag{11.25}$$

where $\lambda = \min\{\mu, \gamma, \eta\}$, and constants $C_{2,\infty}$ and $C_{2,L}$ are dependent on the λth order partial derivatives $D^{\lambda} f(\mathbf{x})$ but independent of the decomposition level j. Specially, if $D_{\lambda}^{\alpha} f(\mathbf{x}) \equiv 0$ for $\alpha = 0, 1, \ldots, \lambda$, there is $C_{2,\infty} = C_{2,L} = 0$. The proof of this theorem is presented in Appendix 11.4

It can be seen from Theorem 11.2 that letting $\eta_l = \gamma$ for $l = N+1, 2, \ldots, N+N_e$ is an appropriate choice if the nodes can meet the corresponding requirement of $f_l(\mathbf{x})$. In the following analysis, $\eta_l = \gamma$ for $l = N+1, 2, \ldots, N+N_e$ is always set to γ if without a special instruction. But, when the distribution of nodes cannot form a uniform rectangular grid for constructing $f_l(\mathbf{x})$ with $\eta_l = \gamma$, the value $1 \le \eta_l < \gamma$ is also acceptable, because the approximation Eq. (11.22) is also convergent and only the local error will increase. As an example of the accuracy to Eq. (11.22), the numerical confirmation is displayed in Fig. 11.2 for the errors of Eq. (11.22) with $\gamma = 4$ to approximate $f(\mathbf{x}) = \sin(\pi x)\sin(\pi y)$ defined in a quarter of a unit circle centered at the origin, where the absolute error $E(\mathbf{x}) = \left| f(\mathbf{x}) - S_L^j f(\mathbf{x}) \right|$ and the L^2-error norm $E_2 = \|E(\mathbf{x})\|_{L^2(\Omega)}$ are used.

11.1.3 Modified Multiresolution Approximation Defined on a Finite Domain

Notice that the multiresolution approximation $S^j f(\mathbf{x})$ defined by Eq. (11.14) always requires a uniform grid. In order to relax this restriction, we define another multiresolution decomposition as

$$P_{j_0}^{j+1} f(\mathbf{x}) = P_{j_0}^{j} f(\mathbf{x}) + Q_{j_0}^{j} f(\mathbf{x}) \text{ for } j \ge j_0 \tag{11.26}$$

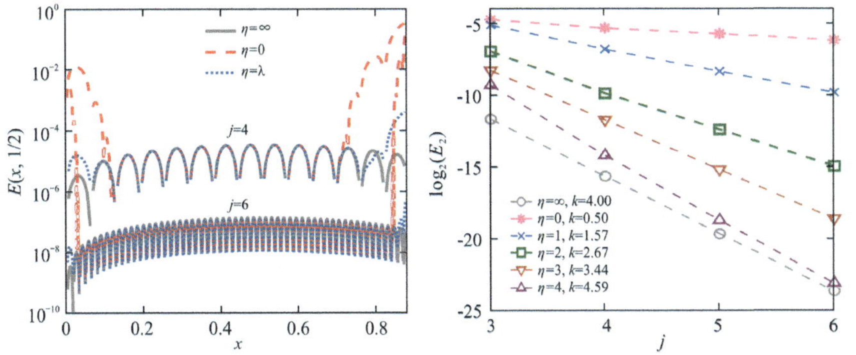

Fig. 11.2 Errors of the approximation Eq. (11.22) with $\gamma = 4$ for $f(\mathbf{x}) = \sin(\pi x)\sin(\pi y)$ defined in a quarter of a unit circle centered at the origin: the absolute error $E(\mathrm{x})$ along x-axis at $y = 1/2$ (Left); the L^2-error norm E_2 varying with the resolution level j (Right) . Here, $\eta = \infty$ means $f(\mathbf{x}_0) = f(\mathbf{x})$ using in Eq. (11.19), and k is the slope of the corresponding line

with

$$Q_{j_0}^j f(\mathbf{x}) = \sum_{(k,l)\in\Re_j} [f(\mathbf{x}_{j+1,k,l}) - P_{j_0}^j f(\mathbf{x}_{j+1,k,l})]\theta_{j+1,k,l}(\mathbf{x}), \tag{11.27}$$

where $\Re_j$ is an arbitrary subset of $\overline{\Re}_j$. Here, the set $\overline{\Re}_j = \{(k,l), k, l \in \mathbb{Z} \text{ and } k \text{ or } l \in Odd\}$ making all nodes, $\mathbf{x}_n \in \Omega$. Then by using the relation of Eq. (11.26) recursively, a multiresolution approximation of a function $f(\mathbf{x}) \in C^0(\Omega)$ is defined by

$$P_{j_0}^J f(\mathbf{x}) = P_{j_0}^{j_0} f(\mathbf{x}) + \sum_{j=j_0}^{J-1} Q_{j_0}^j f(\mathbf{x}) \tag{11.28}$$

with $P_{j_0}^{j_0} f(\mathbf{x}) = S_L^{j_0} f(\mathbf{x})$ given by Eq. (11.22). One can check that the multiresolution approximation Eq. (11.28) holds the following interpolation property and error bound.

Proposition 11.2 There is the relation $P_{j_0}^J f(\mathbf{x}_k) \equiv f(\mathbf{x}_k)$ for all nodes $\mathbf{x}_k$ for $k = 1, 2, \ldots, N$.

The proof of this proposition is presented in Appendix 11.5.

Theorem 11.3 For all functions $f(\mathbf{x}) \in C^\mu(\Omega)$ and $\eta \geq 1$, there are constants $C_{3,\infty}$ and $C_{3,L}$, respectively, satisfying.

$$\left\| f(\mathbf{x}) - P_{j_0}^J f(\mathbf{x}) \right\|_\infty \leq C_{3,\infty} 2^{-j_0\lambda} \tag{11.29}$$

and

$$\left\| f(\mathbf{x}) - P_{j_0}^{J} f(\mathbf{x}) \right\|_{L^2(\Omega)} \leq C_{3,L} 2^{-j_0 \lambda}, \tag{11.30}$$

where $\lambda = \min\{\mu, \gamma, \eta\}$, and constants $C_{3,\infty}$ and $C_{3,L}$ are dependent on the λth order partial derivatives $D^{\lambda} f(\mathbf{x})$ but independent of the decomposition level j_0. Specially, if $D^{\lambda} f(\mathbf{x}) \equiv 0$, there is $C_{3,\infty} = C_{3,L} = 0$.

The proof of this theorem is presented in the Appendix 11.6.

It can be seen from Eq. (11.27) that the distribution of all additional nodes used in $Q_{j_0}^{j} f(\mathbf{x})$ for $j \geq j_0$ can be completely random, since $\mathfrak{R}_j$ can be an arbitrary subset of $\overline{\mathfrak{R}}_j$. And Theorem 11.3 states that these nodes would never reduce the fundamental accuracy which is determined by the uniform distribution of nodes used in $P_{j_0}^{j_0} f(\mathbf{x})$, which means that the addition nodes are completely free at any point for generating a multiresolution interpolation Eq. (11.28), and the accuracy of such an interpolation will never be lower than that of fundamental approximation $P_{j_0}^{j_0} f(\mathbf{x})$.

For example, we apply the multiresolution scheme of Eq. (11.28) with $\gamma = \eta = 2$ to approximate the function $f(\mathbf{x}) = \sin(\pi x/5)\sin(\pi y/5)$ defined in the square domain $[0, 1]^2$. As shown in Fig. 11.3a for the error E_2, the fundamental uniform distribution of nodes consists of four vertexes of square marked by the asterisks, which is corresponding to $j_0 = 0$. When other five nodes are added randomly in the domain, Fig. 11.3b, c show the error distribution of the approximation with two kinds of such additional nodes, while Fig. 11.3d displays the error varying with d_{st}, the standard deviation of distances between any two nodes. From Fig. 11.3a–c, one can see that the additional five nodes can obviously reduce the error near these nodes, since the error at all nodes is zero (see the Proposition 11.2). But the influence region of each node is very limited in most cases. Therefore, the reduction of L^2-error norm E_2 is very slight in most cases, as shown in Fig. 11.3d. This figure shows the error norm E_2 for 1000 random distributions of additional five nodes, in which the dotted lines with labels, $j = 0$ and 1, are the error norm, E_2 of $S_L^j f(\mathbf{x})$, respectively. It can also be seen from Fig. 11.3d that the error norm, E_2, reduces obviously in many cases, but never increases with the addition of local nodes.

By comparing the decompositions of Eqs. (11.12) and (11.26), and considering $\theta_{j,k,l}(\mathbf{x}) \in C_0^{\gamma/2-1}\Omega_{j,k,l}^{\mathrm{inf}}$, when the nodes are added to form a denser uniform distribution $(k/2^j, l/2^j)$ in a local region shown in Fig. 11.4 (Left), the error in this region of multiresolution interpolation Eq. (11.28) can obviously reduce the error formulated by $S_L^j f(\mathbf{x})$, see Fig. 11.4 (Right), where the basic problem is the same as that shown in the Fig. 11.2 except for that the grid for $P_{j_0}^{J} f(\mathbf{x})$ with $j_0 = 4$ is refined to $(k/2^5, l/2^5)$ in the local region $[7/16, 11/16]^2$.

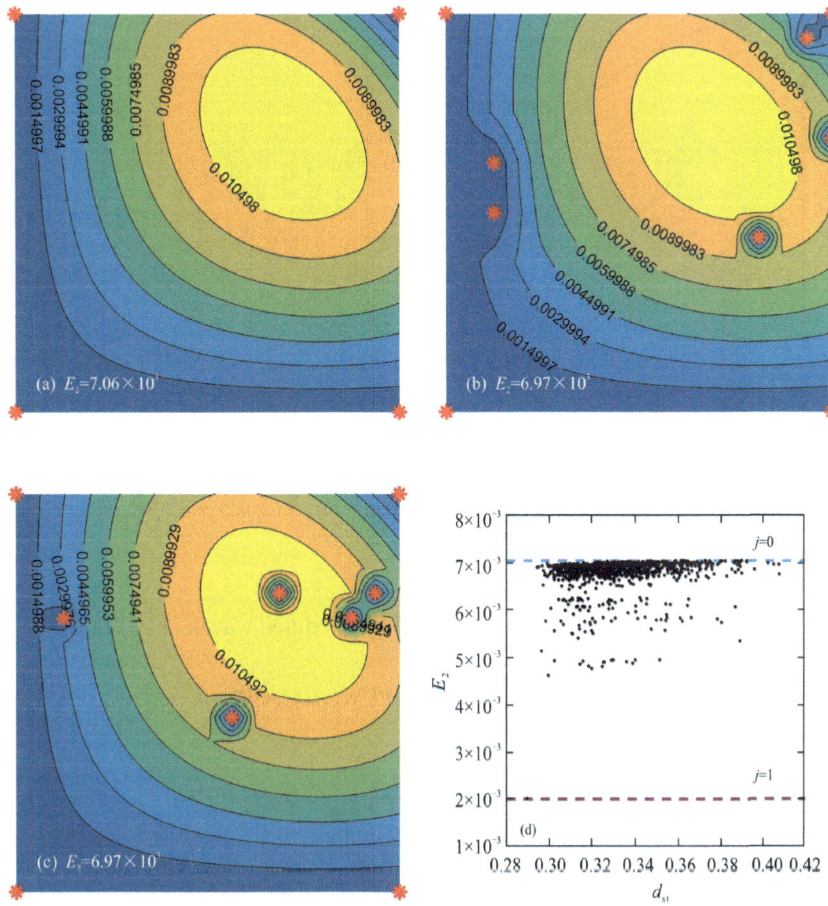

Fig. 11.3 Error distributions of the revised interpolation Eq. (11.28) with $\gamma = \eta = 2$ used to approximate $f(\mathbf{x}) = \sin(\pi x/5)\sin(\pi y/5)$. **a** $E(\mathrm{x})$ of $P_0^0 f(\mathbf{x})$ with four nodes added the vertexes of square, **b** and **c** $E(\mathrm{x})$ of $P_0^0 f(\mathbf{x})$ with two different kinds of other additional five nodes in the square domain $[0, 1]^2$; **d** E_2 varying with the standard deviation d_{st} of distances between any two nodes for 1000 distributions of the random additional five nodes in the domain. Here, asterisks represent additional nodes

11.1.4 Construction of the Wavelet Targeted Interpolation

The wavelet multiresolution approximation Eq. (11.28) can be explicitly written as

$$P_{j_0}^{j_{\max}} f(\mathbf{x}) = \sum_{n\in\Re_{j_0}} f(\mathbf{x}_n)\phi_n(\mathbf{x}) + \sum_{j=j_0+1}^{j_{\max}} \sum_{n\in\Re_j} [f(\mathbf{x}_n) - P_{j_0}^{j-1} f(\mathbf{x}_n)]\phi_n(\mathbf{x}) \quad (11.31)$$

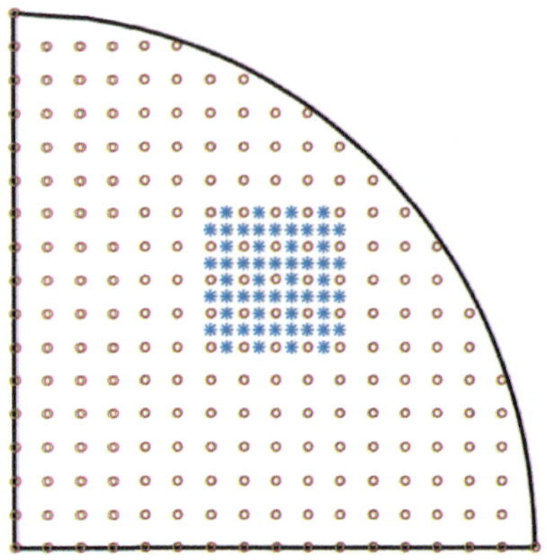

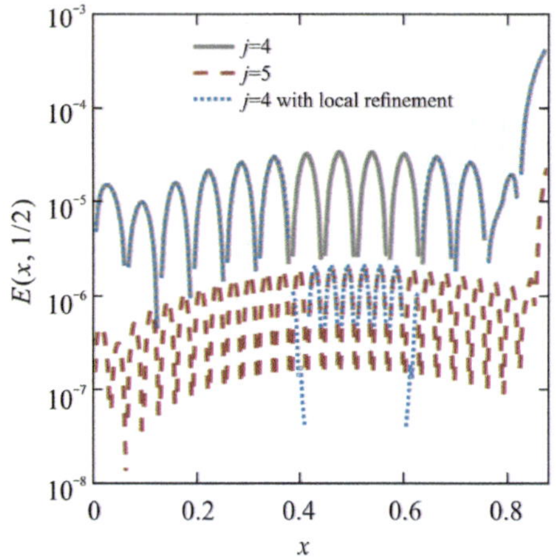

Fig. 11.4 Errors of the interpolation Eq. (11.28) with $\gamma = \eta = 4$ a refined subregion: Distribution of the domain with a refined subregion used to $P_4^J f(\mathbf{x})$ (Left), Absolute error $E(\mathrm{x})$ along x-axis at $y = 1/2$ for $P_4^4 f(\mathbf{x})$, $P_4^J f(\mathbf{x})$ and $P_5^5 f(\mathbf{x})$ (Right)

in which $j_{\max} = \max\{\rho(n),\ \text{for all } \mathbf{x}_n\}$, and the modified wavelet basis functions

$$\phi_n(\mathbf{x}) = \sum_{m\in\mathbb{C}_n} \Pi_n^m(\mathbf{x}_m)\theta_{\rho(m),k(m),l(m)}(\mathbf{x}) \tag{11.32}$$

with the set $\mathbb{C}_n = \Xi_n \cup \{n\}$ and the coefficient $\Pi_n^n(\mathbf{x}_n) \equiv 1$.

One can check that the compact support domain of the modified wavelet basis function $\phi_n(\mathbf{x})$ is $\Omega_n = \cup_{m\in\Xi_n}\Omega_m^{\inf}$ where $\Omega_m^{\inf} = [x_m - d_m, x_m + d_m] \times [y_m - d_m, y_m + d_m]$ with $d_m = 2^{-\rho(m)}(\gamma - 1)$. And there is the relation $\phi_n(\mathbf{x}_m) = \delta_{n,m}$ for all nodes $\mathbf{x}_m$ with $\rho(m) \le \rho(n)$.

Dividing all nodes $\mathbf{x}_n$ into two groups $\mathbb{N}^I = \{\mathbf{x}_n, n \in \Re^I\}$ and $\mathbb{N}^R = \{\mathbf{x}_n, n \in \Re^R\}$ optionally, then the wavelet approximation Eq. (11.31) can be rewritten as

$$\begin{aligned} P_{j_0}^{j_{\max}} f(\mathbf{x}) = &\sum_{n\in\Re_{j_0}^I} f(\mathbf{x}_n)\phi_n(\mathbf{x}) + \sum_{n\in\Re_{j_0}^R} c_n\phi_n(\mathbf{x}) \\ &+ \sum_{j=j_0+1}^{j_{\max}} \sum_{n\in\Re_j^I} [f(\mathbf{x}_n) - P_{j_0}^{j-1} f(\mathbf{x}_n)]\phi_n(\mathbf{x}) + \sum_{j=j_0+1}^{j_{\max}} \sum_{n\in\Re_j^R} c_n\phi_n(\mathbf{x}) \end{aligned} \tag{11.33}$$

in which $\Re_j^I = \Re_j \cap \Re^I$, $\Re_j^R = \Re_j \cap \Re^R$ and the coefficient $c_n = f(\mathbf{x}_n) - P_{j_0}^{\rho(n)-1} f(\mathbf{x}_n)$ with $P_{j_0}^{j} f(\mathbf{x}) \equiv 0$ for $j < j_0$. It can be seen from Eq. (11.33) that the approximation $P_{j_0}^{j_{\max}} f(\mathbf{x})$ is linearly dependent on the nodal values $f(\mathbf{x}_n)$ for $n \in \Re^I$

and coefficients c_n for $n \in \Re^R$. Therefore, Eq. (11.33) can be further expressed as

$$P_{j_0}^{j_{\max}} f(\mathbf{x}) = \sum_{n\in\Re^I} f(\mathbf{x}_n)\Theta_{j_{\max},n}(\mathbf{x}) + \sum_{n\in\Re^R} c_n\Theta_{j_{\max},n}(\mathbf{x}) = \sum_{n\in\Re} a_n\Theta_{j_{\max},n}(\mathbf{x}), \tag{11.34}$$

where the coefficients $a_n = f(\mathbf{x}_n)$ for $n \in \Re^I$ and $a_n = c_n = f(\mathbf{x}_n) - P_{j_0}^{\rho(n)-1} f(\mathbf{x}_n)$ for $n \in \Re^R$.

In Eq. (11.34), the base functions are defined as

$$\Theta_{j_{\max},n}(\mathbf{x}) = \sum_{m\in\mathbb{N}_n} \alpha_{n,m}\phi_m(\mathbf{x}) \tag{11.35}$$

in which the set $\mathbb{N}_n = \{n\} \cup \overline{\mathbb{N}}_n$ with $\overline{\mathbb{N}}_n = \left\{m, \text{ for } m \in \Re^I \text{ and } \beta_{m,n}^{J_I} \neq 0\right\}$ for $n \in \Re^T$ and $\overline{\mathbb{N}}_n = \emptyset$ otherwise, and coefficients $\alpha_{n,n} = 1$ and $\alpha_{n,m} = \beta_{m,n}^{J_I}$ for $m \in \overline{\mathbb{N}}_n$ with $J_I = \max\left\{\rho(n), n \in \Re^I\right\}$. Here, the nodal set $\mathbb{N}^T = \left\{\mathbf{x}_n, n \in \Re^T\right\}$ is consisted of all nodes with $\rho(n) < J_I$, whose compact support domain Ω_n contains at least one node $\mathbf{x}_m$ for $m \in \Re^I$. The matrix $\boldsymbol{\beta}^{J_I}$ verifying the property $\beta_{m,n}^{J_I} = 0$ for $n \notin \Re_{\leq J_I-1}^{T,m}$ can be estimated by

$$\boldsymbol{\beta}^{j+1} = \boldsymbol{\beta}^j - \boldsymbol{\psi}^{j+1} - \overline{\boldsymbol{\psi}}^{j+1}\boldsymbol{\beta}^j \text{ with } \boldsymbol{\beta}^{j_0} = \mathbf{0}, \tag{11.36}$$

where the $N_I \times N_T$ matrix $\boldsymbol{\psi}^j = \{\boldsymbol{\psi}_{m,n}^j = \phi_n(\mathbf{x}_m)$ for $m \in \Re_j^I$ and $n \in \Re_{\leq j-1}^T$, and $\psi_{m,n}^j = 0$ otherwise$\}$ and the $N_I \times N_I$ matrix $\overline{\psi}^j$ $\{\overline{\psi}_{m,n}^j = \phi_n(\mathbf{x}_m)$ for $m \in \Re_j^I$ and $n \in \Re_{\leq j-1}^I$, $\psi_{m,n}^j = 0$ otherwise$\}$ In Eq. (11.36), the sets $\Re_{\leq j}^I = \cup_{n=j_0}^j \Re_n^I$, $\Re_{\leq j}^T = \cup_{n=j_0}^j \Re_n^T$ with $\Re_j^T = \Re^T \cap \Re_j$, and $\Re_{\leq J_I-1}^{T,m} = \Re_{\leq J_I-1}^T \cap \Re_m^\Omega$. Here, the nodal set $\left\{\mathbf{x}_m, m \in \Re_n^\Omega\right\}$ consists of all nodes whose compact support domain Ω_m contains the node $\mathbf{x}_n$. The integers N_I and N_T are the number of nodes in sets $\mathbb{N}^I$ and $\mathbb{N}^T$, respectively. The detailed derivation of Eqs. (11.34)–(11.36) is presented in the Appendix 11.7.

It can be seen from Eq. (11.35) that the shape function $\Theta_{j_{\max},n}(\mathbf{x})$ is a linear combination of the standard wavelet basis functions $\theta_{j,k,l}(\mathbf{x}) = \theta(2^j x - k)\theta(2^j y - l)$. Therefore, it possesses $\gamma/2 - 1$ continuous partial derivatives and the $(\gamma/2)$th order weak partial derivative over the whole domain. Moreover, the shape function $\Theta_{j_{\max},n}(\mathbf{x})$ and its derivatives can be readily calculated by Eqs. (11.34)–(11.36) and $d^n\theta(x)/dx^n$ for $0 \leq n \leq \gamma/2$.

Since the wavelet multiresolution approximation Eq. (11.34) with targeted interpolation is exactly an equivalent expression of the wavelet multiresolution approximation Eq. (11.28), therefore the errors of the approximation Eq. (11.34) also obey Theorem 11.3.

11.1.5 Some Essential Attributes of the Wavelet Targeted Interpolation

One can check that the wavelet multiresolution approximation Eq. (11.34) possesses the following attributes simultaneously, each of which is beneficial for finding a stable and efficient solution to differential equations [9, 15].

(a) *Arbitrary nodal distribution*: The admissible nodal distribution can be arbitrary within reason and can be automatically generated without any difficulty.
(b) *Stability*: No matrix inversion is needed in the construction of the present wavelet approximation. Therefore, the proposed technique is naturally stable and efficient.
(c) *Consistency*: The proposed wavelet approximation holds the capability to locally and exactly reproduce the polynomials with a certain order.
(d) *Delta function property*: The wavelet interpolating basis functions corresponding to some arbitrary specifications can hold the Delta function property, allowing us to directly impose the essential boundary conditions.
(e) *Compact support:* The wavelet interpolating basis function has the compact support, leading to a sparse set of algebraic equations which can significantly save computational cost and storage space.
(f) *Compatibility*: The wavelet interpolating base functions are continuous over whole problem domain.
(g) *Credibility*: There is not any ad hoc parameters in the present wavelet approximation, leading to an excellent usefulness.
(h) *Local enrichment*: The proposed wavelet multiresolution approximation provides an extremely effective and common approach to capture the local characteristic of functions in both time and frequency domains.
(i) *Closedness*: When the nodal set $\mathbb{N}^I = \{\mathbf{x}_n, n \in \Re^I\}$ contains all nodes, the present wavelet interpolation is highly analogous to the wavelet approximation Eq. (3.27) and also holds the property of Eq. (3.29). Therefore, a closed expansion of nonlinear term can be obtained by using the proposed wavelet interpolation, as discussed in the Chap. 2.

11.2 Wavelet Multiresolution Solution to Elasticity Problems

11.2.1 Node Generation and Pre-Processing

An admissible nodal distribution for constructing the wavelet multiresolution approximation Eq. (11.34) can be generated by three completely independent steps, as shown in Fig. 11.5. First, according to the rule that all base nodes $\mathbf{x}_n \in \Omega$ and external auxiliary points $\mathbf{x}_o \notin \Omega$ but $\overline{\Omega}_0 \cap \Omega \neq \emptyset$, where $\overline{\Omega}_0$ is the square with central point $\mathbf{x}_0$ and edge length $d_{j_0} = 2^{1-j_0}(\gamma - 1)$, then the base nodes $\mathbf{x}_n \in \Omega$ and external auxiliary

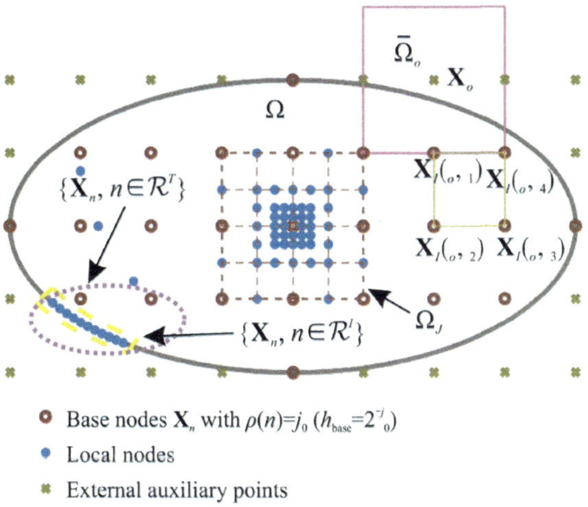

Fig. 11.5 Diagram of the nodal distribution for creating a wavelet multiresolution approximation on Ω when a local region is need to be refined

points $\mathbf{x}_0 \notin \Omega$ can be directly picked out from the points $\mathbf{x}_{j_0,k,l} = (k/2^{j_0}, l/2^{j_0})$ distributed uniformly on the rectangle $(X_{\min} - d_{j_0}, X_{\max} + d_{j_0}) \times (Y_{\min} - d_{j_0}, Y_{\max} + d_{j_0})$. Here, the base resolution level j_0 depends on the required accuracy on most of the problem domain Ω. $X_{\min}$ ($Y_{\min}$) and $X_{\max}$ ($Y_{\max}$) are the minimum and maximum of x (y)-coordinates of the domain Ω, respectively. Second, some local nodes are added into those points at which the essential boundary conditions are imposed. Last, some nodes are created in the region where the accuracy needs to be improved.

After the nodal distribution has been formed, the parameters $\rho(n)$, $k(n)$ and $l(n)$ of all nodes $\mathbf{x}_n$ for $n = 1, 2, \ldots, N$ and external auxiliary points $\mathbf{x}_n \notin \Omega$ for $n = N, N+1, \ldots, N+N_e$ can be calculated by the relation $\mathbf{x}_n = (k(n)/2^{\rho(n)}, l(n)/2^{\rho(n)})$, where $\rho(n) \geq j_0$ is the smallest integer making both $k(n)$ and $l(n)$ being integers. Then, the sets $\mathbb{C}_n$ and the corresponding coefficients $\Pi_n^m(\mathbf{x}_m)$ for all nodes can be obtained based on the definitions of Eqs. (11.23) and (11.32), as well as the compact support domain $\Omega_n = \cup_{m \in \mathbb{C}_n} \Omega_m^{\inf}$. So far, the modified wavelet basis functions defined by Eq. (11.32) have been created.

Then, all nodes $\{\mathbf{x}_n, \text{ for } n \in \Re^I\}$at which essential boundary conditions are imposed are singled out from the global nodal set, where $\mathbf{x}_n$ satisfies $\rho(n) < J_I = \max\{\rho(n), n \in \Re^I\}$, and the compact support domain Ω_n of which contains at least one node $\mathbf{x}_m$ for $m \in \Re^I$ are chosen to form the nodal set $\{\mathbf{x}_n, n \in \Re^T\}$. After that, the nodal sets $\left\{\mathbf{x}_n, n \in \Re_j^T\right\}$ for $j_0 \leq j < J_I$ can be determined through the definition $n \in \Re^T$ and $\rho(n) = j$. Finally, the matrix $\boldsymbol{\beta}^{J_I}$ can be obtained by using Eq. (11.36) iteratively. Further, the shape functions can be constructed directly through its definition of Eq. (11.35).

11.2.2 Variational Formulation of Plane Elasticity Problems

Without losing generality, here, we use the plane elastic problems as an example. From the theory of elastic mechanics, we know that such plane elastic problems can be described by the equilibrium, geometrical, and constitutive equations in the problem domain Ω bounded by $\Gamma = \Gamma_u \cup \Gamma_t$ [9]. The equilibrium equation is written as [9]

$$\nabla_s \boldsymbol{\sigma} + \mathbf{f} = 0, \text{ in } \Omega \tag{11.37}$$

in which $\boldsymbol{\sigma} = \{\sigma_{xx}, \sigma_{yy}, \sigma_{xy}\}^{\mathrm{T}}$ is the stress tensor, $\mathbf{f} = \{f_x, f_y\}^{\mathrm{T}}$ is the body force vector and the differential operator

$$\nabla_s = \begin{bmatrix} \frac{\partial}{\partial x} & 0 & \frac{\partial}{\partial y} \\ 0 & \frac{\partial}{\partial y} & \frac{\partial}{\partial x} \end{bmatrix}. \tag{11.38}$$

Following the linear geometrical and constitutive equations [9], the stress tensor $\boldsymbol{\sigma}$ can be further expressed as

$$\boldsymbol{\sigma} = \mathbf{D} \cdot \boldsymbol{\varepsilon} = \mathbf{D} \cdot \nabla_s^{\mathrm{T}} \mathbf{u}, \tag{11.39}$$

where $\boldsymbol{\varepsilon} = \{\varepsilon_{xx}, \varepsilon_{yy}, \varepsilon_{xy}\}^{\mathrm{T}}$ is the strain tensor, $\mathbf{u} = \{u, v\}^{\mathrm{T}}$ is the displacement vector, and the material matrix $\mathbf{D}$ is of the form

$$\mathbf{D} = \frac{E}{1-\upsilon^2} \begin{bmatrix} 1 & \upsilon & 0 \\ \upsilon & 1 & 0 \\ 0 & 0 & (1-\upsilon)/2 \end{bmatrix} \text{ for plane stress problems.} \tag{11.40}$$

The elastic solid mechanics governed by the above Eqs. (11.37)–(11.40) are referred to the plane stress problems. Once only material constants of the Young modulus, E, and Poisson ratio, υ, in Eq. (11.40) are replaced by $E/(1-\upsilon^2)$ and $\upsilon/(1-\upsilon)$, respectively, the above essential equations behave for a plane strain problem.

The boundary conditions are described by

$$\boldsymbol{\sigma} \cdot \mathbf{n} = \bar{\mathbf{t}} \text{ on the natural boundary } \Gamma_t, \tag{11.41}$$

$$\mathbf{u} = \bar{\mathbf{u}} \text{ on the essential boundary } \Gamma_u \tag{11.42}$$

in which $\boldsymbol{\sigma}$ is the standard stress tensor, $\mathbf{n}$ is the unit outward normal to the domain, and vectors $\bar{\mathbf{t}} = \{\bar{t}_n, \bar{t}_\tau\}^{\mathrm{T}}$ and $\bar{\mathbf{u}} = \{\bar{u}, \bar{v}\}^{\mathrm{T}}$ are given stresses and displacements on boundaries Γ_t and Γ_u, respectively [9].

Corresponding to the above essential equations, the variational form of above equations except for the essential boundary condition of Eq. (11.42) can be written as [9]

$$\int_\Omega \delta(\nabla_s^{\mathrm{T}}\mathbf{u})^{\mathrm{T}} \cdot \mathbf{D} \cdot \nabla_s^{\mathrm{T}}\mathbf{u} d\Omega - \int_\Omega \delta\mathbf{u}^{\mathrm{T}}\mathbf{f} d\Omega - \int_{\Gamma_t} \delta\mathbf{u}^{\mathrm{T}}\bar{\mathbf{t}} d\Omega = 0. \tag{11.43}$$

Based on the wavelet multiresolution approximation Eq. (11.22), the body force $\mathbf{f}(\mathbf{x})$ can be expressed as

$$\mathbf{f}(\mathbf{x}) \approx P_{j_b}^{j_b}\mathbf{f}(\mathbf{x}) = \sum_{m\in\Re_{j_b}} \mathbf{f}(\mathbf{x}_m)\phi_m(\mathbf{x}), \tag{11.44}$$

where a uniform distribution of knots $\mathbf{x}_m$ for $m \in \Re_{j_b}$ are used, so that the expansion coefficients are exactly the function values $\mathbf{f}(\mathbf{x}_m)$ at all knots. In general, one can use the base nodes $\{\mathbf{x}_n, n \in \Re_{j_0}\}$ as the knots $\{\mathbf{x}_m, m \in \Re_{j_b}\}$ for evaluating the body force vector. Thus, we do not need to create these extra knots.

On the other hand, using the wavelet multiresolution approximation Eq. (11.34) to represent the displacement field $\mathbf{u}(\mathbf{x})$, then substituting them and Eqs. (11.44) into (11.43), we get the system of algebraic equations in the matrix form

$$\mathbf{KU} = \mathbf{F} \tag{11.45}$$

in which the unknown nodal vector $\mathbf{U} = \{\mathbf{U}_l\}^{\mathrm{T}} = \{a_l^u, a_l^v\}^{\mathrm{T}}$ with the property $a_l^u = u(\mathbf{x}_l)$ and $a_l^v = v(\mathbf{x}_l)$ for $l \in \Re^I \cup \Re_{j_0}$, the stiffness matrix $\mathbf{K} = \{\mathbf{K}_{k,l}\}$ and load vector $\mathbf{F} = \{\mathbf{F}_k\}^{\mathrm{T}}$ for subscripts $k, l = 1, 2, \ldots, N$. The submatrices can be written as

$$\mathbf{K}_{k,l} = \int_\Omega \mathbf{B}_k^{\mathrm{T}}\mathbf{D}\mathbf{B}_l \, d\Omega \tag{11.46}$$

and

$$\mathbf{F}_k = \mathbf{F}_{k,b} + \mathbf{F}_{k,t} = \sum_{m\in\Re_{j_b}} \mathbf{f}(\mathbf{x}_m) \int_\Omega \Theta_{j_{\max},k}(\mathbf{x})\phi_m(\mathbf{x}) d\Omega + \int_{\Gamma_t} \Theta_{j_{\max},k}(\mathbf{x})\bar{\mathbf{t}} \, d\Gamma, \tag{11.47}$$

where

$$\mathbf{B}_k = \begin{bmatrix} \Theta^{(1,0)}_{j_{\max},k}(\mathbf{x}) & 0 \\ 0 & \Theta^{(0,1)}_{j_{\max},k}(\mathbf{x}) \\ \Theta^{(0,1)}_{j_{\max},k}(\mathbf{x}) & \Theta^{(1,0)}_{j_{\max},k}(\mathbf{x}) \end{bmatrix}. \tag{11.48}$$

In Eq. (11.47), the nodal load vector $\mathbf{F}_{k,t}$ corresponding to the natural boundary condition can be calculated by

$$\mathbf{F}_{k,t} \approx 2^{-j_t} \sum_{l=0}^{[2^{j_t} L]} \Theta_{j_{\max},k}(\mathbf{x}_{j_t,l})\bar{\mathbf{t}}(\mathbf{x}_{j_t,l})[\Phi(2^{j_t} L - l) - \Phi(-l)] \tag{11.49}$$

in which the independent parameter j_t controls the accuracy of the integral, the L is the total arc length of the boundary with $\bar{\mathbf{t}} \neq 0$, the operator $[a]$ returns the largest integer no more than a, and $\mathbf{x}_{j_t,l}$ is the global coordinates of the point with the arc length $s = l/2^{j_t}$, as shown in Fig. 11.6. In Eq. (11.49), the integral $\Phi(x)$ of the autocorrelation function $\theta(x)$ can be obtained by using the method proposed in Chap. 2. When the boundary Γ_t is consisted of several segments, the nodal load vector can be calculated firstly on each segment through Eq. (11.49), and then summing the results gives the final nodal load vector $\mathbf{F}_{k,t}$.

The computation of the stiffness matrix and the nodal load vector $\mathbf{F}_{k,b}$ from the body force will be described in detail in the next subsection. On the other hand, because there is the property $a_l^u \equiv u(\mathbf{x}_l)$ and $a_l^v \equiv v(\mathbf{x}_l)$ for $l \in \Re^l$, the essential boundary condition $\mathbf{u} = \bar{\mathbf{u}}$ on Γ_u can be imposed by directly replacing the corresponding equations in Eq. (11.45), the same as the treatment in the standard FEM [9].

Finally, the unknown nodal vector $\mathbf{U}$ can be obtained by solving the system of linear equations in Eq. (11.45). Then, the displacement field $\mathbf{u}$ can be reconstructed through the approximation Eq. (11.34). Further the fields of strain and stress can

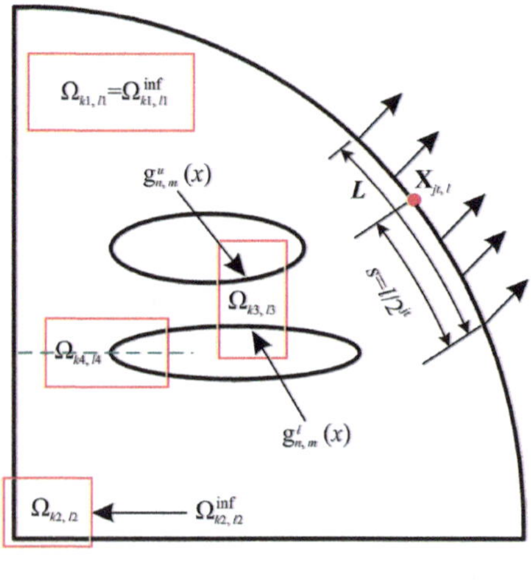

Fig. 11.6 The schematic configuration for estimating the load vector $\mathbf{F}_{k,t}$ and integral Υ_η^K used in next subsection

be calculated directly by the geometrical and material relations of $\boldsymbol{\varepsilon} = \nabla_s^{\mathrm{T}} \mathbf{u}$ and $\sigma = \mathbf{D} \cdot \boldsymbol{\varepsilon} = \mathbf{D} \cdot \nabla_s^{\mathrm{T}} \mathbf{u}$, respectively [9].

11.2.3 Calculation of Stiffness Matrix

To evaluate the stiffness matrix, we first calculate the integral

$$\mathrm{K}_{k,l}^{\kappa} = \int_{\Omega} \frac{\partial^{n_1+m_1} \Theta_{j_{\max},k}(\mathbf{x})}{\partial x^{n_1} \partial y^{m_1}} \frac{\partial^{n_2+m_2} \Theta_{j_{\max},l}(\mathbf{x})}{\partial x^{n_2} \partial y^{m_2}} d\Omega \tag{11.50}$$

in which the multi-index $\kappa = (n_1, m_1, n_2, m_2)$ with either $n_1, m_1, n_2, m_2 = 0$ or 1. Substituting Eqs. (11.35) into (11.50), one can obtain

$$\mathrm{K}_{k,l}^{\kappa} = \sum_{n \in \mathbb{N}_k} \sum_{m \in \mathbb{N}_l} \alpha_{k,n} \alpha_{l,m} \Gamma_{n,m}^{\kappa} \tag{11.51}$$

with the integral

$$\Gamma_{k,l}^{\kappa} = \int_{\Omega} \frac{\partial^{n_1+m_1} \phi_k(\mathbf{x})}{\partial x^{n_1} \partial y^{m_1}} \frac{\partial^{n_2+m_2} \phi_l(\mathbf{x})}{\partial x^{n_2} \partial y^{m_2}} d\Omega \tag{11.52}$$

Further, substitution of Eq. (11.32) into the integral (11.52) yields

$$\Gamma_{k,l}^{\kappa} = \sum_{n \in \Xi_k} \sum_{m \in \Xi_l} \Pi_k^n(\mathbf{x}_n) \Pi_l^m(\mathbf{x}_m) \Upsilon_{\xi}^{\kappa} \tag{11.53}$$

in which the multi-index $\xi = (j_n, k_n, l_n, j_m, k_m, l_m) = (\rho(n), k(n), l(n), \rho(m), k(m), l(m))$ and the integral

$$\Upsilon_{\xi}^{\kappa} = \int_{\Omega} \frac{\partial^{n_1+m_1} \theta_{j_n,k_n,l_n}(\mathbf{x})}{\partial x^{n_1} \partial y^{m_1}} \frac{\partial^{n_2+m_2} \theta_{j_m,k_m,l_m}(\mathbf{x})}{\partial x^{n_2} \partial y^{m_2}} d\Omega. \tag{11.54}$$

Because the wavelet basis function $\theta_{j_n,k_n,l_n}(\mathbf{x})$ is compactly supported on the square domain $\Omega_n^{\inf} = [x_n - d_n, x_n + d_n] \times [y_n - d_n, y_n + d_n]$ with $d_n = 2^{-j_n}(\gamma - 1)$, the integral domain of Eq. (11.54) can be reduced to $\Omega_{n,m} = \Omega_{n,m}^{\inf} \cap \Omega$. Here the domain $\Omega_{n,m}^{\inf} = \Omega_n^{\inf} \cap \Omega_m^{\inf}$ is the rectangle with two diagonal points $(\overline{x}_{n,m}^{\min}, \overline{y}_{n,m}^{\min})$ and $(\overline{x}_{n,m}^{\max}, \overline{y}_{n,m}^{\max})$, where $\overline{x}_{n,m}^{\min} = \max\{x_n - d_n, x_m - d_m\}$, $\overline{x}_{n,m}^{\max} = \min\{x_n + d_n, x_m + d_m\}$, $\overline{y}_{n,m}^{\min} = \max\{y_n - d_n, y_m - d_m\}$ and $\overline{y}_{n,m}^{\max} = \min\{y_n + d_n, y_m + d_m\}$.

It is easy to check that $\Upsilon_\xi^\kappa = 0$ for $\overline{x}_{n,m}^{\min} \geq \overline{x}_{n,m}^{\max}$ or $\overline{y}_{n,m}^{\min} \geq \overline{y}_{n,m}^{\max}$. Especially, when $\Omega_{n,m}^{\inf} \subseteq \Omega$ or $\Omega_{n,m}$ is a rectangular domain such as Ω_{k_1,l_1} and Ω_{k_2,l_2} shown in Fig. 11.6, there is

$$\Upsilon_\xi^\kappa = \overline{\mathrm{T}}_\tau^{n_1,n_2}[x_{n,m}^{\min}, x_{n,m}^{\max}] \times \overline{\mathrm{T}}_\varsigma^{m_1,m_2}[y_{n,m}^{\min}, y_{n,m}^{\max}], \tag{11.55}$$

where $x_{n,m}^{\min}$ ($y_{n,m}^{\min}$) and $x_{n,m}^{\max}$ ($y_{n,m}^{\max}$) are the minimum and maximum of $x(y)$-coordinates of the integral domain $\Omega_{n,m}$, respectively, the multi-indexes $\tau = (j_n, k_n, j_m, k_m)$ and $\varsigma = (j_n, l_n, j_m, l_m)$, and the integral notation is of the form

$$\begin{aligned}\overline{\mathrm{T}}_{j,k,i,l}^{n,m}[a,b] &= \int_a^b \frac{d^n\theta(2^j x - k)}{dx^n}\frac{d^m\theta(2^i x - l)}{dx^m}dx \\ &= \begin{cases} 2^{j(n+m-1)}[\mathrm{T}_{i-j,l-2^{i-j}k}^{n,m}(2^j b - k) - \mathrm{T}_{i-j,l-2^{i-j}k}^{n,m}(2^j a - k)], & \text{for } j \leq i \\ 2^{i(n+m-1)}[\mathrm{T}_{j-i,k-2^{j-i}l}^{m,n}(2^i b - l) - \mathrm{T}_{j-i,k-2^{j-i}l}^{m,n}(2^i a - l)], & \text{for } j > i \end{cases}\end{aligned} \tag{11.56}$$

in which the two-term multiresolution connection coefficient is used by

$$\mathrm{T}_{j,k}^{n,m}(x) = \int_{1-\gamma}^{x} \theta^{(n)}(y)\theta^{(m)}(2^j y - k)dy, \text{ for } 0 \leq n, m \leq \gamma/2, \text{ and } j \geq 0. \tag{11.57}$$

when the nodes $\mathbf{x}_n$ and $\mathbf{x}_m$ are close to the general curved boundary, the corresponding integral domain $\Omega_{n,m}$ will be irregular, such as Ω_{k_3,l_3} and Ω_{k_4,l_4} shown in Fig. 11.6. If the upper boundary $g_{n,m}^u(x)$ and lower boundary $g_{n,m}^l(x)$ of $\Omega_{n,m}$ are single-valued functions, such as Ω_{k_3,l_3}, the integral Υ_ξ^κ can be evaluated by

$$\Upsilon_\xi^\kappa = \sum_{k_3=K_0+1}^{K_1} \overline{\mathrm{T}}_\varsigma^{m_1,m_2}[g_{n,m}^l(\frac{k_3}{2^{j_3}}), g_{n,m}^u(\frac{k_3}{2^{j_3}})]\overline{\Gamma}_\xi^{n_1,n_2}[x_{n,m}^{\min}, x_{n,m}^{\max}] \tag{11.58}$$

in which the independent parameter $j_3 \geq \max\{j_n, j_m\}$ controls the accuracy of integral, $K_0 = [2^{j_3}x_{n,m}^{\min}]$, $K_1 = [2^{j_3}x_{n,m}^{\max}]$, the multi-index $\xi = (j_n, k_n, j_m, k_m, j_3, k_3)$, and the integral

$$\begin{aligned}\overline{\Gamma}_\xi^{n_1,n_2}[a,b] &= \int_a^b \frac{d^{n_1}\theta(2^{j_n}x - k_n)}{dx^{n_1}}\frac{d^{n_2}\theta(2^{j_m}x - k_m)}{dx^{n_2}}\theta(2^{j_3}x - k_3)dx \\ &= \begin{cases} 2^{j_n(n_1+n_2-1)}\Psi_{k_m-2^{j_m-j_n}k_n,k_3-2^{j_m-j_n}k_n}^{n_1,n_2}(j_m - j_n, j_3 - j_n, 2^{j_n}[b:a] - k_n), & \text{for } j_n \leq j_m, \\ 2^{j_m(n_1+n_2-1)}\Psi_{k_n-2^{j_n-j_m}k_m,k_3-2^{j_n-j_m}k_m}^{n_2,n_1}(j_n - j_m, j_3 - j_m, 2^{j_m}[b:a] - k_m), & \text{for } j_n > j_m. \end{cases}\end{aligned} \tag{11.59}$$

In Eq. (11.59), $\Psi_{k,l}^{n,m}(j_1, j_2, 2^j[b : a] - z) = \Psi_{k,l}^{n,m}(j_1, j_2, 2^j b - z) - \Psi_{k,l}^{n,m}(j_1, j_2, 2^j a - z)$ and the three-term multiresolution connection coefficient

$$\Psi_{k,l}^{n,m}(j_1, j_2, x) = \int\limits_{1-\gamma}^{x} \frac{d^n\theta(y)}{dy^n}\frac{d^m\theta(2^{j_1}y - k)}{dy^m}\theta(2^{j_2}y - l)dy, \tag{11.60}$$

for $0 \le n, m \le \gamma/2$, and $j_1, j_2 \ge 0$.

Especially if the upper boundary or lower boundary of $\Omega_{n,m}$ is not a single-valued function, such as Ω_{k_4,l_4}, as shown in Fig. 11.6, a sample partition of $\Omega_{n,m}$ will be implemented firstly.

So far, the computation of the integral $\mathrm{K}_{k,l}^{\kappa}$ has been translated into the evaluation of multiresolution connection coefficients $\mathrm{T}_{j,k}^{n,m}(x)$ and $\Psi_{k,l}^{n,m}(j_1, j_2, x)$. It can be seen from the definitions of Eqs. (11.57) and (11.60) that both $\mathrm{T}_{j,k}^{n,m}(x)$ and $\Psi_{k,l}^{n,m}(j_1, j_2, x)$ are completely independent of the specific problems. Thus, one can build an underlying general database of these connection coefficients, and so the computation of them can be removed in the use of the proposed wavelet method. The algorithm for exactly calculating the connection coefficients $\mathrm{T}_{j,k}^{n,m}(x)$ and $\Psi_{k,l}^{n,m}(j_1, j_2, x)$ has been proposed in Chap. 2.

11.2.4 Error Analysis

As shown in Sect. 11.1.5, the present shape functions of the wavelet approximation possess the global continuity, the linear independence, and the ability to reproduce exactly low-order polynomials. Moreover, the partial differential equations are discretized by using the standard Galerkin method, and the essential boundary conditions are imposed exactly. Thus, the convergence of the proposed wavelet method is guaranteed [9].

In addition, based on Theorem 11.3, the error of the proposed wavelet method for the plane elasticity problems with sufficiently smooth solutions obeys [9]

$$E_d = \|\mathbf{u}_{\text{num}} - \mathbf{u}_{\text{exact}}\|_{L^2(\Omega)} \le C_d h_{\text{base}}^{\lambda} \tag{11.61}$$

and

$$E_e = \left\{\frac{1}{2}\int\limits_{\Omega}(\boldsymbol{\sigma}_{\text{num}} - \boldsymbol{\sigma}_{\text{exact}})^{\mathrm{T}}\mathbf{D}^{-1}(\boldsymbol{\sigma}_{\text{num}} - \boldsymbol{\sigma}_{\text{exact}})d\Omega\right\}^{1/2} \le C_e h_{\text{base}}^{\lambda-1} \tag{11.62}$$

in which E_d and E_e are the displacement and energy error norms, $\mathbf{u}_{\text{exact}}$ and $\boldsymbol{\sigma}_{\text{exact}}$ ($\mathbf{u}_{\text{num}}$ and $\boldsymbol{\sigma}_{\text{num}}$) are the exact (numerical) displacement and stress, C_d and C_e are constants, $h_{\text{base}} = 2^{-j_0}$ is the distance between two adjacent base nodes, $\mathbf{x}_n \in \Re_{j_0}$, and $\lambda = \min\{\gamma, \eta_o\} \geq 2$.

One can see from Eq. (11.61) that the proposed WMIGM possesses γ order convergence rate in the displacement when taking η_o to be γ. Moreover, the above error estimates are in fact rigorous only when a uniform nodal distribution with grid size h_{base} is employed. These errors will be further reduced when using the corresponding irregular nodal distributions created by adding other local nodes.

11.3 Numerical Examples

Based on the previous section for the proposed WMIGM for 2D irregular domain and local refinement of an elastic problem, some examples of the solutions are displayed in this section.

11.3.1 Patch Test

The standard patch test is studied as the first numerical example. Two patches with domain size 2 × 2 are used as shown in Fig. 11.7. One is a patch of nine nodes with one interior node. And the other one is the patch with 15 nodes including seven irregularly distributed interior nodes.

The material parameters are taken as $E = 1.0$ and $\upsilon = 0.3$. The essential boundary conditions on all outside boundaries are extracted from the linear displacement field $u(x, y) = 0.6x$ and $v(x, y) = 0.6y$. Satisfaction of the patch test requires that the

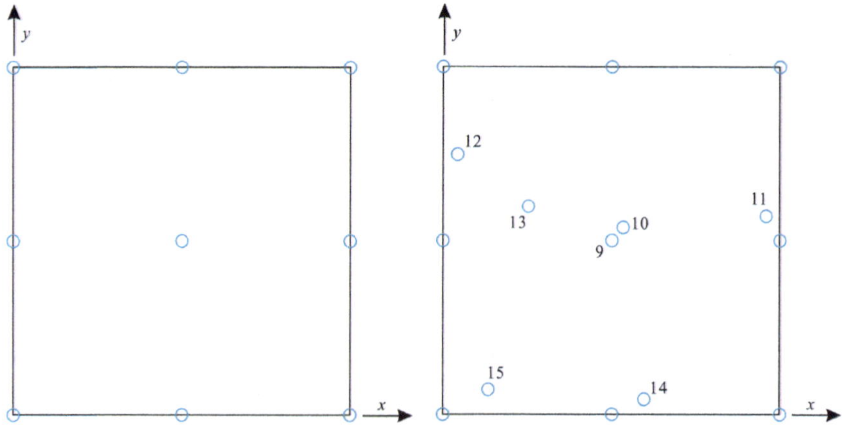

Fig. 11.7 Two test patches with 9 nodes (Left) and with 15 nodes (Right), respectively

Table 11.2 Absolute error at interior nodes for the patch with 15 nodes

No		9	10	11	12	13	14	15
(x, y)		(1, 1)	(17/16, 17/16)	(31/16, 9/8)	(1/16, 3/2)	(1/2, 19/16)	(19/16, 1/16)	(1/4, 1/8)
$\gamma = 2$	u	3.22E-14	2.83E-14	1.78E-15	9.58E-16	1.31E-14	1.78E-15	1.14E-15
	v	2.42E-14	2.12E-14	1.33E-15	9.99E-16	9.88E-15	1.17E-15	8.88E-16
	σ_{xx}	0	4.06E-14	3.13E-14	1.73E-14	2.48E-14	4.44E-15	6.44E-15
	σ_{yy}	0	3.49E-14	1.09E-14	3.77E-15	4.66E-15	2.10E-14	8.10E-15
	τ_{xy}	0	2.04E-14	8.92E-15	3.88E-15	1.37E-15	9.45E-15	4.26E-15
$\gamma = 4$ $\eta = 3$	u	1.11E-12	1.10E-12	1.34E-13	9.59E-14	8.02E-13	1.33E-13	1.26E-13
	v	1.12E-12	1.11E-12	1.37E-13	1.04E-13	8.08E-13	1.26E-13	1.22E-13
	σ_{xx}	1.78E-15	1.89E-13	2.28E-12	1.68E-12	1.08E-12	6.20E-13	7.12E-13
	σ_{yy}	1.78E-15	1.89E-13	7.10E-13	3.69E-13	1.88E-14	2.21E-12	1.07E-12
	τ_{xy}	8.97E-16	1.02E-13	8.04E-13	5.55E-13	2.95E-13	7.57E-13	4.79E-13

displacement of all interior nodes should be given by the same linear functions and that the strain and stress are constant in the whole patch.

The proposed second-order ($\gamma = 2$) and fourth-order ($\gamma = 4$) WMIGM can all pass these two patch testes exactly in the numerical sense. As shown in Table 11.2, for example, the absolute errors of displacement and stress at all interior nodes are listed for the patch with 15 nodes corresponding to Fig. 11.6. From it, one can also see that the WMIGM proposed in this chapter is still very stable and valid even when the node number 9 is very close to node the number 10.

11.3.2 Test of Convergence and Stability Against Irregular Nodes

A plane stress problem with domain $\Omega = [0, 1] \times [0, 1]$ is considered here, in which the material parameters $E = 1.0$ and $\upsilon = 0.3$ are adopted. When the body force is taken by the form $f_x(x, y) = f_y(x, y) = 5\pi^2\{7\cos[\pi(x - y)/2] + 20\sin[\pi(x + y - 1)/2]\}/1456$, and the boundary condition $\mathbf{u}(x, y) = 0$ is enforced on the edge $y = 0$, the exact displacement field $u(x, y) = v(x, y) = \sin(\pi x/2)\sin(\pi y/2)/4$ is employed to give the stress boundary conditions on the other three edges. In this numerical test, the completely uniform distributions of nodes are employed.

Figure 11.8 shows the displacement error norm E_d and the energy error norm E_e as functions of the grid size $h = \Delta x = \Delta y$, as well as the sparsity pattern of the stiffness matrix for $\gamma = 2$ and 4, where the completely uniform distributions of nodes are employed. We can see from Fig. 11.8 that the proposed WMIGM has a satisfactory accuracy and rate of convergence. The rates of convergence of displacement for the second- ($\gamma = 2$) and fourth-order ($\gamma = 4$) method are about 1.96 and 3.84 order,

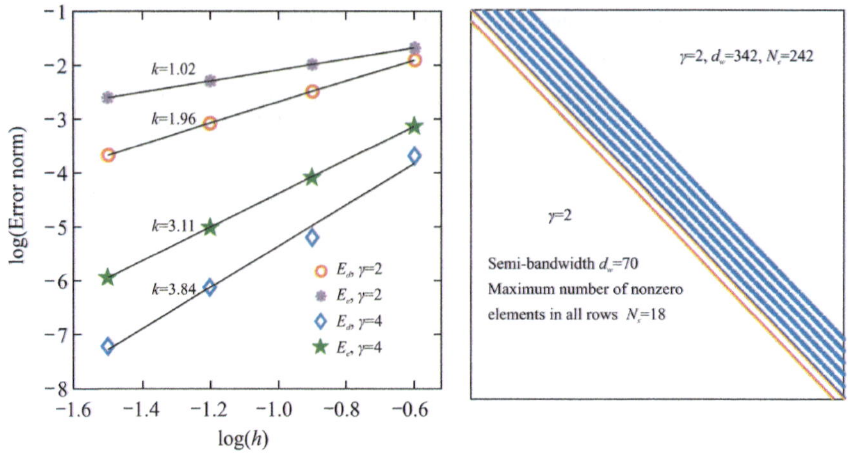

Fig. 11.8 Scaling dependence of accuracy and sparsity pattern of the stiffness matrix of the present wavelet-based solutions with uniform distributions of nodes for solving a plane stress problem with exact solution $u(x, y) = v(x, y) = \sin(\pi x/2)\sin(\pi y/2)/4$: the displacement error norm E_d and energy error norm E_e varying with the grid size $h = \Delta x = \Delta y$ (Left), the sparsity pattern of the stiffness matrix (Right)

respectively, while the rates of convergence of energy for $\gamma = 2$ and 4 are about 1.02 and 3.11 order, respectively. These observations are in good agreement with the theoretical error estimation previously. From Fig. 11.8, it is also found that the stiffness matrix of the present WMIGM for $\gamma = 2$ and 4 are highly regularly banded when using a uniform distribution of nodes. Especially for $\gamma = 2$, the present stiffness matrix is exactly the same as that of the FEM with the plane four-node quadrilateral element [17].

In order to investigate the effect of irregular nodes on the WMIGM, here, four regular addition nodes and either or not five irregular nodes are added as shown in Fig. 11.9, marked by asterisks. Here, the absolute errors in displacement and Von Mises stress for numerical solutions obtained by the proposed second-order ($\gamma = 2$) WMIGM are displayed. According to Fig. 11.9, one sees that properly adding local nodes in the present wavelet method can effectively increase its accuracy both in displacement and stress. For example, the displacement error norm E_d and energy error norm E_e are, respectively, reduced by about 75% and 47% after the five local nodes are added. To obtain a more comprehensive view of such effect of the additional nodes on the distribution of optional nodes, the energy error norm E_e of the present wavelet solutions with 1000 random distributions of these five additional local nodes is plotted in Fig. 11.10, which tells us that these optional nodes can only decrease the error. These results are in good agreement with the error analysis given in Fig. 11.3. Thus, this test example also demonstrates that the proposed WMIGM possesses an excellent stability for irregular additional nodes.

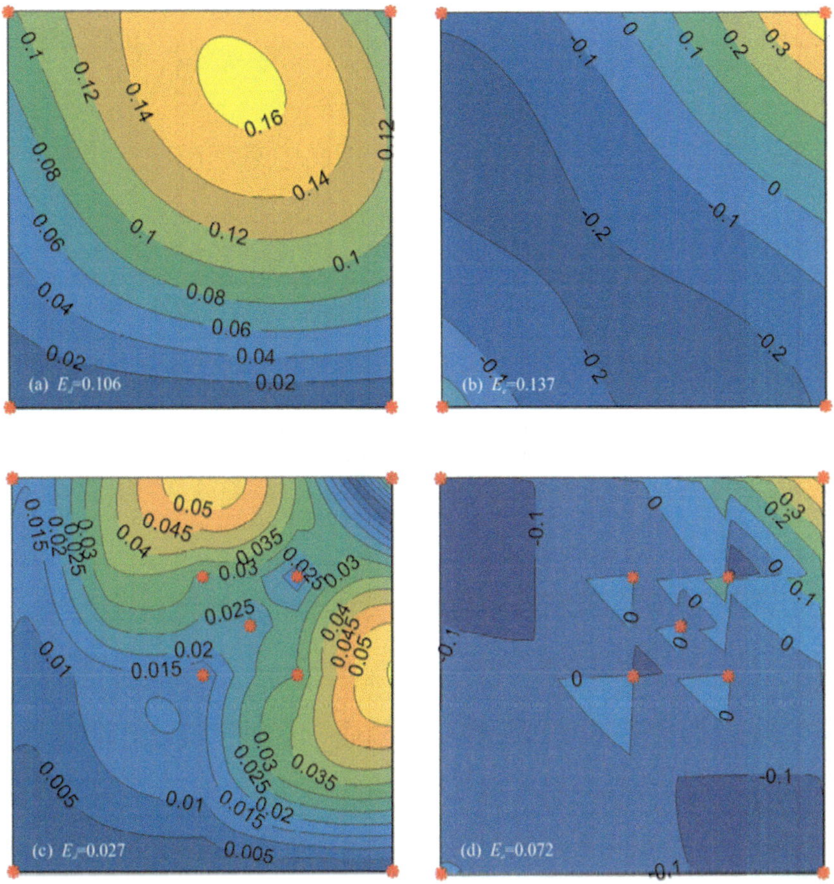

Fig. 11.9 Distributions of the displacement error $E_u(\mathrm{x})$ and the absolute error $E_\sigma(\mathrm{x})$ of the Von Mises stress from the present wavelet solutions with and without five additional local nodes for $\gamma = 2$. **a** and **b** Add four regular distributed nodes; **a** and **d** Add four regular distributed nodes and five other local nodes. Here, the asterisks represent added nodes

11.3.3 Infinite Plate with a Circular Hole

Consider that an infinite plate with a central circular hole of radius $R = 1$ is subjected to a unidirectional uniform tensile load of 1.0 in the x-direction as shown in Fig. 11.11. Here, plane strain problem is studied with Young's modulus $E = 1000$ and Poisson's ratio $\upsilon = 0.3$. To such elastic problem, the exact solution of this problem can be found in the textbook [18], i.e.,

$$\sigma_{xx}(x, y) = 1 - \frac{1}{r^2}\left[\frac{3}{2}\cos(2\theta) + \cos(4\theta)\right] + \frac{3}{2r^2}\cos(4\theta), \tag{11.63}$$

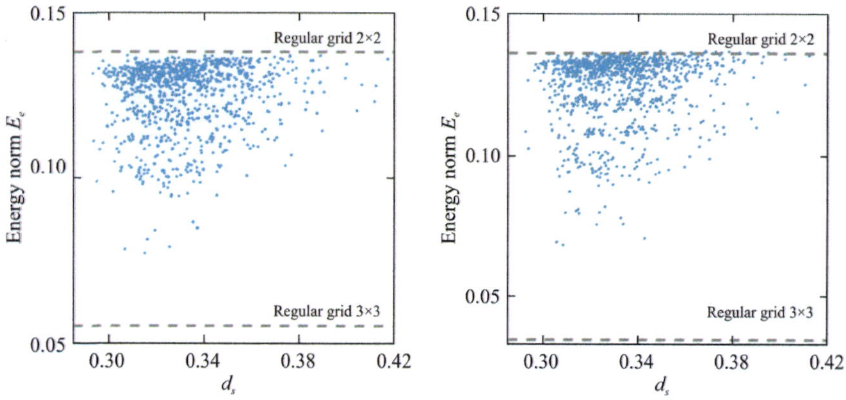

Fig. 11.10 The energy error norm E_e of the numerical solution test from the present wavelet solutions for 1000 random distributions of five additional local nodes when using second-order ($\gamma = 2$) (Left) and fourth-order ($\gamma = 4$) (Right) WMIGM

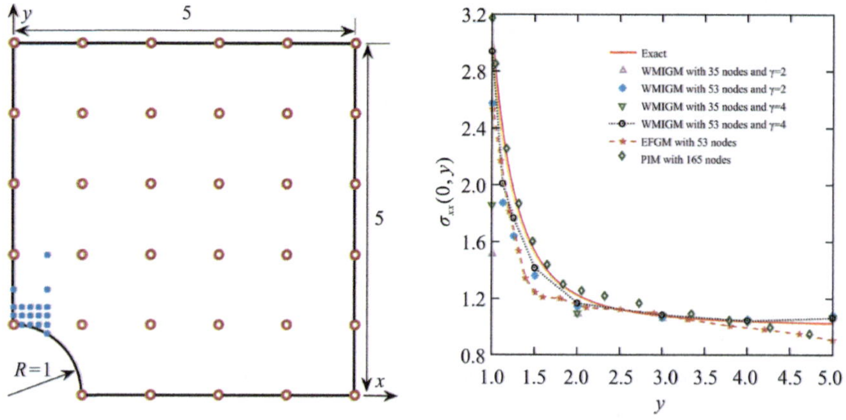

Fig. 11.11 Calculation model of infinite plate with a central circular hole under in-plane unidirectional uniform tensile load and the distribution of noemal stress near the hole: Distribution of nodes of upper right quadrant of the plate used in the wavelet-based method (Left); Distribution of normal stress $\sigma_{xx}(x, y)$ along the left edge ($x = 0$) (Right)

$$\sigma_{yy}(x, y) = -\frac{1}{r^2}\left[\frac{1}{2}\cos(2\theta) - \cos(4\theta)\right] - \frac{3}{2r^2}\cos(4\theta), \tag{11.64}$$

$$\sigma_{xy}(x, y) = -\frac{1}{r^2}\left[\frac{1}{2}\sin(2\theta) + \sin(4\theta)\right] + \frac{3}{2r^2}\sin(4\theta) \tag{11.65}$$

in which $r = \sqrt{x^2 + y^2}$ and $\theta = \arctan(y/x)$. From this set of stress exact solutions, we know that there is a stress concentration at the edge of the hole. Usually, such well-known phenomenon of stress concentration makes extra difficulty in solving this problem and it is always used as an example for testing the accuracy of a numerical solution method.

Due to the symmetry, only the upper right quadrant of the plate is modeled in the present study, as shown in Fig. 11.11. The symmetry condition is imposed on the left ($x = 0$) and bottom ($y = 0$) edges, and the boundary at $R = 1$ is traction free. Moreover, the traction boundary conditions extracted from the exact stresses (11.63)–(11.65) are imposed on the right ($x = 5$) and top ($y = 5$) edges.

A regular nodal distribution (35 nodes in total) and a combination (53 nodes in total) of regular and local nodes shown in Fig. 11.11 are employed. A comparison of the normal stress $\sigma_{xx}(0, y)$ obtained by the present WMIGM, element-free Galerkin method (EFGM) [19] and point interpolation method (PIM) [20] is presented in Fig. 11.11. From it, one sees that the accuracy of the solution near the hole increases obviously by adding these local nodes. And not surprisingly, the accuracy for the fourth-order ($\gamma = 4$) WMIGM is better than that for the second-order ($\gamma = 2$) one, because the solution of this problem is also smooth enough although the stress concentration appears at the edge of the hole.

In addition, it also can be found that the proposed second-order WMIGM using 53 nodes possesses a little better accuracy than the EFGM with $d_{mi} = 4c_i$ using 54 nodes, where d_{mi} is the size of the support for the weight function and c_i is the maximum distance between two adjacent nodes [19]. In Ref. [19], the distribution of nodes used in the EFGM has at least 25 nodes located in the domain of influence of most nodes. On the contrary, only 9 nodes located in the domain of influence of most nodes are employed in the proposed second-order wavelet method. Therefore, the number of nonzero elements in the stiffness matrix for the EFGM is much more than that in the present wavelet method, although the total numbers of nodes used in two methods are almost the same. Moreover, the rate of convergence of wavelet interpolant for approximating curvilinear edge should be faster than the cell structure used in the EFGM [19] in which the implementation of essential boundary conditions in the EFGM is more complex compared to the proposed wavelet method. As a result, the present WMIGM is more efficient than the EFGM. At the same time, Fig. 11.11 also clearly shows that the proposed WMIGM has a higher efficiency than the PIM [20], because the fourth-order WMIGM with only 53 nodes can obtain a more accurate solution compared with the PIM using as many as 165 nodes.

11.3.4 Semi-infinite Plate Subjected to Concentrated Edge Load

As shown in Fig. 11.12a, a concentrated vertical force P acts on a horizontal straight edge of a semi-infinite plate with unit thickness. The distribution of the load P along the thickness of the plate is uniform. For this elastic problem, its analytical solution can be written as [18]

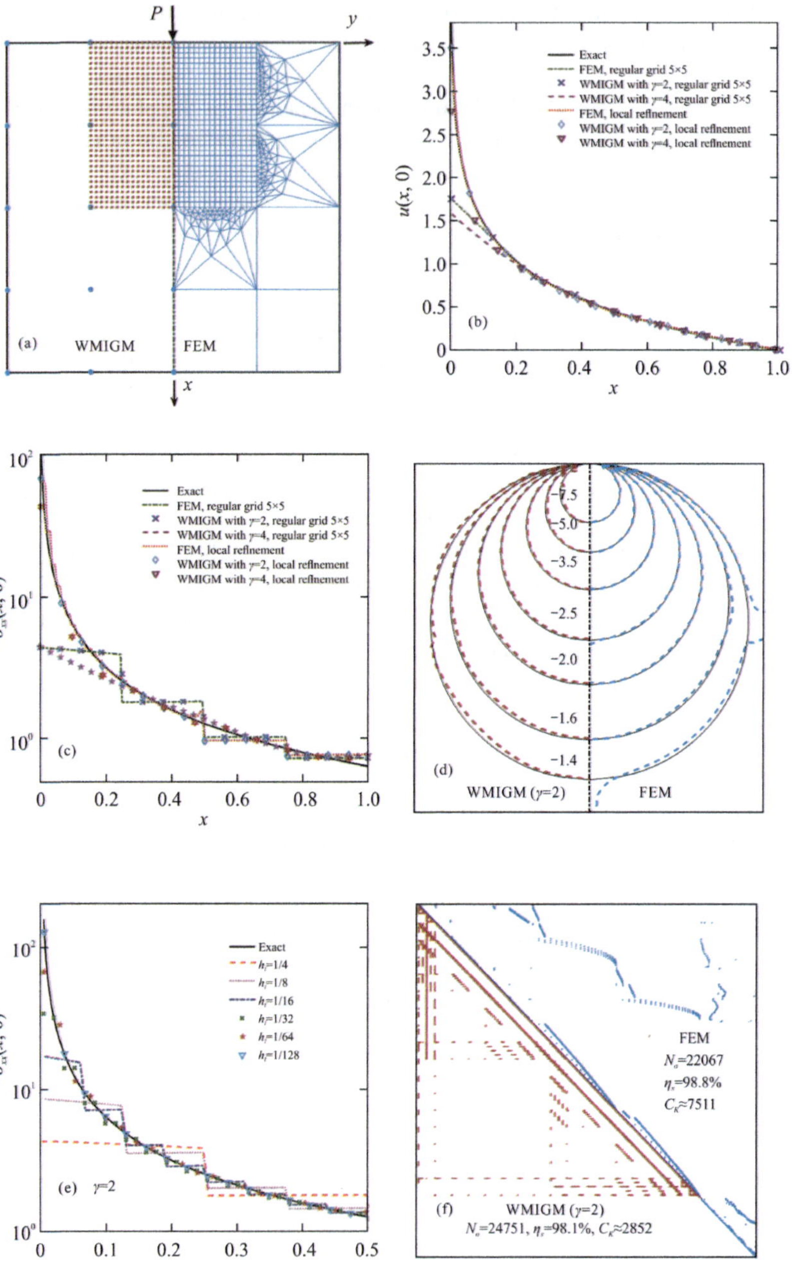

Fig. 11.12 Comparison of the WMIGM and FEM for the semi-infinite plate subjected to an in-plane concentrated load on the horizontal straight edge: **a** the distribution of nodes for WMIGM and the mesh for FEM, **b** the vertical displacement $u(x, 0)$, **c** the normal stress $\sigma_{xx}(x, 0)$, **d** the contour of radial stress $\sigma_r(x, y)$ in polar coordinates, **e** the normal stress $\sigma_{xx}(x, 0)$ for various local grid size h_l; **f** the sparsity pattern of stiffness matrix

$$u(x, y) = -\frac{P[(1+\upsilon)y^2 + (x^2 + y^2)\log(x^2 + y^2)]}{E\pi(x^2 + y^2)}, \tag{11.66a}$$

$$v(x, y) = \frac{P[(1+\upsilon)xy - (1-\upsilon)(x^2 + y^2)\arctan(y/x)]}{E\pi(x^2 + y^2)} \tag{11.66b}$$

and

$$\sigma_{xx}(x, y) = -2Px^3/\pi(x^2 + y^2)^2, \tag{11.67a}$$

$$\sigma_{yy}(x, y) = -2Pxy^2/\pi(x^2 + y^2)^2, \tag{11.67b}$$

$$\sigma_{xy}(x, y) = -2Px^2y/\pi(x^2 + y^2)^2. \tag{11.67c}$$

It can be seen from Eqs. (11.66) and (11.67) that both of the displacement and the stress are singular at the origin, and they change sharply in the region near the origin. The singular point and localized steep gradients will bring severe challenge to numerical methods [9.15].

Figure 11.12 displays a comparison of the present wavelet solution and FEM solution to the semi-infinite plate subjected to an in-plane concentrated load on the horizontal straight edge. In the present numerical simulations, the problem domain is taken as $\Omega = [0, 1] \times [-1/2, 1/2]$. The concentrated force $P = 1$, Young's modulus $E = 1$ and Poisson's ratio $\upsilon = 0.3$ are adopted. The displacement boundary conditions on all edges except for the edge $x = 0$ are extracted from the exact displacement field of Eq. (11.66). The boundary $x = 0$ is traction free except for the origin at which concentrated force P is applied. A nodal distribution including a 5×5 grid ($h = 1/4$) on the whole domain and a refined grid with $h_l = 1/64$ in local region $\Omega_{\text{local}} = [0, 1/2] \times [-1/4, 1/4]$ shown in Fig. 11.12a is employed to study this problem.

From Fig. 11.12b, c, we can see that the accuracy of solutions in Ω_{local} is improved obviously by adding the local nodes, and the second-order ($\gamma = 2$) WMIGM possesses a higher accuracy than the fourth-order ($\gamma = 4$) one for such a problem with singular point and extremely steep gradient. This is reason that the compact support of shape function for the second-order method is smaller than that for the fourth-order method, leading to a better capacity of detecting local mutation for the former [1].

Figure 11.12b–d also shows a comparison between the solutions, respectively, obtained by the second-order WMIGM and the FEM using the plane four-node quadrilateral element and linear triangular element. The nodal distribution in the used finite element mesh is the same as the distribution used in the wavelet method, except for a few extra nodes employed to meet the required in quality of mesh for the FEM, as shown in Fig. 11.12a. This fact shows that the sharp change in gird size is allowable in the proposed WMIGM, but which should be avoided whenever possible in the FEM, as well as most existing methods [9.13]. In order to verify this flexibility, Fig. 11.12e displays the normal stress $\sigma_{xx}(x, 0)$ for various local grid size h_l, which demonstrates that the present wavelet method has good convergence and excellent stability, and its accuracy has been increasing gradually as the decrease of h_l. In the WMIGM, there is no unstable phenomenon even when the maximum ratio between adjacent mesh size is as high as 32. Therefore, the distribution of nodes is more easy to determine, because only the requirement from accuracy should be considered with no need to worry about the stability.

In addition, it can be seen from Fig. 11.12b, c that the accuracy of vertical displacement $u(x, 0)$ and the normal stress $\sigma_{xx}(x, 0)$ obtained by the second-order WMIGM is almost the same as the accuracy of finite element solutions. But, the radial stress $\sigma_r(x, y)$ in polar coordinates achieved from the present WMIGM is slightly better than the finite element solutions, as shown in Fig. 11.12d. Here, the radial stress is defined as

$$\sigma_r(x, y) = \frac{\sigma_{xx} + \sigma_{yy}}{2} + \frac{\sigma_{xx} - \sigma_{yy}}{2}\cos(2\theta) + \sigma_{xy}\sin(2\theta) \tag{11.68}$$

in which $\theta = \arctan(y/x)$. The contour of radial stress $\sigma_r(x, y)$ of this problem is a series of circles except for the singular point shown in Fig. 11.12d.

For the computational efficiency, the sparsity pattern of stiffness matrix in the present second-order WMIGM and FEM is presented in Fig. 11.12f. It can be found that the sparsity of the stiffness matrix in the wavelet method is almost the same as the sparsity for the FEM. Table 11.3 shows a comparison of the time consuming between the second-order WMIGM and FEM, where ten independent sets of computations are performed to eliminate the influence of occasional factors. Here, the exact same computer platform including hardware and software is used for implementing two methods. And the optimized FEM codes proposed by Kattan [17] are employed. It can be seen that the time consuming of two methods are almost the same. We should point out that the time consuming of the wavelet method is spend for the whole analysis including the node generation, but the time consuming of the FEM does not include the cost for generating the mesh. Considering that generating the finite element mesh is usually a time-consuming task [9, 15], therefore, the present second-order WMIGM is more efficient than the FEM for solving the problem with singular point and localized extremely steep gradients.

Table 11.3 The time consuming for the second-order WMIGM and FEM (Note: the consuming time for generating the finite element mesh in FEM does not include)

No	1	2	3	4	5	6	7	8	9	10	Mean
Wavelet	7.313	7.323	7.320	7.335	7.331	7.339	7.316	7.317	7.326	7.320	7.324
FEM	7.115	7.111	7.120	7.126	7.115	7.120	7.120	7.112	7.108	7.127	7.117

11.3.5 Semi-infinite Plate Subjected to a Uniform Local Loading

Consider a semi-infinite plane subjected to a uniform pressure loading $p = 1$ on $[-a, a]$, as shown in Fig. 11.13. To such plane strain problem, the exact displacement and stress of solution are respectively given by [9]

$$\begin{cases} u(x, y) = \frac{p(1-v^2)}{\pi E}\left\{\frac{1-2v}{1-v}[(x+a)\theta_1 - (x-a)\theta_2] + 2y\ln\frac{r_1}{r_2}\right\}, \\ v(x, y) = \frac{p(1-v^2)}{\pi E}\left\{\frac{1-2v}{1-v}\left[y(\theta_1-\theta_2) + 10\arctan\frac{1}{\xi}\right]\right. \\ \left. +2(x-a)\ln r_2 - 2(x+a)\ln r_1 + 2a\ln\left[a^2(1+\xi^2)\right]\right\} \end{cases} \tag{11.69}$$

and

$$\begin{cases} \sigma_{xx}(x, y) = \frac{p}{2\pi}[2(\theta_1-\theta_2) - \sin 2\theta_1 + \sin 2\theta_2], \\ \sigma_{yy}(x, y) = \frac{p}{2\pi}[2(\theta_1-\theta_2) + \sin 2\theta_1 - \sin 2\theta_2], \\ \sigma_{xy}(x, y) = \frac{p}{2\pi}(\cos 2\theta_1 - \cos 2\theta_2) \end{cases} \tag{11.70}$$

in which $\theta_1 = -\arctan[(x+a)/y]$, $\theta_2 = -\arctan[(x-a)/y]$, $r_1 = \sqrt{(x+a)^2+y^2}$, $r_2 = \sqrt{(x-a)^2+y^2}$, and $\xi = 100$.

In the numerical solution simulations, here, we take $E = 100$ and $v = 0.3$, and the domain is taken to be $[0, 1] \times [0, -1]$ due to the symmetry. The displacement $u(0, y) = 0$ is applied on the boundary of symmetry. And the displacements on the bottom and right boundaries are specified by Eq. (11.69). It can be seen from Eq. (11.70) that the point $(a, 0)$ is a jump discontinuity of the stress. Therefore, there is a discontinuity in the first derivative of the displacement filed in the present numerical model with $a \le 1$, and the displacement filed is sufficiently smooth for $a > 1$. Such two cases are all studied here by taking $a = 3/2$ and $a = 1/4$, respectively.

The uniform nodal distribution with grid size h is employed to verify the theoretical error estimate at firstly. Notice that there are two ways to generate a uniform nodal distribution with grid size h in the present WMIGM. One is created using only base

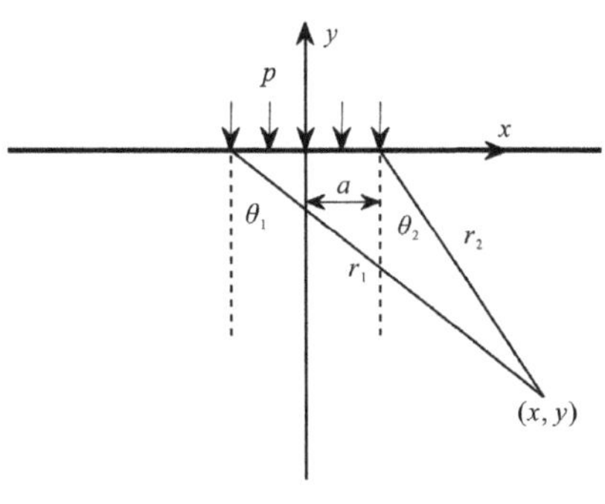

Fig. 11.13 Diagram of semi-infinite plane subjected to a uniform pressure on local edge $[-a, a]$

nodes by taking $h_{\text{base}} = 2^{-j_0} = h$. The other one is assembled by a coarser distribution of base nodes with $h_{\text{base}} > h$ and some other appropriate nodes. Although these uniform nodal distributions with different h_{base} possess exactly the same number of nodes and the same pattern, however, there are two effects on the wavelet methods caused by the different choices of the base size h_{base}. It can be seen from Eq. (11.19) that the distribution of base nodes is also used as the grid for the Lagrange interpolation for obtaining external auxiliary values. Thus, theoretically a larger h_{base} will lead to a decrease in accuracy of the Lagrange interpolation. But, due to the compact support of the wavelet basis function $\theta_{j,k,l}(\mathbf{x})$, the accuracy will only decrease on a very small region near the boundary and can be restored by adding local nodes into this region. On the other hand, the property $a_n = f(\mathbf{x}_n)$ for $n \in \Re^I$ of targeted interpolation shows that the wavelet multiresolution approximation Eq. (11.34) passes exactly through function values at all base nodes. Therefore, a larger h_{base} will relax more shape functions from the Kronecker delta criterion.

The error norms of displacement E_d and energy E_e are shown in Fig. 11.14 as a function of the grid size h of uniform nodal distribution. For this problem with a rectangle domain, both error norms are almost completely independent of the base grid size h_{base}. It can be found from Fig. 11.14 (Left) that when the problem has a sufficiently smooth solution, the convergence rates of displacement and energy for the second-order WMIGM ($\gamma = 2$) are about 2.00 and 1.00 order, respectively, and they are about 4.13 and 3.08 order for the fourth-order ($\gamma = 4$) method. These observations are in excellent agreement with the theoretical error estimates. However, the convergence rates of the fourth-order wavelet method have fell to about 1.97 and 1.01 order for the problems with a jump discontinuity of stresses, as shown in Fig. 11.14 (Right), which are only a little higher than the rates of the second-order method. The error of the former also has increased to about half of the latter when

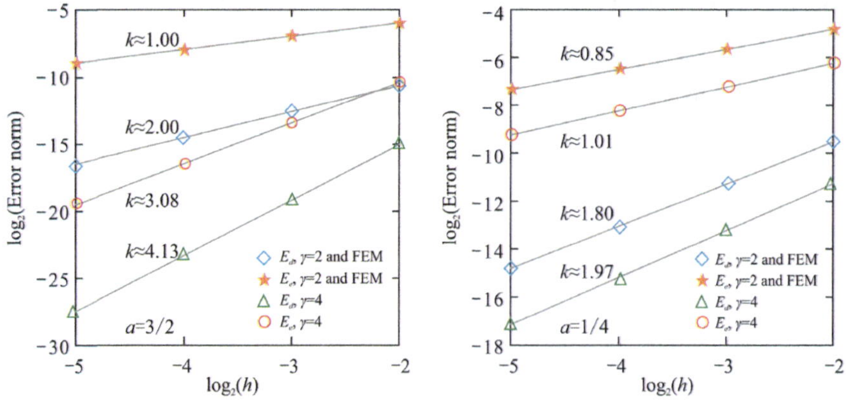

Fig. 11.14 Scaling dependence of the error norms of displacement E_d and energy E_e on the grid size h of uniform nodal distribution for $a = 3/2$ (Left) and $a = 1/4$ (Right) used in the present wavelet solutions of the semi-infinite plane subjected to a uniform pressure on local edge $[-a, a]$

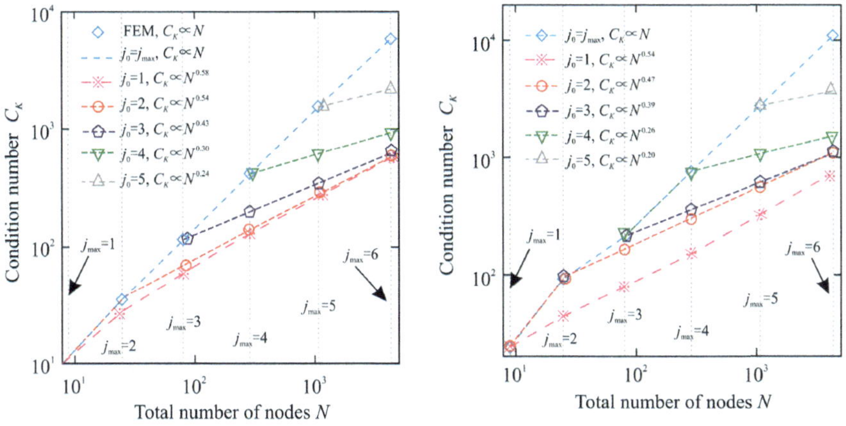

Fig. 11.15 The condition number of stiffness matrix varying with the total number of nodes for the second (Left) and the fourth (Right) order WMIGM by using uniform nodal distributions for solving the semi-infinite plane subjected to a uniform pressure on local edge $[-a, a]$

using the same nodes. In addition, Fig. 11.14 shows that the second-order WMIGM possesses the same accuracy of the FEM using four-node quadrilateral elements.

In Fig. 11.15, the condition number C_K of stiffness matrix for various base grid size $h_{\text{base}} = 2^{-j_0}$ are plotted as a function of the total number N of nodes, from which one can see that when $h_{\text{base}} = h$, the condition number C_K for both the second- and fourth-order WMIGM is proportional to the number of nodes N, as the same as the FEM using four-node quadrilateral element. Especially, the condition number C_K for the second-order wavelet method is almost the same as that for the FEM. In addition, one can also see from Fig. 11.15 that the condition number C_K of stiffness matrix decreases significantly by using a larger h_{base}. More important, the rate of increase of C_K also declines significantly. This might be the result of relaxing more shape functions from the Kronecker delta criterion when a larger h_{base} is used. Such a notable reduction in the condition number C_K of stiffness matrix can lead to a significant decrease of the computational cost for solving the system of linear equations in Eq. (11.45) when using the iterative solver, because there is the relation $n_{iter} \propto \sqrt{C_K}$, where n_{iter} is the number of iterations required to find a converged solution [9]. Thus, this advantage of the proposed WMIGM will be very helpful and attractive in solving large-scale problems.

In the above study, a uniform nodal distribution is employed to solve both the smooth ($a = 3/2$) and non-smooth ($a = 1/4$) problems appeared in the plane strain problem of the semi-infinite plane subjected to a uniform pressure on local edge $[-a, a]$. In fact, for the problem with a jump discontinuity of stress ($a = 1/4$), the error on the region near the point $(a, 0)$ is significantly greater than the error on other regions when using a uniform nodal distribution. To improve the accuracy on the local region, a local enrichment based on the wavelet multiresolution analysis is developed in the proposed wavelet method. As shown in Fig. 11.16, the irregular nodal distributions

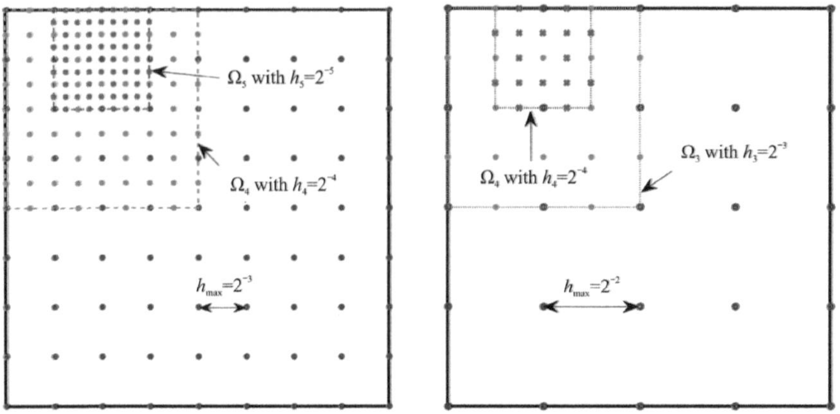

Fig. 11.16 Creation of nodal distributions with local refinement for the second (Left) and the fourth (Right) order WMIGM, respectively, employed in the present wavelet method for solving the semi-infinite plane subjected to a uniform pressure on local edge $[-a, a]$

with local refinement are employed to create a local multiresolution enrichment. For generating the distribution shown in Fig. 11.16 (Left), a set of regular nodes with grid size $h_{\max} = 2^{-3}$ are created on the whole domain Ω first. Then, the set of regular nodes with grid size $h_j = 2^{-j}$ are added into the local region $\Omega_j = [a - 2^{j-2}, a + 2^{j-2}] \times [0, -2^{j-3}]$, respectively, for $j = 4, 5, \ldots, j_{\max}$ with $h_{\min} = 2^{-j_{\max}}$. At the end, merging all nodes is performed to avoid the situation that two or more nodes are at the same point. The nodal distribution shown in Fig. 11.16 (Right) for the fourth-order wavelet method is also generated by using the same process except for $h_{\max} = 2^{-2}$ and $\Omega_j = [a - 2^{j-1}, a + 2^{j-1}] \times [0, -2^{j-2}]$. From them, one can see that in the present WMIGM, creating a nodal distribution with hierarchical refinement is very easy and low cost. By contrast, it is very hard to create a high-quality finite element mesh with hierarchical refinement [9].

Under the refinement nodal distribution shown in Fig. 11.16, the numerical solution results are obtained by the present WMIGM. The relative error of the Von Mises stress at $x = 1/4$ obtained by the second-order WMIGM for various minimum grid size $h_{\min}$ is shown in Fig. 11.17 (Left), which shows that the accuracy near the discontinuous point $(a, 0)$ is improved gradually by enhancing the level of local refinement. In addition, Fig. 11.17 (Left) shows clearly that the present wavelet method has exactly the same accuracy of the full interpolation wavelet method with taking $\Re^I = \Re$. For the computation time, one can see from Fig. 11.17 (Right) that the former possesses a much higher efficiency than the latter, and the condition number of stiffness matrix in the former is far below that of the latter. This may be due to the approximation of displacement field passes exactly through function values only at some specified nodes in the former, but at all nodes in the latter.

The similar observation for the fourth-order wavelet method can be found in Fig. 11.18. By comparing Figs. 11.17 and 11.18, one can find that to achieve the same accuracy for this problem with discontinuity of stress, the fourth-order method

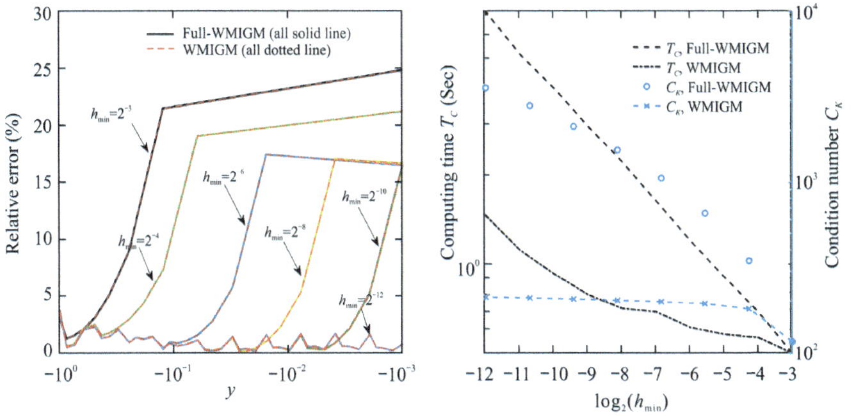

Fig. 11.17 The numerical solution obtained by the second-order WMIGM for solving the semi-infinite plane subjected to a uniform pressure on local edge $[-a, a]$: the relative error of the Von Mises stress at $x = 1/4$ for various minimum grid size $h_{\min}$ (Left); the computation time and condition number of stiffness matrix varying with $h_{\min}$ (Right)

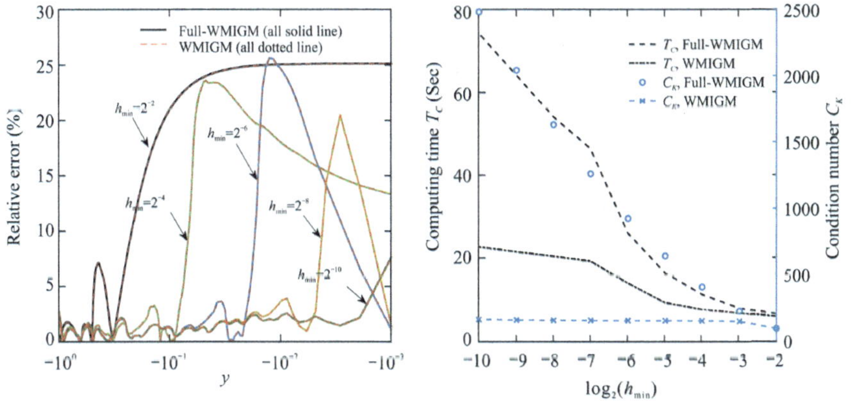

Fig. 11.18 The numerical solution obtained by the fourth-order WMIGM for solving the semi-infinite plane subjected to a uniform pressure on local edge $[-a, a]$: the relative error of the Von Mises stress at $x = 1/4$ for various minimum grid size $h_{\min}$ (Left); the computation time and condition number of stiffness matrix varying with $h_{\min}$ (Right)

has cost more than 10 times computation time of the second one. Therefore, the second-order WMIGM may be more economical than the higher order ones for most problems, because the discontinuity of stress is very common, such as the discontinuous load on the boundary. In the following study, only the second-order WMIGM will be presented by default.

To further study the stability of the proposed WMIGM against irregular nodal distribution, Fig. 11.19 (Left) shows the condition number C_K of stiffness matrix

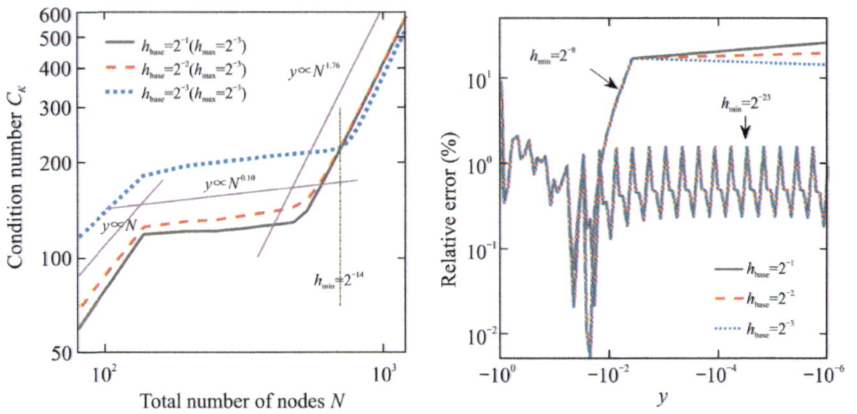

Fig. 11.19 Variation of condition number varying with total node number and relative error distribution of the second-order WMIGM using the nodal distribution shown in Fig. 11.15 (Left): the condition number of stiffness matrix varying with the number of nodes (Left), the relative error distribution of the Von Mises stress at $x = 1/4$ (Right)

as a function of the total number of nodes, which reveals that C_K increases very slowly for irregular nodal distributions with local multiresolution refinement. What is more, the condition number C_K becomes almost a constant in a wide range of the number of nodes. Moreover, one can find from Fig. 11.19 (Left) that the condition number C_K can be effectively decreased by using a larger base grid size h_{base}. As discussed before, such accuracy on a very small region close to the boundary will also decrease when using a larger h_{base}, as shown in Fig. 11.19 (Right) for the relative error distribution of the Von Mises stress at $x = 1/4$. At the same time, Fig. 11.19 (Right) shows clearly that such a losing of local accuracy can be restored by adding local nodes into this region. Thus, the proposed WMIGM can work very well with an extremely irregular nodal distribution in which the ratio of the maximum grid size to the minimum is as high as $h_{\max}/h_{\min} = 2^{-3}/2^{-23} \approx 10^6$. In other words, there is no visible fluctuation in the present numerical solution of Von Mises stress, whose relative error is always less than 2% on almost the whole domain. Such a powerful capacity of local multiresolution enrichment and excellent stability against highly irregular nodal distribution is of great importance in solving problems with singular points or localized extremely steep gradients, such as the crack problems. However, most existing numerical methods are short of a general enrichment technique for capturing local feature of solution, and normally refuse to work with extremely irregular distributions of nodes [9].

11.3.6 Bridge Pier

We consider a bridge pier with unit thickness subjected to a uniform pressure $p = 100\,\text{kN/m}^2$ on the upper surface and its own gravity, as shown in Fig. 11.20. In the numerical solution simulations, the Young modulus $E = 30$ GPa, Poisson ratio $\upsilon = 0.15$, volumetric mass density $\rho_v = 2000\,\text{kg/m}^3$, and gravitational acceleration $g = 9.8\,\text{m/s}^2$ are used. Owing to the symmetry, only the right half of the bridge pier is modeled in the present analysis.

As a comparison, Table 11.4 lists the numerical models gained by the present WMIGM and FEM, respectively, with different nodes or elements. The nodal distributions with 35 and 109 nodes for implementing the wavelet method are presented in Fig. 11.20. And the nodal distribution with 400 (1054) nodes is created by adding other nodes into the distribution with 109 nodes, such that all nodes form a uniform grid with size $h = 1$ ($h = 1/2$). All finite element meshes used have the same topology as shown in Fig. 11.21. In Table 11.4, the reference solution is obtained by the FEM using 140,000 four-node quadrilateral elements. It can be seen from Table 11.4 that when using the same number of nodes, the proposed WMIGM possesses a slightly better accuracy than the EFM using high-quality quadrilateral elements. But the condition number C_K of the present model is far below the finite element one, although the sparsity η_K of stiffness matrix is almost the same in both models.

From the observation of stiffness matrix generated by the present WMIGM in this problem, the band width of the symmetric stiffness matrix is graphically shown in Fig. 11.22, where the band width of FEM is also plotted. From this figure, we see that the stiffness matrix of the wavelet method is a symmetric banded one with a similar sparse pattern of the FEM.

Finally, Fig. 11.23 displays the Von Mises stress obtained by the proposed WMIGM using 400 nodes. From Fig. 11.23, one sees that the present wavelet solution with $h_{\text{base}} = 2$ is in good agreement with the reference one, and there are some slight deviations only near the boundary, which can be removed by taking $h_{\text{base}} = 1$.

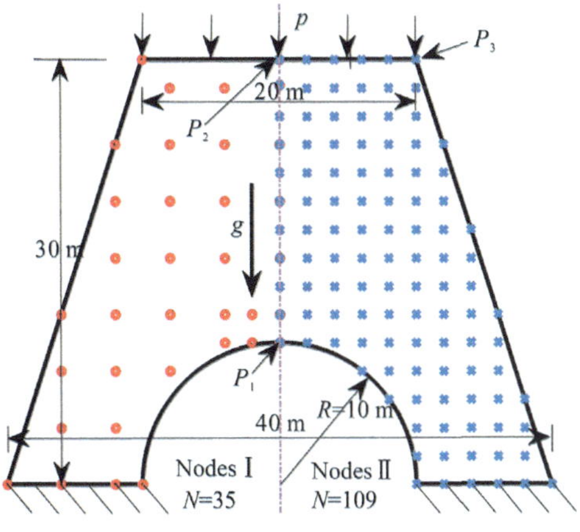

Fig. 11.20 Diagram of two nodal distributions used in the present wavelet method for solving the problem of bridge pier subjected to a uniform pressure on the upper surface and its own gravity

Table 11.4 The comparison of the vertical displacements, total strain energy E_s, the condition number C_K and sparsity η_K of stiffness matrix between the WMIGM and FEM for the problem of bridge pier

Items	$v \times 10^4$ (m) at P_1		$v \times 10^4$ (m) at P_2		$v \times 10^4$ (m) at P_3		E_s (J)		C_K	η_K
	Value	Error	Value	Error	Value	Error	Value	Error		
Reference	5.3453	–	5.8635	–	5.3436	–	1716.9	–	–	–
FEM ($N = 35$)	5.0854	4.86%	5.6750	3.21%	5.3087	0.65%	1663.1	3.13%	385	82.71%
FEM (108)	5.2543	1.70%	5.7973	1.13%	5.3591	0.29%	1698.6	1.07%	1754	93.30%
FEM (400)	5.3229	0.42%	5.8468	0.28%	5.3601	0.31%	1712.7	0.25%	6123	97.98%
FEM (1066)	5.3378	0.14%	5.8580	0.09%	5.3527	0.17%	1715.5	0.08%	14,941	99.20%
WMIGM (35)	5.0680	5.19%	5.6213	4.13%	5.2516	1.72%	1653.4	3.70%	393	79.08%
WMIGM (109)	5.3078	0.70%	5.8335	0.51%	5.3501	0.12%	1709.7	0.42%	917	94.20%
WMIGM (400)	5.3289	0.31%	5.8480	0.26%	5.3437	0.00%	1712.6	0.25%	1245	97.40%
WMIGM (1054)	5.3408	0.08%	5.8589	0.08%	5.3470	0.06%	1715.7	0.07%	1422	99.05%

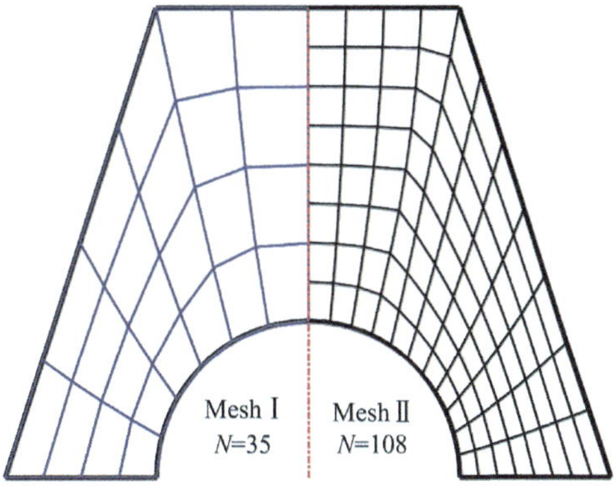

Fig. 11.21 Two meshes used in the finite element analysis for solving the bridge pier subjected to a uniform pressure on the upper surface and its own gravity

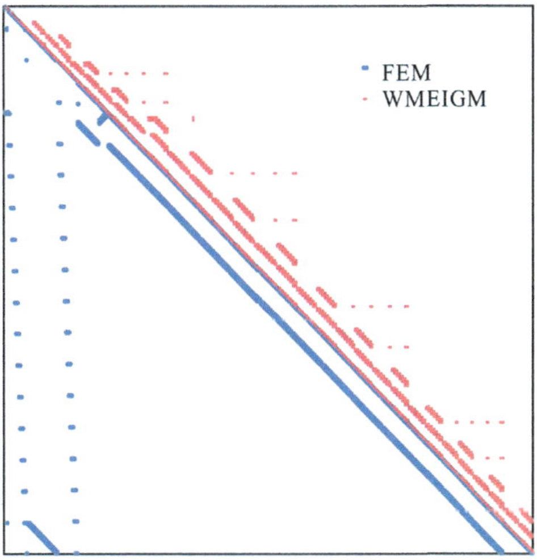

Fig. 11.22 The sparsity pattern of the stiffness matrix in the present wavelet method (red) and FEM (blue) used in solving the bridge pier subjected to a uniform pressure on the upper surface and its own gravity

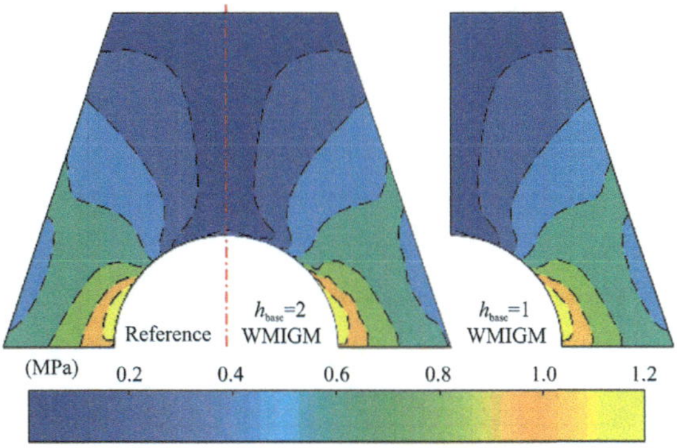

Fig. 11.23 The distributions of the Von Mises stress obtained by the wavelet method using the distribution with 400 nodes and two different base grid sizes h_{base} for solving the bridge pier subjected to a uniform pressure on the upper surface and its own gravity

However, when $h_{\text{base}} = 1$ is taken, the condition number C_K of stiffness matrix will increase to 4493 which is about 3 times of C_K for $h_{\text{base}} = 2$ but still smaller than that in the FEM, as shown in Table 11.4. These results support the preceding analysis that the proposed wavelet method with a larger base grid size h_{base} can effectively reduce the condition number of stiffness matrix, but at the same time will lead to a decrease of accuracy on a very small region close to the boundary.

11.3.7 Corner Brace

Without losing generality, here, we consider the corner brave as shown in Fig. 11.24, where a corner right angle bracket is fixed at two bottom circular holes and subjected to a uniform pressure p on the left half of two top circular holes. In the present numerical solution simulations, we take $p = 100\,\text{kN/m}^2$, the Young modulus $E =$ 200 GPa and Poisson ratio $\upsilon = 0.25$.

Table 11.5 gives a quantitative comparison of the numerical solutions obtained by the WMIGM and FEM. Here, the nodal distribution used in the wavelet method is the combination of a set of regular nodes with grid size h and 48 additional nodes located at two bottom circular holes for imposing the displacement boundary condition, as shown in Fig. 11.24 (Left). In addition, the topology of all finite element meshes used is plotted in Fig. 11.24 (Right). The reference solution in Table 11.5 is obtained by the FEM using 163,200 four-node quadrilateral elements. From Table 11.5, one sees that the accuracy of the proposed WMIGM with $h_{\text{base}} = h$ is better than the FEM using good-quality quadrilateral elements, when the same number of nodes are used in both methods. The error of the former is only about one-third of that

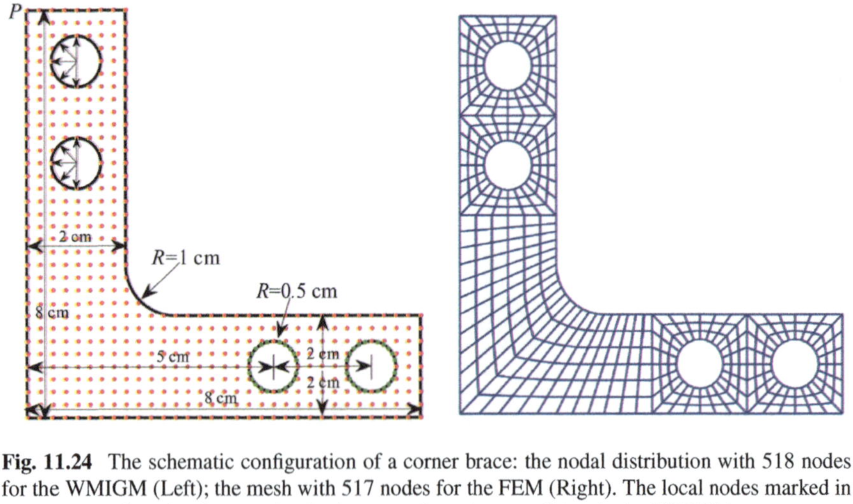

Fig. 11.24 The schematic configuration of a corner brace: the nodal distribution with 518 nodes for the WMIGM (Left); the mesh with 517 nodes for the FEM (Right). The local nodes marked in green are employed to impose the displacement boundary condition in the WMIGM

Table 11.5 The comparison of the displacements at point P, the total strain energy E_s, and the condition number C_K and the sparsity η_K of stiffness matrix between the WMIGM and FEM for the problem of corner brace

Items	$u \times 10^4$ (cm) at P		v $\times 10^4$ (cm) at P		$E_s \times 10^4$ (J)		C_K	η_K
	Value	Error	Value	Error	Value	Error		
Reference	2.5409	–	1.0282	–	1.7235	–	–	–
FEM ($N = 517$)	2.4608	3.15%	1.0054	2.22%	1.6698	3.12%	4.18E + 05	98.57%
FEM (1825)	2.5178	0.91%	1.0217	0.63%	1.7081	0.90%	1.73E + 06	99.56%
FEM (6767)	2.5348	0.24%	1.0265	0.17%	1.7195	0.24%	1.03E + 07	99.87%
WMIGM (518, $h_{\text{base}} = h = 1/4$)	2.5080	1.30%	1.0290	0.07%	1.7032	1.18%	2.85E + 05	98.77%
WMIGM (1789, $h_{\text{base}} = h = 1/8$)	2.5222	0.74%	1.0224	0.57%	1.7113	0.71%	1.09E + 06	99.63%
WMIGM (6735, $h_{\text{base}} = h = 1/16$)	2.5402	0.03%	1.0287	0.05%	1.7234	0.01%	8.84E + 06	99.90%
WMIGM (1789, $h_{\text{base}} = h/2 = 1/4$)	2.5531	0.48%	1.0473	1.86%	1.7341	0.61%	2.94E + 05	99.41%
WMIGM (6735, $h_{\text{base}} = h/4 = 1/4$)	2.5471	0.24%	1.0378	0.93%	1.7290	0.32%	6.20E + 05	99.78%

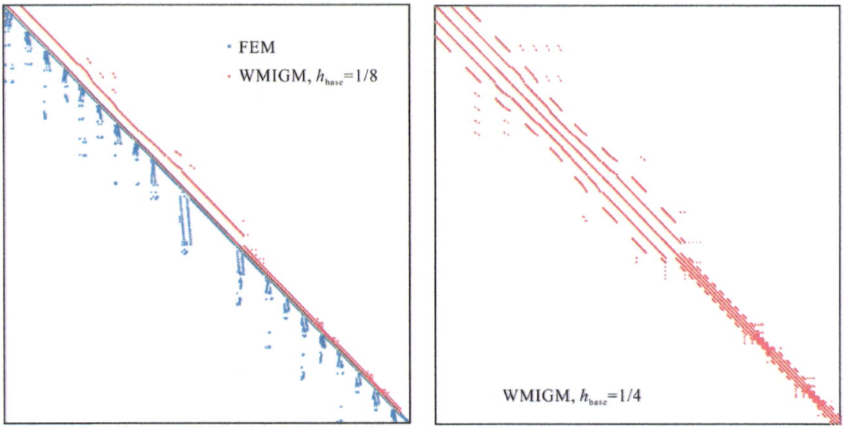

Fig. 11.25 The sparsity pattern of the stiffness matrix in the FEM (blue) and WMIGM with $h_{\text{base}} = 1/8$(red) (Left), as well as the WMIGM with $h_{\text{base}} = 1/4$ (Right)

of the latter. Moreover, one can also see from Table 11.5 and Fig. 11.25 that the stiffness matrix of the proposed wavelet method with $h_{\text{base}} = h$ possesses a smaller condition number, density and bandwidth compared with the FEM. In addition, the results shown in Table 11.5 verify the preceding conclusion again that a larger base grid size h_{base} can significantly reduce the condition number of stiffness matrix but will cause additional errors near the boundary.

Figure 11.26 plots the present numerical results of the Von Mises stress distribution in the corner brace considered. From it, one can see that the proposed wavelet method is of capacity to obtain satisfactory solutions of stress. Especially, the present result using 1789 nodes is in excellent agreement with the reference using solution163,200 elements. What is more, by comparing the results shown in Fig. 11.26b, d, one can also find that the local accuracy of the present WMIGM can be significantly improved by adding more nodes into the corresponding local region needed. The most interesting is that the addition of these local nodes has almost no effect on the condition number of stiffness matrix, as displayed in Fig. 11.26b, d. Moreover, as discussed before, in the proposed WMIGM, the addition of nodes into any local region can be readily implemented without changing existing nodes, which it is not easy to FEM for refining a finite element mesh on a local region and maintaining a good mesh quality simultaneously [9].

In addition, the above results have demonstrated that the proposed wavelet method possesses the capacity for solving the problems with irregular domain, which may be a challenging task for previous wavelet-based methods [3–7].

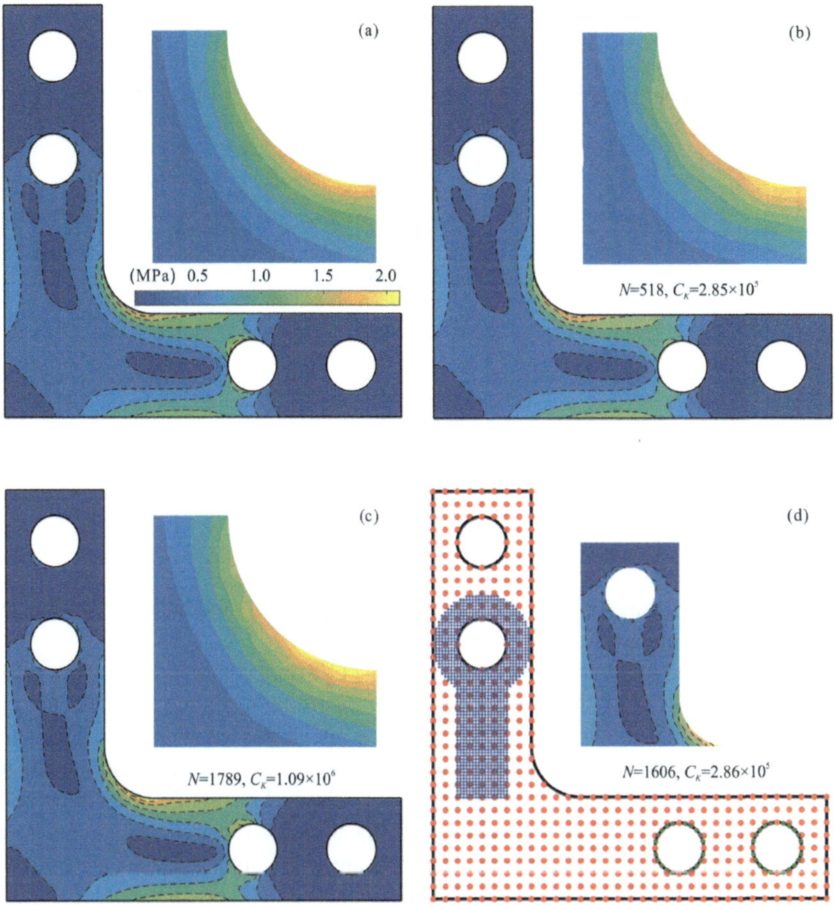

Fig. 11.26 The Von Mises stresses of **a** the reference solution and obtained by the WMIGM, respectively, using uniform distributions with **b** 518 nodes and **c** 1789 nodes, as well as **d** an irregular nodal distribution with local refinement (created based on the nodal distribution with 518 nodes)

11.3.8 Automotive Wheel

As we have known, a typical automotive wheel has a complicated shape as shown in Fig. 11.27, where the wheel is fixed at the center circular hole, and subjected to a uniform pressure p along the local lower arc edge. In the present numerical simulations to this elastic problem with so complex shape, we take $p = 1$ MPa, the Young modulus $E = 200$ GPa and the Poisson ratio $\upsilon = 0.25$ as an example.

Due to the symmetry, only the right half of the wheel is modeled in the present numerical analysis. Although a proper polar coordinate can reduce the complexity of simulation, here, the Cartesian coordinate shown in Fig. 11.27 is still employed

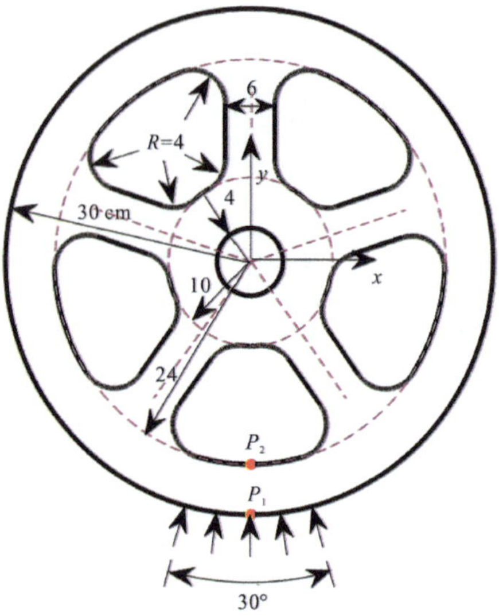

Fig. 11.27 The schematic configuration of an automotive wheel subjected to a local uniform pressure at the lower arc edge

to further demonstrate the capacity of the proposed wavelet method for handling the complicated boundaries.

When the nodal distributions shown in Fig. 11.28 are taken in the present wavelet method, the results of the Von Mises stress distribution obtained by the present WMIGM associated with the FEM reference are displayed in Fig. 11.29. From Fig. 11.29, we find that the present wavelet solution results are in good agreement with the reference solution which is obtained by the FEM using 297,913 three-node triangular elements except for those small local regions near to the points *A*, *B*, and *C*, marked in Fig. 11.29, where the errors are relatively large. This is due to the reason that there is only one node on the horizontal (vertical) grid line passed through

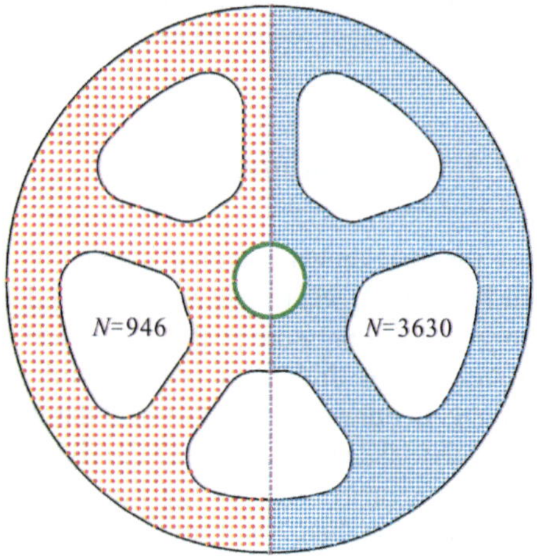

Fig. 11.28 Two nodal distributions for solving the problem of automotive wheel. The local nodes marked in green are employed to impose the displacement boundary condition

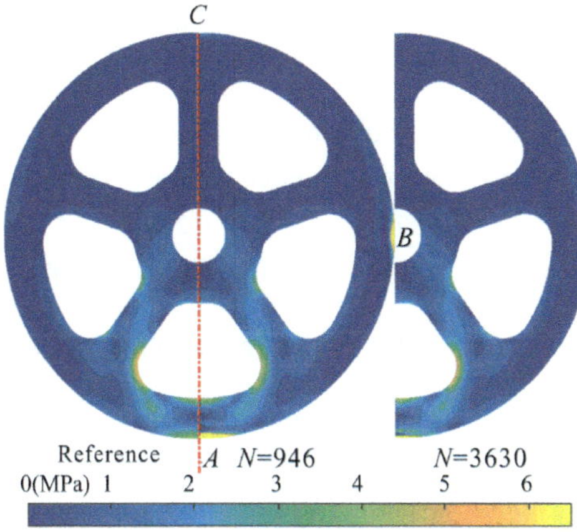

Fig. 11.29 Numerical solution results of the Von Mises stresses obtained by the WMIGM respectively using the nodal distribution shown in Fig. 11.27 comparing with the reference of FEM using 297,913 three-node triangular elements

the point A or C (point B), as shown in Fig. 11.28. In fact, such a problem can be avoided if a polar coordinate is used. After that, the local multiresolution enrichment in the present simulation is employed to restore the accuracy on these local regions. Figure 11.30 clearly shows that the accuracy on local regions is significantly improved by refining the nodal distribution on the corresponding regions. When the number of total nodes is 6366, as shown in Fig. 11.30, the present wavelet numerical solution is in excellent agreement with the reference one shown in Fig. 11.29.

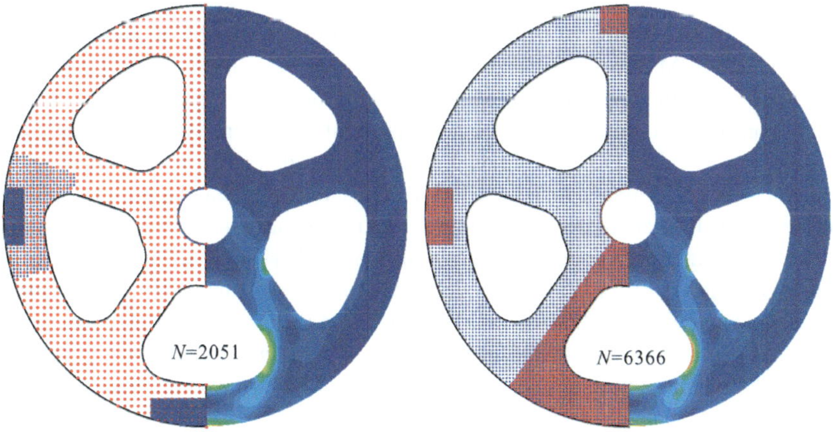

Fig. 11.30 Numerical solutions of the Von Mises stresses obtained by the present WMIGM with multiresolution enhancement through local refinement of nodal distribution

Table 11.6 The comparison of the displacements, total strain energy E_s, and the condition number C_K and sparsity η_K of stiffness matrix between the WMIGM and FEM for the problem of automotive wheel

Items	$v \times 10^4$ (cm) at P_1		$v \times 10^4$ (cm) at P_2		E_s (J)		C_K	η_K
	Value	Error	Value	Error	Value	Error		
Reference	7.0585	–	6.9762	–	0.24483	–	–	–
FEM (N = 950)	6.7090	4.95%	6.6268	5.01%	0.23226	5.14%	1.55E + 04	99.33%
FEM (2103)	6.8993	2.26%	6.8167	2.29%	0.23910	2.34%	3.76E + 04	99.69%
FEM (3756)	6.9698	1.26%	6.8875	1.27%	0.24167	1.29%	7.58E + 04	99.82%
FEM (6383)	7.0065	0.74%	6.9238	0.75%	0.24296	0.77%	1.50E + 05	99.89%
FEM (14,625)	7.0377	0.29%	6.9551	0.30%	0.24407	0.31%	4.37E + 13	99.95%
WMIGM (946)	7.2545	2.78%	7.1694	2.77%	0.25220	3.01%	3.84E + 04	99.21%
WMIGM (2051)	7.1728	1.62%	7.0815	1.51%	0.24875	1.60%	3.85E + 04	99.45%
WMIGM (3630)	7.0141	0.63%	6.9327	0.62%	0.24327	0.64%	1.43E + 05	99.81%
WMIGM (6366)	7.0397	0.27%	6.9566	0.28%	0.24412	0.29%	3.92E + 06	99.85%
WMIGM (14,355)	7.0589	0.01%	6.9767	0.01%	0.24486	0.01%	1.47E + 14	99.95%

A quantitative comparison of the numerical results obtained by the present wavelet method and FEM respectively is given in Table 11.6. Since it is very hard to generate a high-quality finite element mesh using quadrilateral elements for this problem with complicated shape, the three-node triangular element is used in the FEM. According to Table 11.6, one finds that the proposed wavelet method possesses a much better performance than the FEM using triangular elements. For example, the present wavelet method using only 6366 nodes can obtain a little better accuracy than the FEM using 14,625 nodes. And the condition number C_K of stiffness matrix of the latter is as high as about 10^7 times of that of the former. As shown in Fig. 11.31, moreover, the bandwidth of stiffness matrix in the FEM using triangular elements is also much larger than the bandwidth in the proposed wavelet method. Thus, the present WMIGM is more efficient and low cost than the FEM using three-node triangular elements.

In addition, it can be seen from Table 11.6 that the error of the proposed wavelet method decreases gradually with increasing the total number of nodes. When 14,355 nodes are employed, for example, the relative error of the present wavelet solution is as small as 0.01% of the reference solution gained by FEM using about 20 times

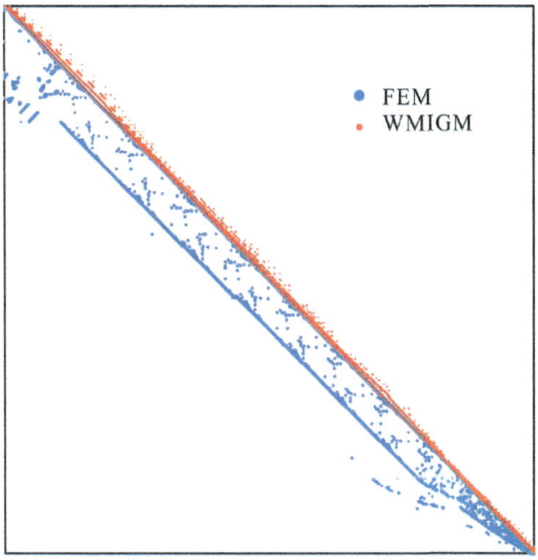

Fig. 11.31 The sparsity pattern of the stiffness matrix in the FEM (blue) and WMIGM (red) used in the numerical solution simulation to the automotive wheel

nodes of the present wavelet solution. These results have verified again that the proposed WMIGM has an excellent convergence and stability, as well as the strong capacity for handling the irregular problem domains with complicated boundaries.

11.3.9 Stress Intensity Factors (SIFs) of Shear Edge Crack

As shown in Fig. 11.32, here, we consider an edge cracked rectangular plate with geometric parameters $H = 16$, $W = 7$, and $a = 3.5$ which is subjected to a uniform

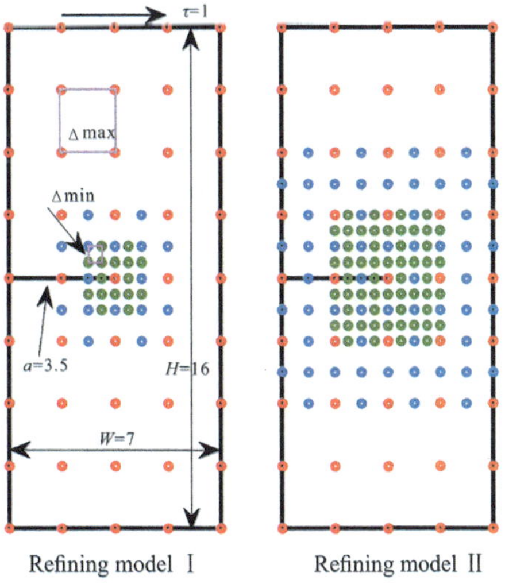

Fig. 11.32 Two nodal distributions generated by nodal refining model I and model II, respectively, for the shear edge crack problem

shear loading $\tau = 1$. The plane strain conditions with $E = 3 \times 10^7$ and $\upsilon = 0.25$ are assumed.

In numerically solving such a problem of fracture mechanics, there are two challenges that deserve special focus. One is how to deal with the strong discontinuity (displacement discontinuity) on the crack face. To handle such a discontinuity, some specialized techniques which will cost additional computational resource is necessary. For example, the mesh must be always conforming to the crack faces in the FEM [9]. In the extended finite element method (XFEM) , a jump function is proposed to modeling the displacement discontinuity [21]. And the so-called visibility criterion is employed in the meshless methods [22]. The other challenge is how to capture the singular stress field near the crack tips. There are two approaches to handle such a singular field. One is creating a high degree of refinement of the grid around the crack tips. However, usually there is obvious numerical oscillation in most methods when such a highly irregular grid is used, leading to a poor accuracy [9, 22]. The other one is incorporating some enrichment basis into the approximation of displacement field [21]. However, the addition of these bases will cause that the numerical method cannot satisfy the partition of unity property in the transition zone connecting the enrichment to the standard approximation [9]. Such a local loss of partition of unity property will lead to a decrease in accuracy, even some potential problems of convergence and stability [9]. In addition, the choose of enrichment basis is based on a priori knowledge about localized solution behavior, which is normally difficult to gain in complicated practical problems.

In the present wavelet-based method, a straightforward technique is developed to handle the displacement discontinuity on the crack face. The detailed instruction on how to implement such a technique can be found in our work [23]. And the multiresolution refinement of nodal distribution around the crack tips is employed to capture the singular stress fields, as shown in Fig. 11.31. In order to efficiently generate a proper local multiresolution refinement of nodal distribution, two nodal refining models are proposed to control the generation of these local nodes, as shown in Fig. 11.33. In the nodal refining model I, all local nodes $\mathbf{x}_{j,k,l} = (k/2^j, l/2^j)$ with resolution level j form a uniform grid with size $\Delta x_j = \Delta y_j = 2^{-j}$ in the square Ω_j whose center is as close to the crack tip as possible and edge length is $d_j^{\mathrm{I}} = 2^{1-j}$, respectively, for $j = j_0 + 1, j_0 + 2, \ldots, j_{\max}$. The only difference between nodal refining model I and model II is that the edge length of the square Ω_j is extended to $d_j^{\mathrm{II}} = 2^{2-j}$ in the latter, as shown in Fig. 11.33. When the crack tip is very close to the boundary of problem domain, some of these additional local nodes will be outside of the problem domain. In this case, these external nodes can be simply thrown away and a higher local resolution level $j_{\max}$ may be employed to restore the accuracy if needed.

Based on the above two nodal refining models, the nodal distributions with local multiresolution refinements can be generated with ease, as shown in Fig. 11.32.

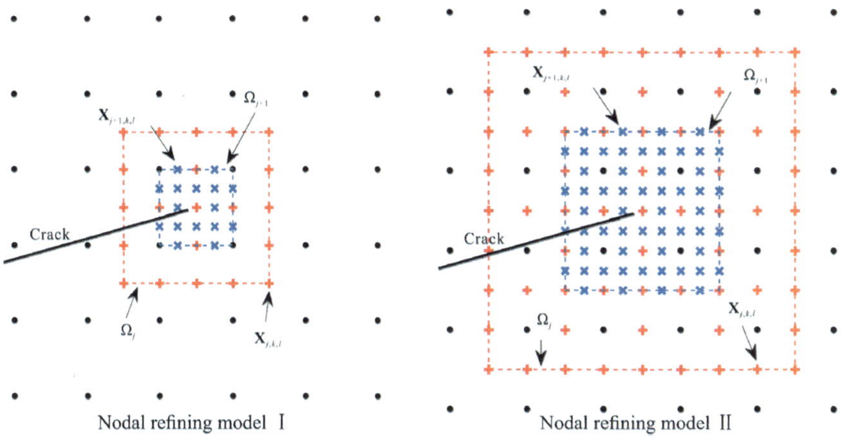

Fig. 11.33 Two nodal refining models of multiresolution enhancement around the crack tip

The relative errors of the SIFs for various grid sizes and J-integral domains are presented in Fig. 11.34, where the grid size $\Delta = \Delta x \times \Delta y$. The reference SIFs are taken to be $K_{\text{I}} = 34.0$ and $K_{\text{II}} = 4.55$ [22]. The calculation method of SIFs based on the J-integral can be found in our work [23]. It can be seen that the errors of SIFs decrease gradually by refining the node distribution, meaning that the present WMIGM has a good convergence. Moreover, one also can find from Fig. 11.34 that the SIFs are almost completely independent of the size of integral domain once it is greater than a critical value for both nodal refining model I and model II. But, such a path-independence for nodal refining model II is much better than that for the model I. Therefore, if the SIFs with a high accuracy (error less than 1%) are required, the nodal refining model II is recommended. Table 11.7 shows a comparison of relative errors of the SIFs obtained, respectively, by various numerical methods, including the proposed WMIGM, the XFEM [21], the standard FEM [24], the cell-based smoothed radial point interpolation method (CS-RPIM) [24], the EFGM [22], the enriched radial point interpolation method (e-RPIM) [25], and the extended radial point interpolation method (XRPIM) [26]. One can see from Table 11.7 that the proposed WMIGM using the same number of nodes can obtain more accurate SIFs compared with many other methods, even some enriched methods with using targeted asymptotic basis functions. Moreover, in the present method the nodal distribution can be automatically and efficiently generated, and all the shape functions are piecewise linear functions, as discussed before. Thus, the proposed WMIGM is more efficient than these methods in estimating the SIFs.

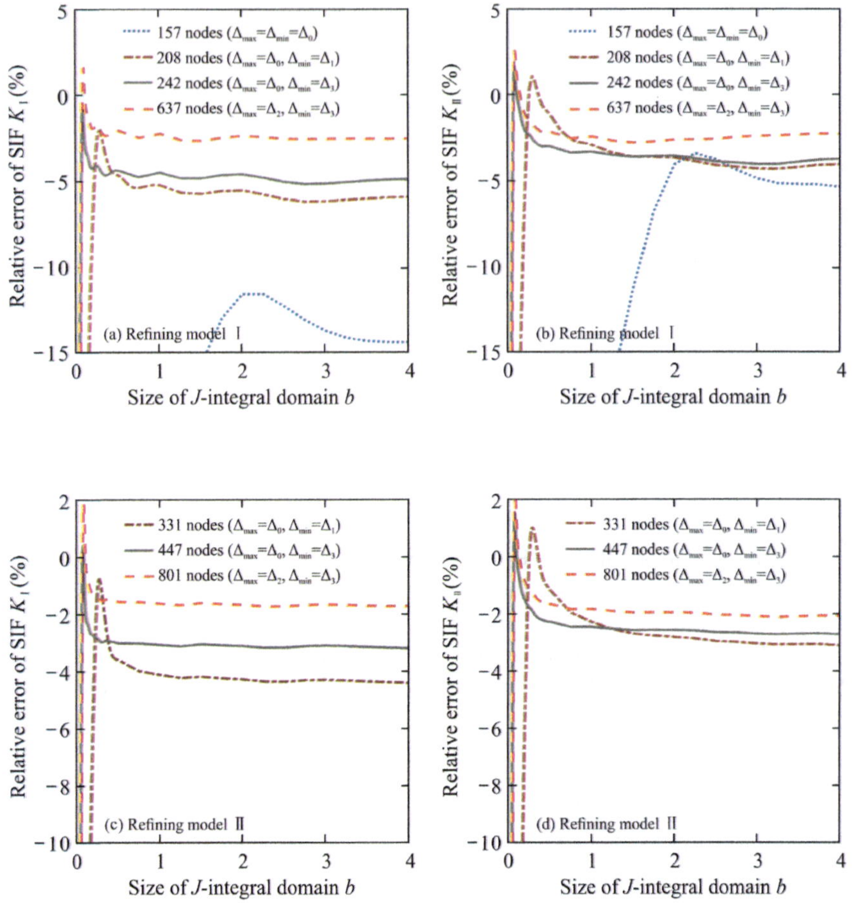

Fig. 11.34 The relative error of SIFs as a function of the size of J-integral domain for the nodal distributions with various grid sizes $\Delta_{\min}$ and $\Delta_{\max}$ generated based on nodal refining (a and b) model I and (c and d) model II. Here $\Delta_0 = 7/8 \times 1$, $\Delta_1 = 7/64 \times 1/4$, $\Delta_1 = 7/16 \times 1/2$ and $\Delta_1 = 7/256 \times 1/16$

Table 11.7 Relative errors of SIFs obtained by various numerical methods

Method	Nodes ($\Delta_{\max}$, $\Delta_{\min}$)	K_I	K_{II}	Method	Nodes	K_I	K_{II}
WMIGM (Refining model I)	208 (Δ_0, Δ_1)	5.96%	4.11%	XFEM [21]	1955	3.00%	1.50%
	242 (Δ_0, Δ_3)	4.93%	3.78%	FEM-T3 [24]	884	11.26%	7.37%
	637 (Δ_2, Δ_3)	2.55%	2.29%	CS-RPIM [24]	884	6.40%	1.85%
WMIGM (Refining model II)	331 (Δ_0, Δ_1)	4.43%	3.11%	EFGM [22]	324	4.38%	4.84%
	447 (Δ_0, Δ_3)	3.21%	2.72%	*e*-RPIM [25]	861	0.14%	8.79%
	801 (Δ_2, Δ_3)	1.74%	2.07%	XRPIM [26]	800	1.68%	3.08%

11.3.10 Crack Propagation in a Rectangular Plate

Here, two cases of crack growth in square plexiglass plates of edge length $W = H = 10$ mm are experimentally studied by Erdogan and Sih [27]. One is that the plate with an initial inclined crack of length $2a = 2$ mm and angle $\alpha = 40°$ is subjected to a uniform tensile loading σ, and the other is that the plate with an initial horizontal crack of length $2a = 2$ mm is subjected to a uniform shear loading τ, as shown in Fig. 11.35 (the solid black line represents the initial crack). The material and loading parameters are taken to be $E = 3300$ MPa, $\upsilon = 0.33$, $\sigma = 1$ MPa and $\tau = 1$ MPa, here.

In the present analysis of crack growth, only the propagation path is traced step by step, without considering the fracture criterion for checking whether the crack grows. In each step, as shown in Fig. 11.36, the increment of the crack length Δa is chosen to be a small value compared with the total crack length. And the direction of crack propagation measured based on the current crack line $\Delta\alpha$ is determined by the maximum circumferential stress criterion [27], which can be expressed as

$$\Delta\alpha = 2\mathrm{Arctan}\left(\frac{1 \pm \sqrt{1 + 8(K_{\mathrm{II}}/K_{\mathrm{I}})^2}}{4(K_{\mathrm{II}}/K_{\mathrm{I}})}\right). \tag{11.71}$$

As the crack grows, the local multiresolution refinement of nodal distribution around the crack tips for capturing the singular stress fields should be updated following the movement of the tips, but without changing other nodes. Since generating such a multiresolution enrichment is independent of other nodes, these local

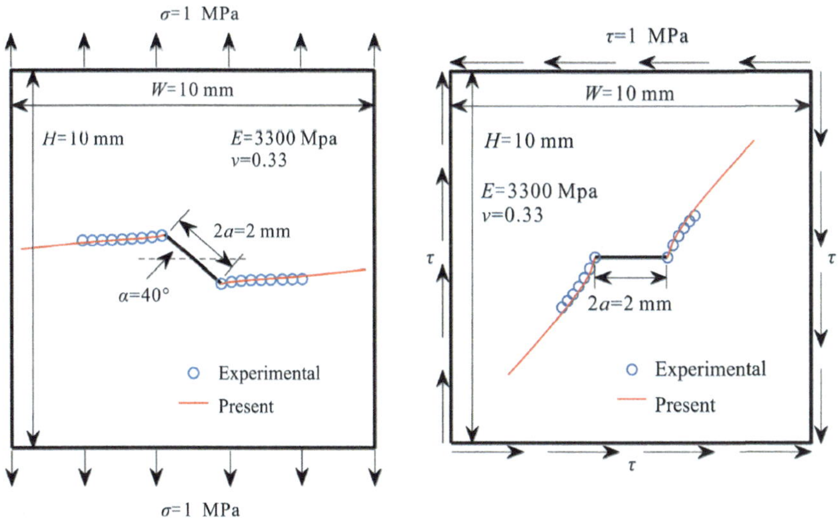

Fig. 11.35 Crack propagation in a rectangular plate subjected to uniform tensile loading (Left) and uniform shear loading (Right), respectively

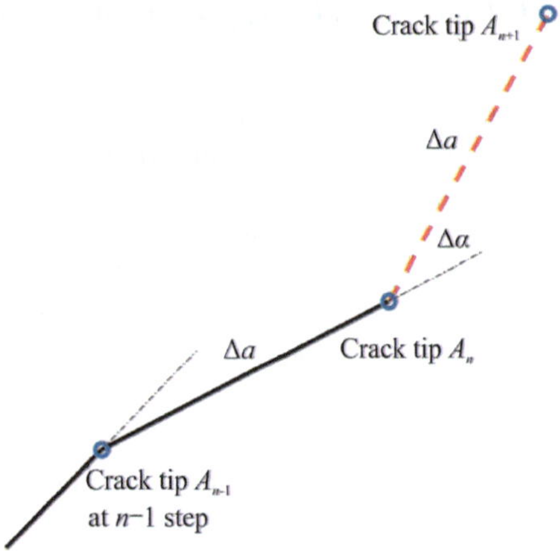

Fig. 11.36 Schematic configuration for tracing crack extension from the step n to the step $n+1$

nodes can be automatically and efficiently removed and regenerated on other regions according to the proposed nodal refining models.

For simulating this crack propagation by using the present WMIGM, a uniform nodal distribution with grid size $h_{\max} = 1/2$ is arranged over whole plate and remains unchanged. Two sets of local nodes with the minimum grid size $h_{\min} = 1/256$ generated by the nodal refining model II shown in Fig. 11.33 are added around two crack tips, respectively. Such sets of local nodes which create multiresolution enhancements around the crack tips should be updated following the movement of tips in each step. The increment of the crack length and the size of J-integral domain are always chosen to be $\Delta a = 0.2$ mm and $b = 0.3$ mm, respectively. A total number of 20 steps are performed in each numerical simulation.

The paths of crack propagation predicted by the present WMIGM are plotted in Fig. 11.35 with the experimental measurements. The excellent agreement shown in Fig. 11.35 illustrates that the proposed WMIGM has an excellent accuracy and stability, and hence is of capacity for tracing the crack propagation.

11.4 Summarized Remarks

A novel wavelet multiresolution approximation in which arbitrary specified nodal coefficients can be exactly sampled out by the function values at the corresponding nodes and the relevant wavelet expansion is established to represent continuous functions defined on two-dimensional finite domain. Using such a wavelet multiresolution approximation to create trial and weighted functions, a wavelet multiresolution interpolation Galerkin method (WMIGM) is developed to solve plane elastic problems. The numerical test results demonstrate that such a wavelet method has a strong capacity for handling irregular domains with complicated shape, especially for a large gradient of those unknown quantities in the local subregion with singularity. Such excellent advantages in the proposed WMIGM can be attributed to the followings:

1. According to a loose and explicit criterion, the nodal distribution required in the present wavelet method can be automatically and efficiently generated on irregular domains with complicated shape. Also, the admissible distribution of nodes possesses a great flexibility to solve various practical engineering problems, in which an extremely large ratio of the maximum grid size $h_{\max}$ to the minimum $h_{\min}$ is permissible. For example, the ratio $h_{\max}/h_{\min} \approx 10^6$ has passed the test in the numerical example. Based on such a flexible nodal distribution, a robust multiresolution enrichment is feasible in our wavelet method for improving local accuracy or for capturing localized steep gradients, even the singular stress field in fracture mechanics.
2. In the present wavelet method, the shape function is constructed explicitly without any matrix inversion or ad hoc parameters, leading to an excellent stability and usefulness.
3. The stiffness matrix can be efficiently obtained through semi-analytic method in the present wavelet method.
4. The present wavelet approximation of field functions, arbitrary specified nodal coefficients can be exactly sampled out by their function values at the corresponding nodes. And at the same time a node can be created at any point. Thus, in the proposed wavelet method the essential boundary condition can be imposed with ease as in the FEM.
5. The computation cost of the second-order wavelet method for achieving the same accuracy can be approximate to the cost (excluding the mesh generation) of the FEM using good-quality four-node quadrilateral elements, and is significantly less than that of the FEM using three-node triangular elements.
6. In the proposed wavelet method, the condition number of stiffness matrix can be reduced significantly by relaxing more shape functions from the Kronecker delta criterion. This property may be very helpful and attractive in solving large-scale problems.

The above advantages lead to a satisfactory accuracy, rate of convergence, stability, confidence level and efficiency for the proposed WMIGM, which have been verified by the numerical results.

Since all the techniques used in the proposed WMIGM, such as the Lagrange interpolation for suppressing numerical instability near the boundary and the multiresolution analysis for constructing shape functions, can be directly extended to three-dimensional space without difficulty, hence, the present wavelet method also can be extended to solve three-dimensional problems. Moreover, some attractive attributes of the present wavelet method in two dimensions, such as the easy implementation of essential boundary condition and robust multiresolution enrichment, also can be expectantly ensured when they are employed in the three-dimensional problems.

Therefore, the present WMIGM, as a truly meshfree method, offers a tremendous potential in numerically solving problems of engineering applications, including the problems of fracture mechanics.

Appendix 11.1 Proof of Essential Properties of Interpolating Wavelet

In order to prove the relation of Eq. (11.8), we first investigate the nth moment $M_n^{\vartheta} = \int x^n \vartheta(x)dx$ of $\vartheta(x)$. Using the definition of Eq. (11.5) and the binomial theorem, we have

$$M_n^{\vartheta} = \int x^n \vartheta(x)dx = \int x^n \int \psi(y)\psi(y-x)dydx = \int \psi(y)[\int (y-x)^n \psi(x)dx]dy \\ = \int \psi(y)[\int \sum_{i=0}^{n} C_n^i y^{n-i}(-x)^i \psi(x)dx]dy = \sum_{i=0}^{n} C_n^i (-1)^i M_{n-i}^{\psi} M_i^{\psi} \quad . \tag{11.72}$$

Then following Eq. (11.3), one can obtain

$$M_n^{\vartheta} = 0, \text{ for } n = 0, 1, \ldots, \gamma - 1. \tag{11.73}$$

From Eqs. (11.72) and (11.73), one can see that the auto-correlation function $\vartheta(x)$ of wavelet function has γ, and only has γ vanishing moments, because $M_\gamma^{\vartheta} = C_\gamma^{\gamma/2}(-1)^{\gamma/2}(M_{\gamma/2}^{\psi})^2 \neq 0$ since $M_{\gamma/2}^{\psi} \neq 0$ [12].

On the other hand, there is the relation $\vartheta(x) = 2\theta(2x) - \theta(x)$ [8, 14]. So, we have

$$M_n^{\vartheta} = \int x^n[2\theta(2x) - \theta(x)]dx = 2^{-n}\int (2x)^n \theta(2x)d(2x) - \int x^n \theta(x)dx = (2^{-n} - 1)M_n^{\theta}. \tag{11.74}$$

Considering Eq. (11.73), we obtain

$$M_n^{\theta} = 0, \text{ for } n = 1, 2, \ldots, \gamma - 1. \tag{11.75}$$

We note that $M_\gamma^{\theta} \neq 0$ since $M_\gamma^{\vartheta} \neq 0$.

When $m = 0$, by using the definition of Eq. (11.5) and the property $M_0^{\phi} = 1$, one can directly obtain

$$M_0^{\theta} = \int \theta(x)dx = \int\int \phi(y)\phi(y-x)dydx = \int \phi(y)\left[\int \phi(y-x)dx\right]dy = 1. \tag{11.76}$$

Thus, the relation of Eq. (11.8) is proved.

For the relation of Eq. (11.9), considering the property $p_{2k} = \delta_{0,k}$ shown in Eq. (11.7), one can directly obtain

$$\sum_{k\in\mathbb{Z}} p_{2k}(2k)^n = \delta_{0,n}. \tag{11.77}$$

On the other hand, by using the two-scale relation of Eq. (11.6) and the binomial theorem, we have

$$\begin{aligned}
M_n^\theta &= \int x^n[\theta(2x) + \sum_{k\in\mathbb{Z}} p_{2k+1}\theta(2x-2k-1)]dx \\
&= 2^{-n-1}\int (2x)^n\theta(2x)d(2x) + \frac{1}{2}\sum_{k\in\mathbb{Z}} p_{2k+1}\int (\frac{x+(2k+1)}{2})^n\theta(x)dx \\
&= 2^{-n-1}M_n^\theta + 2^{-n-1}\sum_{k\in\mathbb{Z}} p_{2k+1}\int\sum_{i=0}^{n} C_n^i(2k+1)^i x^{n-i}\theta(x)dx \\
&= 2^{-n-1}M_n^\theta + 2^{-n-1}\sum_{k\in\mathbb{Z}} p_{2k+1}\sum_{i=0}^{n} C_n^i(2k+1)^i M_{n-i}^\theta.
\end{aligned} \tag{11.78}$$

Considering the relation of Eq. (11.8), for $n = 0, 1, \ldots, \gamma - 1$, Eq. (11.78) can be reduced into

$$M_n^\theta = 2^{-n-1}\delta_{0,n} + 2^{-n-1}\sum_{k\in Z} p_{2k+1}(2k+1)^n = \delta_{0,n}. \tag{11.79}$$

According to Eqs. (11.77) and (11.79), the relation of Eq. (11.9) has been gained. For the relation of Eq. (11.10), applying the binomial theorem, we get

$$\begin{aligned}
g_{n,m}(x) &= \sum_{k\in\mathbb{Z}} (x-k)^n\theta^{(m)}(x-k) \\
&= 2^m\sum_{k\in\mathbb{Z}} (x-k)^n\sum_{l\in\mathbb{Z}} p_l\theta^{(m)}(2x-2k-l) \\
&= 2^m\sum_{k\in\mathbb{Z}}\sum_{i\in\mathbb{Z}} p_{i-2k}(x-k)^n\theta^{(m)}(2x-i) \\
&= 2^{m-n}\sum_{k\in\mathbb{Z}}\sum_{i\in\mathbb{Z}} p_{i-2k}[(2x-i)+(i-2k)]^n\theta^{(m)}(2x-i) \\
&= 2^{m-n}\sum_{i\in\mathbb{Z}}\sum_{o=0}^{n} C_n^o(2x-i)^o[\sum_{k\in\mathbb{Z}} p_{i-2k}(i-2k)^{n-o}]\theta^{(m)}(2x-i).
\end{aligned} \tag{11.80}$$

Here, we have used the relation $\theta^{(m)}(x) = 2^m\sum_{k\in\mathbb{Z}} p_k\theta^{(m)}(2x-k)$ which can be directly obtained from Eq. (11.6). Considering the relation in Eq. (11.9), then, Eq. (11.80) can be reduced into

$$g_{n,m}(x) = 2^{m-n}\sum_{i\in\mathbb{Z}} (2x-i)^n\theta^{(m)}(2x-i) = 2^{m-n}g_{n,m}(2x). \tag{11.81}$$

Defining $G_{n,m}(x) = \int_0^x g_{n,m}(y)dy$ and following Eq. (11.81), we have

$$G_{n,m}(x) = 2^{m-n-1} G_{n,m}(2x). \tag{11.82}$$

On the other hand, further, there is the relation

$$\begin{aligned} G_{n,m}(1) &= \int_0^1 \sum_{k\in\mathbb{Z}} (x-k)^n \theta^{(m)}(x-k)dx = \sum_{k\in\mathbb{Z}} \int_{-k}^{-k+1} x^n \theta^{(m)}(x)dx \\ &= \int x^n \theta^{(m)}(x)dx. \end{aligned} \tag{11.83}$$

For $n < m$, conducting n times integrations by parts to Eq. (11.83), we have

$$\begin{aligned} G_{n,m}(1) &= \int x^n \theta^{(m)}(x)dx = x^n \theta^{(m-1)}(x)\big|_{-\infty}^{+\infty} - n \int x^{n-1} \theta^{(m-1)}(x)dx \\ &= (-1)^n n! \int \theta^{(m-n)}(x)dx = (-1)^n n! \theta^{(m-n-1)}(x)\big|_{-\infty}^{+\infty} = 0, \end{aligned} \tag{11.84}$$

where the property $\theta(x) \in C_0^{\gamma/2-1}[1-\gamma, \gamma-1]$ is used. Similarly, for $n = m$, one can obtain

$$G_{n,m}(1) = (-1)^n n! \int \theta(x)dx = (-1)^n n! M_0^\theta = (-1)^n n!. \tag{11.85}$$

And for $n > m$, doing m times integrations by parts to Eq. (11.83), we have

$$\begin{aligned} G_{n,m}(1) &= \int x^n \theta^{(m)}(x)dx = x^n \theta^{(m-1)}(x)\big|_{-\infty}^{+\infty} - n \int x^{n-1} \theta^{(m-1)}(x)dx \\ &= (-1)^m n! \int x^{n-m} \theta(x)dx/m! = (-1)^n n! M_{n-m}^\theta / m! = 0 \end{aligned} \tag{11.86}$$

for $m = 0, 1, \ldots, \gamma/2 - 1$ and $n = m+1, m+2, \ldots, \gamma - 1$. Eqs. (11.84)–(11.86) give the relation

$$G_{n,m}(1) = (-1)^n \delta_{n,m} n!. \tag{11.87}$$

By performing the same analysis shown in Eqs. (11.83)–(11.87), one can obtain $G_{n,m}(-1) = (-1)^{n+1} \delta_{n,m} n!$ for $n = 0, 1, \ldots, \gamma - 1$ and $m = 0, 1, \ldots, \gamma/2 - 1$. From them, one can see that there is the relation $G_{n,m}(x) \equiv 0$ for $n \neq m$ since $G_{n,m}(1) = G_{n,m}(-1) = 0$ and $G_{n,m}(x) = 2^{m-n-1} G_{n,m}(2x)$. Thus, we have

$$g_{n,m}(x) = \sum_{k\in\mathbb{Z}} (x-k)^n \theta^{(m)}(x-k) = 0 \tag{11.88}$$

for $n = 0, 1, \ldots, \gamma - 1$, $m = 0, 1, \ldots, \gamma/2 - 1$ and $n \neq m$.

When $n = m$, Eq. (11.82) gives $2G_{n,n}(x) = G_{n,n}(2x)$, i.e., $2\int_0^x g_{n,n}(y)dy = \int_0^{2x} g_{n,n}(y)dy$, which means the relation $g_{n,n}(x) = g_{n,n}(2x)$ for all x. Therefore, we have $g_{n,n}(x) = c$ which is a constant, since $g_{n,n}(x)$ is a continuous function. Finally, considering Eq. (11.87), one can obtain

$$g_{n,n}(x) \equiv g_{n,n}(1) = G_{n,n}(1) = (-1)^n n!. \tag{11.89}$$

According to Eqs. (11.88) and (11.89), the relation of Eq. (11.10) can be obtained.

Appendix 11.2 Multiresolution Decomposition of Interpolating Wavelet

Substituting the two-scale relation of Eq. (11.6) into the approximation Eq. (11.11) and using Eq. (11.7), we get

$$\begin{aligned}
S^j f(\mathbf{x}) = &\sum_{k\in\mathbb{Z}}\sum_{l\in\mathbb{Z}} f(\mathbf{x}_{j,k,l})[\theta(2^{j+1}x-2k) + \sum_{n\in Odd} p_n\theta(2^{j+1}x-2k-n)][\theta(2^{j+1}x-2l) \\
&+ \sum_{m\in Odd} p_m\theta(2^{j+1}y-2l-m)] \\
= &\sum_{k\in Even}\sum_{l\in Even} f(\mathbf{x}_{j+1,k,l})\theta_{j+1,k,l}(\mathbf{x}) + \sum_{k\in Even}\sum_{l\in Odd}[\sum_{m\in Odd} p_m f(\mathbf{x}_{j+1,k,l-m})]\theta_{j+1,k,l}(\mathbf{x}) \\
&+ \sum_{k\in Odd}\sum_{l\in Even}[\sum_{n\in Odd} p_n f(\mathbf{x}_{j+1,k-n,l})]\theta_{j+1,k,l}(\mathbf{x}) \\
&+ \sum_{k\in Odd}\sum_{l\in Odd}[\sum_{n\in Odd}\sum_{m\in Odd} p_n p_m f(\mathbf{x}_{j+1,k-n,l-m})]\theta_{j+1,k,l}(\mathbf{x})
\end{aligned} \tag{11.90}$$

On the other hand, for $k \in Even$ and $l \in Odd$, one can obtain

$$\begin{aligned}
S^j f(\mathbf{x}_{j+1,k,l}) &= \sum_{n\in\mathbb{Z}}\sum_{m\in\mathbb{Z}} f(\mathbf{x}_{j,n,m})\theta(k/2-n)\theta(l/2-m) \\
&= \sum_{m\in\mathbb{Z}} f(\mathbf{x}_{j+1,j,k,m}) \sum_{o\in Odd} p_o\theta(l-2m-o) = \sum_{o\in Odd} p_o f(\mathbf{x}_{j+1,k,l-o}),
\end{aligned} \tag{11.91}$$

where Eqs. (11.6) and (11.7) and the property $\theta(k) = \delta_{0,k}$ for $k \in \mathbb{Z}$ are considered. By using the same method, we can also obtain

$$\sum_{o\in Odd} p_o f(\mathbf{x}_{j+1,k-o,l}) = S^j f(\mathbf{x}_{j+1,k,l}) \text{ for } k \in Odd \text{ and } l \in Even, \tag{11.92}$$

$$\sum_{o\in Odd}\sum_{q\in Odd} p_o p_q f(\mathbf{x}_{j+1,k-o,l-q}) = S^j f(\mathbf{x}_{j+1,k,l}) \text{ for } k, l \in Odd. \tag{11.93}$$

Substitution of Eqs. (11.91)–(11.93) into (11.90) yields

$$S^j f(\mathbf{x}) = \sum_{k\in Even}\sum_{l\in Even} f(\mathbf{x}_{j+1,k,l})\theta_{j+1,k,l}(\mathbf{x}) + \sum_{(k,l)\in\Re} S^j f(\mathbf{x}_{j+1,k,l})\theta_{j+1,k,l}(\mathbf{x}) \tag{11.94}$$

in which $\Re$ is the set of all integer pair (k, l) with at least one element is an odd number.

Finally, Eq. (11.13) for $R^j f(\mathbf{x})$ is obtained by subtracting $S^j f(\mathbf{x})$ in Eq. (11.94) from $S^{j+1} f(\mathbf{x})$ defined by Eq. (11.11). Hence, the multiresolution decomposition of Eq. (11.12) is proved.

Appendix 11.3 Error Estimation of the Interpolating Wavelet Approximation Defined on the Whole Space

Here, we first examine the relation of Eq. (11.15). For any point $\mathbf{x}$, one can always find a point $\mathbf{x}_{j,k,l}$ satisfying $\left|\mathbf{x} - \mathbf{x}_{j,k,l}\right| \le 2^{-j}/\sqrt{2}$. Then because both of $f(\mathbf{x})$ and $S^j f(\mathbf{x})$ are continuous functions, there exist two constants ε_0 and ε_1 satisfying

$$\left\| f(\mathbf{x}) - f(\mathbf{x}_{j,k,l})\right\|_\infty \le \varepsilon_0\left|\mathbf{x} - \mathbf{x}_{j,k,l}\right| \le \varepsilon_0 2^{-j}/\sqrt{2}, \tag{11.95}$$

$$\left\| S^j f(\mathbf{x}) - S^j f(\mathbf{x}_{j,k,l})\right\|_\infty \le \varepsilon_1\left|\mathbf{x} - \mathbf{x}_{j,k,l}\right| \le \varepsilon_1 2^{-j}/\sqrt{2}. \tag{11.96}$$

Then, one can obtain

$$\begin{aligned}\left\| f(\mathbf{x}) - S^j f(\mathbf{x})\right\|_\infty &= \left\|[f(\mathbf{x}) - f(\mathbf{x}_{j,k,l})] + [S^j f(\mathbf{x}_{j,k,l}) - S^j f(\mathbf{x})]\right\|_\infty \\ &\le \left\| f(\mathbf{x}) - f(\mathbf{x}_{j,k,l})\right\|_\infty + \left\| S^j f(\mathbf{x}_{j,k,l}) - S^j f(\mathbf{x})\right\|_\infty \le C_{0,\infty}2^j,\end{aligned} \tag{11.97}$$

which gives the relation of Eq. (11.15).

For the relation of Eq. (11.16), by using the Taylor expansion for $f(\mathbf{x})$ at the point $\mathbf{x}$, we have

$$f(\tilde{\mathbf{x}}) = \sum_{n=0}^{\lambda-1}\frac{1}{n!}[(\tilde{x} - x)\frac{\partial}{\partial x} + (\tilde{y} - y)\frac{\partial}{\partial y}]^n f(\mathbf{x}) + \frac{1}{\lambda!}[(\tilde{x} - x)\frac{\partial}{\partial x} + (\tilde{y} - y)\frac{\partial}{\partial y}]^\lambda f(\mathbf{x}_\theta), \tag{11.98}$$

where $\mathbf{x}_\theta$ is on the rectangle with two diagonal apexes $\tilde{\mathbf{x}}$ and $\mathbf{x}$, and

$$[(\tilde{x} - x)\frac{\partial}{\partial x} + (\tilde{y} - y)\frac{\partial}{\partial y}]^n f(\mathbf{x}) = \sum_{m=0}^{n} C_n^m(\tilde{x} - x)^m(\tilde{y} - y)^{n-m} D_n^m f(\mathbf{x}) \tag{11.99}$$

with $D_n^m f(\mathbf{x}) = \left.\frac{\partial^n f}{\partial x^m \partial y^{n-m}}\right|_{\mathbf{x}}$. Assigning $\tilde{\mathbf{x}} = \mathbf{x}_{j,k,l}$ to Eq. (11.98) and then substituting them into Eq. (11.11), one gains

$$
\begin{aligned}
S^j f(\mathbf{x}) &= \sum_{k\in\mathbb{Z}} \sum_{\in\mathbb{Z}} \sum_{n=0}^{\lambda-1} \frac{1}{n!} \sum_{m=0}^{n} C_n^m (\frac{k}{2j} - x)^m (\frac{l}{2j} - y)^{n-m} D_n^m f(\mathbf{x})\theta_{j,k,l}(\mathbf{x}) \\
&+ \sum_{k\in\mathbb{Z}} \sum_{l\in\mathbb{Z}} \frac{1}{\lambda!} \sum_{m=0}^{\lambda} C_n^m (\frac{k}{2j} - x)^m (\frac{l}{2j} - y)^{\lambda-m} D_\lambda^m f(\mathbf{x}_{\theta,k,l}) \\
&= \sum_{n=0}^{\lambda-1} \frac{2^{-jn}}{n!} \sum_{m=0}^{n} [C_n^m \sum_{k\in\mathbb{Z}} (k - 2^j x)^m \theta(2^j x - k) \sum_{l\in\mathbb{Z}} (l - 2^j y)^{n-m} \theta(2^j y - l) D_n^m f(\mathbf{x})] \\
&+ \frac{2^{-j\lambda}}{\lambda!} \sum_{m=0}^{\lambda} [C_\lambda^m \sum_{k\in\mathbb{Z}} \sum_{l\in\mathbb{Z}} (k - 2^j x)^m (l - 2^j y)^{\lambda-m} \theta_{j,k,l}(\mathbf{x}) D_\lambda^m f(\mathbf{x}_{\theta,k,l})] \quad (11.100)
\end{aligned}
$$

in which $\mathbf{x}_{\theta,k,l}$ is on the rectangle with two diagonal apexes $\mathbf{x}_{j,k,l}$ and $\mathbf{x}$. Then, following the property of Eq. (11.10), we have

$$
\sum_{k\in\mathbb{Z}} (k - 2^j x)^q \theta(2^j x - k) = \delta_{0,q} \text{ for } q = 0, 1, \ldots, \gamma - 1. \quad (11.101)
$$

Substituting Eqs. (11.101) into (11.100), and considering $\lambda \le \gamma$, we have

$$
S^j f(\mathbf{x}) = f(\mathbf{x}) + \frac{2^{-j\lambda}}{\lambda!} \sum_{m=0}^{\lambda} [C_\lambda^m \sum_{k\in\mathbb{Z}} \sum_{l\in\mathbb{Z}} (k - 2^j x)^m (l - 2^j y)^{\lambda-m} \theta_{j,k,l}(\mathbf{x}) D_\lambda^m f(\mathbf{x}_{\theta,k,l})]. \quad (11.102)
$$

Since $\Omega_{j,k,l}^{\text{int}} = [2^{-j}(1-\gamma+k), 2^{-j}(\gamma+k-1)] \times [2^{-j}(1-\gamma+l), 2^{-j}(\gamma+l-1)]$ and $\theta_{j,k,l}(\mathbf{x}) \in C_0^{\gamma/2-1} \Omega_{j,k,l}^{\text{inf}}$, we have the relations

$$
\left\| (k - 2^j x)^m (l - 2^j y)^{\lambda-m} \theta_{j,k,l}(\mathbf{x}) \right\|_\infty \le (\gamma - 1)^\lambda \left\| \theta_{j,k,l}(\mathbf{x}) \right\|_\infty \text{ for all } k, l \in \mathbb{Z}, \quad (11.103)
$$

and for any point $\mathbf{x}$

$$
\theta_{j,k,l}(\mathbf{x}) = 0 \text{ when } k \notin (2^j x - \lambda + 1, 2^j x + \lambda - 1) \text{ or } l \notin (2^j y - \lambda + 1, 2^j y + \lambda - 1). \quad (11.104)
$$

Finally, following Eqs. (11.100)–(11.104), one can obtain

$$\left\| f(\mathbf{x}) - S^j f(\mathbf{x}) \right\|_\infty \le 2^{-j\lambda} \frac{(2\gamma-1)^2(\gamma-1)^\lambda}{\lambda!} \left\| \theta_{j,k,l}(\mathbf{x}) \right\|_\infty \sum_{m=0}^{\lambda} C_\lambda^m \left\| D_\lambda^m f(\mathbf{x}_{\theta,k,l}) \right\|_\infty$$
$$\le C_{1,\infty} 2^{-j\lambda} \tag{11.105}$$

in which the properties $\theta_{j,k,l}(\mathbf{x}) \in C^0(R^2)$ and $f(\mathbf{x}) \in C^\lambda(R^2)$ are considered. Specially, if $D_\lambda^m f(\mathbf{x}) \equiv 0$ for $m = 0, 1, \ldots, \lambda$, we have $C_{1,\infty} = 0$ from Eq. (11.105). On the basis of Eq. (11.105), it is very easy to check Eq. (10.17) holding.

Appendix 11.4 Error Estimation of the Interpolating Wavelet Approximation Defined on a Finite Domain

Since $f(\mathbf{x}) \in C^\mu(\Omega)$, one can always find a way to supplement the definition of function $f(\mathbf{x})$ in the domain $\overline{\Omega}$ satisfying $f(\mathbf{x}) \in C^\mu(\Omega \cup \overline{\Omega})$. Then following Eqs. (11.20)–(11.22), we have

$$\begin{aligned} S_L^j f(\mathbf{x}) &= \sum_{l=1}^{N_j} f(\mathbf{x}_l)\theta_{j,kx(\mathbf{x}_l),ky(\mathbf{x}_l)}(\mathbf{x}) + \sum_{l=1}^{N_{\overline{\Omega}}} f(\overline{\mathbf{x}}_l)\theta_{j,kx(\overline{\mathbf{x}}_l),ky(\overline{\mathbf{x}}_l)}(\mathbf{x}) \\ &+ \sum_{l=1}^{N_{\overline{\Omega}}} [\tilde{f}_l(\overline{\mathbf{x}}_l) - f(\overline{\mathbf{x}}_l)]\theta_{j,kx(\overline{\mathbf{x}}_l),ky(\overline{\mathbf{x}}_l)}(\mathbf{x}) \qquad \text{for } \mathbf{x} \in \Omega. \\ &= S^j f(\mathbf{x}) + \sum_{l=1}^{N_{\overline{\Omega}}} [\tilde{f}_l(\overline{\mathbf{x}}_l) - f(\overline{\mathbf{x}}_l)]\theta_{j,kx(\overline{\mathbf{x}}_l),ky(\overline{\mathbf{x}}_l)}(\mathbf{x}) \end{aligned} \tag{11.106}$$

By applying Theorem 11.1 and Eq. (11.106), one can obtain

$$\left\| S_L^j f(\mathbf{x}) - f(\mathbf{x}) \right\|_\infty = \left\| S^j f(\mathbf{x}) - f(\mathbf{x}) \right\|_\infty + \sum_{l=1}^{N_{\overline{\Omega}}} \left\| [\tilde{f}_l(\overline{\mathbf{x}}_l) - f(\overline{\mathbf{x}}_l)]\theta_{j,kx(\overline{\mathbf{x}}_l),ky(\overline{\mathbf{x}}_l)}(\mathbf{x}) \right\|_\infty$$
$$\le C_{1,\infty} 2^{-j\lambda} + \sum_{l=1}^{N_{\overline{\Omega}}} \left\| [\tilde{f}_l(\overline{\mathbf{x}}_l) - f(\overline{\mathbf{x}}_l)]\theta_{j,kx(\overline{\mathbf{x}}_l),ky(\overline{\mathbf{x}}_l)}(\mathbf{x}) \right\|_\infty \tag{11.107}$$

for $\mathbf{x} \in \Omega$. Since $\tilde{f}_l(\mathbf{x})$ is a Lagrange interpolation with order η_l of function $f(\mathbf{x})$, by using the relations $\left|\overline{x}_l - \tilde{x}_{l,i}\right| \le \alpha_1 2^{-j}$ and $\left|\overline{y}_l - \tilde{y}_{l,i}\right| \le \alpha_2 2^{-j}$, we get [16]

$$\left| f(\overline{\mathbf{x}}_l) - \tilde{f}_l(\overline{\mathbf{x}}_l) \right| \le \varepsilon_2 2^{-j\min\{\mu,\eta_l\}}, \tag{11.108}$$

where ε_2 is a constant with property $\varepsilon_2 = 0$ when $D_\lambda^m f(\mathbf{x}) \equiv 0$ for $m = 0, 1, \ldots, \lambda$. Finally, by considering Eq. (11.104) and $\theta_{j,k,l}(\mathbf{x}) \in C_0^{\gamma/2-1}\Omega_{j,k,l}^{\text{inf}}$, Eq. (11.107) can be expressed as

$$\left\| S_L^j f(\mathbf{x}) - f(\mathbf{x}) \right\|_\infty \leq [C_{1,\infty} + (2\gamma - 1)^2 \varepsilon_4] 2^{-j\lambda} = C_{2,\infty} 2^{-j\lambda} \text{ for } \mathbf{x} \in \Omega, \tag{11.109}$$

where $C_{2,\infty} = 0$ when $D_\lambda^m f(\mathbf{x}) \equiv 0$ for $m = 0, 1, \ldots, \lambda$. Based on Eq. (11.109), finally, there is no difficulty for one to check Eq. (11.25) holding.

Appendix 11.5 Proof of the Interpolating Property for the Modified Multiresolution Approximation

If $J = j_0$ in Eq. (11.28), following the delta function property of the modified wavelet basis $\phi_n(\mathbf{x}_k)$ discussed after Eq. (11.23), we can directly obtain

$$P_{j_0}^{j_0} f(\mathbf{x}_k) \equiv S^{j_0} f(\mathbf{x}_k) = f(\mathbf{x}_k) \text{ for all nodes } \mathbf{x}_k. \tag{11.110}$$

Then we assume

$$P_{j_0}^{o} f(\mathbf{x}_{j,n,m}) \equiv f(\mathbf{x}_{j,n,m}) \text{ for } (n, m) \in \Re_j \text{ and } j_0 \leq j \leq o. \tag{11.111}$$

Following Eqs. (11.27) and (11.29), we obtain

$$P_{j_0}^{o+1} f(\mathbf{x}_{j,n,m}) = P_{j_0}^{o} f(\mathbf{x}_{j,n,m}) + \sum_{(k,l)\in\Re_o} [f(\mathbf{x}_{o+1,k,l}) - P_{j_0}^{o} f(\mathbf{x}_{o+1,k,l})]\theta_{o+1,k,l}(\mathbf{x}_{j,n,m}) \tag{11.112}$$

for $(n, m) \in \Re_j$ and $j_0 \leq j \leq o + 1$.

For all nodes $\mathbf{x}_{j,n,m}$, $j_0 \leq j \leq o$, one can gain

$$\theta_{o+1,k,l}(\mathbf{x}_{j,n,m}) = \theta(2^{o+1-j}n - k)\theta(2^{o+1-j}m - l) = 0$$

by considering the interpolation property $\theta(q) = \delta_{0,q}$ for all $q \in \mathbb{Z}$. Because both $2^{o+1-j}n$ and $2^{o+1-j}m$ are even numbers, and at least one of k and l is odd number, therefore, Eq. (11.112) and the assumption (11.111) give the following equation

$$P_{j_0}^{o+1} f(\mathbf{x}_{j,n,m}) = P_{j_0}^{o} f(\mathbf{x}_{j,n,m}) = f(\mathbf{x}_{j,n,m}) \text{ for } j_0 \leq j \leq o. \tag{11.113}$$

For $j = o + 1$ in Eq. (11.112), by using the interpolation property $\theta_{o+1,k,l}(\mathbf{x}_{o+1,n,m}) = \theta(n - k)\theta(m - l) = \delta_{n,k}\delta_{m,l}$, Eq. (11.112) can be reduced into

$$P_{j_0}^{o+1} f(\mathbf{x}_{o+1,n,m}) = P_{j_0}^{o} f(\mathbf{x}_{o+1,n,m}) + \sum_{(k,l)\in\Re_o} [f(\mathbf{x}_{o+1,k,l}) - P_{j_0}^{o} f(\mathbf{x}_{o+1,k,l})]\delta_{n,k}\delta_{m,l}$$

$$= P_{j_0}^{o} f(\mathbf{x}_{o+1,n,m}) + f(\mathbf{x}_{o+1,n,m}) - P_{j_0}^{o} f(\mathbf{x}_{o+1,n,m}) = f(\mathbf{x}_{o+1,n,m}) \quad (11.114)$$

for $j = o + 1$. Hence, Eqs. (11.113) and (11.114) give

$$P_{j_0}^{o+1} f(\mathbf{x}_{j,n,m}) \equiv f(\mathbf{x}_{j,n,m}) \text{ for } (n, m) \in \Re_j \text{ and } j_0 \leq j \leq o + 1. \quad (11.115)$$

Finally, combining the recurrence relations (11.111) and (11.115) with the starting condition of Eq. (11.110), the Proposition 11.2 is proved.

Appendix 11.6 Error Estimation of the Modified Interpolating Multiresolution Approximation

Following the relations of Eqs. (11.26) and (11.27), one can obtain

$$\left\| f(\mathbf{x}) - P_{j_0}^{n} f(\mathbf{x})\right\|_{\infty} \leq \left\| f(\mathbf{x}) - P_{j_0}^{n-1} f(\mathbf{x})\right\|_{\infty}$$
$$+ \sum_{(k,l)\in\Re_{n-1}} \left| f(\mathbf{x}_{j+1,k,l}) - P_{j_0}^{n-1} f(\mathbf{x}_{j+1,k,l})\right| \times \left\|\theta_{j+1,k,l}(\mathbf{x})\right\|_{\infty} \quad (11.116)$$

for $n > j_0$.

By considering $\theta_{j,k,l}(\mathbf{x}) \in C_0^{\gamma/2-1}\Omega_{j,k,l}^{\text{inf}}$ and Eqs. (11.104), (11.116) can be expressed as

$$\left\| f(\mathbf{x}) - P_{j_0}^{n} f(\mathbf{x})\right\|_{\infty} \leq \varepsilon_3 \left\| f(\mathbf{x}) - P_{j_0}^{n-1} f(\mathbf{x})\right\|_{\infty} \quad (11.117)$$

in which ε_3 is a constant. By applying Eq. (11.117) iteratively, and considering the relation $P_{j_0}^{j_0} f(\mathbf{x}) = S_L^{j_0} f(\mathbf{x})$ and Theorem 11.2, one can obtain

$$\left\| f(\mathbf{x}) - P_{j_0}^{J} f(\mathbf{x})\right\|_{\infty} \leq \varepsilon_4 C_{2,\infty} \left\| f(\mathbf{x}) - S_L^{j_0} f(\mathbf{x})\right\|_{\infty} \leq C_{3,\infty} 2^{-j_0\lambda}, \quad (11.118)$$

where ε_4 is a constant, and $C_{3,\infty} = 0$ when $D_{\lambda}^{m} f(\mathbf{x}) \equiv 0$ for $m = 0, 1, \ldots, \lambda$. On the basis of Eq. (11.118), there is no difficulty for one to check Eq. (11.30) holding.

Appendix 11.7 Construction of the Targeted Interpolation Based on Interpolating Wavelet

The method of mathematic induction is employed here to prove the expression of Eq. (11.34). First, for $P_{j_0}^{j_0} f(\mathbf{x})$ one can directly obtain the relation

$$P_{j_0}^{j_0} f(\mathbf{x}) = \sum_{n\in\Re_{j_0}} f(\mathbf{x}_n)\phi_n(\mathbf{x}) = \sum_{n\in\Re_{j_0}} a_n\phi_n(\mathbf{x}) = \sum_{n\in\Re_{j_0}^I} f(\mathbf{x}_n)\Theta_{j_0,n}(\mathbf{x}) + \sum_{n\in\Re_{j_0}^R} c_n\Theta_{j_0,n}(\mathbf{x}), \tag{11.119}$$

where the basis function $\Theta_{j_0,n}(\mathbf{x}) = \phi_n(\mathbf{x})$ for $n \in \Re_{j_0} = \Re_{j_0}^I \cup \Re_{j_0}^R$. And it is also easy to check the relation $P_{j_0}^{j_0} f(\mathbf{x}_n) \equiv f(\mathbf{x}_n)$ held for $n \in \Re_{j_0}^I$ by using the property $\phi_n(\mathbf{x}_m) = \delta_{n,m}$ for $\rho(m) \le \rho(n)$. Based on such a property, in fact, we can further directly obtain the relation $P_{j_0}^{j_{\max}} f(\mathbf{x}_n) \equiv f(\mathbf{x}_n)$ for $n \in \Re_{j_0}$.

Next, we assume that there is the expression

$$P_{j_0}^{j} f(\mathbf{x}) = \sum_{n\in\Re_{\le j}^I} f(\mathbf{x}_n)\Theta_{j,n}(\mathbf{x}) + \sum_{n\in\Re_{\le j}^R} c_n\Theta_{j,n}(\mathbf{x}) = \sum_{n\in\Re_{\le j}} a_n\Theta_{j,n}(\mathbf{x}) \tag{11.120}$$

in which $\Re_{\le j}^I = \cup_{q=j_0}^{j}\Re_q^I$ and $\Re_{\le j}^R = \cup_{q=j_0}^{j}\Re_q^R$. Substitution of Eqs. (11.120) into (11.34) leads to

$$\begin{aligned}
P_{j_0}^{j+1} f(\mathbf{x}) &= P_{j_0}^{j} f(\mathbf{x}) + \sum_{n\in\Re_{j+1}^I} [f(\mathbf{x}_n) - P_{j_0}^{j} f(\mathbf{x}_n)]\phi_n(\mathbf{x}) + \sum_{n\in\Re_{j+1}^R} c_n\phi_n(\mathbf{x}) \\
&= \sum_{n\in\Re_{\le j}^I} f(\mathbf{x}_n)\Theta_{j,n}(\mathbf{x}) + \sum_{n\in\Re_{j+1}^I} f(\mathbf{x}_n)\phi_n(\mathbf{x}) + \sum_{n\in\Re_{\le j}^R} c_n\Theta_{j,n}(\mathbf{x}) + \sum_{n\in\Re_{j+1}^R} c_n\phi_n(\mathbf{x}) \\
&\quad - \sum_{n\in\Re_{j+1}^I}\sum_{m\in\Re_{\le j}^I} f(\mathbf{x}_m)\Theta_{j,m}(\mathbf{x}_n)\phi_n(\mathbf{x}) - \sum_{n\in\Re_{j+1}^I}\sum_{m\in\Re_{\le j}^R} c_m\Theta_{j,m}(\mathbf{x}_n)\phi_n(\mathbf{x}) \\
&= \sum_{n\in\Re_{\le j}^I} f(\mathbf{x}_n)[\Theta_{j,n}(\mathbf{x}) - \sum_{m\in\Re_{j+1}^I}\Theta_{j,n}(\mathbf{x}_m)\phi_m(\mathbf{x})] + \sum_{n\in\Re_{j+1}^I} f(\mathbf{x}_n)\phi_n(\mathbf{x}) \\
&\quad + \sum_{n\in\Re_{\le j}^R} c_n[\Theta_{j,n}(\mathbf{x}) - \sum_{m\in\Re_{j+1}^I}\Theta_{j,n}(\mathbf{x}_m)\phi_m(\mathbf{x})] + \sum_{n\in\Re_{j+1}^R} c_n\phi_n(\mathbf{x}) \\
&= \sum_{n\in\Re_{\le j+1}^I} f(\mathbf{x}_n)\Theta_{j+1,n}(\mathbf{x}) + \sum_{n\in\Re_{\le j+1}^R} c_n\Theta_{j+1,n}(\mathbf{x}) = \sum_{n\in\Re_{\le j+1}} a_n\Theta_{j+1,n}(\mathbf{x}).
\end{aligned} \tag{11.121}$$

By iteratively using the relation of Eq. (11.121) and the starting condition (11.119), one can obtain the formula of Eq. (11.34).

It can be seen from Eqs. (11.119)–(11.121) that the function $\Theta_{j,n}(\mathbf{x})$ for $j_0 \le j \le j_{\max}$ and $n \in \Re$ is a linear combination of modified wavelet basis functions $\phi_n(\mathbf{x})$. Therefore, Eq. (11.120) can be rewritten into the matrix form

$$P_{j_0}^{j} f(\mathbf{x}) = \mathbf{\Phi}(\mathbf{x})\mathbf{B}^{j}\mathbf{a} \tag{11.122}$$

in which the vectors $\mathbf{\Phi}(\mathbf{x}) = \{\phi_n(\mathbf{x}), n = 1, 2, \ldots, N\}$ and $\mathbf{a} = \{a_n, n = 1, 2, \ldots, N\}^{\mathrm{T}}$, and $\mathbf{B}^{j}$ is the $N \times N$ transformation matrix. Based on Eq. (11.119), one can directly obtain

$$\mathbf{B}^{j_0} = \left\{ b_{kk}^{j_0} = 1 \text{ for } k = \Re_{j_0},\ b_{kl}^{j_0} = 0 \text{ otherwise} \right\}. \tag{11.123}$$

Substituting Eqs. (11.122) into (11.121), we get

$$P_{j_0}^{j+1} f(\mathbf{x}) = \mathbf{\Phi}(\mathbf{x})\mathbf{B}^{j}\mathbf{a} + \mathbf{\Phi}(\mathbf{x})\overline{\mathbf{B}}^{j+1}\mathbf{a} - \mathbf{\Phi}(\mathbf{x})\widetilde{\mathbf{B}}^{j+1}\overline{\mathbf{\Phi}}^{j+1}\mathbf{B}^{j}\mathbf{a} = \mathbf{\Phi}(\mathbf{x})\mathbf{B}^{j+1}\mathbf{a} \tag{11.124}$$

in which

$$\overline{\mathbf{B}}^{j+1} = \left\{ \overline{b}_{kk}^{j+1} = 1 \text{ for } k = \Re_{j+1},\ \overline{b}_{kl}^{j+1} = 0 \text{ otherwise} \right\}, \tag{11.125}$$

$$\widetilde{\mathbf{B}}^{j+1} = \left\{ \tilde{b}_{kk}^{j+1} = 1 \text{ for } k = \Re_{j+1}^{I},\ \tilde{b}_{kl}^{j+1} = 0 \text{ otherwise} \right\}, \tag{11.126}$$

$$\overline{\mathbf{\Phi}}^{j+1} = \left\{ \overline{\Phi}_{kl}^{j+1} = \phi_l(\mathbf{x}_k) \text{ for } k \in \Re_{j+1}^{I},\ \overline{\Phi}_{kl}^{j+1} = 0 \text{ otherwise} \right\}. \tag{11.127}$$

By considering the above sparse features of $\widetilde{\mathbf{B}}^{j+1}$ and $\overline{\mathbf{\Phi}}^{j+1}$, it is easy to check the relation $\overline{\mathbf{\Phi}}^{j+1} = \widetilde{\mathbf{B}}^{j+1}\overline{\mathbf{\Phi}}^{j+1}$. Thus, Eq. (11.124) gives

$$\mathbf{B}^{j+1} = \mathbf{B}^{j} + \overline{\mathbf{B}}^{j+1} - \overline{\mathbf{\Phi}}^{j+1}\mathbf{B}^{j}. \tag{11.128}$$

Considering that the wavelet basis function $\phi_n(\mathbf{x})$ has the compact support domain Ω_n and the function $P_{j_0}^{j} f(\mathbf{x})$ defined by Eq. (11.120) is completely independent of $\phi_n(\mathbf{x})$ with $\rho(n) > j$, there is the property $\overline{\Phi}_{kl}^{j+1} = 0$ for $k \notin \Re_{j+1}^{I}$ or $l \notin \Re_{\le j}^{T,k}$ for the matrix $\overline{\mathbf{\Phi}}^{j+1}$ with $\Re_{\le j}^{T,k} = \Re_{\le j}^{T} \cap \Re_{k}^{\Omega}$. Here the nodal set $\left\{\mathbf{x}_n, n \in \Re_{\le j}^{T}\right\}$ is consisted of all those nodes with $\rho(n) \le j$, whose compact support domain Ω_n contains at least one node $\mathbf{x}_m$ for $m \in \Re^{I}$. And the nodal set $\left\{\mathbf{x}_m, m \in \Re_{k}^{\Omega}\right\}$ is consisted of all nodes whose compact support domain Ω_m contains the node $\mathbf{x}_k$.

By using the above property of $\overline{\mathbf{\Phi}}^{j}$, the transformation matrix $\mathbf{B}^{j}$ can be expressed as

$$\mathbf{B}^j = \mathbf{B}_d^j + \mathbf{B}_s^j \tag{11.129}$$

in which the incomplete identity matrix $\mathbf{B}_d^j = \left\{ b_{d,kk}^j = 1 \text{ for } k \in \Re_{\leq j},\ b_{d,kl}^j = 0 \text{ otherwise} \right\}$ and the sparse matrix $\mathbf{B}_s^j$ are verified for all elements $b_{s,kl}^j = 0$ for $k \notin \Re^I$, $l \notin \Re_{\leq j-1}^{T,k}$ or $k = l$.

Based on the definition in Eq. (11.123) of $\mathbf{B}^{j_0}$, it is very easy to obtain $\mathbf{B}_d^{j_0} = \mathbf{B}^{j_0}$ and $\mathbf{B}_s^{j_0} = \mathbf{0}$. Then, substituting Eqs. (11.129) into (11.128), we gain

$$\mathbf{B}^{j+1} = (\mathbf{B}_d^j + \overline{\mathbf{B}}^{j+1}) + \mathbf{B}_s^j - \overline{\boldsymbol{\Phi}}^{j+1} B_d^j - \overline{\boldsymbol{\Phi}}^{j+1} \mathbf{B}_s^j = \mathbf{B}_d^{j+1} + \mathbf{B}_s^{j+1}. \tag{11.130}$$

By directly examining the definition of $\overline{\mathbf{B}}^{j+1}$ given in Eq. (11.125), it is easy to check that the matrix $\mathbf{B}_d^{j+1} = \mathbf{B}_d^j + \overline{\mathbf{B}}^{j+1}$ is exactly the incomplete identity matrix defined in Eq. (11.129). By applying the properties $\overline{\Phi}_{kl}^{j+1} = 0$ for $k \notin \Re_{j+1}^I$ or $l \notin \Re_{\leq j}^{T,k}$, $b_{s,kl}^j = 0$ for $k \notin \Re^I$ or $l \notin \Re_{\leq j-1}^{T,k}$, and the definition of $\mathbf{B}_d^j$ given in Eqs. (11.129), (11.130) gives

$$\mathbf{B}_s^{j+1} = \mathbf{B}_s^j - \overline{\boldsymbol{\Phi}}^{j+1} - \overline{\boldsymbol{\Phi}}^{j+1} \mathbf{B}_s^j \tag{11.131}$$

which verifies the sparsity pattern $b_{s,kl}^{j+1} = 0$ for $k \notin \Re^I$, $l \notin \Re_{\leq j-1}^{T,k}$ or $k = l$.

In Eq. (11.131), the matrix $\overline{\boldsymbol{\Phi}}^{j+1}$ has been redefined as

$$\overline{\boldsymbol{\Phi}}^{j+1} = \left\{ \overline{\Phi}_{kl}^{j+1} = \phi_l(\mathbf{x}_k) \text{ for } k \in \Re_{j+1}^I \text{ and } l \in \Re_{\leq j}^T,\ \overline{\Phi}_{kl}^{j+1} = 0 \text{ otherwise} \right\}, \tag{11.132}$$

where the properties $\overline{\Phi}_{kl}^{j+1} = 0$ for $l \notin \Re_{\leq j}^T$ is applied.

In addition, since $b_{s,kl}^{j+1} = 0$ for $k \notin \Re^I$, there is the relation $\overline{\boldsymbol{\Phi}}^{j+1} \mathbf{B}_s^j = \widehat{\Phi}^{j+1} \mathbf{B}_s^j$ with the sparse matrix $\widehat{\Phi}^{j+1} = \left\{ \widehat{\Phi}_{kl}^{j+1} = \phi_l(\mathbf{x}_k) \text{ for } k \in \Re_{j+1}^I \text{ and } l \in \Re_{\leq j}^I,\ \widehat{\Phi}_{kl}^{j+1} = 0 \text{ otherwise} \right\}$.

Therefore, the final transformation matrix can be expressed as $\mathbf{B}^{j_{\max}} = \mathbf{I} + \mathbf{B}_s^{J_I}$ with the identity matrix $\mathbf{I}$ and the sparse matrix $\mathbf{B}_s^j$ satisfying the recurrence relation

$$\mathbf{B}_s^{j+1} = \mathbf{B}_s^j - \overline{\boldsymbol{\Phi}}^{j+1} - \widehat{\Phi}^{j+1} \mathbf{B}_s^j \text{ with } \mathbf{B}_s^{j_0} = \mathbf{0}, \tag{11.133}$$

where the property $\overline{\boldsymbol{\Phi}}^j = \mathbf{0}$ for $j > J_I$ with $J_I = \max\{\rho(n), n \in \Re^I\}$ has been considered.

By directly reducing the dimensions of matrices $\mathbf{B}_s^j$, and according to the sparsity features $\overline{\mathbf{\Phi}}^j$ and $\widehat{\Phi}^j$ in Eq. (11.133) displayed above, finally, one can obtain the shape functions $\left\{\Theta_{j_{\max},n}(\mathbf{x}),\ n \in \Re\right\} = \mathbf{\Phi}(\mathbf{x})\mathbf{B}^{j_{\max}}$ defined by Eq. (11.35).

References

1. Meyer Y (1992) Wavelets and operators:, vol 1. Cambridge University Press, London
2. Li B, Chen X (2014) Wavelet-based numerical analysis: a review and classification. Finite Elem Anal Des 81:14–31
3. Chen XF, Yang SJ, Ma JX, He ZJ (2004) The construction of wavelet finite element and its application. Finite Elem Anal Des 40:541–554
4. Liu XJ, Wang JQ, Zhou YH, Wang JZ (2017) Wavelet methods and applications in nonlinear mechanics problems. Chin J Solid Mech 38:287–310 (in Chinese)
5. Liu Y, Liu Y, Cen Z (2011) Multi-scale Daubechies wavelet-based method for 2-D elastic problems. Finite Elem Anal Des 47:334–341
6. Vasilyev OV, Kevlahan NKR (2005) An adaptive multilevel wavelet collocation method for elliptic problems. J Comput Phys 206:412–431
7. Sannomaru S, Tanaka S, Yoshida K, Bui TQ, Okazawa S, Hagihara S (2017) Treatment of Dirichlet-type boundary conditions in the spline-based wavelet Galerkin method employing multiple point constraints. Appl Math Model 43:592–610
8. Donoho DL (1992) Interpolating wavelet transforms. Preprint, Department of Statistics, Stanford University 2:1–54
9. Liu GR (2009) Meshfree methods: moving beyond the finite element method. Taylor and Francis
10. Liu XJ, Liu GR, Wang JZ, Zhou YH (2019) A wavelet multiresolution interpolation Galerkin method for targeted local solution enrichment. Comput Mech 64:989–1016
11. Liu XJ, Liu GR, Wang JZ, Zhou YH (2020) A wavelet multi-resolution enabled interpolation Galerkin method for two-dimensional solids. Eng Anal Boundary Elem 117:251–268
12. Daubechies I (1988) Orthonormal bases of compactly supported wavelets. Commun Pure Appl Math 41:909–996
13. Beylkin G (1992) On the representation of operators in bases of compactly supported wavelet. SIAM J Numer Anal 29:1716–1740
14. Beylkin G, Saito N (1992) Wavelets, their autocorrelation functions, and multiresolution representation of signals. SPIE Proc 1826:39–50
15. Liu GR (2016) An overview on meshfree methods: for computational solid mechanics. Int J Comput Methods 13:1630001
16. Kincaid D, Kincaid DR, Cheney EW (2009) Numerical analysis: mathematics of scientific computing. Amer Math Soc
17. Kattan P (2003) MATLAB guide to finite elements: an interactive approach. Springer-Verlag, New York
18. Timoshenko S, Goodier JN (1951) Theory of elasticity. McGraw-Hill book Company
19. Belytschko T, Lu YY, Gu L (1994) Element-free Galerkin methods. Int J Numer Meth Eng 37:229–256
20. Liu GR, Gu YT (2001) A point interpolation method for two-dimensional solids. Int J Numer Meth Eng 50:937–951
21. Moes N, Dolbow J, Belytschko T (1999) A finite element method for crack growth without remeshing. Int J Numer Meth Eng 46:131–150
22. Fleming M, Chu YA, Moran B, Belytschko T (1997) Enriched element-free Galerkin methods for crack tip fields. Int J Numer Meth Eng 40:1483–1504

23. Liu XJ, Liu GR, Wang JZ, Zhou YH (2020) A wavelet multiresolution interpolation Galerkin method with effective treatments for discontinuity for crack growth analyses. Eng Fract Mech 225:106836
24. Liu GR, Jiang Y, Chen L, Zhang GY, Zhang YW (2011) A singular cell-based smoothed radial point interpolation method for fracture problems. Comput Struct 89:1378–1396
25. Li S, Liu WK, Rosakis AJ, Belytschko T, Hao W (2002) Mesh-free Galerkin simulations of dynamic shear band propagation and failure mode transition. Int J Solids Struct 39:1213–1240
26. Nguyen NT, Bui TQ, Zhang C, Truong TT (2014) Crack growth modeling in elastic solids by the extended meshfree Galerkin radial point interpolation method. Eng Anal Boundary Elem 44:87–97
27. Erdogan F, Sih GC (1963) On the crack extension in plates under plane loading and transverse shear. J Basic Eng 85:519–525

Chapter 12
Brief Introduction in Applications of Other Groups

From previous Chaps. 2–11, the wavelet-based solution method and its applications are detail introduced to what were conducted by the author and his colleagues. In this chapter, we give a brief introduction of those applications done by else groups using the generalized Coiflets and relevant method proposed by the author's group. When their evaluations on the efficiency of the method are directly cited, for simplicity, the reference numbers in their original references are renewed by replacing ones in order here, where some relevant necessary research background and main conclusion are cited and the details of researches are omitted. Once a reader is interested in them, one can find them following the cited references here. From the cited applications, one can also see that the generalized Coiflets and relevant methods play a powerful role in their applications.

12.1 Deep Improvement for Homotopy Analysis Method (HAM) by the Generalized Coiflets

The early concept of homotopy method can be tracked back to Poincaré [1]. Leray and Schauder [2] refined this mathematical tool and Lahaye [3] used it to solve nonlinear algebraic equations. In such method, the solution of a nonlinear equation can be obtained by solving a constructed homotopy operator with a tracking parameter varying from 0 to 1, where the original nonlinear equation is contained. When the tracking parameter is taken as zero, some solution for a linear equation should be relatively easy gained. When the tracking parameter increases step by step, the solution for the homotopy operator can be gained by substituting the solution of former step, like a perturbation method. Finally, the solution for the original nonlinear problem can be gained when the tracking parameter approaches to 1. As the solution methods for nonlinear problems were substantively attracted by the scientific committee, this

Y.-H. Zhou, *Wavelet Numerical Method and Its Applications in Nonlinear Problems*, Engineering Applications of Computational Methods 6,
https://doi.org/10.1007/978-981-33-6643-5_12

method was generalized to solving the nonlinear differential equations with either boundary conditions or initial conditions or both, e.g., [4–6]. In theory, this method can solve any kind of nonlinear problems, either weakling or strongly.

In order to get some analytical solution of this method to a nonlinear problem, Liao proposed the homotopy analysis method (HAM) by means of a perturbation method where the solution is expended by a set of power series of the tracking parameter and the original equations in the homotopy method is reduced into a set of linear equations on unknown functions with orders from zero to infinite, for example, see [7, 8]. Following them, the essential of HAM can be displayed as follows.

Based on the basic concept of homotopy, the HAM describes a continuous deformation or variation from a topological space to another. Therefore, the most important step is to construct a homotopy between the initial guess $V_0(x, y)$ and the final solution $V(x, y)$. Let $q \in [0, 1]$ denote an embedding parameter for homotopy, c_0 the convergence-control parameter, $\mathcal{L}$ an auxiliary linear operator and $\mathcal{N}$, the nonlinear operator, respectively. Then, a homotopy (or continuous variation) from the initial guess $V_0(x, y)$ to the solution $V(x, y)$ can be built by

$$(1-q)\mathcal{L}[\varphi(x, y; q) - V_0(x, y)] = c_0 q \mathcal{N}[\varphi(x, y; q)] \tag{12.1}$$

subject to the boundary conditions, e.g., of the form in a 2D problem.

$$\varphi(0, y; q) = \varphi(1, y; q) = \varphi(x, 0; q) = \varphi(x, 1; q) = 1. \tag{12.2}$$

Here $\mathcal{N}[\varphi(x, y; q)]$ is a nonlinear operator relevant to the original differential equation when $q = 1$. For example,

$$\mathcal{N}[\varphi(x, y; q)] = \varphi(x, y; q)\nabla^2\varphi(x, y; q) - \nabla\varphi(x, y; q) \cdot \nabla(x, y; q) - \frac{\lambda}{2} \tag{12.3}$$

Equation (12.1) tells us that when $q = 0$, we have $\mathcal{L}[\varphi(x, y; 0) - V_0(x, y)] = 0$ or $\varphi(x, y; 0) = V_0(x, y)$, and $\varphi(x, y; 1) = V(x, y)$ when $q = 1$. In other words, $\varphi(x, y; q)$ denotes a continuous variation from the initial guess $V_0(x, y)$ to the solution $V(x, y)$ as q increases from 0 to 1. Then, $\varphi(x, y; q)$ can be expanded in Maclaurin series with respect to the embedding parameter q, i.e.,

$$\varphi(x, y; q) = V_0(x, y) + \sum_{k=1}^{+\infty} V_k(x, y) q^k \tag{12.4}$$

where

$$V_k(x, y) = \mathscr{D}_k[\varphi(x, y; q)] = \frac{1}{k!}\frac{\partial^k \varphi(x, y; q)}{\partial q^k}\Big|_{q=0} \tag{12.5}$$

Substituting Eq. (12.4) into Eq. (12.1) and comparing the same order of power of q, one can obtain a set of linear governing equations on the unknowns $V_k(x, y)$ ($k = 1, 2, \ldots$). When $V_0(x, y)$ is guessed, then substituting it into the governing equation on $V_1(x, y)$. Then repeating the similar iteration, the higher order of $V_k(x, y)$ can be gained successively.

Liao and his group used the HAM to solve many problems of nonlinear differential equations. Although the HAM exhibits more advantages of higher efficiency, such a method is strongly dependent on how the auxiliary linear operator and the base functions were chosen. In other words, to a given nonlinear problem, the HAM should pay a long attention trying to find a suitably auxiliary linear operator associated with a set of base functions. Even when such selection is OK for a given nonlinear problem, however, such selection is usually not suitable to other kinds of nonlinear problems. In order to conquer this key problem, Yang and Liao [9, 10] employed the generalized or revised Coiflets base scaling functions of our group in their HAM to solve the nonlinear ordinary and partial differential equations, respectively. In the background introduction of their researches, they [9] pointed out that "the base function plays a significant role for efficiently obtaining a good approximation of a highly nonlinear equation. Our purpose is to develop such a new HAM approach that:

(A) It is insensitive to the choice of the auxiliary linear operator so that it can be more flexible and adaptive to choose the auxiliary linear operator;
(B) The exponential expansion of the right-hand side term in high-order deformation equation can be overcome in a convenient way;
(C) It can provide a convenient way to balance the computational efficiency and approximation accuracy so that an acceptable approximation can be obtained efficiently.

For these purposes, the generalized Coiflet -type orthogonal wavelet developed by Wang and Zhou [11–15] is applied as base functions to develop such a new HAM approach" and "The early applications of wavelet in differential equations could be traced back to 1990s [16–18]. However, as the theory of wavelet method for differential equations was not established, many basic problems such as boundary leaping and computation of connection coefficients were needed to solve [11, 19, 20], Wang and Zhou et al. [11–15] developed the generalized Coiflet -type orthogonal wavelet and proposed a new boundary extension technique [11, 15, 20–22] similar to Taylor's expansion ."

When they selected a set of base functions, they compared the efficiency among three different kinds of existing wavelet methods. They [10] gave the main reason why our wavelet method is selected, that is "Briefly speaking, there are three kinds of wavelet methods: wavelet finite element method [23–25], wavelet collocation method [26–29], and Wavelet-Galerkin method [13, 30–33]. Comparing these wavelet-based methods, it is found that the wavelet type plays a very important role for the convergence property as well as the efficiency. So, how could we choose a "proper wavelet" to solve nonlinear BVPs? Such kind of "**proper wavelet**" should possess the following properties: **orthogonality**, **compact support**, **interpolation properties** as well as **high algebraic accuracy** and the **ability to represent functions**

at different levels of resolution [20, 34]. **The generalized Coiflet-type wavelet possesses almost all of these properties.**" Hereafter, the bond mark in the citation is made by the author of this book.

In these two essential papers, they told us that "In Section 2, some related theories of the generalized Coiflet-type orthogonal wavelet are introduced" [9] and "In Section 2, using the generalized Coiflet-type wavelet developed by Wang and Zhou et al. [11–15], the base functions for solutions with two variables are constructed. More importantly, a section-based wavelet approximation to approximate partial derivatives is proposed, which greatly improves the efficiency" [10]. In such situation, they named such improved HAM as wHAM.

After the examples of solving nonlinear problems were numerically displayed in [9, 10], the prominent advantages of the improved method were given in their conclusions, which are directly cited as follows:

In the paper [9], the relevant conclusion part is "a HAM-based wavelet approach for nonlinear ordinary differential equations, namely, the wavelet homotopy analysis method (wHAM), is proposed, which successfully combines the homotopy analysis method, the generalized Coiflet-type orthogonal wavelet and the Galerkin method together. The one-dimensional Bratu's equation is used as an example to describe the basic ideas of the wHAM and to illustrate its validity. By means of choosing a proper value of the so-called convergence-control parameter, the wavelet approximations given by the wHAM with **various auxiliary linear operators** converge quickly to the exact solution. It is found that the wHAM not only keeps the advantages of the normal HAM, but also possesses some new merits. First of all, the convergence-control parameter, which plays a significant role in the normal HAM, still provides us a convenient way to guarantee the convergence of the wavelet approximation in the wHAM. Secondly, since the wavelet approximation is based on multiresolution analysis, the resolution level provides us a **convenient way** to balance the computational efficiency and approximative accuracy so that an accurate enough approximation can be obtained efficiently. Note that the convergent solutions for all cases considered in this paper can be gained in less than one second CPU times by means of the wHAM. This indicates that the wHAM possesses the high computational efficiency. Besides, different from the normal HAM, the CPU time of the wHAM just increases linearly with respect to the approximation order. Furthermore, in the frame of the wHAM, the base function is always the wavelet function and besides **the validity of the wHAM is not sensitive to the choice of the auxiliary linear operator**, so that the wHAM provides us **larger freedom** than the normal HAM to choose the auxiliary linear operator."

In the paper [10], the relevant conclusion part is "Compared to the normal HAM, the wHAM has some obvious advantages. Firstly, the wHAM provides us **larger freedom** to choose the auxiliary linear operator than the normal HAM. Secondly, the wHAM possesses **much higher computational efficiency** than the normal HAM. In addition, it is found that the iteration approach of the wHAM can greatly accelerate the convergence. Therefore, the iteration wHAM is strongly suggested in practice.

Moreover, it is found that **the convergence of solution series and the computational efficiency is not sensitive to the choice of the auxiliary linear operator**, which makes it easy for users to use the wHAM."

Following the above wHAM, Liao and his colleagues solved other nonlinear problems, e.g., the mixed convection flow [35], the lid-driven cavity flow and heat transfer with non-uniform boundary conditions [36], the stream function–vorticity formulation of Navier–Stokes equations [37], the extreme large bending deflection of a rectangular plate on non-uniform elastic foundations [38], and the channel flow due to orthogonally moving porous walls [39], etc.

12.2 Applications of the Generalized Coiflets and Relevant Method in Random Dynamic Problems

After the main results of the generalized Coiflets and relevant methods of the Laplace inverse transform were published, the new methods were also used in solving the nonlinear random dynamic problems, which displays a powerful merit comparing with those convention methods. As Ray and Gupta [40] pointed that "Zhou et al. [21] present an **efficient** wavelet-based algorithm for solving a class of fractional vibration, diffusion and wave equations with strong nonlinearities. For that purpose, they first suggest a wavelet approximation for a function defined on a bounded interval, in which expansion coefficients are just the function sampling at each nodal point. They use Laplace transform to convert fractional differential equations containing strong nonlinear terms and singular integral kernels into the second type Voltera integral equations with non-singular kernels. They use certain property of the integral kernel and the ability of **explicit wavelet approximation** to the nonlinear terms of the unknown function in the equation which enables to **numerically decouple spatial and temporal dependencies** during solution of those equations. They proposed an efficient numerical method **without involving any matrix inversions** for numerically solving the nonlinear fractional vibration, diffusion and wave differential equations."

The early applications of the generalized Coiflets-based Laplace transform was conducted by Koziol and his colleagues [41] for solving the problems of the bending waves in bending on viscoelastic random foundation [41] and the vibration of surface due to a load moving in the layer [42]. For example, one of their equations of the beam vertical motions can be written by using the Euler–Bernoulli theory as

$$EI\frac{\partial^4 W}{\partial x^4}+\rho_B\frac{\partial^2 W}{\partial t^2}=P(t)\delta(x-Vt)+a[\sigma_{zz}(x,h^-,t)-\sigma_{zz}(x,h^+,t)], \quad (12.6)$$

where $P(t)$ is the vertical point load, $W(x,t)$ is the vertical displacement of the beam, $\sigma_{zz}(x,z,t)$ is the vertical stress, EI and ρ_B are the bending stiffness and the mass per unit length of the beam, $\delta(\cdot)$ is the Dirac delta function, a is a characteristic length associated with the length of the structure in y-direction (the thickness of the beam in y-direction).

In Ref. [41], for example, they gave how and what method were used, that is "To solve this problem two different approaches were used: Bourret's approximation [43] and Adomian's decomposition [44]. It has been made a comparison of numerical results calculated by using these two methods. To investigate the properties of the solution a new method of finding inverse Laplace transform based on the wavelet theory has been adopted [45, 12]. **This method allows to find solutions especially for more complicated cases of correlation functionsfor which traditional methods are inadequate**."

After the generalized Coiflets and relevant Laplace transform were employed in Section 6 of [41], they got the numerical test results. In its conclusion, they pointed out "The new method of finding the inverse Laplace transform based on the wavelet theory has been adopted to carry out the parametric study. The proposed method allows to calculate and compare the solutions for the arbitrarily chosen correlation function. **The numerical tests indicate that the Coiflet-based method is very efficient and relatively easy to implement in practical calculations**."

Also in Ref. [42], Koziol et. al. used the Coiflet-based Laplace transform and pointed out "The scaling function based on these coefficients **has good low-pass characteristic** [12]. The difference between the energy spectrum of the perfect low-pass filter and the energy of the chosen filter is less than 0.3% [12] in the frequency domain $[0, \pi/2]$. **This filter is much more accurate than** the one used previously in [41] for investigating the same problem. Therefore, the results obtained in this paper are **more precise** in the sense that the **faster convergence** of the approximation (25) has been achieved. It should be noted that using this more precise filter preserves the effectiveness of numerical calculations." Then, they concluded that "The numerical tests indicate that the Coiflets-based method is **very efficient** and **relatively easy** to implement in practical calculations allowing also different parametric studies. The effectiveness of the Coiflets-based approximation grows with the level of accuracy of the Coiflets filters. The **complexity** of numerical calculations **is independent** on the choice of the filter and **the amount of time** needed for calculations **does not change significantly**. Possible further applications of presented semi-analytical wavelet-based method exist for the problems **where the Fourier and the Laplace transforms cannot be obtained** by the methods of direct numerical calculations."

Following above successful applications, Koziol and his colleagues used such method to solve more complex problems, e.g., the vibratory analysis of surface due to a load moving in the layer [46], the beam-soil structure response for fast moving train [47], the vibration analysis of fast moving load on a viscoelastic medium [48], the dynamic response of Rayleigh beam on nonlinear foundation due to moving load [49], the dynamic response of a beam resting on a nonlinear foundation to a moving load [50], the dynamic response and optimization for multilayered infinite medium subject to a moving load: Dynamic response [51], the vibration of a beam on a nonlinear viscoelastic foundation due to moving load [52], the nonlinear problem of a double-beam response subjected to a series of moving loads [53] and the dynamic response of railway track subjected to a moving train [54]. For example, in their publication [47], they pointed out "In order to **avoid numerical instabilities** appearing for some systems of parameters, a wavelet approximation method [12] and a specially chosen Coiflets filter [55] are used for derivation of the inverse Fourier transform. This method is **very effective** for high velocities and high load frequencies and allows to reduce numerical problems for calculations that prevent effective analysis of the system due to the big amount of time needed for calculations [46]. The problem of viscoelastic soil vibrations generated by a load moving along a beam resting on the surface was analyzed in two cases: with finite thickness of supporting layer and with half-space under the beam. A special wavelet method was adopted for the derivation of vertical displacements at the surface. This method allowed to carry out the analysis in the area of high velocities and low frequencies leading to the approximation of the point load constant in time. The applied estimation allowed to **alleviate the problem of singularities** appearing in the integrated formulas when the constant load was considered. The modified Coiflets filter improved the accuracy of wavelet approximation compared to previously published results."

Other two groups, leaded by Kennedy in UK and Shi in China, conducted the similar mechanics problems using the generalized Coiflets and relevant Laplace or Fourier transform [56–59]. In Ref. [56], for example, the conventional and the wavelet-based methods were given in its Sections 4 and 5, respectively, where a comparison was given in itself Section 5 by "Section 4 gives analytical solutions, but they are difficult to obtain accurate numerical results in practical applications because they are in integral form. Additionally, the integrand in Eq. (23) in [56] is singular and highly oscillatory, and the integration interval is infinite. Moreover, computational time is an **important** consideration because large number of frequency points must be computed. In this section, the **wavelet approach** is introduced, which is **very accurate** and **highly efficient** [12, 55, 60, 61]. The wavelet approach is not only suitable to solve linear problems [46, 48, 62, 63] but is also **robust enough** to address nonlinear problems [50, 52, 53]. Here, we use the approach of numerical integration related to a beam-soil structure subjected to a moving load presented by [46, 48, 50, 52, 53]." In the conclusion of this paper, it pointed out that "The stochastic vibration analysis of a beam-soil structure subjected to a moving random load has been studied. Based on the pseudo-excitation method, analytical solutions for the **non-stationary** power spectral density and standard deviation of vertical displacement are derived in integral form. **A wavelet approach is introduced to calculate the integrand**. Numerical results for the power spectral density locate the major frequency band at which large vibrations occur at different load velocities. The plots of standard deviation against time show the general trend of vibration, while the maximum standard deviations indicate the velocities worthy of attention. Furthermore, a parametric study is conducted to study the mechanism of the critical velocity of the system. **The hybrid method presents a practical and efficient approach** for studying the random responses of beam-soil structures subjected to moving random loads."

12.3 Applications of the Generalized Coiflets in Dynamic Control Systems and Others

The researchers conducted by the author's group using the wavelet-based method were original about 20 years ago [12, 64–70], containing the relevant method and applications to static deflection and dynamic control of thin beam and plate structures. Following them and subsequent results, this method has been also used by else groups in different areas, e.g., the simulation of temperature field of copier paper [71], the vibration control of large space smart structures [72], and the construction wavelet-based beam element [73]. All of them displayed high efficiency.

For example, Si and Li [72] used the wavelet-based method proposed in Refs. [65, 70] to design the piezoelectric sensors and actuators bounded on the large space truss structures with scale of 0.3m × 0.3m × 9m consisted by 5 active bars and 389 ordinary bars as well 124 connect nodes. As pointed out in this paper, when the conventional modal control method is employed to such large space structures, there are two situations to make the calculations loss efficiency: one is to solve the observation signals of charges or electric currents, and other is to solve the inversion matrix of modal displacement or velocity. When the matrix is in singularity, such method leads to inefficiency of the observation signals of displacement or velocity. At the same time, it is unavoidable that there are some leaking of observation and control signals in the conventional modal control method. Such coupled leaking signals between sensors and actuators in the connectional feedback control system sometimes make the control instability, i.e., some higher order modal of vibration is inspired out when the low-order modal of vibration is suppressed. At present, the researchers have been paid high attention to these problems in researches and designs of such kinds of smart structures. Due to the ability of a wavelet scaling function filtering high frequency components, the control system employed by a wavelet scaling function makes the phenomenon of control instability disappeared. After they gained the numerical simulation results, that gave a conclusion of the form "From the simulation results, it is known that the vibration of such large space truss structures can be suppressed in 5 s by the wavelet-based method, while the suppression duration is 7.5 s when the conventional modal control method is employed. Thus, the control method employed in this paper is significantly effective. Since the approximation of wavelet scaling function has the ability of automatically filtering high frequency components, this control method does not indeed appear the phenomenon of control instability generated by the coupling of leaking signals between the observation and control, which overcomes the difficulty original from the low frequency and dense frequency in the smart control system of large space structures."

The similar review to the dynamic control with wavelet-based method established in Ref. [69] was also given by Fenik and Starek [74], which are those that "Furthermore, in the vibration control of structures with varying dynamical properties it is difficult to achieve good results without some level of self-identification or adaptive control design [75, 76]. Among recent works concerned with velocity feedback a new approach was proposed in Zhou et al. [69], where the authors used wavelet transform scaling functions for approximation of the deflection of the controlled plate. Control law with negative deflection and velocity feedback was chosen and the wavelet weighted residual method was used to determine applied control voltages. Due to the fact that the scaling function transform is like a low-pass filter, which can **automatically filter out high-order signals** of vibration or disturbance from the measurement, **this control approach does not lead to the undesired phenomenon of control instability**."

Recently, Ma et al. [77] used our wavelet-based method in their solution program for the coupled nonlinear governing equations for longitudinal and transverse vibrations of axially moving strings. As pointed out by them: "A generalized Coiflets scaling function $\varphi(x)$ developed by [11, 32, 79, 80] are adopted in this paper." In this paper, the typical governing equations for a uniform axially moving string was given by

$$\rho\frac{\partial^2 w}{\partial t^2}+2\rho v\frac{\partial^2 w}{\partial x\partial t}+\rho v^2\frac{\partial^2 w}{\partial x^2}-T\frac{\partial^2 w}{\partial x^2} -\frac{\partial}{\partial x}\left[EA\left(\frac{\partial w}{\partial x}+\frac{1}{2}\left(\frac{\partial u}{\partial x}\right)^2\right)+f_1(w)h_1(u)\right]=0, \tag{12.7}$$

$$\rho\frac{\partial^2 u}{\partial t^2}+2\rho v\frac{\partial^2 u}{\partial x\partial t}+\rho v^2\frac{\partial^2 u}{\partial x^2}-T\frac{\partial^2 u}{\partial x^2} -\frac{\partial}{\partial x}\left[\frac{1}{2}EA\left(\frac{\partial w}{\partial x}+\frac{1}{2}\left(\frac{\partial u}{\partial x}\right)^2\right)\frac{\partial u}{\partial x}+f_2(w)h_2(u)\right]=0, \tag{12.8}$$

where w (x, t) and u (x, t) are longitudinal and transverse displacement, respectively, at axial coordinate x and time t. The parameters of ρ, v, T, E, and A are, respectively, density per unit length, moving speed, pre-tension, Young's modulus, and the cross-sectional area of the string. Besides, the origin of the coordinate is set at the fixed end of the string and the instantaneous length of the string is l.

From Eqs. (12.7) and (12.8), it is obvious that such governing equations are nonlinear and coupled. After the numerical results were gained in [32, 78], They gave that "the following conclusions can be obtained:

(1) The proposed method with wavelet–Galerkin followed by SDC has **higher accuracy** and **efficiency** compared to the traditional methods with a classic FEM followed by Runge–Kutta or SDC, wavelet–Galerkin followed by Runge–Kutta.
(2) The proposed method with wavelet–Galerkin followed by SDC is **valid** in solving coupled nonlinear string vibrations problems.

It has been shown that the proposed method is **very efficient** to solve **coupled nonlinear** string vibrations problems. Since there are only longitudinal and transverse vibrations for axially moving strings, all of the vibrations equations discussed in this work are one or two dimensional. However, the proposed method can be used to solve more general nonlinear coupled vibrations problems, e.g., the three-dimensional vibrations problems of belt considering longitudinal, transverse, and torsional vibrations."

Wang [13] used the generalized Coiflets-based Laplace transform to solve those typical mathematical equations, e.g., the fractional Riccati differential equations [80], a class of nonlinear Volterra–Fredholm integral equations [82], and fractional Bagley–Torvik equation [82].

For the purpose of clarity, here, we cite the essential governing equations as follows.

(1) the fractional Riccati differential equations is of the nonlinear form [80]

$$D^{\alpha} y(t) = 1 - [y(t)]^2, \quad t > 0,\ 0 < \alpha < 1 \tag{12.9}$$

with initial conditions

$$y(0) = 0, \quad (0 < \alpha < 1) \tag{12.10}$$

and the exact solution of Eq. (12.9) is $y(t) = (e^{2t} - 1)/(e^{2t} + 1)$ when $\alpha = 1$.

(2) The nonlinear Volterra–Fredholm integral equation is employed by [81]

$$y(x) = f(x) + \lambda_1 \int_a^{h(x)} K_1(x,t) G_1(y(t))dt + \lambda_2 \int_a^{b} K_2(x,t) G_2(y(t))dt, \tag{12.11}$$

where $f(x)$, $h(x)$, $K_1(x,t)$, and $K_2(x,t)$ are known functions, $a \le x, t \le b$, $a \le h(x) < \infty$; a, b are known constants; $y(x)$ is the unknown function. $G_1(x,t)$ and $G_2(x,t)$ are composite functions or nonlinear functions. It can be easily seen that, when $h(x)$ is a first-order polynomial, Eq. (12.11) is a functional integral equation with proportional delay.

(3) The fractional Bagley–Torvik equation employed in [84] is of the form

$$Ay''(t) + BD^{3/2}y(t) + Cy(t) = f(t), \quad t > 0 \tag{12.12}$$

with initial conditions

$$y(0) = y_0, \quad y'(0) = y_1. \tag{12.13}$$

After the above mathematical equations are numerically solved, the prominent advantages of the generalized Coiflets-based method are also revealed in [80–82]. At follows, the conclusions given in these three publications are directly used as the end of this book except for the reference number renewed in order here.

In this paper [80], “a numerical method based on the Coiflets wavelet operational method is applied to solve the fractional differential equations. In this method, the equation with fractional differential order is transferred to an integral equation of convolution type by the Laplace transform and then the solution is approximated by the modified wavelet approximate scheme. This simple method was established by Zhou et al. [21] and has been applied to solve nonlinear equations of vibration, diffusion, and wave equations. Compared with the results for solving fractional Riccati differential equation by the other numerical methods [83–87], the results for numerical examples demonstrate that the present method can give a **high accurate** approximation in a larger region. This is also the advantage of the present method.”

By using the Bagley–Torvik equation as an example in [81], “we find that this wavelet algorithm has a convergence rate and shows a **very high precision** comparing with many other existing numerical methods. The examples show **simplicity** and **effectiveness** of this method. Moreover, such an approach can be applied to any reasonable function categories and it is **not necessary to know the properties** of original function in advance.”

In this paper [82], “by combining techniques of boundary extension and Coiflet-type wavelet expansion, an approximation scheme of the function defined on a finite interval is proposed. With such a modified Coiflet-type wavelet approximation scheme, any nonlinear term containing unknown function can be explicitly expressed by a single-point sampling of the function successfully. Using Galerkin method, finally, based on this method, the original nonlinear integral equation was transformed into the solving of some simple nonlinear algebraic equations. Since it does not need to calculate the connection coefficients, it **avoids numerical errors** and **reduces the computational complexity** of the connection coefficients. Moreover, **numerical errors** of the present method are **not sensitive to the nonlinear intensity** of the equations. Also in the future, the method proposed in this paper is expected to be further applied to solve other nonlinear problems in other fields.”

References

1. Poincaré H (1882) Sur les courbes definies par les equations differentielles. Journal De Mathématiques Pures Et Aplliquées 3(8):251–286
2. Leray J, Schauder J (1934) Topologie et équations fonctionnelles. Annales Scientifiques De l'École Normale Supérieure. 51:45–78
3. Lahaye E (1934) Une méthode de résolution d'une catégorie d'équations transcendantes. Comptes rendus des séances de l'Académie des sciences. Vie académique 197:1840–1842
4. Hilton PJ (1953) An introduction to homotopy theory. Cambridge University Press, Cambridge
5. Sen S (1983) Topology and geometry for physicists. Academic Press, Florida
6. Alexander JC, Yorke JA (1978) The homotopy continuation method: numerically implementable topological procedures. Trans Ameri Math Soc 242:271–284
7. Liao SJ (1992) A second-order approximate analytical solution of a simple pendulum by the process analysis method. J Appl Mech 59:970–975
8. Liao SJ (2004) An analytic approximate approach for free oscillations of self-excited systems. Int J Non-Linear Mech 39(2):271–280

9. Yang ZC, Liao SJ (2017) A HAM-based wavelet approach for nonlinear ordinary differential equations. Commun Nonlinear Sci Numer Simul 48:439–453
10. Yang ZC, Liao SJ (2017) A HAM-based wavelet approach for nonlinear partial differential equations: two dimensional Bratu problem as an application. Commun Nonlinear Sci Numer Simul 53:249–262
11. Wang JZ (2001) Generalized theory and arithmetic of orthogonal wavelets and applications to researches of mechanics including piezoelectric smart structures. Lanzhou University; 2001. PhD Thesis. (advised by Zhou Y H)
12. Wang JZ, Zhou YH, Gao HJ (2003) Computation of the Laplace inverse transform by application of the wavelet theory. Commun Numer Methods Eng 19(12):959–975
13. Wang JZ, Wang XM, Zhou YH (2012) A wavelet approach for active-passive vibration control of laminated plates. Acta Mech Sin 28(2):520–531
14. Liu XJ, Wang JZ, Zhou YH (2013) A wavelet method for solving nonlinear time-dependent partial differential equations. Computer Modeling in Engineering and Science. 94:225–238
15. Liu XJ, Zhou YH, Zhang L, Wang JZ (2014) Wavelet solutions of Burgers' equation with high Reynolds numbers. Sci China Technol Sci 57(7):1285–1292
16. Kosloff D, Kessler D, Filho A, Tessmer E, Behle A, Strahilevitz R (1990) Solution of the equations of dynamic elasticity by a Chebychev spectral method. Geophysics 55(6):734–748
17. Jaffard S (1992) Wavelet methods for fast resolution of elliptic problems. SIAM J Numerical Anal 29(4):965–986
18. Alpert B, Beylkin G, Coifman R, Rokhlin V (1993) Wavelet-like bases for the fast solution of second-kind integral equations. SIAM J Scient Comput 14(1):159–184
19. Xu JC, Shann WC (1992) Galerkin-wavelet methods for two-point boundary value problems. Numer Math 63(1):123–144
20. Liu XJ (2014) A wavelet method for uniformly solving nonlinear problems and its application to quantitative research on flexible structures with large deformation. Lanzhou University; 2014. PhD Thesis. (Supervised by Zhou Y H)
21. Zhou YH, Wang XM, Wang JZ, Liu XJ (2011) A wavelet numerical method for solving nonlinear fractional vibration, diffusion and wave equations. Comput Model Eng Sci 7(2):137–160
22. Zhang L, Liu XJ, Zhou YH, Wang JZ (2013) Influence of vanishing moments on the accuracy of a modified wavelet Galerkin method for nonlinear boundary value problems. AIP Conference Proceedings. Amer Inst Phys 1558(1): 942–945
23. Ko J, Kurdila AJ, Pilant MS (1995) A class of finite element methods based on orthonormal, compactly supported wavelets. Comput Mech 16(4):235–244
24. Ma JX, Xue JJ, Yang SJ, He ZJ (2003) A study of the construction and application of a Daubechies wavelet-based beam element. Finite Elem Anal Des 39(10):965–975
25. Han JG, Ren WX, Huang Y (2006) A spline wavelet finite-element method in structural mechanics. Int J Numer Meth Eng 66(1):166–190
26. Ray SS (2012) On Haar wavelet operational matrix of general order and its application for the numerical solution of fractional Bagley Torvik equation. Appl Math Comput 218(9):5239–5248
27. Ray SS, Gupta AK (2014) A two-dimensional Haar wavelet approach for the numerical simulations of time and space fractional Fokker-Planck equations in modelling of anomalous diffusion systems. J Math Chem 52(8):2277–2293
28. Gupta AK, Ray SS (2015) Numerical treatment for the solution of fractional fifth-order Sawada-Kotera equation using second kind Chebyshev wavelet method. Appl Math Model 39(17):5121–5130
29. Ray SS, Gupta AK (2016) Numerical solution of fractional partial differential equation of parabolic type with Dirichlet boundary conditions using two-dimensional Legendre wavelets method. J Comput Nonlinear Dyn 11(1):011012
30. Amaratunga K, Williams JR, Qian S, Weiss J (1994) Wavelet-Galerkin solutions for one-dimensional partial differential equations. Int J Numer Meth Eng 37(16):2703–2716
31. Avudainayagam A, Vani C (2000) Wavelet-Galerkin method for integro-differential equations. Appl Numer Math 32(3):247–254

32. Liu XJ, Zhou YH, Wang XM, Wang JZ (2013) A wavelet method for solving a class of nonlinear boundary value problems. Commun Nonlinear Sci Numer Simul 18(8):1939–1948
33. Zhang L, Wang JZ, Zhou YH (2015) Wavelet solution for large deflection bending problems of thin rectangular plates. Arch Appl Mech 85(3):355–365
34. Chen MQ, Hwang C, Shih YP (1996) The computation of wavelet-Galerkin approximation on a bounded interval. Int J Numer Meth Eng 39(17):2921–2944
35. Yu Q, Xu H, Liao SJ (2018) Analysis of mixed convection flow in an inclined lid-driven enclosure with Buongiorno's nanofluid model. Int J Heat Mass Transf 126:221–236
36. Yu Q, Xu H (2018) Novel wavelet-homotopy Galerkin technique for analysis of lid-driven cavity flow and heat transfer with non-uniform boundary conditions. Appl Math Mech 39(12):1691–1718
37. Yu Q, Xu H, Liao SJ, Yang ZC (2019) A novel homotopy-wavelet approach for solving stream function-vorticity formulation of Navier-Stokes equations. Commun Nonlinear Sci Numer Simul 67:124–151
38. Yu Q, Xu H, Liao SJ, Yang ZC (2018) Nonlinear analysis for extreme large bending deflection of a rectangular plate on non-uniform elastic foundations. Appl Math Model 61:316–340
39. Chen QB, Xu H (2020) Coiflet wavelet-Homotopy solution of channel flow due to orthogonally moving porous walls governed by the Navier-Stokes equations. J Math 2020:1–12
40. Ray SS, Gupta AK (2013) On the solution of Burgers-Huxley and Huxley equation using wavelet collocation method. Comput Model Eng Sci 91:409–424
41. Koziol P, Hryniewicz Z (2006) Analysis of bending waves in bending on viscoelastic random foundation using wavelet technique. Int J Solids Struct 43:6965–6977
42. Koziol P, Mares C, Esat I (2006) A wavelet approach for the analysis of bending waves in a beam on viscoelastic random foundation. Appl Mech Mater 5:239–246
43. Soong TT (1973) Random differential equations in science and engineering. Academic Press, New York
44. Adomian G (1983) Stochastic systems. Academic Press, New York
45. Daubechies I (1993) Orthonormal bases of compactly supported wavelets II. Variations on a theme. SIAM J Math Analysis 24(2):499–519
46. Koziol P, Mares C, Esat I (2008) Wavelet approach to vibratory analysis of surface due to a load moving in the layer. Int J Solids Struct 45:2140–2159
47. Koziol P, Hryniewicz Z, Mares C (2009) Wavelet analysis of beam-soil structure response for fast moving train. J Phys: Confer Ser. 181(1):012052. IOP Publishing
48. Koziol P, Mares C (2010) Wavelet approach for vibration analysis of fast moving load on a viscoelastic medium. Shock Vib 17(4, 5):461–472
49. Hryniewicz Z (2011) Dynamics of Rayleigh beam on nonlinear foundation due to moving load using Adomian decomposition and Coiflet expansion. Soil Dyn Earthq Eng 31(8):1123–1131
50. Koziol P, Hryniewicz Z (2012) Dynamic response of a beam resting on a nonlinear foundation to a moving load: coiflet-based solution. Shock Vibr 19(5):995–1007
51. Koziol P, Neves MM (2012) Multilayered infinite medium subject to a moving load: dynamic response and optimization using coiflet expansion. Shock Vibr 19(5):1009–1018
52. Hryniewicz Z, Kozioł P. Wavelet-based solution for vibrations of a beam on a nonlinear viscoelastic foundation due to moving load. Journal of Theoretical and Applied Mechanics. 2013, 51.
53. Koziol P (2014) Wavelet approximation of Adomian's decomposition applied to the nonlinear problem of a double-beam response subject to a series of moving loads. J Theor Appl Mech 52(3):687–697
54. Koziol P (2016) Experimental validation of wavelet based solution for dynamic response of railway track subjected to a moving train. Mech Syst Signal Process 79:174–181
55. Beylkin G, Coifman R, Rokhlin V (1991) Fast wavelet transforms and numerical algorithms I. Commun Pure Appl Math 44(2):141–183
56. Si LT, Zhao Y, Zhang YH, Kennedy D (2016) A hybrid approach to analyze a beam-soil structure under a moving random load. J Sound Vib 382:179–192

57. Lu PM, Shi CJ (2012) Analysis solution study of steady response on asphalt road based on the wavelet theory. Chinese J Comput Mech 29(5):806–810 (in Chinses)
58. Chun-Juan SHI, Peng-min L (2013) Study on the dynamic response of asphalt pavement based on the nonlinear viscoelastic model. Eng Mech 2:47 (in Chinses)
59. Liu XY, Shi CJ (2012) Random characteristics and reliability analysis of asphalt pavement under vehicle random load. China J Highway Transport 25(6):49–55 (in Chinses)
60. Meyer Y (1992) Wavelets and operators. Cambridge University Press
61. Monzón L, Beylkin G, Hereman W (1999) Compactly supported wavelets based on almost interpolating and nearly linear phase filters (coiflets). Appl Comput Harmonic Anal 7(2):184–210
62. Lieb M, Sudret B (1998) A fast algorithm for soil dynamics calculations by wavelet decomposition. Arch Appl Mech 68(3–4):147–157
63. Grundmann H, Lieb M, Trommer E (1999) The response of a layered half-space to traffic loads moving along its surface. Arch Appl Mech 69(1):55–67
64. Zhou YH, Wang JZ, Zheng XJ (1998) Applications of wavelet Galerkin FEM to bending of beam and plate structures. Appl Math Mecha 19(8):745–755
65. Zhou YH, Wang JZ (1998) A dynamic control model of piezoelectric cantilevered beam-plate based on wavelet theory. Acta Mech Sin 30(6):719–727 (in Chinese)
66. Wang JZ, Zhou YH (1998) Error estimation of the generalized wavelet Gaussian integral method. J Lanzhou Univ 34:26–30. (in Chinese)
67. Zhou YH, Wang JZ (1999) Generalized Gaussian integral methods for calculations of scaling function transform of wavalets and its applications. Acta Math Scientia 19(2):293–300 (in Chinese)
68. Zhou YH, Wang JZ, Zheng XJ (2001) A wavelet-based approach for dynamic control of intelligent piezoelectric plate structures with linear and non-linear deformation. IUTAM Symposium on Smart Structures and Structronic Systems. Springer, Dordrecht, pp 179–187
69. Zhou YH, Wang JZ, Zheng XJ, Jiang Q (2000) Vibration control of variable thickness plates with piezoelectric sensors and actuators based on wavelet theory. J Sound Vib 237(3):395–410
70. Zhou YH, Wang JZ, Zheng XJ (2001) A numerical inversion of the Laplace transform by use of the scaling function transform of wavelet theory. Acta Math Scientia 21A(1):86–93 (in Chinese)
71. Yang SJ, Ma JX, Xue JJ, He ZJ (2002) Simulation of temperature field of copier paper based on wavelet finite element method. Acta Simulate Systematica Sinica 14(9):1243–1245 (in Chinese)
72. Si HW, Li DX (2003) Researches on wavelet-based vibration control of large space smart structures. J National Univ Defense Technol 25(3):14–18 (in Chinese)
73. Ma JX, Xue JJ, Yang SJ, HE ZJ (2004) Study of the construction and application of Daubechies wavelet-based beam element. Minimicro Syst-ShenYang 25:663–666
74. Fenik S, Starek L (2010) Optimal PI controller with position feedback for vibration suppression. J Vib Control 16(13):2023–2034
75. Hegewald T, Inman DJ (2001) Vibration suppression via smart structures across a temperature range. J Intell Mater Syst Struct 12(3):191–203
76. Rew KH, Han JH, Lee I (2002) Multi-modal vibration control using adaptive positive position feedback. J Intell Mater Syst Struct 13(1):13–22
77. Ma XL, Wu B, Zhang JH, Shi X (2019) A new numerical scheme with wavelet-Galerkin followed by spectral deferred correction for solving string vibration problems. Mech Mach Theory 142:103623
78. Wang JZ, Zhang L, Zhou YH (2018) A simultaneous space-time wavelet method for nonlinear initial boundary value problems. Appl Math Mech 39(11):1547–1566
79. Zhang L, Wang JZ, Liu XJ, Zhou YH (2017) A wavelet integral collocation method for nonlinear boundary value problems in physics. Comput Phys Commun 215:91–102
80. Wang XM (2014) A Coiflets-based wavelet Laplace method for solving the Riccati differential equations. J Appl Math 2014:257049

81. Wang XM (2014) A new wavelet method for solving a class of nonlinear Volterra-Fredholm integral equations. Abstract Appl Analy 2014:975985
82. Wang XM (2014) A wavelet method for solving Bagley-Torvik equation. Comput Model Eng Sci 102(2):169–182
83. Momani S, Shawagfeh N (2006) Decomposition method for solving fractional Riccati differential equations. Appl Math Comput 182(2):1083–1092
84. Li Y, Sun N, Zheng B, Wang Q, Zhang Y (2014) Wavelet operational matrix method for solving the Riccati differential equation. Commun Nonlinear Sci Numer Simul 19(3):483–493
85. Cang J, Tan Y, Xu H, Liao SJ (2009) Series solutions of non-linear Riccati differential equations with fractional order. Chaos Solitons Fractals 40(1):1–9
86. HosseinNia SH, Ranjbar A, Momani S (2008) Using an enhanced homotopy perturbation method in fractional differential equations via deforming the linear part. Comput Math Appl 56(12):3138–3149
87. Odibat Z, Momani S (2008) Modified homotopy perturbation method: application to quadratic Riccati differential equation of fractional order. Chaos Solitons Fractals 36(1):167–174

Index

Y.-H. Zhou, *Wavelet Numerical Method and Its Applications in Nonlinear Problems*, Engineering Applications of Computational Methods 6,
https://doi.org/10.1007/978-981-33-6643-5

9789813366428